Agachi, Cristea, Csavdári, Szilágyi
Advanced Process Engineering Control
De Gruyter Graduate

Also of Interest

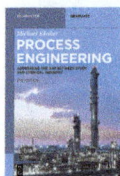

Process Engineering.
Addressing the Gap between Study and Chemical Industry
Kleiber, 2023
ISBN 978-3-11-102811-8, e-ISBN 978-3-11-102814-9

Sustainable Process Engineering
Szekely, 2021
ISBN 978-3-11-071712-9, e-ISBN 978-3-11-071713-6

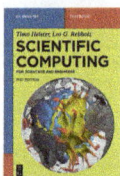

Scientific Computing.
For Scientists and Engineers
Heister, Rebholz, 2023
ISBN 978-3-11-099961-7, e-ISBN 978-3-11-098845-1

Process Systems Engineering.
For a Smooth Energy Transition
Zondervan (Ed.), 2022
ISBN 978-3-11-070498-3, e-ISBN 978-3-11-070520-1

Integrated Chemical Processes in Liquid Multiphase Systems.
From Chemical Reaction to Process Design and Operation
Kraume, Enders, Drews, Schomäcker, Engell, Sundmacher (Eds.), 2022
ISBN 978-3-11-070943-8, e-ISBN (OA) 978-3-11-070985-8

Paul Șerban Agachi, Mircea Vasile Cristea,
Alexandra Ana Csavdári, Botond Szilágyi

Advanced Process Engineering Control

2nd, Revised and Extended Edition

DE GRUYTER

Authors

Prof. Dr. Paul Şerban Agachi
Faculty of Chemistry and Chemical Engineering
Babeş-Bolyai University
Arany Janos Street 11
400028 Cluj Napoca
Romania
serban.agachi@ubbcluj.ro
and
Chemical, Materials and Metallurgical Engineering
Department
Botswana International University of Science and
Technology
Private Bag 16
Palapye
Botswana
agachip@biust.ac.bw

Prof. Dr. Mircea Vasile Cristea
Faculty of Chemistry and Chemical
Engineering
Babeş-Bolyai University
Arany Janos Street 11
400028 Cluj-Napoca
Romania
mircea.cristea@ubbcluj.ro

Assoc. Prof. Dr. Alexandra Ana Csavdári
Faculty of Chemistry and Chemical Engineering
Babeş-Bolyai University
Arany Janos Street 11
400028 Cluj-Napoca
Romania
alexandra.csavdari@ubbcluj.ro
and
Department of analytical, colloidal chemistry and
technology of rare elements
Faculty of Chemistry and Chemical Technology
Al-Farabi Kazakh National University
71 Al-Farabi Avenue
050040 Almaty
Republic of Kazakhstan

Dr. Botond Szilágyi
Department of Chemical and Environmental
Process Engineering
Faculty of Chemical Technology and Biotechnology
Budapest University of Technology and Economics
Műegyetem rkp. 3
1111 Budapest
Hungary
szilagyi.botond@vbk.bme.hu

ISBN 978-3-11-078972-0
e-ISBN (PDF) 978-3-11-078973-7
e-ISBN (EPUB) 978-3-11-078986-7

Library of Congress Control Number: 2023941358

Bibliographic information published by the Deutsche Nationalbibliothek
The Deutsche Nationalbibliothek lists this publication in the Deutsche Nationalbibliografie;
detailed bibliographic data are available on the Internet at http://dnb.dnb.de.

© 2024 Walter de Gruyter GmbH, Berlin/Boston
Cover image: Janaka Dharmasena/Hemera/Thinkstock
Typesetting: Integra Software Services Pvt. Ltd.
Printing and binding: CPI books GmbH, Leck

www.degruyter.com

Preface to the second edition

Six years ago, when we published *Advanced Process Engineering Control*, we did not know the editor will propose quite soon, a second edition. In historical terms, 6 years is almost a blink of the eye, but if we look at the tremendous changes which occurred, especially in the computer industry with important consequences in process industries, we were thinking that yes, the book deserves a second edition – an edition to also capture the progress in control of other nontraditional industries, beyond the chemical ones, from where we started.

This work is intended to be the continuation of the authors' *Advanced Process Engineering Control* published by De Gruyter in 2016. The first edition presented the holistic approach of synthesis and separation processes, wherever they occur. Titles containing the concept of process engineering were deliberately chosen to suggest the inclusion, within the same approach, of processes other than the traditional chemical engineering ones. These come from outside the traditional fields of chemistry and coal/petrochemistry: the spheres – wastewater management, water purification, construction material industry, food processing, and mineral processing. We have added one new chapter, Chapter 15, containing several applications of control of processes in various industries: cement manufacturing (such an important construction material!), fluid catalytic cracking (deserving this place as the process is the core of the refinery), biowaste and coal-based processing industries, and mineral processing industry are very promising in Botswana where I had the opportunity to live in the last 8 years. Botswana has the second largest coal reserve in the world and has also important reserves of salt, rare in the region, and reserves deserve more than raw exploitation. A special case study is quality by control, a quasi-new concept used especially in the pharma industries.

On the other hand, new developments in the nowadays already "traditional control fields" such as optimal, feedforward, predictive, and adaptive control asked more attention from us, adding new information. Some very recent developments were reported in the series of seminars kept by ASPEN (The Self-Optimizing Plant: A New Era of Autonomy, Powered by Industrial AI[1]). It is about enterprises operating in a global world, characterized by extreme volatility, uncertainty, complexity, and ambiguity (VUCA). In this environment, more and more "autonomous and semiautonomous processes augmented by the latest advances in artificial intelligence (AI)" will start operating. A quite new concept[2] of self-optimizing plant has the ability of meeting the dynamic requests of the customers, operating safely at the limits of stability, thus increasing the profitability in the harsh competition on the market.

---

1 Antonio Pietri, https://www.aspentech.com/en/resources/executive-brief/the-self-optimizing-plant-a-new-era-of-autonomy-powered-by-industrial-ai, ASPENTECH seminar, February 2022.
2 We had introduced the concept of Wise Machinery in our first book in the series, *Basic Process Engineering Control*.

https://doi.org/10.1515/9783110789737-202

During use and development of automatic control systems, control analysis and control system design for process industries have followed the traditional unit operation approach. It means that all control loops are established individually for each unit or piece of equipment in the plant, and that the final plant-wide control system represents the sum of the individual parts. The disadvantage of this method is the difficulty in stabilizing potential conflicts among individual loops. One very handy method of avoiding these interactions (and this approach is not new at all, coming from the practice of the field engineers) is the different tuning of control loops: those controlling the most important parameters are tuned tight and the others loose. Another way to improve the behavior of the whole system is the decoupling of the loops, which is largely presented in the book. Despite any process complexity, the unit operation approach provided reasonable results and remains in use on a large scale for designing control systems. Consequently, the book follows this traditional approach but provides updates to new industrial achievements.

Modern process plants are designed for flexible production and maximization of energy and material savings, especially within the frame of globalization and strong competition among manufacturers. Additionally, according to the fourth paradigm of process/ chemical engineering, the processes have to fulfill tight environmental constraints. Industrial plants become more complex and have therefore strong interactions between process units. As a consequence, the failure of one unit might have a negative effect both on overall productivity and on environmental performances. This situation raises important control issues. A significant example is that of the thermally integrated plants, a concept born during the global energy crisis that started in 1973. Energy recovery became a priority for the industry and at the same time a scientific challenge. The necessity of redesigning industrial processes in terms of energetic efficiency was identified. Moreover, it was discovered that energy saving can be achieved using retrofit and recovery of extra energy from all secondary sources of a process. These aspects posed complex control problems because of the weak process controllability (effects of all disturbances are collected at the end of the process and reintroduced as enhanced disturbances at the input).

The emergence and continuous development of advanced control techniques provided solutions for plant-wide control at any level of process complexity and in the abovementioned conditions. According to Willis and Tham, a definition of the advanced process control can be formulated as "a systematically studied approach for the choice of pertinent techniques and their integration into a co-operative management and control system that will significantly enhance plant operation and profitability". Applied on complex chemical processes, advanced control can improve product yield, reduce energy consumption, increase plant capacity, improve product quality and consistency, enhance process safety, and reduce environmental impact. The benefits of the advanced control implementation are noticeable in the overall operating costs of a plant. These can decrease by 2% to 6%. Another benefit is the reduction of process variability. Consequently, a plant can be operated at its designed capacity.

This book is structured into two parts. Part I comprises Chapters 1–7 and defines advanced control as any control system that surpasses simple and conventional loops.

This could mean either smarter control configuration (cascade, feedforward, ratio, inferential, digital, or multivariable control) or improved regulator features (fuzzy, model predictive, or optimal control). Approaches for the design of plant-wide control systems are presented, and a new paragraph of combined configurations (e.g., cascade–feedforward, ratio–feedforward, and inferential–MPC) was elaborated. Part II includes Chapters 8–15 and refers to control solutions for the so-called unit operations: reaction and separation processes (distillation, absorption-desorption, extraction, evaporation, drying, crystallization, and filtration). The sections desorption and filtration control were added, considering their importance in process engineering. In the new Chapter 15, the studies are about treating complex manufacturing processes: cement manufacturing, fluid catalytic cracking, biowaste/coal pyrolysis, mineral processing control, pharmaceutical manufacturing. The reader can check his or her level of comprehension by solving the problems and exercises proposed in the updated Chapter 16. These cover the entire list of discussed topics.

The authors hope that by including many industrial examples and applications, as well as their own and other researcher's experience accumulated over many years within the Group of Computer Aided Process Control, this work will be useful for all interested parties in process engineering and process control: students in electrical, chemical, or process engineering; specialists in chemical, petrochemical, or automation companies; professionals of water or natural gas management; engineers in food or pharma industries; and so on.

The idea of this book series describing the main aspects of modern process engineering as applied to (not only) chemical industry belongs to Prof. Dr. Paul Şerban Agachi. He initiated the manuscripts, developed their structures, and coordinated the authors. More than 25 years ago, he recognized the ever-increasing importance of the subject and founded the *Group of Computer Aided Process Engineering* at the Faculty of Chemistry and Chemical Engineering of the *Babeş-Bolyai University* in Cluj-Napoca, Romania, and continued this approach at *Botswana International University of Science and Technology*. Many professionals emerged from it, and the two younger authors of this work have also started their carriers in this group. Although writers have exchanged ideas and discussed all topics of this book, work was distributed in agreement with individual strengths, experience, and competencies: Prof. Dr. Paul Şerban Agachi was in charge of Chapters 1, 4, 12, and 13; Prof. Dr. Mircea Vasile Cristea shared his experience in Chapters 2, 3, and 5–7; Assoc. Prof. Dr. Alexandra Csavdári wrote Chapters 8, 11 and coauthored with Research Engineer Botond Szilágyi Chapters 9, 10, and 14 with assistance from P. Ş. Agachi. Chapter 15 and the list of problems and exercises in Chapter 16 are the result of a joint effort.

Finally, within the framework of this laborious enterprise, the authors gratefully acknowledge graduate students Keabetswe Mbayi, Thapelo Shomana, Daniel Eric Botha, Mmoloki Makoba and Leonard Akofang for their dedicated and valuable help. Many thanks to the editing team, Ria Sengbusch, Melanie Goetz, Cherline Daniel, Ishwarya Mathavan with whom we wonderfully cooperated. The last but not the least, many, many thanks to Karin Sora, the Chief Editor, without whom, we would have not done this much improved Second Edition.

Palapye, July 2023   *Paul Șerban Agachi*

Contents

Part II: Applied Process Engineering Control

Part I: Advanced Process Control

The first part of the book, entitled *Advanced Process Control*, refers, in the view of the authors, to all levels of complexity of automatic control systems beyond the simple feedback control loop, the backbone of the industrial process control.

Advanced control emerges from the human need to make groups of phenomena behave in a desired way, meeting productivity, quality, efficiency, safety, and environmentally friendly objectives. Usually, this assembly of physical, chemical, and biological phenomena shows complicated behavior and asks for complex control systems.

Within this context, in Part I (Advanced Process Control), the authors have treated first the nonconventional control systems, such as cascade, feed-forward, ratio, and inferential control. The following chapters deal with the most important and significant control techniques for advanced, performing process industries: model predictive control, multivariable control including decoupling, fuzzy control, optimal control, plantwide control as well as the method of applying control algorithms in practice.

Model predictive control is undoubtedly the most industrially applied advanced control algorithm. Its roots are in the practice-proven native applications, followed by the theoretical developments based on the open-loop optimal control and receding horizon control. Model predictive control makes possible the intelligent embedment of the process behavior described by the model while taking advantage of its prediction capability and coupling it with solving the constrained optimization to find the best control solution.

Fuzzy logic, with its extension to fuzzy control, can provide human consistency to the way control action is conceived. As human experts are able to solve complex control problems on the basis of the rule-based approach, the fuzzy controllers may achieve the same task, being able to embed in the controller the human expertise in a systematic way.

Multivariable control strives to consider the complex system as a whole and accordingly aims to simultaneously achieve all desired control objectives, despite the possibly interacting effects of the multiple controls. Decentralized and centralized control are the two facets of the multivariable control approach featuring particular motivations from the practice applying perspective.

Optimal control is practically devoted to optimizing one objective function such as the quality control criterion (e.g. integral absolute error IAE or integral of time weighted absolute error ITAE), relevant for the quality of the control action, the economic benefit, the return of investment, the duration of a batch process, or the conversion rate of a raw material, relevant for the overall control/optimization of the industrial process. The optimal control is approached here both from the point of view of the process and of the control engineer. The steady-state control is practically an optimization of the process that uses the control elements as instruments. This approach is extended to batch and continuous processes as well.

Plantwide control is still an emerging field of research and development. It strives to find the harmony of coupling the regulatory layer of the control hierarchy with the

https://doi.org/10.1515/9783110789737-001

supervisory/advanced control layers. However, a set of guidelines for the development of a systematic plantwide control approach are available and they are useful for the control system design.

The theoretical concepts of part I are the fundaments for the control applications presented in part II (Applied Process Engineering Control).

1 Complex and nonconventional control systems

We consider all automatic control systems (ACS) that exceed as complexity, the traditional feedback control loops as being complex and nonconventional or advanced. As Agachi and Cristea [1] have stated, the controllability of a process can be improved by using more complex systems as cascade, feedforward, predictive, multivariable, adaptive, or optimal control systems. The first two, at which we add two special configurations of ratio and inferential control, are classified in this book as complex and nonconventional. In this second edition, the authors added some combinations of the previously mentioned types of complex systems: cascade–feedforward, cascade–inferential, etc.

1.1 Cascade control systems

1.1.1 Processes in series

This type of control system is next in complexity after the simple feedback control. The better performances of the cascade reside in capturing the effects of the disturbances inside the process and not at its end (Fig. 1.1). In this way, the controller can intervene earlier, being more efficient in cancelling the effects of the disturbance.

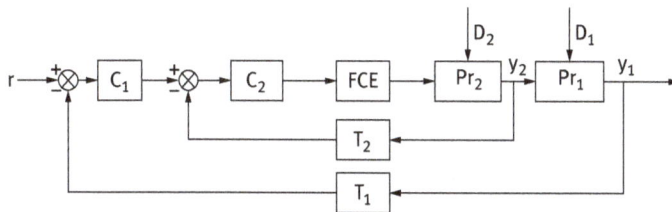

Fig. 1.1: Cascade ACS.

The process can be "decomposed" into two subprocesses, one faster in response to the intervention of the controller (Pr_2) and the other slower (Pr_1). As a matter of fact, the process is a whole, but the system captures an intermediate-controlled variable by means of the inner loop transducer (T_2).

The cascade consists of two nested loops, the "fast" inner one with the index 2 and the "slow" outer one with the index 1. The temperature control of a continuous stirred tank reactor (CSTR) is given here as an example of applying the cascade control (Fig. 1.2). The inner loop is formed of the "faster" process, the heat transfer in the jacket (Pr_2), the resistance thermometer in the jacket (T_2), the controller C_2, and the final control element, which operates for both nested loops, manipulating the cooling or heating agent flow. The outer loop is formed of the "slower" process, the heat trans-

https://doi.org/10.1515/9783110789737-002

fer in the reactor (Pr_1), the resistance thermometer in the jacket (T_1), the controller C_1, which gives the setpoint to be followed to C_2.

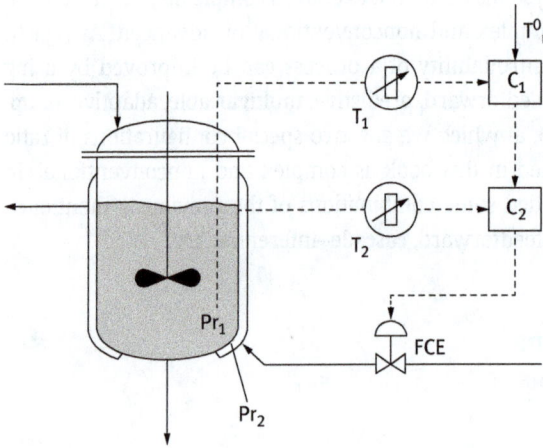

Fig. 1.2: Cascade ACS for the temperature of a nonisothermal CSTR.

The configuration is most efficient when the frequent disturbances are in the cooling/heating agent circuit because the ACS does not wait for the inner temperature of the reactor to change, but intervenes after the slightest change of temperature in the jacket.

Another example of applying successfully the cascade control is represented in Fig. 1.3.

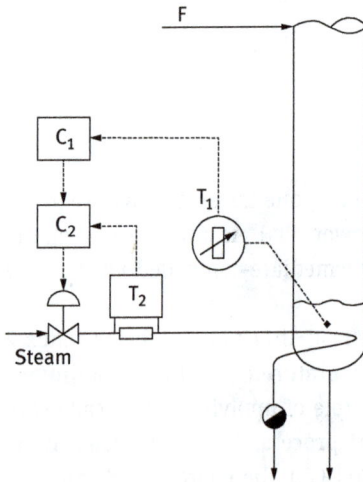

Fig. 1.3: Temperature cascade control at the bottom of a distillation column.

In this case, the steam flow to the reboiler is considered the inner process variable, subjected to the "faster" changes than the temperature at the bottom. The inner loop manages the steam flow, whereas the outer loop manages the temperature.

Generally, the cascade control is most efficient when disturbances of the type D_2 are most frequent and important as magnitude. Since the interaction between Pr_2 and Pr_1 is not only one way, from left to right, the efficiency of the cascade should still be considered when type D_1 disturbances occur. The efficiency is the highest when

$$3 \le \frac{T_1}{T_2} \le 10 \tag{1.1}$$

where T_1 and T_2 are the time constants of the two subprocesses.

Structure of the controllers and tuning the controllers' parameters

The main goal of the cascade ACS is the tight control of the output y_1 and less accurate control of the intermediate variable y_2. This is why the inner controller C_2 can have a simple P structure [2]. The external controller C_1, function of the slowness of Pr_1, can have a PI or PID structure (PID is designed for very slow processes, as heat or component transfer) [2]. In the case of PID structure of the external controller, the presence of the D element in the control algorithm of C_2 has to be avoided since it amplifies any sudden small change of disturbance signals.

Tuning the controllers' parameters has to take into consideration the strong interaction between the two controllers, C_1 imposing the setpoint value for C_2.

Experimentally, tuning has the following steps:
- One of the experimental tuning methods is chosen (e.g., Ziegler-Nichols or Cohen-Coon) [3].
- With both controllers on the manual mode of operation, the inner controller is fixed at the beginning to PB_{2max} (P structure); the chosen method is applied and PB_{2opt} is found.
- With the C_2 controller on automatic mode of operation, the inner loop becomes a dynamic element in the outer loop (Fig. 1.4).
- The procedure of tuning the parameters of C_1 is repeated for the "new" control loop with the inner loop as dynamic element of the outer loop.

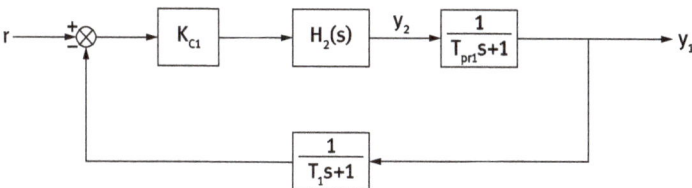

Fig. 1.4: The "new" structure of the cascade reduced to feedback control loop.

Example 1.1

Tuning the parameters of the controllers in a series cascade system.

The control system in Fig. 1.5 has two controllers, one, the internal, with an already chosen gain, $K_{c2} =$ 12, and the other, which must be chosen properly.

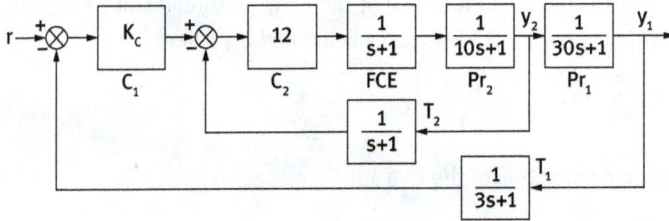

Fig. 1.5: Block diagram of the cascade control system corresponding to the CSTR in Fig. 1.2.

The obtained values of the C_1 controller should be compared with those of a simple feedback loop controller, controlling the same process. Analysis must be made to compare the values obtained in both cases and explain why the cascade is superior. All constants are expressed in minutes. The feedback control loop of y_1 is presented in Fig. 1.6.

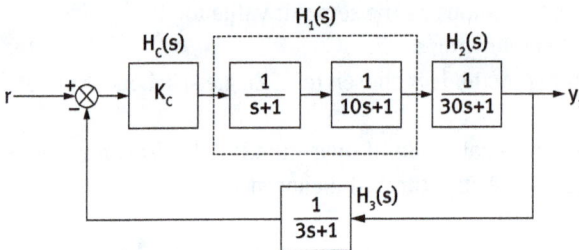

Fig. 1.6: Block diagram of the feedback control system controlling the heat transfer process in the CSTR from Fig. 1.2.

According to section 12.4 of [2], the crossover frequency condition for tuning the controller is

$$\frac{360}{2\pi}\left(-\tan^{-1}(1\omega) - \tan^{-1}(10\omega) - \tan^{-1}(30\omega) - \tan^{-1}(3\omega)\right) = -180° \quad (1.2)$$

and

$$\omega_{osc} = 0.18\,\text{rad/min} \quad (1.3)$$

that is, one oscillation at every 35 min.

From relationship (11.44) from [3],

$$K_{C1} \times \frac{1}{\sqrt{1+1\times0.18^2}} \cdot \frac{1}{\sqrt{1+10^2\times0.18^2}} \cdot \frac{1}{\sqrt{1+30^2\times0.18^2}} \cdot \frac{1}{\sqrt{1+3^2\times0.18^2}} = 0.5 \quad (1.4)$$

it results that:

$$K_{Copt} = 6.5 \quad (1.5)$$

According to Fig. 1.4, the "simplified" control system has as $T_{pr1} = 30$ min and $T_{T1} = 3$ min and an additional element with the transfer function:

$$H_1(s) = \frac{12 \cdot \frac{1}{s+1} \cdot \frac{1}{10s+1}}{1+12 \cdot \frac{1}{s+1} \cdot \frac{1}{10s+1} \cdot \frac{1}{s+1}} \quad (1.6)$$

To determine the parameters of the C_1 controller, we may use the Black-Nichols diagrams (Fig. 1.7) [4].

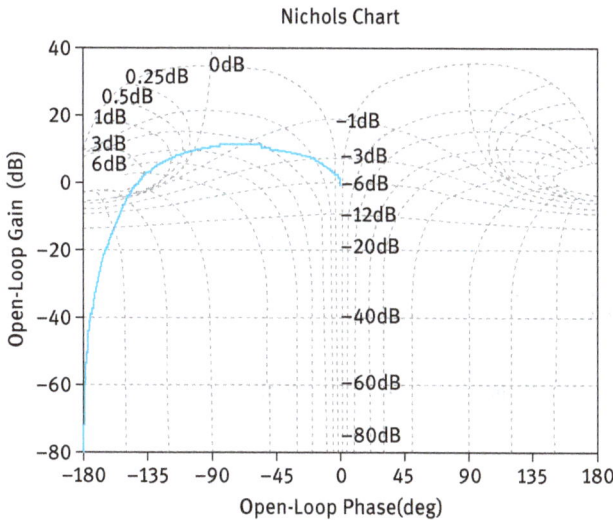

Fig. 1.7: Black-Nichols diagram used to determine the closed-loop gain and the phase angle when we know both open-loop values.

Using the open-loop phase angle

$$\varphi_{ol}(w) = 0 - \tan^{-1}(1\cdot\omega) - \tan^{-1}(10\cdot\omega) - \tan^{-1}(1\cdot\omega) \quad (1.7)$$

and the open-loop module

$$M_{ol}(\omega) = 12 \cdot \frac{1}{\sqrt{1+1\cdot\omega^2}} \cdot \frac{1}{\sqrt{1+10^2\cdot\omega^2}} \cdot \frac{1}{\sqrt{1+1\cdot\omega^2}} \tag{1.8}$$

we obtain the following closed-loop values φ_{cl} and M_{cl} from the Black-Nichols plots:

ω (rad/min)	φ_{ol} (deg)	M_{ol}	φ_{cl} (deg)	M_{cl}	φ_{tot} (deg)
0.1	−56	8.40	−5	0.93	−93.0
0.2	−86	5.16	−11	0.97	−122.5
0.3	−104	3.48	−19	1.03	−144.6
0.4	−119	2.10	−28	1.17	−160.6
0.5	−12	1.87	−32	1.32	−174.5
0.6	−142	1.38	−40	1.60	−197.0

that is, that the crossover frequency for the cascade is $\omega_{osc} = 0.53$ rad/min or an oscillation at every 12 min.

The alternative calculus can be done by processing the transfer function directly from eq. (1.5).

The frequency function is

$$H_1(j\omega) = \frac{12 \cdot \frac{1}{j\omega+1} \cdot \frac{1}{10j\omega+1}}{1 + 12 \cdot \frac{1}{j\omega+1} \cdot \frac{1}{10j\omega+1} \cdot \frac{1}{j\omega+1}}$$

$$= \frac{12 - 1440\omega - 3158\omega^2 + 5040\omega^3 - 40{,}068\omega^4 - 37{,}320\omega^5 - 12{,}000\omega^8}{1 - 264\omega + 50\omega^2 - 1964\omega^3 - 1059\omega^4 - 82{,}460\omega^5 - 19{,}220\omega^6 + 6200\omega^7}$$

$$+ j\frac{-12 - 120\omega - 18{,}371\omega^2 - 4584\omega^3 + 17532\omega^4 - 36120\omega^5 - 1200\omega^6 - 39720\omega^7}{1 - 264\omega + 50\omega^2 - 1964\omega^3 - 1059\omega^4 - 82{,}460\omega^5 - 19{,}220\omega^6 + 6200\omega^7} \tag{1.9}$$

and from it, $M_{cl}(\omega)$ and $\varphi_{cl}(\omega)$ for different values of ω are obtained directly, but in a much more laborious manner.

From eq. (11.3), K_{c1} can be calculated as

$$K_{c1} \cdot M_{c1}(\omega = 0.53) \cdot \frac{1}{\sqrt{1+30^2 \times 0.53^2}} \cdot \frac{1}{\sqrt{1+3^2 \times 0.53^2}} = 0.5 \tag{1.10}$$

resulting in $K_{c1} = 10.6$.

If we compare the results of the controllers' parameter tuning in both cases (feedback versus cascade), we obtain

ACS feedback		ACS cascade	
ω_{osc}	K_C	ω_{osc}	K_{C1}
(rad/min)	–	(rad/min)	–
0.16	6.5	0.53	10.6

results that are in the favor of the cascade:
- The crossover frequency is pushed for the cascade in the range of higher frequencies, meaning that the chance of a "normal" disturbance (of lower frequency) to destabilize the loop is lower; thus, the loop is more stable.
- The response of the system is three times faster in its intervention to eliminate the disturbance effect ($P_{osc\ casc} < P_{osc}$ fb); the result is also a shorter settling time in the case of the cascade ACS.
- The gain of the controller is higher, with the result of a smaller offset of the controlled variable.

The practical and measurable economic results are visible from Example 1.2 [4].

Example 1.2

The temperature control of a dryer is made both with a feedback control and cascade control systems (Fig. 1.8). Data of the process and elements of the ACS are given in Fig. 1.8c.

Considering a load disturbance of 10% (the variation of the steam quality), the response of both systems is shown in Fig. 1.9 with the tuning of both controllers, PB = 30% and T_i = 35s (a) and PB$_1$ = 100%, T_{i1} = 10s and PB$_2$ = 15%.

One may observe the following facts:
- The overshoot is smaller in the case of the cascade (20% versus 30%).
- The settling time is much smaller, decreasing from 150 s to 90 s; these two first facts show an increased stability of the system.
- The ratio of the areas under the response curve, signifying the energy consumption in both cases of control, is $S_b = \frac{1}{4}S_a$, implying a very important economy of consumed energy. The area under the temperature curve expresses, proportionally, the energy consumption. The smaller the area is, the smaller the energy consumption is.

Even when the disturbance is not of the type D_2, but D_1 (Fig. 1.10), the cascade is still very economically efficient ($S_b = \frac{1}{2}S_a$). The disturbance could be from higher humidity of the material introduced in the dryer.

It should be noted that the performance of the cascade control in the latter case is worse since the ACS does not profit on the fast response of the internal loop, but only on the inverse interaction between the dryer and its heater.

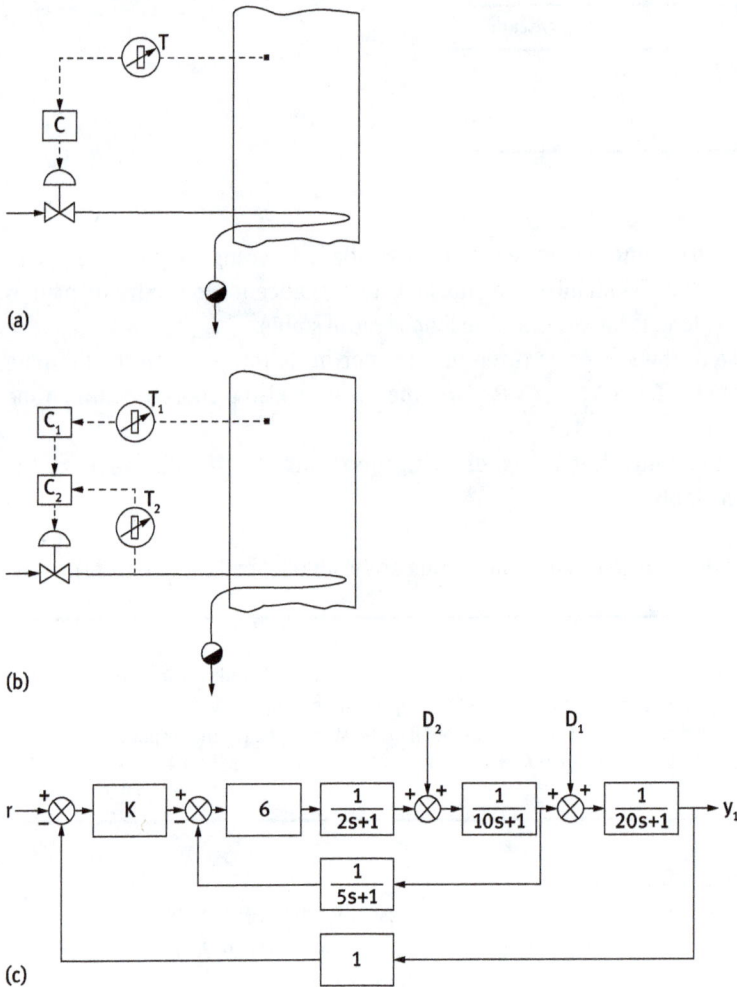

(a)

(b)

(c)

Fig. 1.8: Temperature control of a dryer: (a) feedback temperature control; (b) cascade temperature control; and (c) block diagram of the control system with the data of the components of the system (time constants are measured in seconds).

1.1.2 Processes in parallel

The first who defined the concept was Luyben in 1973 [5]. Luyben referred to the systems for which the manipulated variable influences two processes in parallel, with two output variables (Fig. 1.11). One example is the influence of both the top tray temperature and the composition at the top of a distillation column, via its reflux. The primary loop of the cascade is that controlling the overhead composition and the secondary one is that controlling the temperature of the top tray. The manipulative variable (reflux

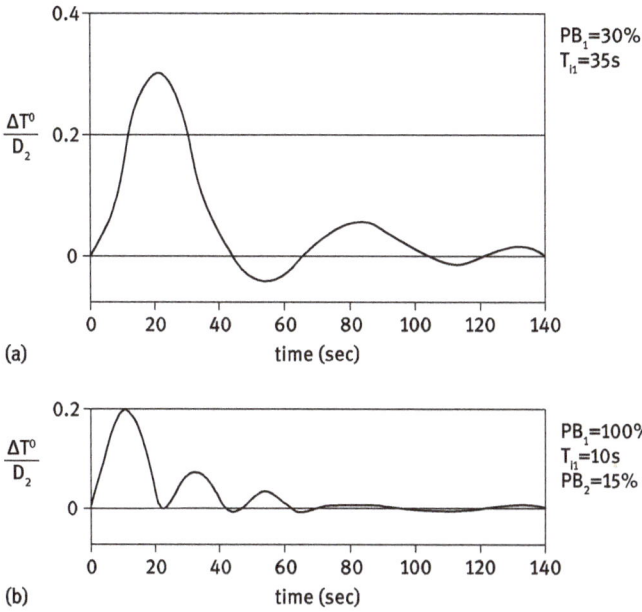

Fig. 1.9: The response of feedback and cascade control systems at a load disturbance (type D_2): (a) response of the cascade and (b) response of the feedback control.

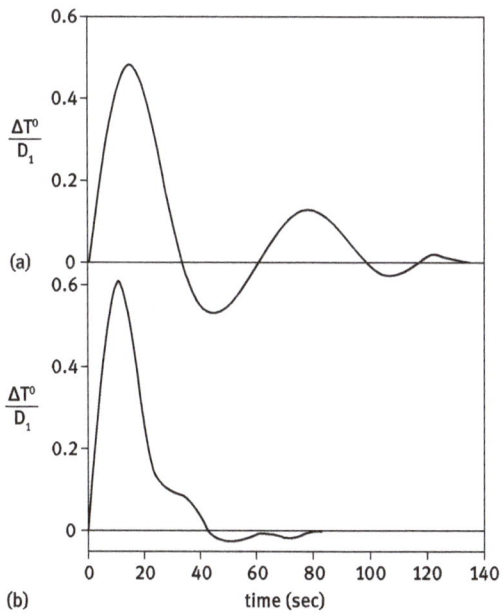

Fig. 1.10: The response of the temperature control systems (feedback and cascade) in the case of a disturbance of type D_1: (a) feedback control and (b) cascade control.

flow) affects overhead composition and tray temperature through two parallel process transfer functions, H_{pr1} and H_{pr2}. According to the original notations, these correspond to GM and GS, respectively.

Another example was given by Rao et al. [8] for a liquefied petroleum gas (GPL) splitter model; its structure is given in Fig. 1.12.

Pr_1 and Pr_2 are the two processes in parallel, the primary and the secondary ones; the secondary loop is much faster than the primary one. H_{C1} and H_{C2} are the controllers of the loops. H_{D1} and H_{D2} are the transfer functions of the channels of the disturbance in the process.

The cascade control is especially beneficial when the secondary loop is much faster than the primary one, as in the series cascade systems.

Tuning the controllers starts with the C_2 controller by using the inner open-loop transfer function.

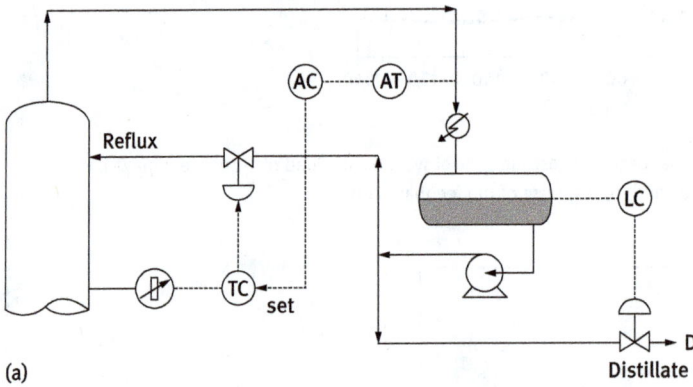

(a)

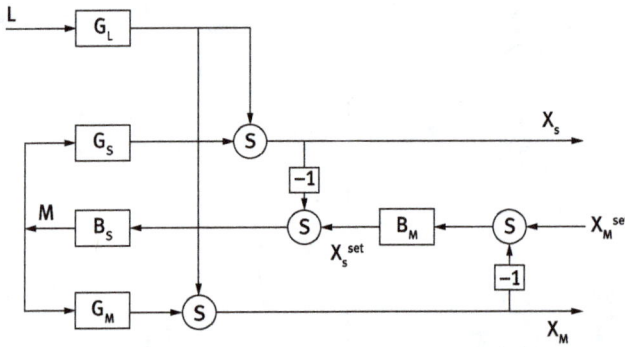

(b)

Fig. 1.11: Parallel cascade control: (a) top composition and top tray temperature control for a distillation process and (b) block diagram of the parallel cascade ACS. The master loop is characterized by G_M – process transfer function; B_M – feedback controller's transfer function; and X_M – process output. The similar characteristics of the slave loop are G_M, B_M, and X_M, respectively, whereas G_L stands for the load transfer function.

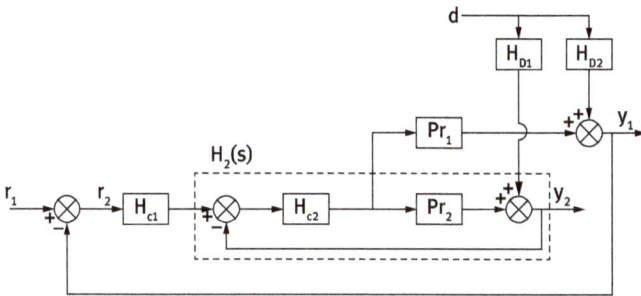

Fig. 1.12: Parallel cascade control structure.

$$H_2(s) = \frac{H_{C2}(s) \cdot H_{AD}(s) \cdot H_{pr2}(s) \cdot H_{T2}(s)}{1 + H_{C2}(s) \cdot H_{AD}(s) \cdot H_{pr2}(s) \cdot H_{T2}(s)} \qquad (1.11)$$

with the consequent $\varphi_{ol}(\omega)$ and $M_{ol}(\omega)$ from [4]. For simplicity reasons, the value of the actuator transfer function $H_{AD}(s)$ as well as those of the transducers' transfer functions $H_{T1}(s)$ and $H_{T2}(s)$ in Fig. 1.12 is considered to be equal to 1.

The controller C_1 is tuned, considering that function $H_{pr2}(s)$ is placed on the reaction of loop 2 (Fig. 1.13).

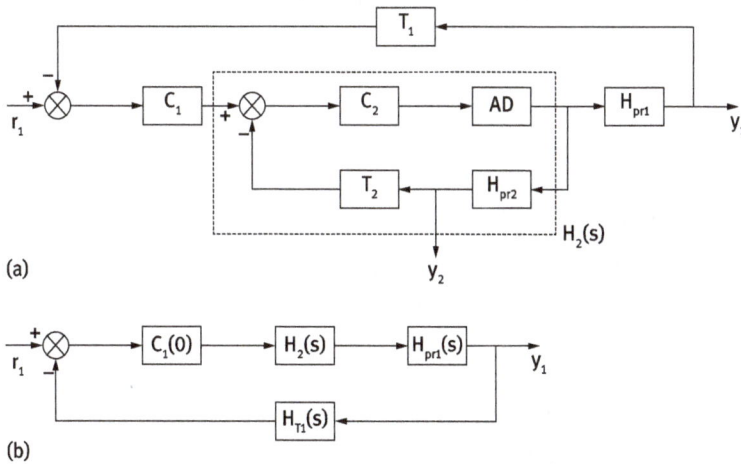

(a)

(b)

Fig. 1.13: The block diagram of the tuning of parallel cascade: (a) the "restructured" control system and (b) the equivalent system for tuning the C_1 controller.

The transfer function of the external loop is

$$H_1(s) = \cfrac{H_{C1} \cdot \cfrac{H_{C2}(s) \cdot H_{AD}(s) \cdot H_{pr2}(s) \cdot H_{T2}(s)}{1 + H_{C2}(s) \cdot H_{AD}(s) \cdot H_{pr2}(s) \cdot H_{T2}(s)} \cdot H_{pr1}(s)}{1 + H_{C1} \cdot \cfrac{H_{C2}(s) \cdot H_{AD}(s) \cdot H_{pr2}(s) \cdot H_{T2}(s)}{1 + H_{C2}(s) \cdot H_{AD}(s) \cdot H_{pr2}(s) \cdot H_{T2}(s)} \cdot H_{pr1}(s) \cdot H_{T2}(s)} \tag{1.12}$$

ⓘ Example 1.3

Tuning the parameters in a parallel cascade system.

Let us take the example given by Luyben [5]. He defined the master loop (loop 1 in previous notation) as that of composition and slave loop (loop 2 in previous notation) as that of temperature with the consequent notations. For clarity reasons, we decided to retain Luyben's notations.

Considering the system in Fig. 1.11, Luyben particularized the transfer functions of the process as (Fig. 1.14):

$$H_{pr1}(s) = H_{prM}(s) = \frac{1}{(T_M s + 1)^2} \text{ and } H_{pr2}(s) = H_{prS}(s) = \frac{1}{(T_S s + 1)^2} \tag{1.13}$$

by giving only one value for $T_S = 1$ min and several values for $T_M = 0.5, 1, 2, 4$ min, respectively; T_S and T_M stand for the time constants of the process transfer functions of slave and master loops, respectively.

For the master loop, additional dead time was considered $D = 0.5, 1, 2$ min to notice the difference in calculating the gain of the master controller, GCM. The controllers are considered proportional, and GCS was set to 1 to obtain a damping ratio of 0.707.

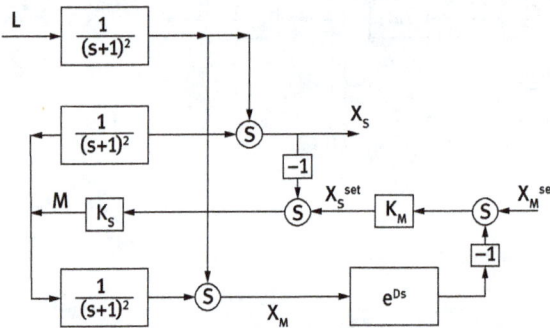

Fig. 1.14: Particular block diagram for the alternatives feedback and parallel cascade system in Fig. 1.11. In the figure, $T_S = T_M = 1$ min. Meaning of notations is: L – load disturbance; D – dead time; M – manipulative variable; $X_M{}^{set}$ and $X_M{}^{set}$ – setpoints of master and slave controllers; K_M and K_S – gains of master and slave controllers.

The tuning of the master controller is given in Tab. 1.1. Luyben used the root locus method for tuning [6, 7]. But other tuning methods, as presented in [3], give the same results.

Tab. 1.1: The results of the master P controller tuning for different ACSs (feedback, series, and parallel cascade) for different time constants and dead times.

T (min)	K_M ACS feedback		K_M ACS cascade		Additional dead time τ (min)	K_M ACS feedback	K_M ACS cascade
	Series	Parallel	Series	Parallel		Series	Parallel
0.5	0.7	3	0.6	2.5	0.5	2.500	2.500
1.0	0.7	3	1	2	1	1.350	1.560
2.0	0.7	3	2	4	2	0.862	1.136
4.0	0.7	3	3	13		$T = 1$ min	

$K_S = 1$ and the damping coefficient of the slave loop is 0.7.

The supplementary dead time can belong to a gas chromatograph in the composition loop, meaning that the transducer T_1 (T_M in Luyben's version) has the transfer function e^{-Ds}, where D takes different values (from 0.5, 1, or 2 min) to see the influence of this element on the tuned values of the master controller: the larger the D is, the smaller the controller gain is, due to the reduced controllability of the process plus transducer.

More recently, there have been attempts to revise the tuning techniques by proposing procedures of auto-tuning that are extensively discussed in [8].

The results reported in [8–11] are encouraging the use of parallel cascade control, with amendments to the original thinking of Luyben as adding a dead-time compensator for the dead time existing in the outer loop (Figs. 1.1 and 1.15). The proposed control structure [11] uses the outer-loop controller G_{C1} in the feedback path. As a result, although G_{C1} is meant to reject the load disturbance, it contributes to the stabilization of the process in the outer loop.

Actually, as Luyben affirms, one cannot truly appreciate the performances of different types of cascade control but only in comparison with the noncascade situation (see Fig. 1.16). It is obvious that the performance of the cascade control loop is better (shorter settling time and smaller steady-state error). In the cascade system, the manipulative variable immediately begins its intervention when load disturbance occurs, whereas in the feedback system action starts later with the time elapse of the dead time.

If cascade control is not efficient enough, a more complex control system should be used.

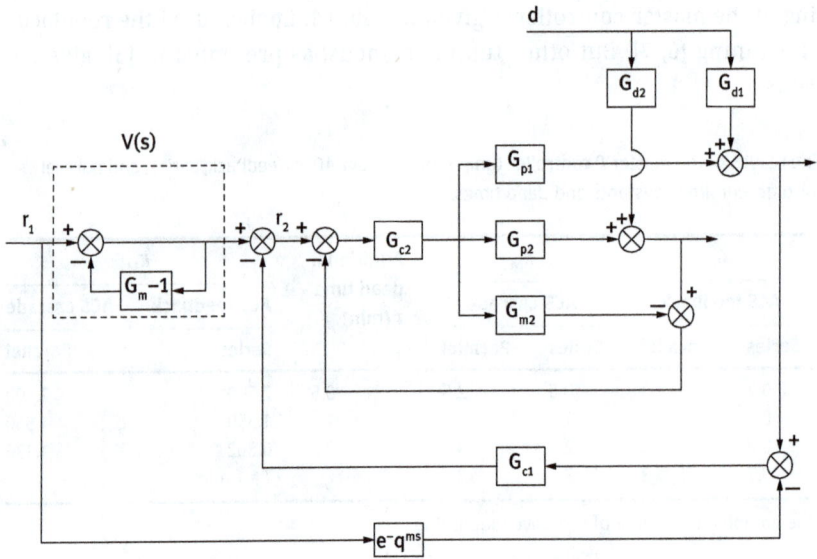

Fig. 1.15: Parallel cascade with dead-time compensator [11]. G_{p1} and G_{p2} stand for transfer functions of the primary and secondary processes; G_{m1} and G_{m2} for those of the primary and secondary process models, and G_{d1} and G_{d2} for the disturbance paths in the primary and secondary loops, respectively.

1.2 Feedforward control systems

In [1], when the controllability of the process is discussed, for a poor controllability, a superior organization of the control system is needed: either a feedforward or a model predictive structure. In this section, we propose to examine the feedforward control.

Feedforward control (Fig. 1.17) is one of the most advanced control structures used on a large scale in industry.

It captures the most important disturbance of the process, measures, and processes it in a disturbance controller (DC) in such a way that when the disturbance action is propagated through process D path to the exit, it is totally annulled by the counteraction on the path "DT-DC-AD-process m" (Fig. 1.18).

The DC must be so "intelligent" that its action should be perfectly coordinated in time and magnitude with the action of the main disturbance D in the process to completely counteract it. Thus, it has to contain all the information related to the gains and delays in all elements involved: process D, process m, DT, and AD. As observed from Fig. 1.16, the result of the feedforward control is the sum of the symmetric actions on both process D and "DT-DC-AD-process m" paths, meaning that the disturbance does not in fact disturb the process at all. To obtain such a result, one must know the dynamic behavior of all elements, meaning that their models are

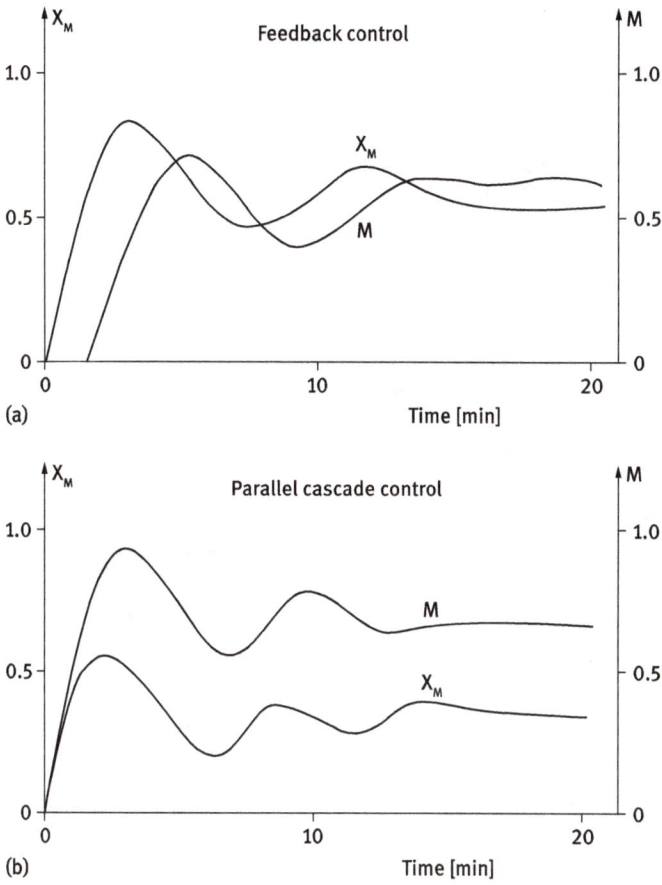

Fig. 1.16: The comparative results of controlling the output of a 2-min dead time for the primary process: (a) feedback control and (b) parallel cascade control. X_M stands for the controller output and M for the manipulative variable [5].

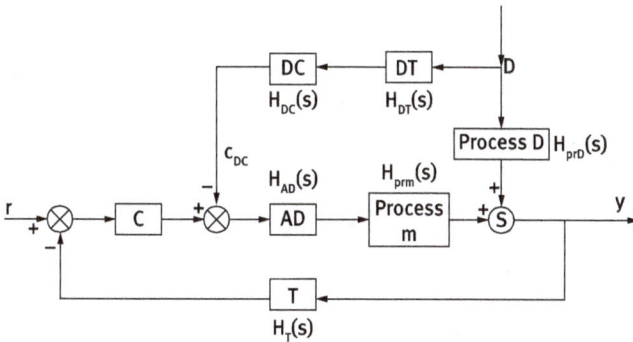

Fig. 1.17: Feedforward ACS block diagram.

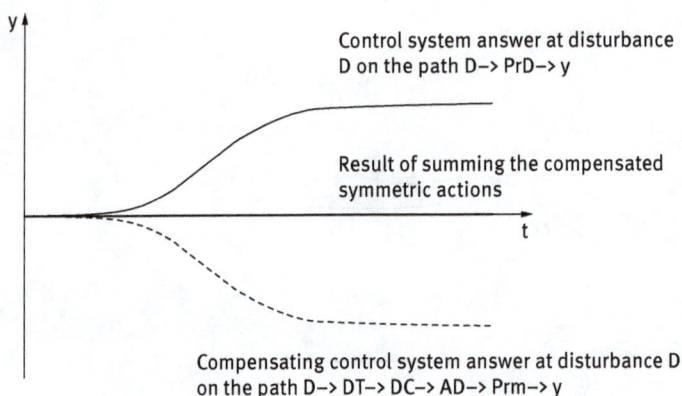

known. As D is not the only disturbance on the process, but the main one, the effect of the other disturbances is rejected through a regular feedback ACS containing the regular controller (C), transducer (T), and actuating device (AD).

The synthesis of the DC is realized based on the desired behavior of the system expressed mathematically as follows:

$$\frac{\Delta y(t)}{\Delta D(t)} = 0 \text{ or } \frac{Y(s)}{D(s)} = 0 \tag{1.14}$$

This means that the effect of the disturbance on the process output-controlled variable is 0.

Thus,

$$Y(s) = [C(s) - CD(s)] \cdot H_{AD}(s) \cdot H_{prm}(s) + D(s) \cdot H_{prD}(s) \tag{1.15}$$

with

$$C(s) = H_C(s)[R(s) - H_T(s) \cdot Y(s)] \text{ and } CD(s) = H_{DC}(s) \cdot H_{DT}(s) \cdot D(s) \tag{1.16}$$

Equation (1.14) becomes

$$Y(s) = R(s)\frac{C(s) \cdot H_{AD}(s) \cdot H_{pr\,m}(s)}{1 + H_T(s) \cdot H_C(s) \cdot H_{AD}(s) \cdot H_{prm}(s)}$$
$$+ D(s)\frac{H_{prD}(s) - H_{DC}(s) \cdot H_{DT}(s) \cdot H_{AD}(s) \cdot H_{pr\,m}(s)}{1 + H_T(s) \cdot H_C(s) \cdot H_{AD}(s) \cdot H_{pr\,m}(s)} \tag{1.17}$$

To have no effect of the disturbance on the process, the numerator of the second term in eq. (1.17) has to be 0,

$$H_{prD}(s) - H_{DC}(s) \cdot H_{DT}(s) \cdot H_{AD}(s) \cdot H_{pr\,m}(s) = 0 \tag{1.18}$$

and from which the DC transfer function can be calculated:

$$H_{DC}(s) = \frac{H_{prD}(s)}{H_{DT}(s) \cdot H_{AD}(s) \cdot H_{prm}(s)} \tag{1.19}$$

It is remarkable that the DC contains all the information needed to compensate perfectly the effect of the disturbance in the process; all gains and delays on the propagation paths are embedded in its transfer function.

Example 1.4

Consider a distillation column subjected to the main disturbance, the feed flow. This is very often true in a
train of columns where the feed flows cannot be controlled. The controlled variable is the light component
concentration (or the temperature) in the bottom of the column. Figure 1.19 describes the technological
process of the column.

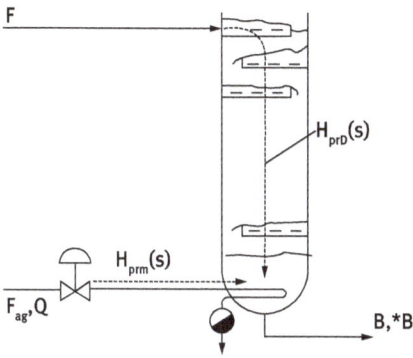

Fig. 1.19: Distillation process subjected to
feedforward control.

According to the mathematical model of the binary distillation column described in
[12], each tray can be described as having a capacitive behavior. Since we have k
trays on the concentration section of the column, the transfer function of transfer
path process D is

$$H_{prD}(s) = \frac{K_{eD}e^{-\tau_{eD}s}}{1 + T_{eD}s} = \sum_{1}^{k} \frac{K_i}{1 + T_i s} \tag{1.20}$$

where K_{eD}, T_{eD}, and τ_{eD} are the equivalent gain, time constant, and dead time, respectively, on the transfer path $F \to x_B$; K_i and T_i are the gain and the time constant for the
tray i (series of trays each with capacitive behavior), respectively.

On the path $m \to x_B$, the process has practically only one element, the bottom of
the column (including the reboiler or the steam serpentine). It exhibits a capacitive
behavior and has a transfer function of

$$H_{prm}(s) = \frac{K_B}{1 + T_B s} \tag{1.21}$$

where K_B and T_B are the gain and the time constant of the bottom, respectively.

The other elements of the feedforward control are the disturbance transducer (flow transducer), the actuating device (control steam valve) and the additional transducer (temperature transducer). All their delays (in comparison with the delays in the column) are insignificant; therefore, one can approximate their behavior as being proportional with the gains K_{DT}, K_{AD}, and K_T, respectively.

Thus, the transfer function of the feedforward controller for this case study is:

$$H_{DC}(s) = \frac{\dfrac{K_{eD}e^{-\tau_{eD}s}}{1+T_{eD}s}}{K_{DT} \cdot K_{AD} \cdot \dfrac{K_B}{1+T_B s}} = \frac{K_{eD}e^{-\tau_{eD}s} \cdot (1+T_B s)}{K_{DT} \cdot K_{AD} \cdot K_B \cdot (1+T_{eD}s)} \qquad (1.22)$$

and the corresponding feedforward control system is presented in Fig. 1.20.

Equation (1.22) can be very easily transformed in a practical feedforward control algorithm by using the Z-transform [13]:

$$c_{DC}(nT^*) = \frac{T_{eD}}{\beta} \cdot c_{DC}[(n-1)T^*] + \frac{K\alpha}{\beta} \cdot x_{DT}[(n-2)T^*] - \frac{KT_B}{\beta} \cdot x_{DT}[(n-3)T^*] \qquad (1.23)$$

where T^* is the sampling time chosen so that $\tau_{eD} = kT^*$; $K = K_{DT} \cdot K_{AD} \cdot K_B$; $\alpha = TB + T^*$ and $\beta = T_{eD} + T^*$; x_{DT} is the signal of the disturbance transmitter at different sampling times.

In some works of Niesenfeld [14, 15], the authors mentioned an experiment worth citing: they measured the performances of a feedforward control system applied to a distillation column, exactly in the format mentioned in Example 1.4. The results are exposed in Tab. 1.2.

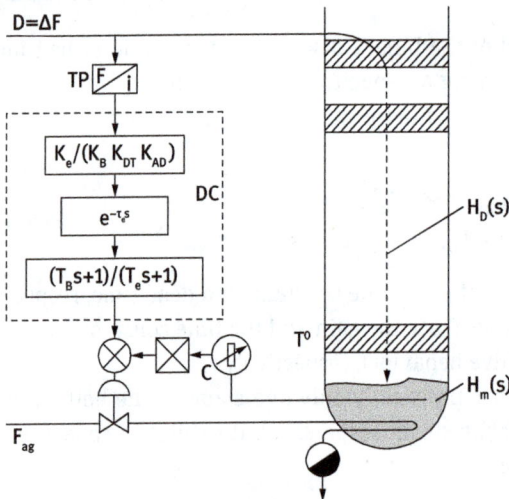

Fig. 1.20: Feedforward control system of the distillation column from Fig. 1.17.

Tab. 1.2: The results of the feedforward control of the distillation column in [14, 15].

Disturbance ΔF (%) from F_{max}	ACS structure	Oscillations damping time (min)
60–70%	PID	10
	Feedforward	No oscillations
70–80%	PID	Maintained oscillations
	Feedforward	3

It is worth mentioning that due to the increased stability of the process, a 15% saving in energy consumption was estimated.

Another experiment carried out in 1987 at the Brazi Refinery in Romania by a group of professors from the Petroleum and Gas Institute, Ploieşti, Romania [16] reported 20% energy savings when using a benzene–toluene column with 90 trays and a 2-m diameter. The usually reported consumption of steam was in the average of 4 t/h.

To give an idea on the significance of a 20% savings, let us make a cost estimate. At the average price of US $70/Gcal and a total operating timespan of 7200 h/year, with a latent heat of 525 kcal/kg, the total amount of consumption for the mentioned column would be:

$$Q_{year} = 7200 \text{ h} \cdot 4000 \text{ kg/h} \cdot 525 \text{kcal/kg} = 15{,}120 \text{Gcal/year}$$

The monetary savings per year are MS = 0.2 × 15,120 Gcal/year × US $70/Gcal = US $211,680.

The investment for such an industrial application can be estimated at a total of US $123,000 by including:

Human resources (two process engineers to work on process models and computer algorithms) ~ US $8000/month = US $96,000/year.
A process computer ~ US $10,000.
Control equipment (controller, transducers, control valve) ~ US $15,000.
Electric and air fittings, rack, manpower ~ US $2000.

This means that the investment can be recovered in about 7 months.

Furthermore, a process computer, the major infrastructure additional investment, can manage not only one but several applications.

There are situations when more important disturbances are significantly influencing the process and they have to be considered in the feedforward control [17–21], together with more advanced techniques as the model predictive control.

Example 1.5
Consider a CSTR (Fig. 1.21) in which a first-order exothermic reaction $A \rightarrow B$ takes place. k and ΔH_r are the rate constant and heat of reaction, respectively. The process is disturbed essentially by C_{Ai} and $T_i°$.

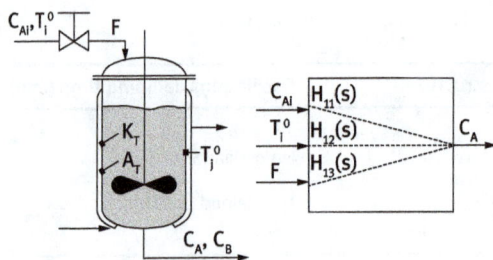

Fig. 1.21: CSTR subjected to two measurable main disturbances. T_j° is the jacket temperature; K_T and A_T are the heat transfer characteristics.

The output-controlled variable is C_A and the manipulative variable is F. The difference from the previous example of feedforward control is that there are two important disturbances instead of one, which have as consequence the existence of two disturbance controllers, D_{C1} for C_{Ai} and D_{C2} for T_i°. The two process transfer paths for the two disturbances are described by the following equations:

$$V\frac{dC_A}{dt} = F(C_{Ai} - C_A) - VkC_A \tag{1.24}$$

$$Vpc_p\frac{dT0^\circ}{dt} = Fpc_p\left(T_i^\circ - T_j^\circ\right) - \Delta H_r VkC_A - K_T A_T\left(T^\circ - T_j^\circ\right) \tag{1.25}$$

$$k = k_0 e^{-\frac{E}{RT^\circ}} \tag{1.26}$$

The linearized equations, by following the procedure from [22] to obtain linear transfer functions, are presented in the following equations:

$$\frac{dC_A}{dt} = a_{11}C_A + a_{12}T^\circ + a_{13}C_{Ai} + a_{15}F \tag{1.27}$$

$$\frac{dT^\circ}{dt} = a_{21}C_A + a_{22}T^\circ + a_{24}T_i^\circ + a_{25}F + a_{26}T_j^\circ \tag{1.28}$$

In these orders 1–6 of the correlation coefficients' indexes are allocated to the variables C_A, T°, C_{Ai}, T_i°, F, and T_j°, respectively.

$$a_{11} = -\frac{F_n}{V} - k_n; \; a_{12} = -\frac{C_{An}Ek_n}{RT_n^{\circ 2}}; \; a_{13} = \frac{F_n}{V}; \; a_{15} = \frac{C_{Ai} - C_{An}}{V}$$

$$a_{21} = -\frac{\Delta H_r K_n E}{RT_n^{\circ 2}pc_p} - \frac{F_n}{V} - \frac{K_T A_T}{Vpc_p}; \; a_{24} = \frac{F_n}{V}; \; a_{25} = \frac{T_i^\circ - T_n^\circ}{V}; \; a_{26} = \frac{K_T A_T}{Vpc_p}$$

The variables with the index n are at steady state and have steady-state values. Density, specific heat, and volume are considered constant.

The block diagram of the feedforward control system is presented in Fig. 1.22.

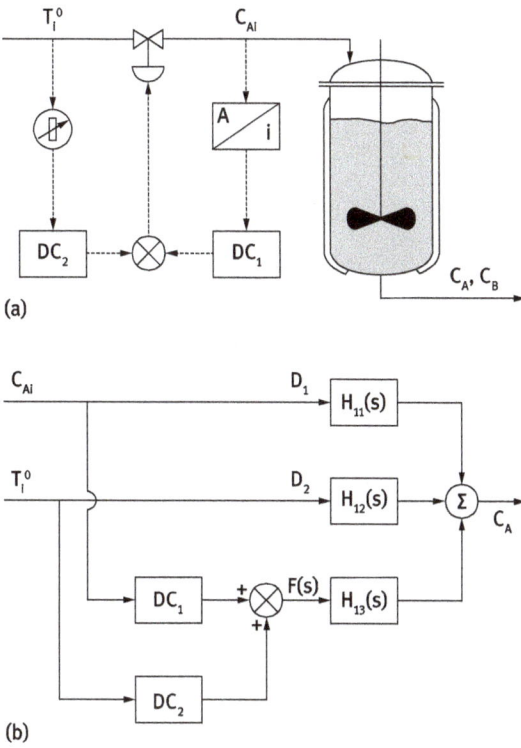

(a)

(b)

Fig. 1.22: Block diagram of the CSTR feedforward control of both main disturbances: C_{Ai} and T_i^o: (a) feedforward scheme without feedback control and (b) block diagram of the feedforward CSTR control system with two identified disturbances.

Thus, the transfer functions which describe the system are:

$$C_A(s) = H_{11}(s)C_{Ai}(s) + H_{12}(s)T_i(s) + H_{13}(s)F(s) + H_{14}(s)T_j(s) \tag{1.29}$$

$$T(s) = H_{21}(s)C_{Ai}(s) + H_{22}(s)T_i(s) + H_{23}(s)F(s) + H_{24}(s)T_j(s) \tag{1.30}$$

and from eqs. (1.27)–(1.28), the following is obtained:

$$sC_A(s) = a_{11}C_A(s) + a_{12}T(s) + a_{13}(s)C_{Ai}(s) + a_{15}(s)F(s) \tag{1.31}$$

$$sT(s) = a_{21}C_A(s) + a_{22}T(s) + a_{24}T_i(s) + a_{23}F(s) + a_{24}T_j(s) \tag{1.32}$$

Finally,

$$C_A(s) =$$

$$\frac{a_{13}(s - a_{22})}{s^2 - (a_{11} + a_{22})\, s + a_{11}a_{22} - a_{12}a_{21}}\, C_{Ai}(s) + \frac{a_{12}a_{24}}{s^2 - (a_{11} + a_{22})\, s + a_{11}a_{22} - a_{12}a_{21}}\, T_i(s) + \quad (1.33)$$

$$\frac{a_{12}a_{23} + a_{15}(s - a_{22})}{s^2 - (a_{11} + a_{22})\, s + a_{11}a_{22} - a_{12}a_{21}}\, F(s) + \frac{a_{12}a_{26}}{s^2 - (a_{11} + a_{22})\, s + a_{11}a_{22} - a_{12}a_{21}}\, T_j(s)$$

and

$$T(s) =$$

$$\frac{a_{13}a_{21}}{s^2 - (a_{11} + a_{22})\, s + a_{11}a_{22} - a_{12}a_{21}}\, C_{Ai}(s) + \frac{a_{24}(s - a_{11})}{s^2 - (a_{11} + a_{22})\, s + a_{11}a_{22} - a_{12}a_{21}}\, T_i(s) + \quad (1.34)$$

$$\frac{a_{15}a_{21} + a_{25}(s - a_{11})}{s^2 - (a_{11} + a_{22})\, s + a_{11}a_{22} - a_{12}a_{21}}\, F(s) + \frac{a_{26}(s - a_{11})}{s^2 - (a_{11} + a_{22})\, s + a_{11}a_{22} - a_{12}a_{21}}\, T_j(s)$$

The condition for invariability of the output concentration is $C_A(s) = 0$, thus:

$$C_A(s) = H_{11}(s)C_{Ai}(s) + H_{12}(s)T_i(s) + H_{13}(s)F(s) = 0 \quad (1.35)$$

and then, from (1.19):

$$H_{DC1}(s) = \frac{H_{11}(s)}{H_{13}(s)} \text{ and } H_{DC2}(s) = \frac{H_{12}(s)}{H_{13}(s)} \text{ or}$$

$$H_{DC1}(s) = \frac{a_{13}(s - a_{22})}{a_{12}a_{23} + a_{15}(s - a_{22})} \text{ and } H_{DC2}(s) = \frac{a_{12}a_{24}}{a_{12}a_{23} + a_{15}(s - a_{22})} \quad (1.36)$$

The efficiency of the feedforward control with a change of input flow on the bottom concentration is shown in Fig. 1.23.

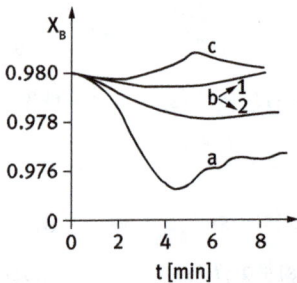

Fig. 1.23: The efficiency of the feedforward control with an input change in the molar feed flow of a distillation column: (a) feedback control; (b) feedforward control only (1 – for process 1; 2 – for process 2); and (c) combined feedforward and feedback.

One important drawback is that linearization decreases the degree of accuracy. Research has been done in the direction of nonlinear feedforward control.

To summarize, the feedforward control is a very successful control strategy in the case of poor controllability processes. However, it has some drawbacks:
- The main disturbances should be measurable and captured.
- It is quite difficult to elaborate good dynamic models for the entire process; in addition, in case of linear feedforward control, linearization reduces the accuracy of the model.
- High level of competency is requested from process engineers.

For a long time, the interest in development of feedforward control was somehow tempered and determined especially by the fact that in many situations the ideal algorithms elaborated from process models could not be implemented in practice. Recent developments including internal model control and model predictive control, or "neuromorphic" control by using artificial neural networks and robust nonlinear controllers are nowadays presented in the literature [23–25].

1.3 Ratio control systems

In many situations of process engineering, one has to keep constant the ratio between two flows of material, especially when one of the flows cannot be controlled from the beginning: this is the situation of keeping constant the ratio between reflux and distillate in a distillation column, or the ratio of flows of the reactants in a CSTR, or the ratio of flows of primary and secondary solvents in a liquid-liquid extraction process, or the air-gas ratio in a burner. The general block diagram of the ratio control systems is given in Fig. 1.24.

The noncontrolled, independent flow is sometimes named "wild feed" and the controlled, dependent flow, closely following the noncontrolled one, is called "controlled feed". In version (a) of the ratio control system, the two flows are measured with flow transducers and the values of their signal are divided inside a ratio relay; the controller becomes a ratio controller, with the setpoint equal to the desired ratio between the two flows. The controller acts by modifying the controlled feed stream in such a way that the ratio between the wild and the controlled streams is kept at the setpoint.

$$r = \frac{F_B}{F_A} = r_{set} \tag{1.37}$$

In this case, the control system reacts slowly when the value of the controlled stream is relatively large and acts faster when the value of the controlled stream is relatively small.

In version (b) of the ratio control system, the wild feed is multiplied with a constant r in the multiplication element and the result is fed as the setpoint to the controller; the result is that the controlled stream is kept by the flow control system to the setpoint (variable) value that varies with flow A:

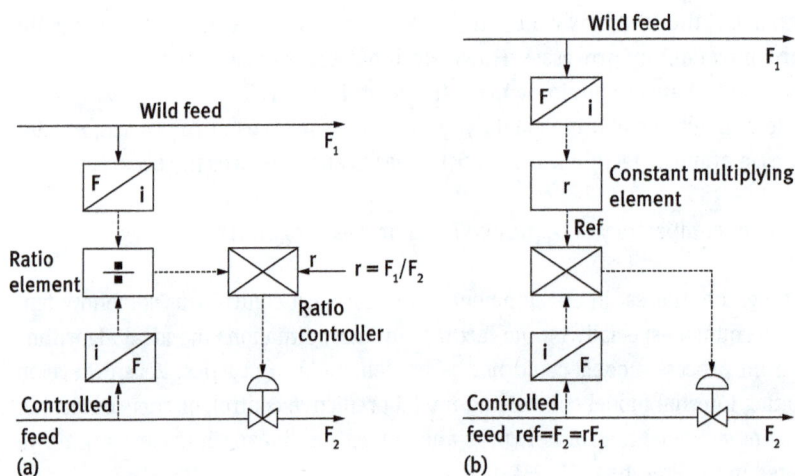

Fig. 1.24: Ratio control systems in two possible versions: (a) ratio control based on variable ratio error and (b) ratio control based on error of the flow subjected to ratio constraint.

$$F_B = F_{B\,set} = rF_A \qquad (1.38)$$

In this version, the action of the control system is independent from the relative amount of the controlled feed stream.

Example 1.6

One good example of the application of ratio control systems is that of a reversible reaction $aA + bB \Leftrightarrow cC$ taking place in a CSTR (see Fig. 1.25); to maximize the production of C, one has to move the equilibrium of the reaction to the right, keeping an excess of B over A. It means that whenever the flow of A increases, the ratio control system automatically increases the flow of B, and keeps constant a certain excess ratio value (Fig. 1.22). In manual control operation mode of a CSTR, the highest excess of B over A is often used by fixing the controlled flow at maximum. This is feasible, yet the disadvantages are the following:

- In the case of economically natural recycling of B, the stage following the CSTR is the separation whose costs depends on the molar fraction of B in the mixture $B - C$. The higher the fraction of B, the higher the separation costs.
- In the case of not recycling B due to its very low price, either there is no separation stage and B severely impurifies C because of its high quantity, or there exists a separation stage with the adjacent costs so that a solution has to be found for depositing and further using B.

The advantages of the automatic ratio control are thus obvious.

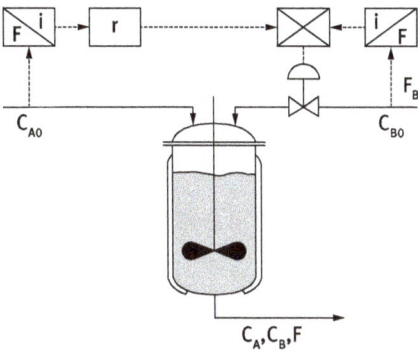

Fig. 1.25: CSTR with ratio control. F_A and F_B are the volumetric reactants flow rates; C_{Ai} and C_{Bi} are the input molar concentrations of the reactants.

From chemical kinetics information [26], the excess factor that has to be respected is

$$\gamma = \frac{aC_{B0}}{bC_{A0}} \tag{1.39}$$

where C_{A0} and C_{B0} are the initial concentrations of both reactants after mixing their flows:

$$C_{A0} = \frac{F_A \cdot C_{Ai}}{F_A + F_B} \text{ and } C_{B0} = \frac{F_B \cdot C_{Bi}}{F_A + F_B} \tag{1.40}$$

Keeping in mind that the ratio control system operates with volumetric flows and keeps constant their ratio, the value of setpoint ratio r calculated from eqs. (1.38)–(1.40) is

$$r = \frac{F_B}{F_A} = \gamma \frac{bC_{Ai}}{aC_{Bi}} \tag{1.41}$$

By supposing that in the CSTR the following reaction takes place: $FeCl_3 + 3H_2O \Leftrightarrow Fe(OH)_3 + 3HCl$, with the mass concentration of the ferric chloride solution of 20%, by knowing the molecular weights of feed solution chemicals $FeCl_3$ and H_2O (162,2 and 18 kg/kmol, respectively), their densities (1135 and 1000 kg/m³, respectively), and finally by using eqs. (1.38)–(1.40), the molar concentrations of feed solutions can be calculated. These are 36.77 kmol/m³ and 55.6 kmol/m³. The values yield $\gamma = 0.22$ and $r = 0.43$, and can be fixed on the ratio relay.

The approximate variation of the flows in the ratio control system is given in Fig. 1.23. The tuning of the controller parameters is based on the process response curve method [4] to $K_{P\,opt} = 0.35\,K_{pr}\frac{T_{pr}}{\tau_{pr}}$ and $T_{i\,opt} = 1.2\tau_{pr}$.

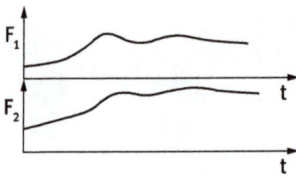

Fig. 1.26: Evolution of flows in the ratio control system. The controlled flow F_2 follows with a small delay in the wild feed F_1.

1.4 Inferential control systems

Sometimes it is useful to not control directly one variable, but a calculated one. For example, it is better to control the mass flow instead of volumetric flow of liquids or gases, especially when large variations of temperature occur.

It is known that the mass flow can be calculated as $F_m = F_v \rho = F_v \frac{pMg}{RT^\circ}$, where the volumetric flow F_v is measured with an orifice plate transducer. If one limits the measurement at the flow transducer, the mass content remains unknown. This is why in inferential control, measurements of pressure, temperature, and additional computing elements are added (Fig. 1.27).

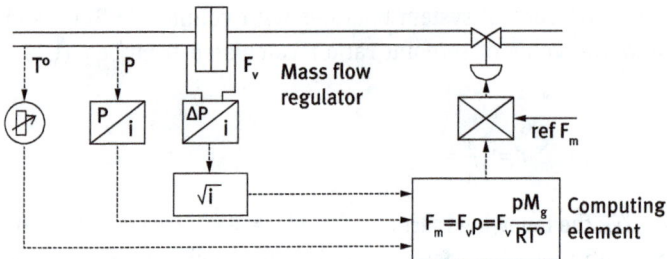

Fig. 1.27: Inferential control of the mass flow of a gas.

Another interesting example is the cascade control of heat content of the reboiler in the bottom of a distillation column (Fig. 1.28).

The controlled variable of the inner loop of the cascade is the heat flow depending on the input and output temperatures of the steam in and out streams from the reboiler (T_1° and T_2°). Thus, the controlled variable in the inner loop is $Q = F\rho \left[c_p \left(T_1^\circ - T_2^\circ \right) + l_v \right]$, where c_p, the specific heat, and l_v, the latent heat of vaporization, are considered constant in the usual range of temperature variation.

Thus, the final controlled parameter is the bottom temperature, T_o, and the intermediate controlled one is Q.

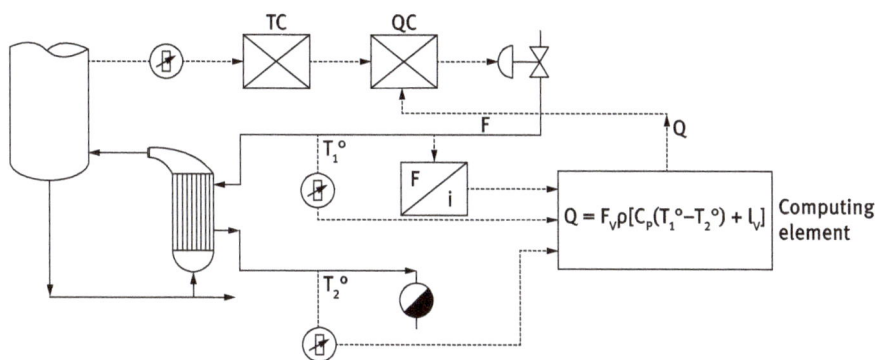

Fig. 1.28: Bottom temperature cascade control with the inner controlled variable Q.

1.5 Selective control systems

There are situations when the appropriate functioning of a control loop depends not only on the controlled variable, but also on other variables. Usually, the level at the bottom of a separation column is controlled through the bottom flow B. The temperature profile in the stripping section can be kept constant through the heat flow into the reboiler. In start-up or shut-down situations or on strong disturbances, the level at the bottom may decrease even when the bottom control valve is completely closed. This happens because the vapor flow is higher than the reflux flow from the first tray. If there is no attempt to reduce the heat entering the reboiler, then the column could work with an empty bottom, which can in return destroy it. If an operator observes the event, he/she shifts the operating mode of the level control loop from AUTOMATIC to MANUAL mode and closes the steam valve of the reboiler. This operation requires the operator to have high professional expertise and qualification. A selective control system (Fig. 1.29) can solve this problem.

The selective system has in its composition a selector switch that can choose which control signal is predominant (T_B° or L index B) in manipulating the control valve of the reboiler. If the bottom is nearly empty (supposing a level transducer with the output signal 4–20 mA) at a transducer signal of 5 mA, the value selector VS takes the decision and conveys the smallest signal received: either the temperature signal or the level signal. This way, the steam valve of the reboiler is closed at higher-than-normal temperature, and if this is not the case, at the minimum level signal. It must be mentioned that in this case the bottom flow is reduced to 0.

Similar selective control systems are designed to locate the "hot spot" in plug flow reactors and to select which is the important signal to be considered in the control loop (Fig. 1.30). The position of the "hot spot" is not always the same; it depends on other operating conditions as well.

The temperatures measured along the reactor are fed to a selector switch and the highest value is transmitted to the controller, in order to be controlled.

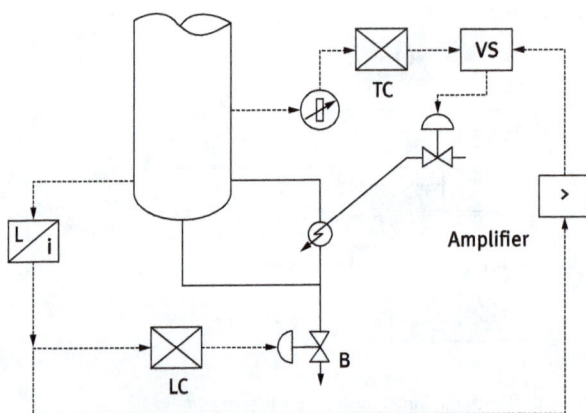

Fig. 1.29: Selective control system for the temperature and level of the bottom of the column.

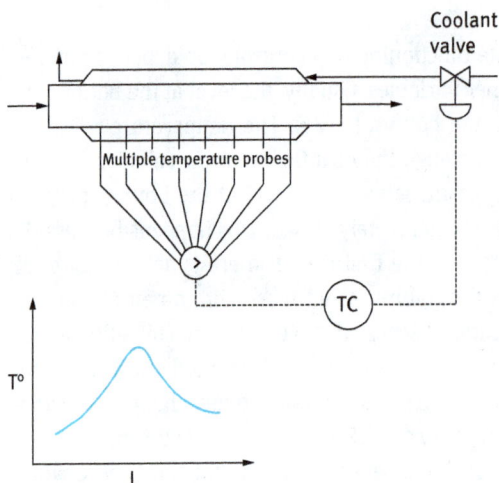

Temperature profile (qualitative representation along the length of a tubular reactor)

Fig. 1.30: Selective control loop of the highest temperature of a tubular reactor.

1.6 Combinations of different nonconventional CS

1.6.1 Cascade–feedforward–ratio control [26]

In some "difficult-to-control" situations [26], a combination of the three previous configurations is proposed. There are situations in industry, when, the change of the ratio setpoint is critical and the change has to be done in minimum time in order to affect the
process as little as possible.

The authors applied this combination to a wastewater treatment process for ammonium removal. Such a process is described in [27].

The example is given in the very unstable process of controlling the dissolved oxygen (DO) in the biomass. The reactors in discussion are those with biomass retention (details about wastewater treatment control are found in Chapter 15. Wastewater treatment systems are roughly bioreactors in which some parameters are important to be kept in tight limits).

Ammonium (the cation NH_4^+) removal from wastewater can be done either using conventional nitrification-denitrification over nitrite or using anaerobic ammonia oxidation. Nitrification is a two-step reaction in which the ammonium is oxidized into nitrite by ammonium-oxidizing bacteria (AOB) and further on, goes to nitrate using nitrite-oxidizing bacteria (NOB). In [26], to remove ammonium, the authors opted for partial nitration coupled with ammonium anaerobic oxidation (Anammox) process. The scope is to convert half of the ammonium contained in the influent stream into nitrite, to allow both (ammonium and nitrate) to be converted to nitrogen through Anammox reaction. To avoid transformation of the nitrite in nitrate, AOB must be in excess over NOB. In this process, the DO concentration plays an important role, the slave/internal controller of the cascade keeping its concentration (C_{O_2}) constant. This suppresses unwanted NOB, encourages the wanted AOB, and keeps a suitable ratio between nitrite and ammonium (setpoint for the ratio computing element) needed for coupling with the Anammox reaction.

Considering the cascade control loop (Fig. 1.31), the master controller aiming to regulate the ammonia/nitrite ratio generates the setpoint for the DO concentration slave controller. The master control is a combination of the feedforward controller giving the optimal oxygen concentration for the current influent ammonium and nitrite concentrations, and a feedback controller based on the measurement of the effluent nitrite/ammonium concentrations ratio.

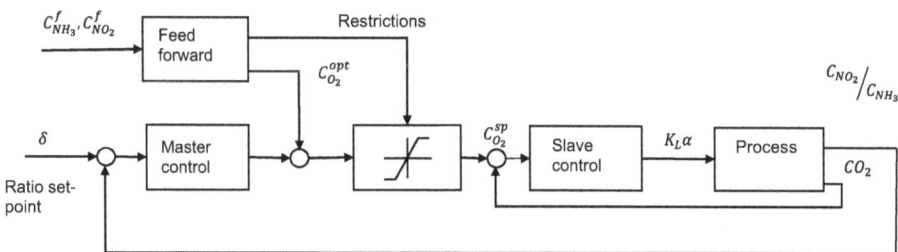

Fig. 1.31: Proposed control strategy for setting up the ratio between ammonia and nitrate. Based on the process inputs, the feedforward block calculates the DO constraints and the optimal value $C_{O_2}^{opt}$. The master controller compensates the errors in $C_{O_2}^{opt}$ to obtain the desired effluent ratio C_{NO_2}/C_{NH_3} (δ). The corrected value is compared with the constraints imposed, and DO setpoint $C_{O_2}^{sp}$ is obtained. The slave-controller tracks the DO set point by adjusting the $k_L a$.

The feedforward controller calculates the optimal DO concentration based on the influent wastewater stream flowrate and concentrations of ammonia and nitrate. The calculated DO concentration setpoint is compared with the constraints to really assure the suppression of unwanted NOB and the appropriate quantity of AOB. For this reason, a saturation function is applied to the setpoint before being applied to the slave controller. The conditioned setpoint is the one making the controller's output equal to the saturation value. The feedforward controller contributes to a faster response to the disturbances in the influent, but it is not absolutely necessary. The output of the combined control loop is the necessary value of the ratio of nitrite/ammonium concentrations. The simulations (Fig. 1.32) showed the importance of the control scheme in the two cases of the high and low ammonium influent concentration scheme (left and right).

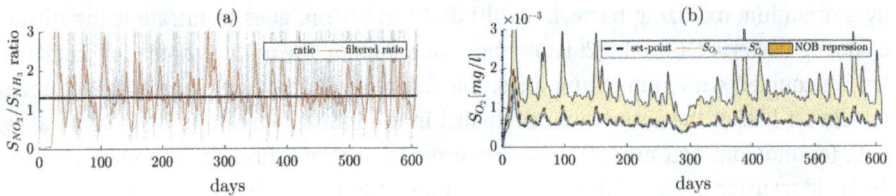

Fig. 1.32: The ratio values are different in the case of high ammonium influent concentration (a) and low ammonium influent concentration (b) (reprinted with the permission of Elsevier [26]). Note: In the figures reproduced, the concentrations are denoted by S.

1.6.2 Inferential cascade control [28]

The application is in the powder milk manufacturing, more precisely in the evaporation section of the powder milk process. The present example is a good representation of application of advanced process control in a process engineering field other than chemical engineering, that of food and beverage industry. The traditional control of the evaporation is treated in Chapter 12 of this book.

Food and dairy industries use widely multieffect falling film evaporators. A falling film evaporator [29] is a heat exchanger that uses a shell and tube design to evaporate heat-sensitive liquids (the proteins in milk are very temperature-sensitive). Evaporators are fed at the top and then the feed is uniformly dispersed throughout the heating tubes. Partially evaporated, the liquid flows through tubes in a thin layer on the tube walls, with a high heat exchange coefficient. The heating agent is usually steam. Under gravity, the liquid and vapor move downhill. The vapor flow in the co-current direction aids the liquid's downward descent. At the bottom of the falling film evaporator unit, the concentrated product and its vapor are separated from one another in the separators, which are vapor traps (Fig. 1.33).

Due to large time delay induced in the chain of evaporators and because of the process disturbances (variable steam pressure, cooling water temperature depending on climate or season, and concentration of the feed) it is difficult to control the total solid concentration (TSC) in the final product in a robust and accurate way. The classic con-

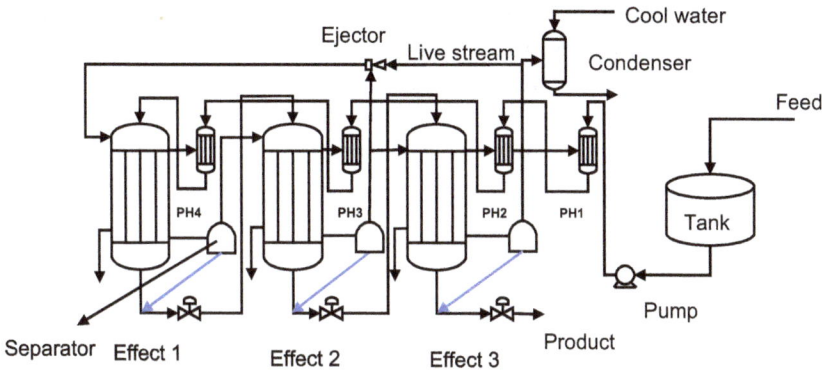

Fig. 1.33: Three-effect falling film evaporator.

trol scheme (Figure 12.2) has as main purpose the control of the concentration of total solids in the presence of disturbances. But it has the disadvantage of inducing transitional delays causing important deviations from the quality required from the product.

The important parameters to be controlled are the TSC, which are difficult-to-measure, and are replaced by temperatures measurements and, additionally, the bottom liquid levels for the sake of keeping mass balances constant and protection of the evaporators.

Using inferential cascade control, the control of three-effect falling film evaporator in Isfahan (Iran) milk powder factory was discussed. In Fig. 1.34, the block diagram of the TSC control is presented.

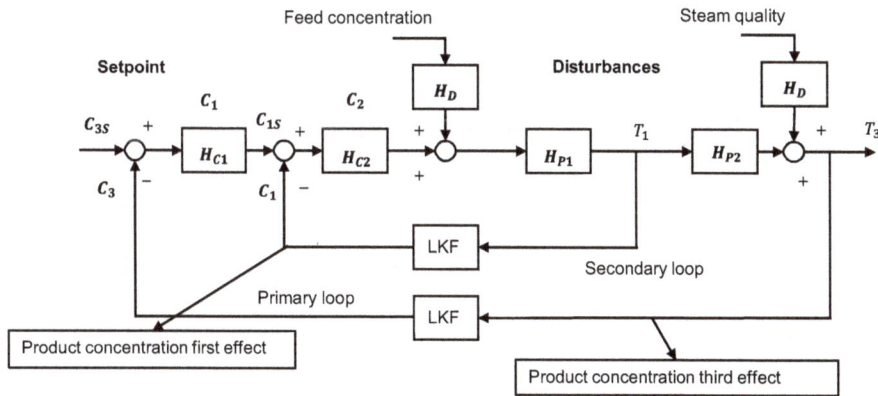

Fig. 1.34: Block diagram of inferential cascade control for TSC control loop.

Instead of measuring the solid concentration in a complicated way, the authors opted to measure the "easy-to-measure" temperatures and to estimate the concentrations.

Using Linear Kalman Filter (LKF) and process measurements, TSC of first effect product can be significantly estimated. The estimated state (Fig. 1.35) was used in inferential cascade control as secondary measurement.

The alternative could have been the estimation of the concentrations based on a steady state mathematical model. But as the authors point out, the estimation was poorer. From Fig. 1.35 one may see that the LKF estimation is good and can be used as a proxy for TSC measurement.

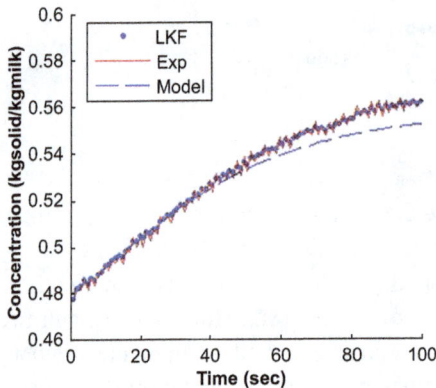

Fig. 1.35: Estimation through LKF versus model. Measurements were done on the real plant of powder milk in Isfahan, Iran (reprinted with the permission of Elsevier [28]).

The goal of the evaporator battery is to obtain a certain TSC after the third effect. Thus, the setpoint of the loop is C_{3S}. The master controller (CC$_1$) sets the setpoint for the slave controller (CC$_2$) which manipulates the steam valve of the first effect evaporator (transfer function H_{P1}). The transfer function H_{P2} stands for the effects 2 and 3. The disturbances are the feed concentration and the steam quality, the main one being the feed concentration. Therefore, it is identified and included in the feedforward loop (feedforward controller transfer function is H_D). The internal loop controls the first effect solid product concentration, estimated by LKF based on temperature (T_1°) measurement (see Tab. 1.3 [26]).

Tab. 1.3. Measurements and estimations.

Tab. 1.3: Output measurements and estimated states of the falling film three-effect evaporator.

Output measurements	Estimated states
First effect temperature	First effect TSC
Second effect temperature	Second effect TSC
Third effect temperature	Third effect TSC

Due to feedforward approach, the feed disturbance is immediately identified, and the system corrects the effects in advance, not waiting for through propagation of the signal through the whole plant.

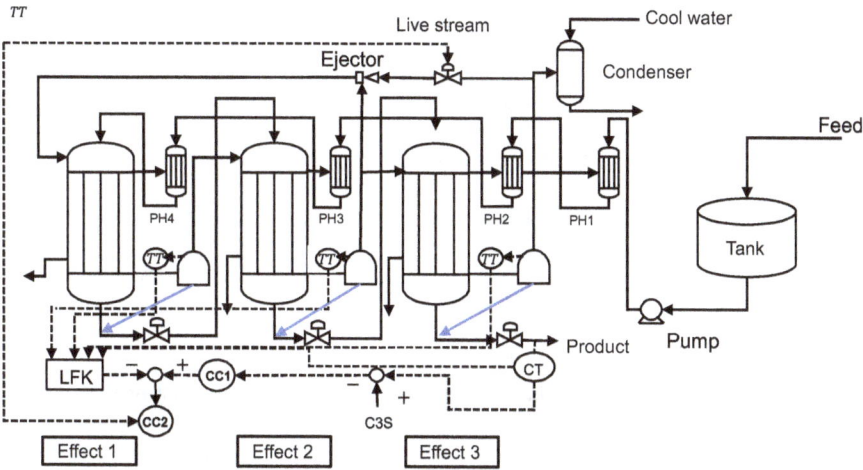

Fig. 1.36: The inferential cascade control applied to the three-effect evaporator.

The results of this approach are shown in Fig. 1.37. In both situations the deviations and the response time are reduced significantly, resulting in a better quality of the product and reduced steam consumption.

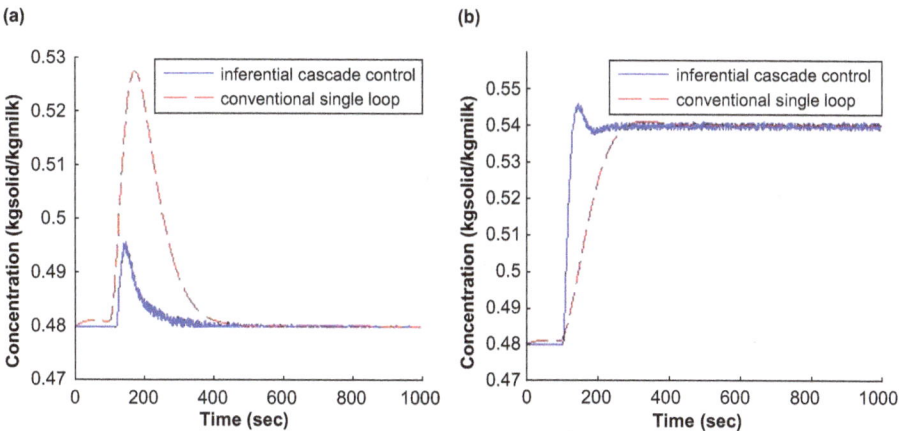

Fig. 1.37: The results of inferential feedforward control of the total solid concentration in milk, using Kalman filter (reprinted with the permission of Elsevier [28]). (a) The performance of the conventional versus inferential control loop on concentration at 18% change in feed concentration; (b) the performance of the conventional versus inferential control loop on TSC concentration at 6% change in setpoint change.

Many other combinations of traditional complex control schemes and other more advanced combinations with adaptive, optimal, predictive, fuzzy, internal model control schemes [30–33] were imagined and applied in process industries, but due to the limited space available, we just mention them.

References

[1] Agachi, P.S., Cristea, V.M., *Basic Process Engineering Control*, p. 142, Walter de Gruyter Gmbh, Berlin/ Boston, 2014.

[2] Agachi, P.S., Cristea, V.M., *Basic Process Engineering Control*, Walter de Gruyter Gmbh, Berlin/Boston, chapter 12, 2014.

[3] Agachi, P.S., Cristea, V.M., *Basic Process Engineering Control*, Walter de Gruyter Gmbh, Berlin/Boston, chapter 11, 2014.

[4] Agachi, S.. *Automatizarea proceselor chimice (control of chemical processes)*, p. 185, Casa Cărții de Știință, Cluj – Napoca, 1994.

[5] Luyben, W., *Parallel cascade control*, Industrial and Engineering Chemistry Fundamentals, 12, 463–467, 1973.

[6] Lumkes, J.H., Jr, *Control Strategies for Dynamic Systems. Design and Implementation*, p. 207, CRC Press.

[7] Baker, G., *PID tuning of plants with time delay, using root locus*, master theses and graduate research, Masteral Thesis Paper 4036, San Jose Scholar Works, San Jose University, http://scholarworks.sjsu. edu/etd_theses?utm_source=scholarworks.sjsu.edu%2Fetd_theses%2F4036&utm_medium= PDF&utm_campaign=PDFCoverPages, 2011.

[8] Rao, A.S., et al., *Enhancing the performance of parallel cascade control using smith predictor*, ISA Transactions, 48, 220–227, 2009.

[9] Bharati, M., et al., *Auto tuning of parallel cascade control using setpoint relay*, International Journal of Computer Applications, Special Issue on "Evolutionary computation for optimization techniques", Ecot, 2, 57–61, 2010.

[10] Vilanova, R., Visioli, A., *PID Control in the Third Millennium: Lessons Learned and New Approaches*, Springer, the Netherlands, chapter 8, 2012.

[11] Padhan, D.G., Majhi, S., *Synthesis of PID Tuning for a New Parallel Cascade Control Structure*, fr.a2, in *IFAC Conference on Advances in PID Control, PID'12*, Brescia (Italy), March 28–30, 2012.

[12] Agachi, P.S., Cristea, V.M., *Basic Process Engineering Control*, p. 87, Walter de Gruyter, Berlin/Boston, 2014.

[13] Agachi, S.. *Automatizarea proceselor chimice (chemical process control)*, p. 330, casa casa cărții de știință, Cluj Napoca, 1994.

[14] Nisenfeld, A., Miyasak, R., *Applications of feedforward control to distillation columns*, Automatica, 9, 319–327, 1973.

[15] Nisenfeld, A.E., *Reflux or distillate. which to control?* Chemical Engineering, 78(21), 169–171, 1969.

[16] Marinoiu, V., Paraschiv, N., Patrascioiu, C., *Conducerea cu calculatorul a unei coloane de distilare (Distillation column computer control)*, Contract Report, No. 18, CP Brazi, 1987.

[17] Congalidis, J., Richards, J., Harmon Ray, W., *Feedforward and feedback control of a solution copolymerization reactor*, AIChE Journal, 35(6), 891–907, 1989.

[18] Brauner, N., Lavie, R., *Feedforward quality control through inventory manipulation in periodically perturbed mixed accumulators*, Computer and Chemical Engineering, 14(9), 1025–1029, 1990.

[19] Rovaglio, M., Ranzi, E., Biardi, G., Fontana, M., Domenichini, D., *Rigorous dynamics and feedforward control design for distillation processes*, AIChE Journal, 36(4), 576–586, 1990.

[20] Sheffield, R.E., *Integrate process and control system design*, Chemical Engineering Progress, 88(10), 30–35, 1992.

[21] Biao, H., *Feedforward plus feedback controller performance assessment of MIMO systems*, Control Systems Technology, IEEE Transactions, 8(3), 580–587, 2002.

[22] Agachi, P.S., Cristea, V.M., *Basic Process Engineering Control*, Walter de Gruyter, Berlin/Boston, 2014.

[23] Nandong, J., *A unified design for feedback-feedforward control system to improve regulatory control performance*, International Journal of Control and Automation, 13(1), 91–98, 2015.

[24] Cus, F., Zuperl, U., Balic, J., *Combined feedforward and feedback control of end milling system*, Journal of Achievements in Materials and Manufacturing Engineering, 45(1), 79–88, 2011.

[25] Rusli, E., Drews, T.O., Ma, D., Alkire, R., Braatz, R., *Robust nonlinear feedback–feedforward control of a coupled kinetic monte carlo–finite difference simulation*, Journal of Process Control, 16, 409–417, 2006.

[26] Jamilis, M., Garelli, F., Battista, H.D., Volcke, E., *Combination of cascade and feed-forward constrained control for stable partial nitration with biomass retention*, Journal of Process Control, 95, 55–66, 2020.

[27] Zhang, X., Wu, P., Xu, L., Ma, L., *A novel simultaneous partial nitritation, denitratation and anammox (SPNDA) process in sequencing batch reactor for advanced nitrogen removal from ammonium and nitrate wastewater*, Bioresource Technology, 343, 126105, 2022.

[28] Karimi, M., Jahanmiri, A., Azarmi, M., *Inferential cascade control of multi-effect falling-film evaporator*, Food Control, 18, 1036–1042, 2007.

[29] Hongfei, Z., *Solar Energy Desalination Technology*, pp. 173–258, Elsevier inc, Chapter 3, 2017.

[30] Da silva, B., Dufour, P., Sheibat-Othman, N., Othmhman., S., *Inferential MIMO predictive control of the particle size distribution in emulsion polymerization*, Computer Chemical Engineering, 38, 115–125, 2012.

[31] Visioli, A., Hägglund, T., *Minimum-time feedforward control in ratio control systems*, IFAC Papers OnLine, 53-2, 11818–11823, 2020.

[32] Mohan, V., Pachauri, N., Panjawni, B., Kamath, D., *A novel cascaded fractional fuzzy approach for control of fermentation process*, Bioresource Technology, 357, 127377, 2022.

[33] Chen, F., Jiao., J., Hou, Z., *Robust polymer electrolyte membrane fuel cell temperature tracking control based on cascade internal model control*, Journal Power Sources, 479, 229008, 15 December 2020.

2 Model predictive control

2.1 Introduction

Currently, the most prominent representative of the model-based control strategies is considered to be the model predictive control (MPC) algorithm. This consideration relies on the success MPC has demonstrated in industrial applications, associated with the interest of control researchers for developing its design over the last four decades. Emerged from practice-driven needs for high-performance control, the MPC design theory brings out a natural relationship between the behavior of the process to be controlled, revealed by the model, and the design of the controller. MPC may be considered a member of the internal model control family with whom it shares the direct use of the process model when building the controller, but with added new capabilities such as prediction, optimization, and handling constraints.

MPC strategy relies on the use of the process model to make predictions of the future behavior of the process, as a result of both the past, but known, and the future, but unknown inputs, to find the best future input sequence aimed at optimizing a control performance index, having associated constraints on inputs, states, and outputs. The constrained optimization feature of the MPC control strategy is most appreciated, as it allows the controller to overcome the traditional stabilizing capability by involving optimization in the control design while incorporating the constraints that always accompany the control problem. The multivariable approach that MPC methodology is able to cope in a straightforward way is also appreciated for the control of processes, with interactions between input and output variables.

The MPC applications cover a very large area of fields, with thousands of successful reported industrial implementations [1–2]. The literature shows that almost all practice and research studies, where control is involved, may benefit form the MPC approach [2]. Some of them are: refining and production of petrochemicals [3–5], scheduling semiconductor production [6], PVC reactor [7], polyethylene reactor [8], water gas-shift reactor [9], thermal regenerator [10], autoclave composite processing [11], drying [12–13], heat exchanger network [14], drainage and irrigation channels [15–16], waste water treatment [17–18], solar air conditioning and desalination plants [19–20], cruise control [21], flight control [22–23], robotics [24], and medical applications [25]. The trend in the expansion of MPC fields is continuously widening, with applications for large-scale systems, fast dynamic systems, and low-cost systems [26].

https://doi.org/10.1515/9783110789737-003

2.2 MPC history

The MPC history may be considered to have its roots in the linear quadratic regulator (LQR) developed by Kalman in the early 1960s, which described the proportional controller emerging from the minimization of a performance function that penalizes the squared deviation of the states and inputs from the origin, working on the basis of the state feedback, and computed as solution of a Riccati equation [27–28]. But the LQR approach lacked the ability of handling constraints.

More than five decades ago, the work of Zahed and Whalen [29] revealed the potential of coupling the minimum time optimal control with linear programming, and Propoi [30] showed the benefits of the receding horizon approach. They were the first two ingredients for the birth of the originally called *open-loop optimal control*. However, it had to pass some time, until the 1970s, when the potential of this control approach was proven by industrial applications. The results presented by Richalet showed that model predictive heuristic control (MPHC) was a powerful control strategy, as its software implementation, IDCOM, was successfully used for identification and control [31]. The main characteristics of MPHC were the use of linear impulse response models for making predictions used in a quadratic performance function, considered on a finite prediction horizon, to which input and output constraints were associated. The computation of the optimal inputs was based on a heuristic iterative algorithm. Soon, Cutler et al. [4] and Prett et al. [32] elaborated the new control version by the so-called *dynamic matrix control* (DMC) and successfully applied it for the control of the challenging fluid catalytic cracking process. IDCOM and DMC are considered the first generation of the newborn industrial MPC technology [1].

The first generation of the industrial MPC technology lacked the systematic handling of constraints but this problem was addressed by Cutler et al. [33], who posed the DMC as a quadratic program (QP) with constraints directly included in the optimization problem, and the optimal inputs emerging as solution to a QP. The new quadratic dynamic matrix control (QDMC), broadly described by Garcia et al. [34], refined the concept of hard and soft constraints and may be considered the second generation of the industrial MPC technology [1].

The second generation of the industrial MPC technology could not handle one important problem occurring in QDMC applications, i.e., tackling the situation of getting an infeasible solution and recovering from infeasibility. The newly developed IDCOM-M algorithm uses a controllability supervisor to identify ill-conditioned regions and has a mechanism for their avoidance, operates with two quadratic objective functions (one for the output and one for the input), makes control for a single future point (denoted as coincidence point on the reference trajectory), and distinguishes hard constraints by ranking them in the order of priority [35–36]. This is considered as a representative of the third generation of the industrial MPC technology [1].

It is a recognized fact that at the beginning of the MPC technology development, MPC industrial applications have preceded the MPC theoretical design, but the order has been reversed in the last few decades. During the 1980s to the 1990s, the MPC theory mostly focused on the quantitative analysis of the industrial MPC algorithm's performance. Since the 1990s, the MPC qualitative synthesis theory has predominated over the industrial applications. A large interest was shown and effort made by the scientific community during the last decades to develop a sound theoretical theory for dealing with different MPC design aspects that emerged form process requirements, such as stability, tuning, optimization, robustness, uncertainty, etc. with the aim of building MPC controllers that are able to guarantee both general and specific plant performance [26]. In the future, these theoretical developments need to be validated by industrial applications. Practical solutions are expected to match the two aspects of the MPC development's directions, such as reducing the computational burden, simplifying the theoretical approach, and rendering physical meaning to the complex MPC theoretical research results. The costs for maintenance and training should be kept as low as possible and the need for low-cost MPC controllers should also be satisfied [26].

The MPC technology is far from reaching its maximum potential, and further developments are expected both from theory and practice. The approach of MPC control for large-scale, fast dynamic, and low-cost systems is the current and a challenging area of research for large communities of researchers and practitioners. One of the major research directions remains the way process nonlinearity may be handled both in MPC applications and in the development of an efficient but comprehensive nonlinear model predictive control (NMPC) theoretical framework. Adaptive, robust (tube-based), distributed, embedded, cooperative, economic, and stochastic (scenario based) MPC are just some of the several hot topics of research waiting for new ideas and solutions.

2.3 Basics of MPC control strategy

In the following, a tutorial on linear MPC is presented in a simple but intuitive way. At its root, the MPC strategy may be considered to have the same basic structure as the traditional feedback-feedforward control loop, as it is presented comparatively in Fig. 2.1.

As shown in Fig. 2.1, there is no difference between the arrangements of the elements building the two control loops, i.e., the process, controller, measuring devices, and the final control element have the same role and position in the feedback (plain line) and feedforward (dotted line) control configurations. The meaning of the signals in the loops remains the same too, i.e., the setpoint (reference), output (controlled variable), measurements (both for the output and for the disturbance), control (input) variable, and the manipulated variable carry the same information. The difference is the way the control (input) variable is computed. The classical controller is replaced by an optimization algorithm that runs online.

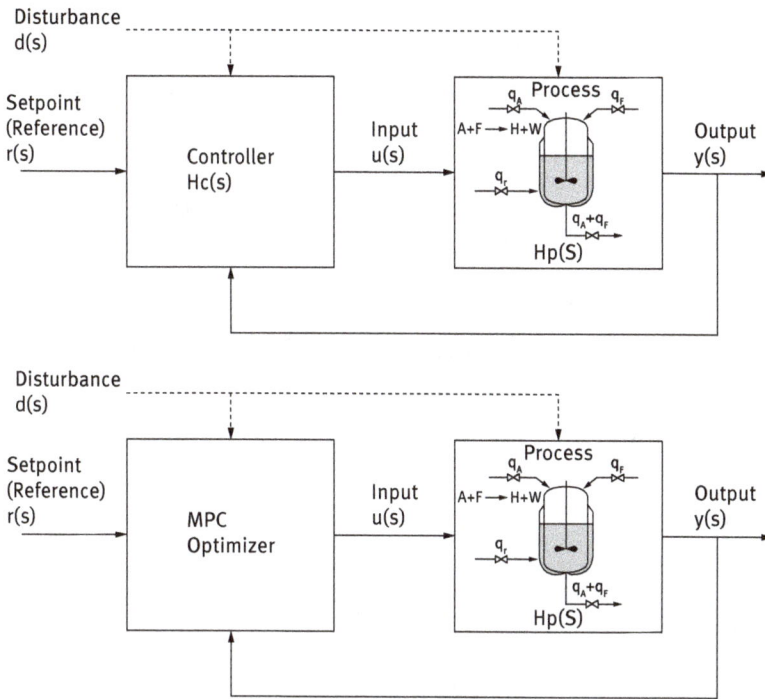

Fig. 2.1: (a) Traditional and (b) MPC control structures.

For the MPC control loop, the control variable is generated by the MPC controller as a result of solving an optimization problem. The performance index of the optimization problem consists of a function that evaluates the square difference between the desired reference and the predicted output variable over a future time interval, denoted as the prediction horizon. As the performance index depends on the future predicted variable, it implicitly also depends on the future values of the control variable. This makes the independent (unknown) variable of the performance index to be, in fact, the future control values, which will be computed by solving the minimization problem.

A process model is needed for making predictions on the future output variable evolution, as they depend on the future changes of the input variable. Mathematically, the process model includes in the performance index, the relationship between the future values of the controlled variable and the future values of the input variable (the unknown of the optimization problem).

But what emerges from solving this minimization problem is only an open-loop optimal control. There is an important weakness in the open-loop optimal control, i.e., no feedback is included yet. This is introduced by the so-called receding horizon approach, which will be further described. As the MPC algorithm is implemented on a computer, the discrete time approach is usually the one that is assumed for the MPC control strategy. The discrete approach introduces the so-called future input sequence, having values changing only at the sampling time moments and remaining constant in between (i.e., zero order hold, ZOH, assumption).

ℹ Example 2.1

To better understand the MPC control approach, we may consider a trivial but intuitive tutoring example. It consists of the way a car (or bike) is controlled in traffic. The analogy of controlling a process and driving a car on a road will reveal the natural way that MPC control strategy is conceived. Let us consider the task of driving a car on the road, from a starting point to a destination point, as is schematically presented in Fig. 2.2.

Fig. 2.2: Analogy between driving a car and the MPC control.

The driver (controller) has to manipulate the steering wheel, the accelerator (throttle pedal), and the brakes to stay in lane, obey speed limitations, avoid over acceleration or over braking, etc. The position of the car in the lane may be considered a controlled variable and the middle of the lane trajectory as reference (setpoint) for the car system.

Reaching the destination point on the best possible trajectory, i.e., keeping the position of the car, as much as possible, in the middle of the lane all along the starting-destination travel, may be planned before leaving the starting point. This means planning the best control actions for the steering wheel, the accelerator, and the brakes for all future moments of the planned road trip. Finding these optimal control actions is done by the driver who knows not only the desired trajectory but also has knowledge about the behavior of his car when he manipulates the steering wheel, accelerator, and brakes. This means the driver has a mental model of his car. Only on the basis of this model, the driver is able to make the best plan, aiming to keep the minimum distance between the car's position and the reference trajectory. This best plan, done before leaving the departure point, represents the optimal open-loop control of the car. It is worthy to mention that it relies on the model of the car and predicts its future position all along the time of the trip (time horizon), while computing the best future sequence for the control actions.

As may be observed from the driving example, the control problem may be formulated on the basis of a performance index to be minimized, i.e., the deviation (possibly square) distance between the current (and future) position of the car and the reference trajectory, while conforming to constraints such as the obeying speed limitations and avoiding over acceleration or over braking.

Unfortunately, although this optimal open-loop control could be considered of interest, it shows to have major problems. These problems are related to the unknown events that may appear during the trip, making ineffective the optimal plan. Such unpredictable events are other vehicles moving in the same lane at different speeds or stopping (as the truck is in Fig. 2.1), coming in contact with a wet track with low adhesion, people or animals crossing the road in front of the car, etc. All these events may be considered unknown (unmeasured) disturbances. Such disturbances cannot be handled by the open-loop optimal control hypothetical plan made by the driver before starting the trip.

There is still a solution to these problems, while keeping the prediction and optimization benefits introduced by the open-loop optimal control strategy. The driver should make his plan on a shorter time interval and apply it. Then, the driver should resume building a new plan upon reaching the end of this time interval, taking into account, at that moment, the unexpected event that happened in the meantime. The new plan should be applied for a new time interval. And the succession of building new plans based on current information and their application during their subsequent time interval should be repeated all along the trip, up to its end. This manner of driving a car is in fact very close to the way the driver acts by making a plan for his best future driving actions for a short period of time, considering all available information about the environment, applying it, and repeating this planning and execution.

The general theoretical approach for coping with the open-loop optimal control weakness is to split the time intended for controlling the system into small time intervals and compute an open-loop optimal control on each of these time periods. Every new open-loop optimal control plan should be performed when the system reaches the end of the previous time interval and using the information in control on the outputs (states) of the system at this moment (possibly affected by unmeasured disturbances). What results is a succession of optimal plans computed and applied all over the period of control. Resuming the design of a new control plan is denoted as the receding horizon control. It is a necessary ingredient of the MPC strategy as it brings feedback to the control approach by considering, when planning, the current values of the outputs or states. This is obviously needed to counteract the effect of unmeasured disturbances. As the discrete time approach is commonly used for implementing MPC, the moments of building new optimal open-loop control plans are set at *each of the sampling time moments*. It is important to note that the length of a new open-loop optimal control period of time spans over multiple sampling time intervals. Although the plan is made over a large future time period, only the first of the computed optimal control variable is sent (applied) to the controlled process. The optimization is done again at the very next sampling moment and a new open-loop optimal control plan is rebuilt. This feature of the MPC control approach is presented in Fig. 2.3.

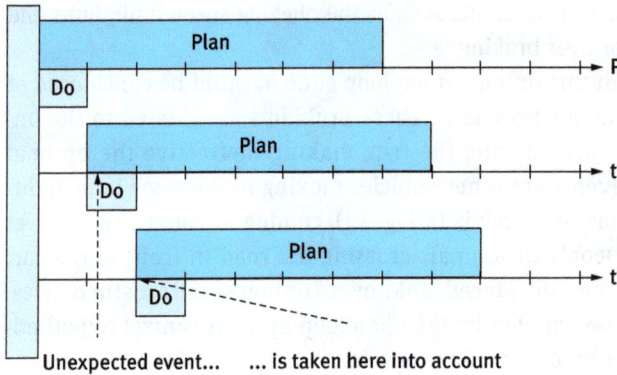

Fig. 2.3: Receding horizon feature of MPC.

As presented in Fig. 2.3, an unexpected event is taken into consideration in a new optimization plan at the very next sampling time moment, following its occurrence, when this feedback information is used for the new prediction.

The future time period (multiple of the sampling time) over which the output is predicted when performing the open-loop control optimization, i.e., future plan duration with respect to the output, is denoted as the *prediction (output) horizon*. This prediction horizon is shifted one sampling time-step ahead into the future at each sampling period when the optimization is resumed. This is also shown in Fig. 2.3.

At first sight, it might be considered the same length for the future time period over which the optimal control variable sequence is computed. This is usually not the case for practical MPC implementations when this future time period associated with the optimal control variable is considered to have a smaller length, i.e., the control variable is allowed to change only for a shorter future period of time, and then remains constant for the rest of the time up to the prediction horizon end. This approach is due to reasons of sparing the computation resources, since the number of unknowns of the optimization problem is reduced and so is the computation load. The future time period (multiple of the sampling time) over which the open-loop control variable is computed when performing the open-loop control optimization, i.e., future plan duration with respect to the control (input) variable, is denoted as the *input horizon*.

Recalling the driving car example previously presented, the prediction horizon may be illustrated as the corresponding time needed by the car during night traffic to cover the distance its headlights are revealing, i.e., how long (in time) the driver sees in front of the road to make his open-loop optimal control plan.

In a nutshell, it may be concluded that MPC control strategy makes a repeated open-loop optimization of a performance index with respect to the future control actions, on the basis of the predictions made for the output variable, using the model of the process. The family of the MPC methodology has several versions, such as: IDCOM [31], IDCOM-M [35], DMC [4], MAC [3], QDMC [34], GPC [37–38], SPC [39], or UPC [40].

The representation of the MPC horizons for an arbitrary current moment k and considering the input variable emerging from a ZOH is presented in Fig. 2.4.

An important element of the MPC strategy is the *model* of the process. As presented before, the prediction made at any current (sampling) moment used in the optimization performance index relies on the model. Implication of the *prediction* in the control algorithm allows the MPC strategy to consider the future behavior of the controlled process, based on the current information available and take early counteracting measures before the output is substantially diverged from the reference trajectory.

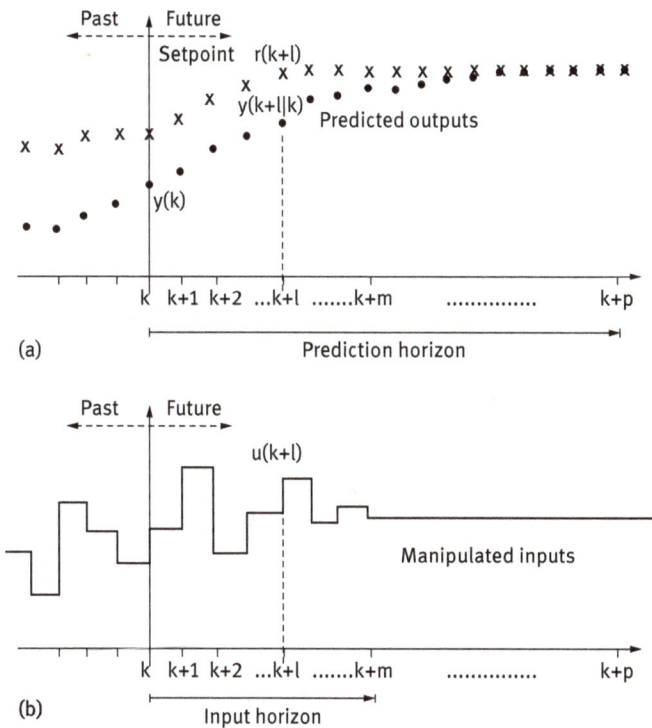

Fig. 2.4: Significance of the prediction and output horizons.

It may be considered that taking into account the future reference changes, MPC prediction allows "feedforward" control with respect to the setpoint future evolution. This "look ahead" or "preview" capability brings efficiency to the algorithm. Coming back to the driving car example, the control based on the traditional feedback control (such as PID) would mean driving the car only by looking at the lateral rear windows (or rear-view mirror). Control based on looking ahead ability would mean driving the car looking forward through the front window and making control, based on knowing the future road reference trajectory.

Nevertheless, the classical feedforward control made on the basis of knowing the measured disturbances is also possible in the MPC framework. This is also based on a model of the process, which, in this case, must consider the disturbance-output path. It is sufficient to include in the output prediction of the optimization index, the measured disturbance effect associated with the input (control variable) effect to obtain a combined feedback-feedforward MPC controller.

As the control problem is discrete, the optimization problem is performed in a discrete framework. A typical form of the MPC optimization problem, with its associated performance index, for the input-output process model description has the following form:

$$\min_{\Delta u(k)\ldots\Delta u(k+m-1)} \{J(u,k)\} = \min_{\Delta u(k)\ldots\Delta u(k+m-1)} \left\{ \sum_{l=1}^{p} ||Y\mathrm{wt}_l[r(k+l) - y(k+l|k)]||^2 + \right.$$
$$\left. + \sum_{l=1}^{m} ||U\mathrm{wt}_l[\Delta u(k+l-1)]||^2 \right\} \qquad (2.1)$$

where
$J(u, k)$ is the performance index (cost function),
$r(k)$ is the reference trajectory vector,
$y(k)$ is the process output vector,
$u(k)$ is the process input (control signal) vector (i.e., manipulated variable),
$\Delta u(k)$ is the process input (control) incremental signal (i.e., manipulated variable move) $\Delta u(k) = u(k) - u(k-1)$, p is the prediction horizon,
m is the input (control) horizon,
$Y\mathrm{wt}_l$ is the output (error) weighting factor for the moment $(k+l)T$,
$U\mathrm{wt}_l$ is the input (control) weighting factor for the moment $(k+l)T$,
$y(k+l, |k)$ is the prediction of $y(k+l)$ for the generic moment "$k+l$", based on information available at time "k",
$||y(k)||^2 = y(k)^2$ for scalar case ($||y(k)||^2 = y^T(k) \cdot y(k)$ for vector case),
$||\Delta u(k)||^2 = u(k)^2$ for scalar case ($||\Delta u(k)||^2 = \Delta u^T(k) \cdot \Delta u(k)$ for vector case).

As presented in eq. (2.1), the performance index consists of two terms [41–42]. The first term penalizes the reference tracking error vector, multiplied by the output weighting factor $Y\mathrm{wt}_l$, and the second term penalizes the input (control) move vector, multiplied by the input weighting factor $U\mathrm{wt}_l$.

Another form of the input-output MPC optimization formulation is framed in the generalized model predictive control (GPC) with a simplified performance index [37–38]:

$$\min_{\Delta u(k)\ldots\Delta u(k+m-1)} \{J(u,k)\} = \min_{\Delta u(k)\ldots\Delta u(k+m-1)} \left\{ \sum_{l=p_1}^{p_2} ||[r(k+l) - P(q)y(k+l|k)]||^2 + \right.$$

$$\left. + \sum_{l=1}^{m} ||\lambda^2[\Delta u(k+l-1)]||^2 \right\} \qquad (2.2)$$

where $Ywt_l = 1$ and $Uwt_l = \lambda^2$, p_1 is the minimum prediction (output) horizon, p_2 is the prediction (output) horizon, and $P(q)$ in a polynomial with the desired closed-loop poles, $P(q) = a_r q^{-r} + a_{r-1}q^{-r+1} + \cdots + a_1 q^{-1} + 1$ (in the simplest case $P(q) = 1$).

Despite the fact that the input-output MPC formulation of the performance index is more intuitive and directly involves the practice-emerged signals, the input-state-output model is very frequently used due to its comprehensive approach. The space-based MPC optimization may be formulated as [43–44]

$$\min_{\Delta u(k)\ldots\Delta u(k+m-1)} \{J(u,k)\} = \min_{\Delta u(k)\ldots\Delta u(k+m-1)} \left\{ \sum_{l=1}^{p} Q||x(k+l|k)||^2 + \right.$$

$$\left. + \sum_{l=1}^{m} R||\Delta u(k+l-1)||^2 \right\} \qquad (2.3)$$

where $x(k)$ is the vector of the states, $\Delta u(k)$ is the vector of the inputs move, Q is the state weighting matrix, and R is the input (control) weighting matrix.

Most of the MPC performance index formulations are based on the square norm $||\cdot||^2$, as presented before. The 1-norm (based on the sum of the vector absolute values) or the infinity-norm (based on the maximum absolute value of vector's elements) is rarely used.

Constraints are always associated with the MPC optimization problem defined by eqs. (2.1) to (2.3). Solving the constrained optimization problems is the most important gain of MPC. The capability to take into account constraints in a systematic way during the design and implementation of the controller makes it the most popular advanced control algorithm. Simple constraints such as the final control element saturation, or more complex constraints on the states or on the controlled outputs such as those coming out from equipment, technological, safety, environmental or economic reasons are placing challenges on the MPC-based control. What results is the best control sequence of the control variable changes that brings the predicted output as close as possible to the reference while satisfying all input states and output constraints. For linear models with linear constraints and the performance index as presented in eq. (2.1), the solution to the optimization problem can be found using QP algorithms. If a 1-norm or infinity-norm performance index is used, linear programming algorithms will provide the solution. Both algorithms are convex and show convergence.

MPC may be successfully used for the control of processes showing large time constants, large pure time delay, nonminimum phase or even instability, and having the same systematic approach both for SISO and MIMO systems. It is also valued that the MPC has the ability to work with unequal (excess) number of manipulated *vs.* the number of controlled variables. As the model is explicitly used in the controller design, MPC has inherently robust properties as against the model-process mismatch and, to a certain extent, may also cope with structural modifications of the process, as long as they are revealed by the prediction model.

2.4 Types of MPC process models

The optimal input (control) sequence $\{u(k), u(k+1), \ldots, u(k+l), \ldots, u(k+m-1)\}$ minimizes the square error (first term) of the optimization index, possibly associated with the input variable change (second term) and the constraints. The error term is computed as the difference between the future reference and the predicted sequences. The predicted sequence is calculated on the basis on the process model. The explicit use of the model for designing the MPC controller differentiates the model-based controller design, compared to the traditional PID design. Different types of models may be used for making predictions [39]. A typical linear, discrete, and time-invariant model of the process has the form:

$$y(k) = H_p^*(q)u(k) + d(k) = H_p(q)\Delta u(k) + H_{PD}(q)\Delta d(k) = H_p(q)\Delta u(k) + H_{PD}(q)e(k) \quad (2.4)$$

where $y(k)$ denotes the output controlled variable, $u(k)$ is the input (control or manipulated) variable; $d(k)$ is the disturbance variable; $\Delta u(k)$ is the input variable incremental change ($\Delta u(k) = u(k) - u(k-1)$); $\Delta d(k)$ is the disturbance variable incremental change ($\Delta d(k) = d(k) - d(k-1)$), considered to be the white noise signal with zero mean $e(k)$; $H_p(q)$ is the discrete transfer function of the process on the input-output path; $H_{PD}(q)$ is the discrete transfer function of the process on the disturbance-output path; and q is the forward shift operator. The output-input increment model assumes that the disturbance in not changing. As a result, in the prediction, it will not be necessary to include the model of the disturbance estimate as it is already included in the incremental variables.

2.4.1 Impulse and step-response models

The most intuitive models that emerged from direct process identification are the step- and impulse-response models. They have been used as the first MPC models at the beginning of the MPC applications development, but were later extended to the more elaborate first-principle models.

For the SISO case, the impulse response h_i and the unit step response s_i of a linear time-invariant process are related by the well-known equations:

$$h_i = s_i - s_{i-1}, s_i = \sum_{j=1}^{i} h_j \quad (2.5)$$

The discrete transfer functions are defined by

$$H_p^*(q) = \frac{y(q)}{u(q)} = \sum_{i=1}^{\infty} h_i q^{-1} \tag{2.6}$$

$$H_p(q) = \frac{y(q)}{\Delta u(q)} = \sum_{i=1}^{\infty} s_i q^{-1} \tag{2.7}$$

and, as a result,

$$H_p(q) = \frac{y(q)}{\Delta u(q)} = \frac{y(q)}{(1-q^{-1})u(q)} = H_p^*(q)\frac{1}{(1-q^{-1})} = \sum_{i=1}^{\infty} h_i \frac{q^{-i}}{(1-q^{-1})} \tag{2.8}$$

For practical reasons, the truncated impulse- and step-response models (up to a multiple of n sampling times, nT, i.e., the time the process needs to settle and reach almost the steady state) are used instead of the infinite response models. They have the following form in the discrete time formulation:

$$y(k) = \sum_{i=1}^{n} h_i u(k-i) + d(k) \tag{2.9}$$

$$y(k) = \sum_{i=1}^{n-1} s_i \Delta u(k-i) + s_n u(k-n) + d(k) \tag{2.10}$$

For the MIMO case, when the system has a v-dimensional input vector $u(k) = [u_1(k)\ u_2(k)\ldots u_v(k)]^T$ and a w-dimensional output vector $y(k) = [y_1(k)\ y_2(k)\ldots y_w(k)]^T$, the truncated response models have the forms:

$$y(k) = \sum_{i=1}^{n} H_i u(k-i) + d(k) \tag{2.11}$$

$$y(k) = \sum_{i=1}^{n-1} S_i \Delta u(k-i) + S_n u(k-n) + d(k) \tag{2.12}$$

where the impulse-response matrix is

$$H_i = \begin{bmatrix} H_{11}(i) & H_{12}(i) & \cdots & H_{1v}(i) \\ H_{21}(i) & H_{22}(i) & \cdots & H_{2v}(i) \\ \vdots & \vdots & \vdots & \vdots \\ H_{w1}(i) & H_{w2}(i) & \cdots & H_{wv}(i) \end{bmatrix} \tag{2.13}$$

For this MIMO case, the step-response matrix is related to the impulse-response matrix by

$$S_i = \sum_{l=1}^{i} H_l \tag{2.14}$$

or

$$H_i = S_i - S_{i-1} \tag{2.15}$$

i **Example 2.2**
The unit impulse- and step-response models may be obtained from experimental identification data. Consider the unit impulse input and its corresponding truncated unit impulse response presented in Figs. 2.5 and 2.6.

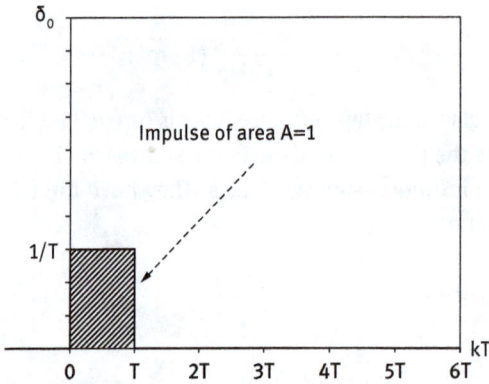

Fig. 2.5: Unit impulse input signal.

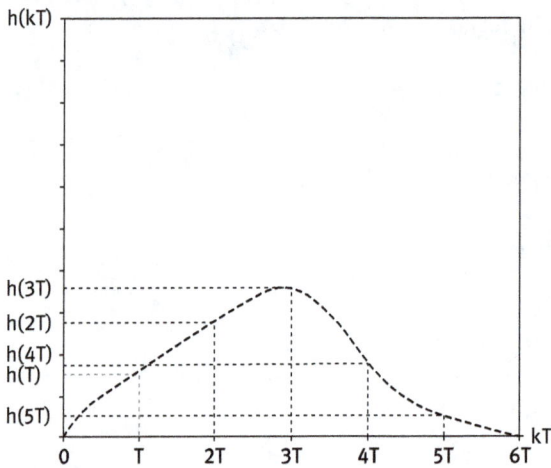

Fig. 2.6: Truncated unit impulse response signal.

and unit step input with its corresponding truncated step response presented in Figs. 2.7 and 2.8.

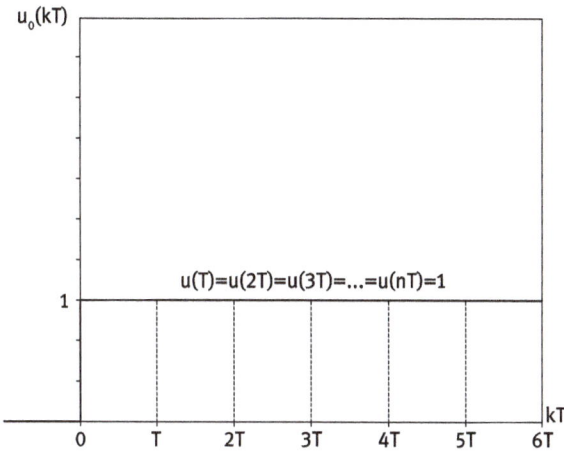

Fig. 2.7: Unit step input signal.

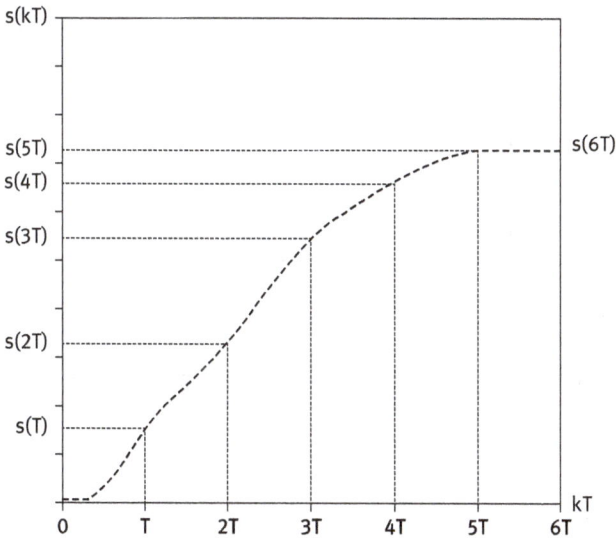

Fig. 2.8: Truncated unit step response signal.

Based on the homogeneity and additivity properties of the linear systems implicit in the convolution model presented in eq. (2.9) and the unit impulse response parameters presented in Fig. 2.6, it is possible to compute the response of the system to an arbitrary input. A trivial but tutorial example is the computation of the system's response to the unit step input signal using the information emerging from the unit impulse-truncated response. The graph representation of this computation is given in Fig. 2.9 for the input step signal of the same form as presented in Fig. 2.7.

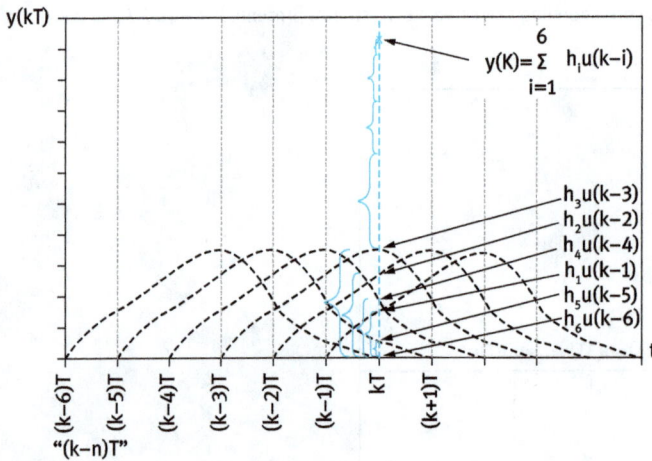

$$y(K)=\sum_{i=1}^{6} h_i u(k-i)$$

Fig. 2.9: Computation of the unit input step response signal based on the truncated impulse response model.

Fig. 2.9 shows the computation by summation of the step response $y(k)$ at the generic time moment $t = kT$ using the truncated impulse-response model presented in eq. (2.9).

Example 2.3

In this example, the way the output (response) of the system can be determined is shown on the basis of the truncated step model presented in eq. (2.10) and the unit step response parameters presented in Fig. 2.8. In this example, the unit ramp input signal is considered and the response of the system is computed. The input ramp signal and its equivalents (obtained by decomposition) are presented in Fig. 2.10. The graph representation of the response computation is given in Fig. 2.11.

Figure 2.11 shows the computation of the step response $y(k)$ at the generic time moment $t = kT$ using the truncated step-response model presented in eq. (2.10). Both examples 2.2 and 2.3 assume that no disturbance is present and $d(k) = 0$.

Examples 2.2 and 2.3 reveal the natural way the impulse- or step-response models may be used for computing the process output for an arbitrary input using the truncated impulse or step-response equations. The latter will be used for making the MPC predictions.

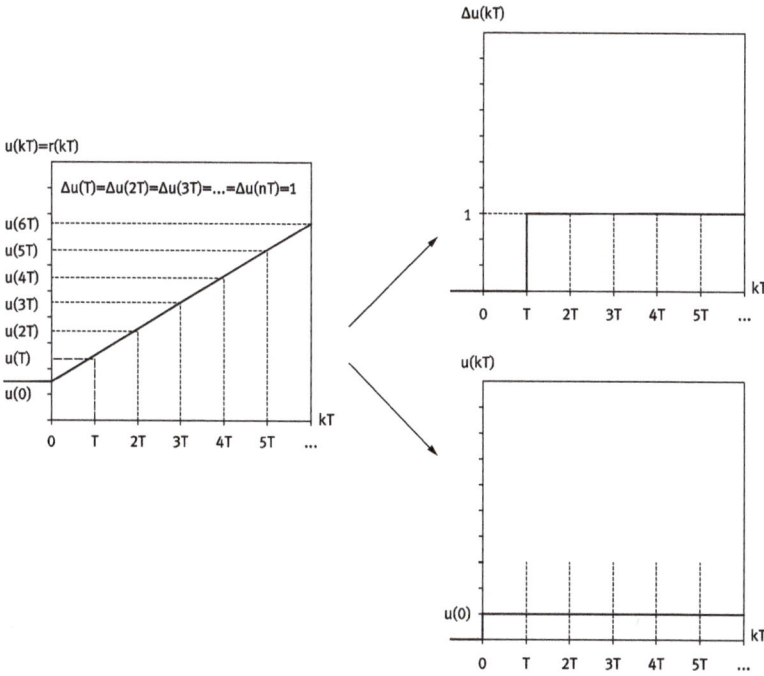

Fig. 2.10: Unit ramp input signal and its equivalent signals.

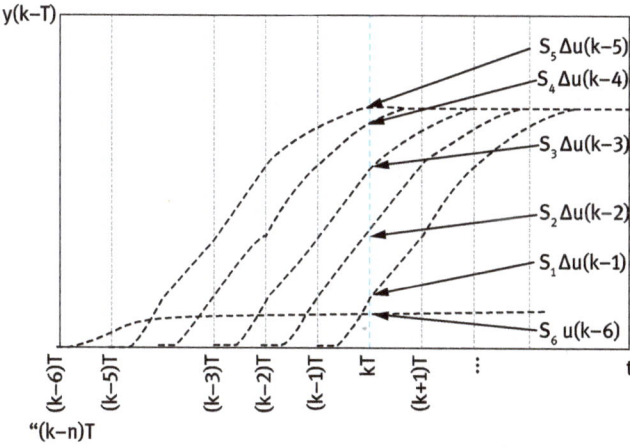

Fig. 2.11: Computation of the unit input ramp response signal based on the truncated unit step-response model.

2.4.2 State-space models

The time-invariant state-space description of the system offers a comprehensive representation of the dynamic and steady-state behavior, since the internal system's variables (states) are also revealed and can be controlled. The state-space realization shows to be appropriate both for SISO and MIMO system description, using the same mathematical form. The general discrete form of the linear time-invariant state-space model has the following representation:

$$\begin{cases} x(k+1) = Ax(k) + B\Delta u(k) + Ke(k) \\ y(k) = Cx(k) + e(k) \end{cases} \tag{2.16}$$

where the matrices A, B, C, and K may be directly computed from their continuous corresponding ones, i.e., the matrices of the continuous linear time-invariant state-space description.

Nevertheless, the input-output models may be also converted to the state space form, the latter being also preferred due to its inherent numerical computations reliability. The relationship between the input-output description and the state-space formulation may be directly obtained from eqs. (2.16) using the Z transform. Using the forward shift operator, the relationship between the two types of models may be described by

$$H_p(q) = C(qI - A)^{-1}B \tag{2.17}$$

$$H_{PD}(q) = C(qI - A)^{-1}BK + I \tag{2.18}$$

It is worth mentioning that due to the fact that usually not all of the states are directly measureable, a state-space estimator is necessary for constructing the initial (current moment) state from the available measured outputs.

2.4.3 Time series models

The CARIMA (controlled autoregressive integrated moving average) type of models has been also used to make MPC prediction [37–39]. Also known as transfer function models, CARIMA models have gained large acceptance especially for SISO systems. The development of the MIMO CARIMA models is straightforward, but becomes cumbersome for applications to large systems.

Its largest popularity among other transfer function models is due to its capability of including uncertainty in a simple way and being suitable to reveal the slowly changing disturbances that exhibit nonzero steady state.

The discrete equation describing the model is

$$A(q)y(k) = B(q)u(k) + C(q)\frac{\xi(q)}{\Delta(q)} \tag{2.19}$$

In eq. (2.19), the shift operator q^{-1} was used and $\Delta(q) = 1 - q^{-1}$. $A(q)$, $B(q)$, $C(q)$ are polynomials in the time shift operator q^{-1}. The factor $\xi(q)$ considers the uncertainty on the input-output path. $\xi(q)$ is assumed to have zero mean and finite variance.

Based on the above equation, the following form of the difference equation can be derived:

$$A(q)\Delta(q)y(k) = B(q)\Delta(q)u(k) + C(q)\xi(q) \tag{2.20}$$

and as a result, the transfer functions of eq. (2.4) get the following forms:

$$H_P(q) = \frac{B(q)}{A(q)\Delta(q)} \tag{2.21}$$

$$H_{PD}(q) = \frac{C(q)}{A(q)\Delta(q)} \tag{2.22}$$

2.5 Predictions for MPC

At each current moment of time kT, the output of the process is considered over the future time horizon pT by making a set of p-step ahead predictions of the output: $\{y(k+1|k), y(k+2|k), \ldots, y(k+l|k), \ldots, y(k+p|k)\}$. The predictions are based on information available at current time step kT, and depend on both the past and future values (with respect to current time) of the input variable $u(k)$ or on its increments $\Delta u(k)$. However, the effect of the past input variable on the predicted output may be directly computed. This computation is based on knowing all the past *input (control) variable values* and the *model* of the process. The effect of the future input variable sequence $\{u(k), u(k+1), \ldots, u(k+l), \ldots, u(k+m-1)\}$ or the sequence of input variable move $\{\Delta u(k), \Delta u(k+1), \ldots, \Delta u(k+l), \ldots, \Delta u(k+m-1)\}$ on the predicted output is not yet known. It will be the task of minimizing the performance index to find this latter sequence, as the performance function penalizes the square error (square difference between the reference trajectory and the predicted output).

From the cause-effect viewpoint, the prediction has two components. One component is due to the action of the past and the future input variable moves $H_P(q)\Delta u(k)$ and the second is due to the action of the disturbances $d(k) = H_{PD}(q)e(k)$ [39]. The assumption that $e(k)$ is the zero-mean white noise will be considered further and the best prediction for future values of $e(k+l) = \Delta d(k+l) = d(k+l) - d(k+l-1)$ will be $e(k+l|k) = \Delta d(k+l|k) = 0$ for $l > 0$.

The development of the prediction equations for SISO case will be presented both for the input-output and the state-space model approaches, but the extension to the

MIMO case is straightforward [39]. A simple form of the prediction equation may be obtained if the prediction p and the control horizon m are considered equal, $p = m = N$.

The vector of predicted values of the output variable based on the information available at current time step kT is defined as

$$\hat{y}(k) = [y(k+1|k)y(k+2|k)...y(k+l|k)...y(k+N|k)]^T \qquad (2.23)$$

and the vector of future values of the control variable move:

$$\hat{u}(k) = [\Delta u(k)\ \Delta u(k+1)...\Delta u(k+l)...\Delta u(k+N-1)]^T \qquad (2.24)$$

A prediction equation may be developed with the same form for both the input-output and the state-space model approaches. Its formulation is

$$\hat{y}(k) = M\hat{u}(k) + \hat{y}_0(k) \qquad (2.25)$$

where M is the so-called *predictor matrix* and describes the effect of the future unknown input moves on the predicted output and $\hat{y}_0$ is the so-called *free run output* and describes the effect of past input moves (before the current time step, kT) on the predicted output.

When *step response models are used for prediction*, the predicted output variable emerges from the step-truncated-response model presented in eq. (2.10), where the generic moment lT is considered all along the prediction horizon $(0 < l < N)$. The prediction is described for the SISO case by [45]

$$y(k+l|k) = \sum_{i=1}^{l} s_i\,\Delta u(k+l-i) + \sum_{i=l+1}^{n-1} s_i\,\Delta u(k+l-i) + s_n\,u(k+l-n) + d(k+l|k) \quad (2.26)$$

In eq. (2.26), the first term considers the effects of the sequence of future control moves, $\{\Delta u(k),\ \Delta u(k+1),...,\ \Delta u(k+l-1)\}$, on the predicted output variable $y(k+l|k)$. The second term of eq. (2.26) considers the effect of the past control moves, $\{\Delta u(k-1),\ \Delta u(k-2),...,\ \Delta u(k-n+1)\}$, on the predicted output variable $y(k+l|k)$. The third term in eq. (2.26) considers the furthermost input that affects the predicted output $y(k+l|k)$, according to the length of the truncation time, nT. The last term considers the prediction of future-assumed disturbances on the same predicted output variable $y(k+l|k)$ As presented before, it is also assumed that $e(k+l|k) = \Delta d(k+l|k) = 0$ and the computation of the disturbance effect is given by

$$d(k+l|k) = d(k|k) = y_m(k) - y(k|k) \qquad (2.27)$$

where $y_m(k)$ is the measured value of the output at the current time kT and $y(k|k)$ is the predicted value of the output at the same moment of time, provided by the model:

$$y(k|k) = \sum_{i=1}^{n-1} s_i\,\Delta u(k-i) + s_n\,u(k-n) \qquad (2.28)$$

This last term in eq. (2.26) is the one bringing feedback in the MPC control algorithm.

For the MIMO case, the prediction is described in a similar way by [2]

$$y(k+l|k) = \sum_{i=1}^{l} S_i \, \Delta u(k+l-i) + \sum_{i=l+1}^{n-1} S_i \, \Delta u(k+l-i) + S_n \, u(k+l-n) + d(k+l|k) \quad (2.29)$$

and

$$d(k+l|k) = d(k|k) = y_m(k) - y(k|k) = y_m(k) - \sum_{i=1}^{n-1} S_i \, \Delta u(k-i) + S_n \, u(k-n) \quad (2.30)$$

Assuming the length of the prediction horizon to be larger than the step-response-truncation time, $N \geq n$, and $s_{n+i} = s_n$ for $0 \leq i \leq N - n$, the predictor matrix of eq. (2.25) gets the following form (SISO case):

$$M = \begin{bmatrix} s_1 & 0 & 0 & \cdots & 0 \\ s_2 & s_1 & 0 & \cdots & 0 \\ s_3 & s_2 & s_1 & \cdots & 0 \\ \vdots & \vdots & \vdots & \ddots & \vdots \\ s_N & s_{N-1} & s_{N-2} & \cdots & s_1 \end{bmatrix} \quad (2.31)$$

As a result, the free run output $\hat{y}_0$ of eq. (2.25) may be described by

$$\hat{y}_0(k) = \begin{bmatrix} s_2 & s_3 & \cdots & s_N \\ s_3 & s_4 & \cdots & 0 \\ \vdots & & \ddots & \vdots \\ s_N & 0 & \cdots & 0 \end{bmatrix} \begin{bmatrix} \Delta u(k-1) \\ \Delta u(k-2) \\ \vdots \\ \Delta u(k-N+1) \end{bmatrix} + \begin{bmatrix} s_N \cdot u(k-N+1) \\ s_N \cdot u(k-N+2) \\ \vdots \\ s_N \cdot u(k-1) \end{bmatrix} + \begin{bmatrix} I \\ I \\ \vdots \\ I \end{bmatrix} (y_m(k) - y(k|k))$$

$$(2.32)$$

Example 2.4

ℹ

Based on the truncated step model presented in eq. (2.10) and the unit step response parameters presented in Fig. 2.8, the computation of the predictions for MPC and an intuitive graphic representation of this computation are presented in the following. The example considers the step response model truncated to $n = 6$ sampling time-steps. The input for which the prediction will be computed is presented in Fig. 2.12.

Figure 2.12 shows the arbitrary input signal for which the prediction will be performed, revealing both its past and future values (with respect to the current time kT). The increment (input move) values are also presented.

The predicted output for the current moment of time is computed with eq. (2.26), for $l = 0$, or with eq. (2.28) and has the following analytical form:

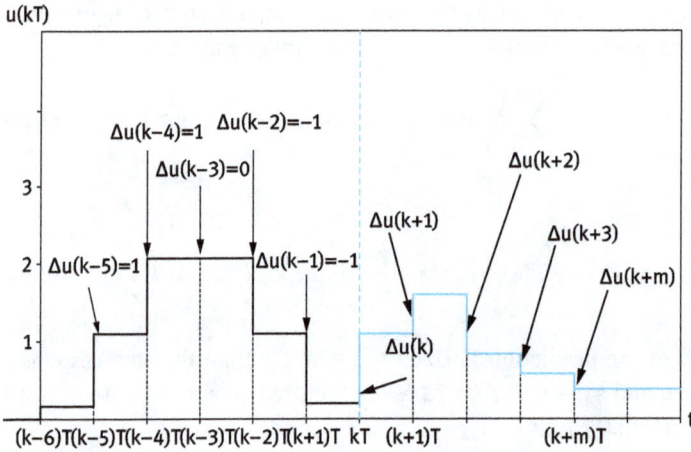

Fig. 2.12: Input signal with arbitrary form used for MPC prediction.

$$y(k|k) = \sum_{i=1}^{0} s_i \Delta u(k-i) + \sum_{i=1}^{5} s_i \Delta u(k-i) + s_6 u(k-6) = s_1 \Delta u(k-1) + s_2 \Delta u(k-2) +$$

$$+ s_3 \Delta u(k-3) + s_4 \Delta u(k-4) + s_5 \Delta u(k-5) + s_6 u(k-6) =$$

$$= s_1 \cdot (-1) + s_2 \cdot (-1) + s_3 \cdot 0 + s_4 \cdot (1) + s_5 \cdot (1) + s_6 u(k-6) =$$

$$= -s_1 - s_2 + 0 + s_4 + s_5 + s_6 u(k-6) \tag{2.33}$$

As expected, the prediction of the output only depends on the past known inputs.

Figure 2.13 shows the graphical computation of the predicted response $y(k|k)$ at the current time moment $t = kT$. The segments corresponding to the components of the prediction, which should be graphically added to build $y(k|k)$, are individually revealed in the same figure. They are considered with a plus sign above the abscissa and with a negative sign below it.

The computation of the prediction, based on the summation of step-response-oriented segments is intuitive and simple.

Continuing the same computation approach, prediction for the following sampling time moment, $y(k+1|k)$, $l = 1$, with the same eq. (2.26), has the following analytical form:

$$(k+1|k) = \sum_{i=1}^{1} s_i \Delta u(k+1-i) + \sum_{i=2}^{5} s_i \Delta u(k+1-i) + s_6 u(k+1-6) =$$

$$= s_1 \Delta u(k) + s_2 \Delta u(k-1) +$$

$$+ s_3 \Delta u(k-2) + s_4 \Delta u(k-3) + s_5 \Delta u(k-4) + s_6 u(k-5) =$$

$$= s_1 \Delta u(k) + s_2 \cdot (-1) + s_3 \cdot (-1) + s_4 \cdot 0 + s_5 \cdot (1) + s_6 u(k-5) =$$

$$= s_3 \Delta u(k) - s_2 - s_3 + 0 + s_5 + s_6 u(k-5) \tag{2.34}$$

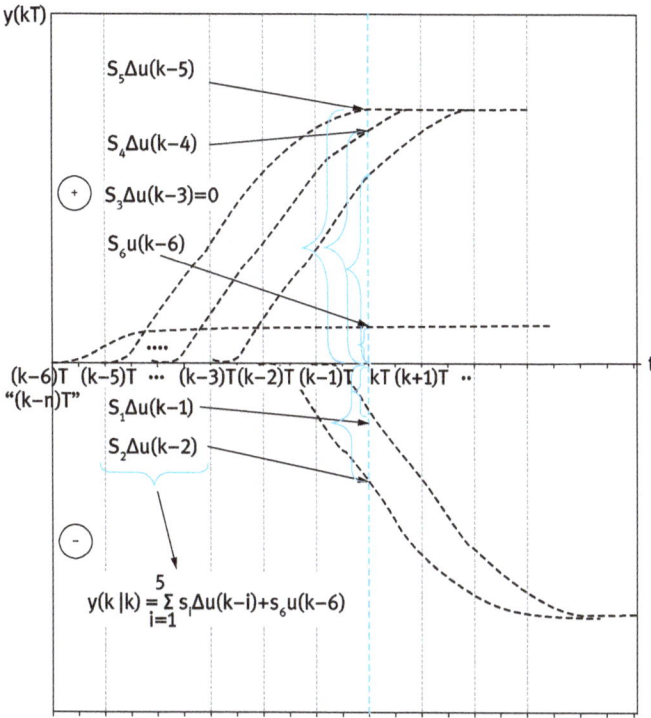

Fig. 2.13: Graphical representation of the components that build by the summation the prediction for $y(k|k)$, based on the truncated step-response model.

Prediction of the output for the future moment of time $(k+1)T$ depends on the known past inputs and the future (unknown) input variable move, $\Delta u(k)$.

Fig. 2.14 shows the graphical computation of the predicted response $y(k+1|k)$, for the time moment $(k+1)T$, on the basis of information known at the current time moment $t = kT$. The segments corresponding to the components of the prediction, which should be graphically added to build $y(k+1|k)$, are individually revealed in the Fig. 2.14. Using the same representation, they are considered with a plus sign above the abscissa and with a negative sign below it.

As noticed, the contribution of the input variable move $\Delta u(k)$ is not represented in Fig. 2.14 because it is not known at the current moment kT. It will be computed at the optimization step of the MPC algorithm.

Based on the same procedure, the computation of the prediction for $y(k+2|k)$, $l = 2$, has the following analytical form:

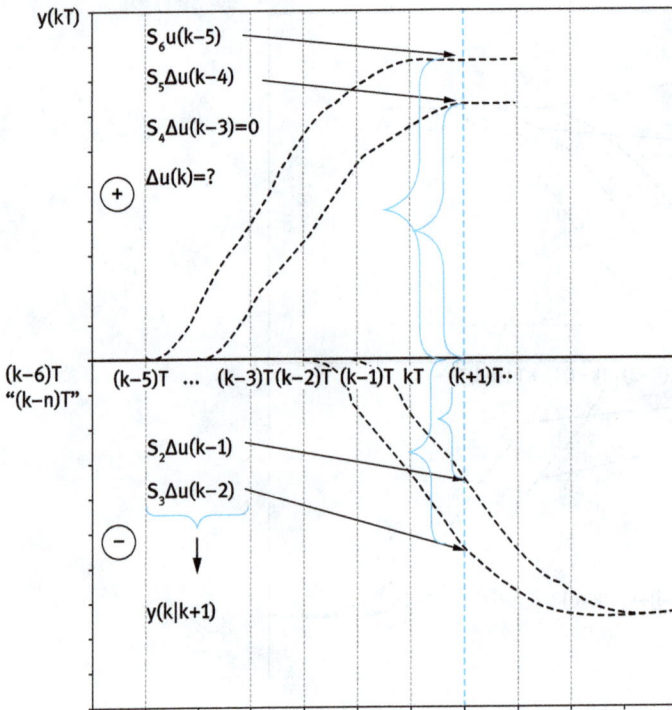

Fig. 2.14: Graphical representation of the components that build by summation the prediction for $y(k+1|k)$, based on the truncated step-response model.

$$y(k+2|k) = \sum_{i=1}^{2} s_i \Delta u(k+2-i) + \sum_{i=3}^{5} s_i \Delta u(k+2-i) + s_6 u(k+2-6) =$$

$$= s_1 \Delta \boldsymbol{u}(\boldsymbol{k}+1) + s_2 \Delta \boldsymbol{u}(\boldsymbol{k}) +$$

$$+ s_3 \Delta u(k-1) + s_4 \Delta u(k-2) + s_5 \Delta u(k-3) + s_6 u(k-4) =$$

$$= s_1 \Delta \boldsymbol{u}(\boldsymbol{k}+1) + s_2 \Delta \boldsymbol{u}(\boldsymbol{k}) + s_3 \cdot (-1) + s_4 \cdot (-1) + s_5 \cdot 0 + s_6 u(k-4) =$$

$$= s_1 \Delta \boldsymbol{u}(\boldsymbol{k}+1) + s_2 \Delta \boldsymbol{u}(\boldsymbol{k}) - s_3 - s_4 + s_5 \cdot 0 + s_6 u(k-4) \qquad (2.35)$$

Prediction of the output for the future moment of time $(k+2)T$ depends on the known past inputs and the future (unknown) input variable moves, $\Delta \boldsymbol{u}(\boldsymbol{k})$ and $\Delta \boldsymbol{u}(\boldsymbol{k}+1)$.

Figure 2.15 shows the graphical computation of the predicted response $y(k+2|k)$ using the same approach of adding the oriented segments.

Considering $m = p = N = n = 6$, the output predictions for all future time moments, $(k+l)T$, $l = 1, \ldots, 6$ depend on the known past input moves, $\{\Delta u(k-1), \Delta u(k-2), \ldots, \Delta u(k-5)\}$, on the past input $u(k-6)$, and the future and the present (**unknown**) input variable moves, $\{\Delta \boldsymbol{u}(\boldsymbol{k}), \Delta \boldsymbol{u}(\boldsymbol{k}+1), \ldots, \Delta \boldsymbol{u}(\boldsymbol{k}+5)\}$. They may be computed in a

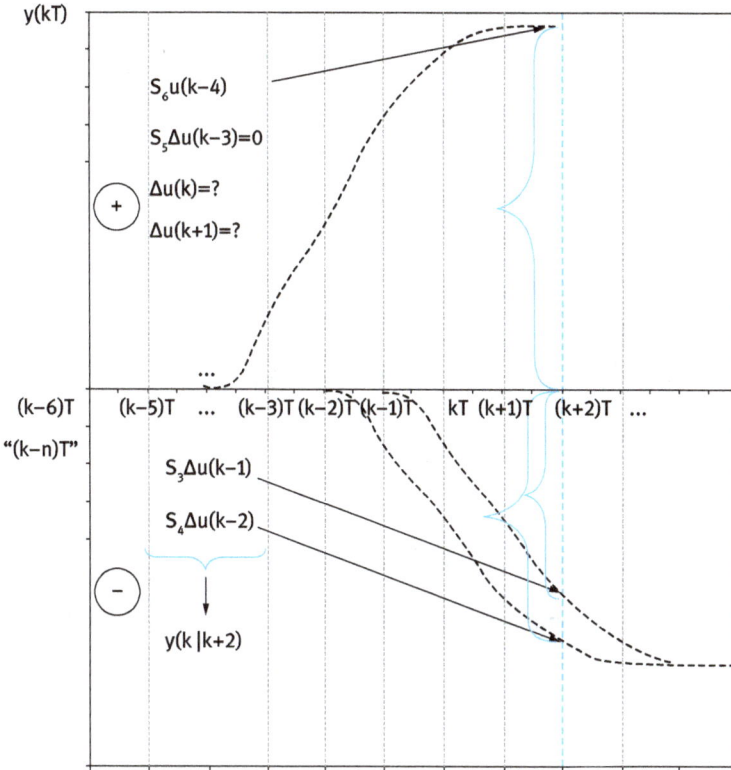

Fig. 2.15: Graphical representation of the components that build by summation of the prediction for $y(k + 2|k)$, based on the truncated step-response model.

similar manner as presented before. At the very core of the prediction computation stays the convolution property of the linear time-invariant systems.

When state-space models are used for prediction, the predicted state and output variables, both for SISO and MIMO case, emerge from the state-space model presented in eq. (2.16). They have the following form:

$$x(k+1|k) = Ax(k|k) + Bu(k) + Ke(k) \tag{2.36}$$

$$y(k+1|k) = Cx(k+1|k) + e(k+1|k) \tag{2.37}$$

where the prediction of the disturbance will be considered $e(k + l|k) = 0$.

Based on eq. (2.36), the state prediction $x(k + l|k)$ for future time steps, $l > 0$, may be obtained as

$$x(k+l|k) = A^k x(k) + \sum_{i=1}^{l} A^{i-1} B \, \Delta u(k+l-i) + A^{k-1} K \, e(kT) \tag{2.38}$$

and the predicted output $y(k + l|k)$ as

$$y(k+l|k) = Cx(k+l|k) = C\left[A^k x(k) + \sum_{i=1}^{l} A^{i-1} B \, \Delta u(k+l-i) + A^{k-1} K \, e(kT) \right] \quad (2.39)$$

The corresponding predictor matrix M and the free-run term $\hat{y}_0$ from eq. (2.25) may be obtained as

$$M = \begin{bmatrix} CB & 0 & 0 & \cdots & 0 \\ CAB & CB & 0 & \cdots & 0 \\ CA^2 B & CAB & CB & \cdots & 0 \\ \vdots & & & \ddots & \vdots \\ CA^{N-1}B & CA^{N-2}B & CA^{N-3}B & \cdots & CB \end{bmatrix} \quad (2.40)$$

$$y_0(k) = \begin{bmatrix} CA \\ CA^2 \\ \vdots \\ CA^N \end{bmatrix} x(k|k-1) + \begin{bmatrix} CK \\ CAK \\ \vdots \\ CA^{N-1}K \end{bmatrix} [y_m(k) - Cx(k|k-1)] \quad (2.41)$$

It may be observed that the relationship between the step-response model S_i (of the model $H^*_p(q)$) and the A, B, C matrices is given by

$$S_i = CA^{i-1}B \quad (2.42)$$

For the measured disturbances, feedforward MPC may be directly performed on the basis of predictions for the measured disturbance effect on the predicted output. Both truncated step/impulse-response models and state-space models may be used. They have forms similar to those already presented for the manipulated (input) variable.

A particular but commonly used form of the state-space formulation may be built using the step-response parameters of the process $s_1, \ldots, s_n$ to the input u and the step-response parameters of the process $s_1^d, \ldots, s_n^d$ to the measured disturbance d [2, 46].

$$Y(k) = M^* Y(k-1) + \begin{bmatrix} S_1 \\ S_2 \\ \vdots \\ S_n \end{bmatrix} \Delta u(k-1) + \begin{bmatrix} S_1^d \\ S_2^d \\ \vdots \\ S_n^d \end{bmatrix} \Delta d(k-1) \quad (2.43)$$

where

$$Y(k) = [y(k) \, y(k+1) \ldots y(k+n-1)]^T \quad (2.44)$$

and

$$M^* = \begin{bmatrix} 0 & 1 & 0 & \cdots & 0 \\ 0 & 0 & 1 & \cdots & 0 \\ 0 & 0 & 0 & \cdots & 0 \\ \vdots & \vdots & \vdots & \ddots & 1 \\ 0 & 0 & 0 & \cdots & 1 \end{bmatrix}$$ (2.45)

Based on this model, the output prediction made at moment kT for the future N sampling time-steps. $Y(k+1|k)$, is

$$Y(k+1|k) = M^{**}y(k|k) + S^u \Delta U(k) + S^d \Delta d(k)$$ (2.46)

where $S^u = M$, eq. (2.31), is the matrix containing the step-response parameters to the input variable u and S^d consists of the step-response parameters to the measured disturbance variable d:

$$S^d = \begin{bmatrix} s_1^d & 0 & 0 & \cdots & 0 \\ s_2^d & s_1^d & 0 & \cdots & 0 \\ s_3^d & s_2^d & s_1^d & \cdots & 0 \\ \vdots & \vdots & \vdots & \ddots & \vdots \\ s_N^d & s_{N-1}^d & s_{N-2}^d & \cdots & s_1^d \end{bmatrix}$$ (2.47)

The input move vector is defined by

$$\Delta U(k) = [\Delta u(k) \ \Delta u(k+1) \ldots \Delta u(k+N-1)]^T$$ (2.48)

and the matrix M^{**} is given by

$$M^{**} = \begin{bmatrix} 0 & 1 & 0 & 0 & 0 & \cdots & 0 \\ 0 & 0 & 1 & 0 & 0 & \cdots & 0 \\ 0 & 0 & 0 & 1 & 0 & \ddots & 0 \\ \vdots & \vdots & \vdots & \ddots & \ddots & \ddots & 0 \\ 0 & 0 & 0 & \vdots & 1 & \cdots & 0 \end{bmatrix}$$ (2.49)

For the state-space-based prediction, it may be necessary to use a state observer to compute the states that are not directly measurable. This computation is based on the control and measured output variables associated with the process model.

When CARIMA time series models are used for prediction, the predicted output variable emerges from solving two Diophantine equations [37–39]:

$$C(q) = E_j(q)\, A(q)\, \Delta(q) + q^{-j} F_j(q) \tag{2.50}$$

$$E_j(q) B(q) = G_j(q)\, C(q) + q^{-j} L_j(q) \tag{2.51}$$

The predictor matrix becomes

$$M = \begin{bmatrix} s_1^* & 0 & 0 & \cdots & 0 \\ s_2^* & s_1^* & 0 & \cdots & 0 \\ s_3^* & s_2^* & s_1^* & \cdots & 0 \\ \vdots & \vdots & \vdots & \ddots & \vdots \\ s_N^* & s_{N-1}^* & s_{N-2}^* & \cdots & s_1^* \end{bmatrix} \tag{2.52}$$

where s_i^* are the step-response values of $H_P^*(q)$. They are given by the coefficients of the $G_j(q)$ polynomial.

The free run output of eq. (2.25) becomes

$$\hat{y}_0(k) = \begin{bmatrix} L_1 u^*(q) \\ L_2 u^*(q) \\ L_3 u^*(q) \\ \vdots \\ L_N u^*(q) \end{bmatrix} + \begin{bmatrix} F_1 y^*(q) \\ F_2 y^*(q) \\ F_3 y^*(q) \\ \vdots \\ F_N y^*(q) \end{bmatrix} \tag{2.53}$$

where $u^*(q)$ and $y^*(q)$ are

$$u^*(q) = C^{-1}(q)\Delta u(k-1) \tag{2.54}$$

$$y^*(q) = C^{-1}(q) y(k) \tag{2.55}$$

2.6 Optimization for MPC

As presented in eqs. (2.1)–(2.3), the MPC algorithm relies on solving a receded optimization problem for computing the open-loop optimal control sequence at each time step, from which only the first control move is sent to the process input. Usually, the optimization index penalizes the sum of the square error (difference between the future reference trajectory and the predicted output) and the square input variable move, subject to constraints. The performance index is quadratic in the unknown input move sequence $\{\Delta u(k), \Delta u(k+1), \ldots, \Delta u(k+l), \ldots, \Delta u(k+m-1)\}$.

For the MIMO case, with the prediction equations (2.29) and (2.30), the optimization problem has the following formulation:

$$\min_{\Delta u(k)\ldots\Delta u(k+m-1)}\{J(u,k)\} = \min_{\Delta u(k)\ldots\Delta u(k+m-1)}\left\{\sum_{l=1}^{p}Y_{wt}||y(k+l|k)-r(k+l)||^2 + \right.$$

$$\left. +\sum_{l=1}^{m}U_{wt}||\Delta u(k+l-1)||^2\right\} \tag{2.56}$$

Although the *quadratic* performance index in the state-space formulation is preferred and extensively applied, the linear performance index may be also used. Despite the fact that when using the linear performance index, the optimization problem is solved in an easier manner, compared to the quadratic performance index, the first one shows some weaknesses, making it less attractive. Among them, the possibility of obtaining not-unique solutions, showing slow control action far away from origin and a lot of action close to origin, difficulty in tuning, and always activating some constraints may be mentioned. The use of the quadratic performance index is harder to solve, but leads to a unique solution while showing a lot of control action far away from origin and smooth action close to origin.

When the quadratic optimization MPC problem is not constrained, an analytical optimal solution may be obtained (good resemblance to the classical LQG regulation). However, when constraints are associated with the optimization problems (2.1–2.3), the optimal solution may be obtained using QP, sequential programming, or nonlinear programming (NLP) numerical solution techniques, as no analytical solution exists. QP refers to an optimization problem having quadratic performance index and linear constraints.

The constraints are always present in control applications, having equipment, technological, safety, environmental, or economic origin. MPC is highly appreciated for its capability to handle constraints in a systematic way, compared to traditional control, such as the PID control. In the latter case, constraints are handled by detuning the controller with negative effect on the control performance, and by ad hoc solutions of using lead-lag and high-low logic blocks [1]. Process requirements may ask for the control task to keep a process variable as close as possible but below a specified constrained value. It is, for example, the case of the reactor riser or regenerator temperature control for the fluid catalytic cracking unit, aimed to keep the temperature in the very close vicinity of the desired temperature value (where optimal economic results are obtained), but below it and with no overshoot [5, 14, 49].

Constraints act on the input (control), output, state, and input move variables, as presented by:

$$u_{min} \le u(k) \le u_{max}$$
$$y_{min} \le y(k) \le y_{max}$$
$$x_{min} \le x(k) \le x_{max} \tag{2.57}$$
$$\Delta u_{min} \le \Delta u(k) \le \Delta u_{max}$$

for any moment of time kT.

Based on the output prediction equation (2.25) and the state prediction equation (2.38), the constraints on predicted inputs and states may be reformulated as constraints on the future control variables $\hat{u}(k)$:

$$\hat{u}_{\min} \leq C_{\hat{u}}\hat{u} \leq \hat{u}_{\max} \tag{2.58}$$

There are cases when the control variable has to be kept close to an optimal value, while it is used to counteract the effect of disturbances and finally brought to this optimal value. This situation may be met when an optimization layer, working on the top of the MPC control layer, sends optimum values for the control variables. Typical circumstances may arise when a higher number of control variables are used, compared to a smaller number of controlled outputs. Handling this situation is possible by adding a supplementary term in the performance index. This term penalizes the deviation of the control variable from its reference value, as presented by the following formulation of the optimization problem:

$$\min_{\Delta u(k)\ldots\Delta u(k+m-1)} \{J(u,k)\} = \min_{\Delta u(k)\ldots\Delta u(k+m-1)} \left\{ \sum_{l=1}^{p} ||Y\mathrm{wt}_l[r(k+l) - y(k+l|k)]||^2 + \right.$$
$$\left. + \sum_{l=1}^{m} ||U\mathrm{wt}_l\Delta u(k+l-1)||^2 + \sum_{l=1}^{m} ||U\mathrm{wt}_l^{\mathrm{ref}}[u(k+l-1) - u^{\mathrm{ref}}(k+l-1)]||^2 \right\} \tag{2.59}$$

where u^{ref} is the reference input variable and $U\mathrm{wt}^{\mathrm{ref}}_l$ is the input reference weighting factor.

The incentive of this approach over constrained optimization consists of putting less restriction on $\Delta u(k)$ for obtaining the optimal control move. This will also enlarge the feasible region of the optimization problem, provided that there are enough degrees of freedom to bring the value of the performance index to zero, i.e., the steady-state outputs be brought to their desired reference values.

Situations may arise when the constrained optimization problem does not have a feasible solution. In such cases, it might be suitable to allow the constraints to be violated, but in a less possible extent. This approach is called constraints softening and may be achieved by the help of a new variable $\delta > 0$, denoted as *slack variable*. This variable allows the enlargement of the hard constraints limits, as presented in the following inequalities:

$$\hat{u}_{\min} - U_{\min}\,\delta \leq C_{\hat{u}}\hat{u} \leq \hat{u}_{\max} + U_{\max}\delta \tag{2.60}$$

where $U_{\min}$ and $U_{\max}$ are the minimum and maximum relaxation nonnegative constant vectors, respectively.

The slack variable δ also becomes an unknown variable of the optimization problem, and its optimal value will minimize the constraints violation. In this case, the form of the optimization problem for MPC gains a new term and gets the formulation:

$$
\min_{\Delta u(k)\ldots\Delta u(k+m-1,\delta)}\{J(u,k)\} = \min_{\Delta u(k)\ldots\Delta u(k+m-1)}\left\{\sum_{l=1}^{p}||Y\mathrm{wt}_l[r(k+l)-y(k+l|k)]||^2 + \right.
$$

$$
\left. + \sum_{l=1}^{m}||U\mathrm{wt}_l\Delta u(k+l-1)||^2 + \sum_{l=1}^{m}||U\mathrm{wt}_l^{\mathrm{ref}}[u(k+l-1)-u^{\mathrm{ref}}(k+l-1)]||^2 + \rho_\delta\delta\right\} \quad (2.61)
$$

where ρ_δ is the weighting factor for penalizing the violation of constraints.

The analytical solution of the MPC unconstrained optimization problem may be obtained as the solution of a least squares problem [37–39]. This solution can be demonstrated for the optimization problem described in eq. (2.2), considering $P(q) = 1$, $p_1 = 1$, $p_2 = N$, $m < N$, and introducing the notation $\hat{r}(k)$ for the future reference vector:

$$
\hat{r}(k) = [r(k+1)\ldots r(k+l)\ldots r(k+N)]^T \quad (2.62)
$$

The vector of future unknown values of the control variable move is described by

$$
\hat{u}(k) = [\Delta u(k)\ \Delta u(k+1)\ldots\Delta u(k+l)\ldots\Delta u(k+m-1)]^T \quad (2.63)
$$

The MPC minimization problem may be sequentially formulated as

$$
\min_{\Delta u(k)\ldots\Delta u(k+m-1)}\{J(u,k)\} = \min_{\Delta u(k)\ldots\Delta u(k+m-1)}\left\{\sum_{l=1}^{N}||[\hat{r}(k+l)-y(k+l|k)]||^2 + \right.
$$

$$
\left. + \sum_{l=1}^{N}||\lambda^2\Delta u(k+l-1)||^2\right\} = \min_{\Delta u(k)\ldots\Delta u(k+m-1)}\left\{||\hat{r}(k)-\hat{y}(k)||^2 + \lambda^2||\hat{u}(k)||^2\right\} =
$$

$$
= \min_{\Delta u(k)\ldots\Delta u(k+m-1)}\{[\hat{r}(k)-\hat{y}(k)]^T[\hat{r}(k)-\hat{y}(k)] + \lambda^2\hat{u}^T(k)\hat{u}(k)\} =
$$

$$
= \min_{\Delta u(k)\ldots\Delta u(k+m-1)}\left\{\hat{u}^T(k)\hat{A}^T\hat{A}\hat{u}(k) + 2\hat{b}\hat{A}\hat{u}(k) + \hat{b}^T\hat{b}\right\} \quad (2.64)
$$

where the last form of the performance index was obtained by replacing $\hat{y}(k)$ with its equal term, presented by the general prediction equation (2.25), and $\hat{A}$ matrix and $\hat{b}$ vector have the following form:

$$
\hat{A} = \begin{bmatrix} -M \\ \lambda I \end{bmatrix} \quad (2.65)
$$

and

$$
\hat{b} = \begin{bmatrix} \hat{r}(k)-\hat{y}_0(k) \\ O \end{bmatrix} \quad (2.66)
$$

The solution to this minimization problem in the least squares sense, having $\hat{u}(k)$ vector as unknown, is given by

$$\hat{u}(k) = -(\hat{A}^T\hat{A})^{-1}\hat{A}^T\hat{b} = (M^TM + \lambda^2 I)^{-1}M^T(\hat{r}(k) - \hat{y}_0(k)) \tag{2.67}$$

Example 2.5

The derivation of the solution presented in eq. (2.67) emerges from solving a linear system of equations having the form:

$$Ax + b = 0 \tag{2.68}$$

where x is the vector of unknown (dim(x) = $m \times 1$), A is a matrix of constant values (dim(A) = $N \times m$, rank(A) = m), and b is a vector of constant free terms (dim(b) = $N \times 1$), $m < N$. As the number of unknowns of the linear system is higher than the number of equations, the eq. (2.68) may only be solved in the least squares sense, i.e., by minimizing the square of the residue $\rho = b + Ax$:

$$\min_x \{\rho^T\rho\} = \min_x \{(Ax + b)^T(Ax + b)\} \tag{2.69}$$

The least squares solution may be found by computing the derivative of the minimization index from eq. (2.69) and making it equal to zero. The solution is obtained successively in the following:

$$\frac{d(\rho^T\rho)}{dx} = 2\left(\frac{d\rho}{dx}\right)^T\rho = 2A^T(Ax + b)$$

$$2A^T(Ax + b) = 0 \tag{2.70}$$

$$A^TAx = -A^Tb$$

$$x = -(A^TA)^{-1}A^Tb$$

Note that the second derivative of the residue is positive:

$$\frac{d^2(\rho^T\rho)}{dx^2} = 2AA^T \geq 0 \tag{2.71}$$

which makes the optimal value to be a minimum.

For the MIMO case of the optimization problem presented in eq. (2.56), the optimal control sequence becomes:

$$\hat{u}(k) = (M^TY_{wt}{}^TY_{wt}M + U_{wt}{}^TU_{wt})^{-1}M^TY_{wt}{}^TY_{wt}(\hat{r}(k) - \hat{y}_0(k)) \tag{2.72}$$

The forms of the optimal MPC control moves, presented in eqs. (2.67) and (2.72), may be reduced to the following well-known form:

$$\hat{u}(k) = K_{MPC}E(k + 1|k) \tag{2.73}$$

where K_{MPC} is the controller gain matrix and $E(k + 1|k)$ is the vector of future predicted errors for zero future-manipulated variable moves (future error produced by the *free run output*).

Due to the receding horizon control approach, only the first of the control variable move:

$$\Delta u(k) = [1\ 0\ 0\ \dots\ 0]\hat{u}(k) \tag{2.74}$$

is extracted from the optimal control vector and sent to the controlled process. The resulting explicit control law is linear and time-invariant, provided the weighting factors are also time-invariant.

2.7 MPC tuning

Tuning the MPC controller is not straightforward, but some general guidelines may be depicted to provide the desired reference tracking, no steady-state offset, and disturbance rejection [2, 39]. MPC has good inherent robustness to model-pant mismatch. Meanwhile, there are no simple mathematical conditions for ensuring stability.

There are several MPC tuning parameters [40, 47]. They are: sampling time T, model horizon n, prediction horizon p (including both p_1 and $p_2 = n$ parameters of the MPC formulation described in eq. (2.2)), control horizon m, penalty weighting factor λ or matrices Y_{wt} and U_{wt}, and filters.

The sampling period T: This must be chosen such as to describe the dynamic behavior without losing the relevant quick process variable changes, but at the same time, to avoid overloading the numerical computation implied by finding the solution to the constrained optimization problem [48]. It is the Shannon theorem that states the compromise for selecting the best value of the sampling time, i.e., half of the period corresponding to the highest considerable angular frequency ω_s present in the harmonic decomposition of the sampled signal $T \le \pi/\omega_s$. This consideration may be reduced to the very simple rule, $T = 0.1(\tau_d + T_d)$, where τ_d and T_d are the pure time delay (dead time) and the dominant time constant of the open-loop system.

Model horizon n: This should be selected such that nT exceeds 95% of the open-loop system settling time. It may also be the recommended value for n, such as the last considered value of the impulse-response parameter h_n is of the order of magnitude comparable to the measurement error for the output variable. Typical values may vary between 20 and 70, but it may also depend on the available computation resources.

Prediction horizon p: The product of the prediction horizon and the sampling time pT should cover the time necessary for the closed-loop system to achieve steady state. Typical values vary between 20 and 30. Short prediction horizon produces large control variable moves, ending in instability. Long prediction horizon produces less control variable moves and slower response. There is a critical minimum horizon length to achieve a stable closed-loop system; the setting $p = n + m$ is suggested. The minimum horizon $p_1 T$ of the performance index (2.2) should be chosen one sampling step larger than the pure time delay, $p_1 = \tau_d + 1$.

Control horizon m: This should be chosen about one-fourth to one-third of the prediction horizon. Typical values vary between 1 and 4. Short control horizon performs good control due to smaller control variable moves and reduced computation resources. When m is increased, the control variable moves become larger, with less robustness and increased computational load, also having an increased degree of freedom in computing the control moves. For the case of using the CARIMA prediction models (or equivalents), the control horizon may be chosen equal to the order of polynomial $A(q)$.

Penalty weighting factor λ: This should be chosen as small as possible (but positive). For $\lambda = 0$, the control action is not penalized. Increasing λ will make the control action less aggressive.

Penalty weighting matrices Y_{wt} and U_{wt}: These are usually diagonal, positive definite matrices. The values of the diagonal elements in the Y_{wt} matrix are measures of the importance of the control effort assigned for each of the controlled variables (MIMO case). As one element of Y_{wt} is increased, the deviation from the reference (set-point) value of the corresponding controlled variable is decreased. Adding in the performance index, the term for penalizing the movement of the control (manipulated) variables, weighted by U_{wt}, reduces the excessive manipulated variable move. For this reason, U_{wt} is also called as "move suppression factor". Increasing the value of one element from U_{wt} decreases the corresponding manipulated variable change, with effect on the degradation of the control performance, but increasing its robustness. The relative magnitudes of Y_{wt} and U_{wt} have to be considered as the magnitudes for the controlled and manipulated variables may be of different orders of magnitude. The weighting matrices Y_{wt} and U_{wt} may be considered nondiagonal, as interactions between the controlled variables are important, but tuning their values is not straight-forward. The weighting matrices Y_{wt} and U_{wt} may be considered time-varying when the control performance has to be dynamically changed, but choosing their appropriate time-dependent functions is difficult. A possible first choice for choosing the values of the weighting matrices is selecting each of their diagonal elements as the inverse of the maximum allowed change in the corresponding output or input variables. These initial guess values should be refined by an iterative simulation procedure [49].

Feedback filter: Introducing a suitable filter on the feedback signal may provide good disturbance rejection. The $P(q)$ polynomial of the performance index formulation (2.2) may also be used to obtain the desired closed-loop control performance by choosing the location of its poles.

Scaled dynamic sensitivity of the MPC performance index with respect to the tuning parameters may reveal their importance on the control performance, suggesting tuning quantitative measures [50].

2.8 MPC stability

Development of the necessary and sufficient conditions for ensuring linear MPC stability is not a trivial task. Most of the developed theoretical conditions for guaranteeing stability are sufficient conditions. One of the most important properties that MPC algorithm must also satisfy is feasibility, i.e., capability to find solution for the optimization problem.

During the last two decades, researchers have focused in their studies on two important ways for guaranteeing stability and feasibility, with significant developments [51]. The first one addresses the development of conditions imposed on the terminal performance function and on supplementary constraints for the terminal states or outputs [39]. *Terminal* values refer to values at the end of the prediction or control horizon. The second direction has been aimed to find conditions for guaranteeing stability by using sufficiently large (infinite) prediction and control horizons.

Infinite control and prediction horizons, $m = p = \infty$, make the unconstrained MPC algorithm similar to the optimal LQ solution. But handling the constraints on an infinite horizon formulation is extremely difficult (or even impossible), especially from the computation effort point of view. One potential approach allows the prediction horizon to be infinite but makes the control horizon finite. In this way, constraints handling becomes a finite-dimensional problem. Conditions may be developed to obtain the closed-loop stability for the monotonic performance index.

The terminal point equality constraint adds a supplementary constraint to the constrained MPC by imposing the condition that the controlled output reaches the reference value or the controlled state reaches the origin (for the state-space MPC performance index formulation) at the end of the prediction horizon:

$$y(k+p+j) = r(k+p), \quad j=1,\dots, N$$
$$x(k+p+j) = \mathbf{0}, \quad j=1,\dots, N \tag{2.75}$$

Such sufficient conditions for guaranteeing asymptotic stability may exhibit feasibility problems if the control horizon is also small, due to the limited degrees of freedom. One solution is to replace the constraint on the terminal point with the constraint of keeping the terminal outputs or states inside a terminal region Ω, where stability is fulfilled:

$$y(k+p+j) \in \Omega_y, \quad j=1,\dots, N$$
$$x(k+p+j) = \Omega_x, \quad j=1,\dots, N \tag{2.76}$$

A terminal penalty term, $E(x(k+p))$, may also be used for obtaining stability by adding it to the optimization performance index:

$$\min_{\Delta u(k)\dots\Delta u(k+m-1)} \{J(u,k)\} = \min_{\Delta u(k)\dots\Delta u(k+m-1)} \left\{ \sum_{l=1}^{p} Q\|x(k+l|k)\|^2 + \right.$$
$$\left. + \sum_{l=1}^{m} R\|\Delta u(k+l-1)\|^2 + E(x(k+p)) \right\} \tag{2.77}$$

Contraction constraints can also offer guarantees for MPC algorithm stability. Such a constrained contraction has the form:

$$\|x(k_0 + p|k)\| \le \varepsilon \|x(k_0)\| \tag{2.78}$$

where ε is a less than one but a positive contraction factor [52].

Recursive feasibility and stability conditions have been developed by research-ers [53, 54], but their practical application is still lacking due to the mathematical complexity [26].

2.9 Nonlinear MPC

During the last two decades, a large effort has been devoted both by academia and in-dustry to develop and apply the Nonlinear Model Predictive Control NMPC techniques [54–62]. Significant results have been obtained but this MPC research of top interest is yet a challenging subject. Although the linear MPC does have well-developed methodol-ogies for obtaining the desired control performance while satisfying the stability re-quirement, the NMPC techniques do not have such general and mature construction.

Essentially, the NMPC techniques are striving to compute the control law as solution to a receding open-loop optimization problem using the predictions of the output or state variable, based on the nonlinear model of the plant. Founded on linearized process models, the linear MPC can handle control of nonlinear processes, especially when the goal is to keep the constrained process variables close to the desired nominal steady-state points. In this case, the use of a linear model for predictions, associated with a qua-dratic performance index, result in a convex QP optimization problem, which can be solved online with the available numerical algorithms. The need for NMPC becomes nec-essary when the control task asks for tracking a changing reference, which implies the covering of a pronounced nonlinearity domain. A typical case is the batch, start-up or shut down operation. Another set of control instances asking for NMPC is the distur-bance rejection control for processes with strong nonlinearities and for important changes of disturbances.

There are different types of models that may be used in the NMPC framework. The most valued are the mechanistic (first principle) models, but statistical black box models are also much appreciated. The first category of models takes advantage of their sound physical meaning and capacity to extrapolate beyond the available pro-cess-measured data or any particular known operating point. The second category al-lows the identification of the model from process measurements and this may be performed with less time and human resources compared to the former category. But black box models lose the physical interpretation and make extrapolation question-able. However, NMPC applications are still dependable to a large extent on the models that emerge from process identification.

From the time variable point of view, both continuous and discrete time models are used in NMPC. Due to the only-on digital computer implementation of the NMPC applications and despite the difficulty in transforming the continuous time nonlinear models in their corresponding discrete time form (unlike the continuous time linear models case), the discrete nonlinear model formulation is preferred. From the contin-uous or discrete time type of the signals, the theoretical approach for NMPC has developed several methodologies for the association between the model used and the control signal obtained. They are: continuous-time model with continuous control signal, continuous-time model with discrete-time control signal, and discrete-time model with discrete-time control signal formulations [39].

The mathematical formulation of the NMPC optimization problem with associated constraints for the discrete-time model with discrete-time control signal is presented in the following:

$$\min_{u(k)\ldots u(k+m-1)} \{J(u,k)\} = \min_{u(k)\ldots u(k+m-1)} \left\{ \sum_{l=1}^{p} ||Y_{\text{wt}}[r(k+l) - y(k+l|k)]||^2 + \right.$$

$$\left. + \sum_{l=1}^{m} ||U_{\text{wt}}u(k+l-1)||^2 \right\} \tag{2.79}$$

subject to

$$x(k+1) = f(x, u, k) \tag{2.80}$$

$$y(k) = g(x, k) \tag{2.81}$$

$$g_1(x) = 0 \tag{2.82}$$

$$x(k) = x_{\text{est}}(k) \tag{2.83}$$

$$u_{\min}(k+l) \le u(k+l) \le u_{\max}(k+l) \tag{2.84}$$

$$u(k+1-1) - \Delta u_{\max} \le u(k+l) \le u(k+l-1) + \Delta u_{\max} \tag{2.85}$$

$$u(k+l) = u(k+m-1), \ l = m-1,\ldots, p \tag{2.86}$$

$$x_{\min}(k+l) \le x(k+l) \le x_{\max}(k+l) \tag{2.87}$$

$$y_{\min}(k+l) \le y(k+l) \le y_{\max}(k+l) \tag{2.88}$$

Equations (2.80) and (2.81) describe the nonlinear discrete state-space model, where both f and g functions are nonlinear. Equation (2.82) reveals the nonlinear algebraic equations of the model. Equation (2.83) shows the use of the estimated state in the discrete state-space model as initial state. The inequality constraint equations (2.84) and (2.85) correspond to the constraints on the control variable and the control variable moves. The additional constraint on the control variable eq. (2.86) enforces the condition to keep the control variable values unchanged for the time interval between the end of the input horizon and the end of the prediction horizon. Constraints from eqs. (2.87) and (2.88) correspond to the constraints on the state and the output variables.

For the case of NMPC presented in eqs. (2.79)–(2.88), where discrete time models are used, solving the constrained optimization is simplified, compared to NMPC using continuous models due to the fact that solution of the optimization problem might be performed without the need for numerical integration for finding the states associated with the next moment of time. Alternation of the optimal control variables' computation, followed by substitution, makes the algorithm of finding the solution to the constrained optimization less demanding. Anyway, even for the case of using the continuous models, the solving of the model differential equations is also based on the numerical approach, i.e., the discretization is inherently performed but the computational effort is increased.

Solving the nonlinear programming (NLP) problem described by eqs. (2.79)–(2.88) is generally nonconvex and requires good NLP software for finding the online solution. There are some approaches for finding solution to the NLP of NMPC [2].

One simple approach is to use a *linearized form of the model* equations around an operating point and successively update the linearizing as the operating point is changing or on a time scheduling basis. If the operating point cannot be directly found by measurements, it may be the estimated from available measurements. This approach transforms NPMC into linear MPC and for the latter, the QP may be used for obtaining the solution if a quadratic performance index is considered. An improved successive linearizing approach consists of using the linearized model for computing only the component of the predictions due to future control moves and the nonlinear model for computing the prediction component due to the past (known) control moves.

Another approach for solving the NPMPC optimization problem is the *sequential algorithm* [2, 63, 64]. This is a two-step method where the sequential model solution and optimization are performed. It is supposed that a continuous model is used for prediction and it consists of a set of ordinary differential equations (ODEs). The control inputs are computed by the NLP solver and, subsequently, the ODE numerical solver computes the solution to the model equations. A schematic representation of this algorithm is presented in Fig. 2.16.

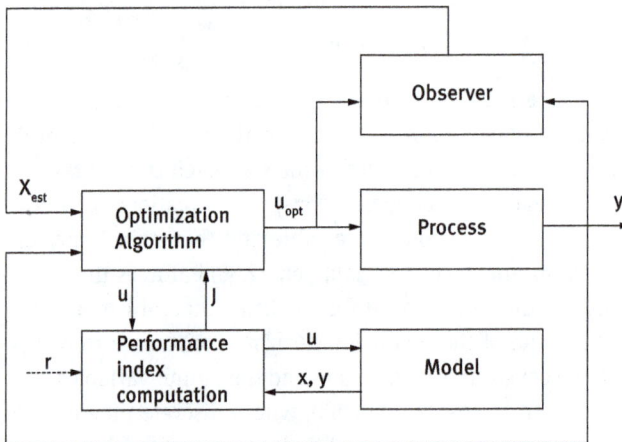

Fig. 2.16: Schematic representation of the sequential NMPC algorithm.

A sequence of input control is first considered and the system of model ODEs is solved to get the states and outputs. With these computed values, the performance index is evaluated and the optimization problem is solved by establishing a new input and an improved control sequence. This succession of optimization and model solution is performed up to the finding the optimal control input sequence.

The *simultaneous algorithm* is another solving approach to the NPMC optimization problem [2, 65]. It is schematically presented in Fig. 2.17.

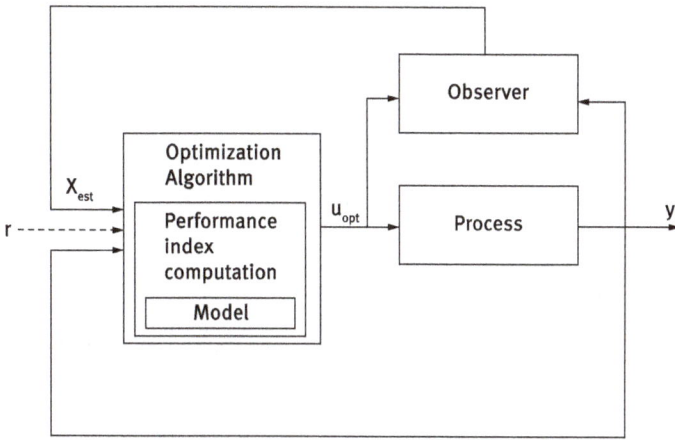

Fig. 2.17: Schematic representation of the simultaneous NMPC algorithm.

In this approach, the model equations, in discretized form, are considered as constraints to the optimization problem. An NLP solver is used to solve the constrained optimization problem.

Further research is devoted today to extend the MPC theory and applications. Some of the directions for these developments are MPC for large-scale systems, either with distributed or hierarchical configurations, MPC for rapid dynamic systems, MPC for uncertain systems, stochastic MPC, adaptive MPC, embedded MPC, low-cost MPC, economic MPC, output MPC, and nonlinear MPC.

References

[1]   Qin, S.J., Badgwell, T.A., *A survey of industrial model predictive control technology*, Control Engineering Practice, 11(7), 733–764, 2003.
[2]   Agachi, P.S., Nagy, Z.K., Cristea, V.M., Imre-Lucaci, A., *Model Based Control – Case Studies in Process Engineering*, Wiley-VCH, Weinheim, 2006.
[3]   Prett, D.M., Gillette, R.D., *Optimization and Constrained Multivariable Control of a Catalytic Cracking Unit*, in *AIChE National Meeting*, Houston, TX, 1979.

[4]   Cutler, C.R., Ramaker, B.L., *Dynamic matrix control-a computer control algorithm*, in *Proceedings of the Joint Automatic Control Conference*, 1980.

[5]   Cristea, M.V., Agachi, S.P., Marinoiu, M.V., *Simulation and model predictive control of a UOP fluid catalytic cracking unit*, Chemical Engineering and Processing, 42, 67–91, (2003).

[6]   Wang, W.L., Rivera, D.E., Kempf, K.G., *Model predictive control strategies for supply chain management in semiconductor manufacturing*, International Journal of Production, Economics, 107(1), 56–77, (2007).

[7]   Nagy, Z., Agachi, S., *Model predictive control of a PVC batch reactor*, Computers and Chemical Engineering, 21(6), 571–591, (1997).

[8]   Van Brempt, W., Backx, T., Ludlage, J., Van Overschee, P., De Moor, B., Tousain, R., *A high performance model predictive controller: application on a polyethylene gas phase reactor*, Control Engineering Practice, 9, 829–835, (2001).

[9]   Wright, G.T., Edgar, T.F., *Nonlinear model predictive control of a fixed-bed water-gas shift reactor: an experimental study*, Computers and Chemical Engineering, 18, 83–102, (1994).

[10] Muske, K.R., Howse, J.W., Hansen, G.A., Cagliostro, D.J., *Model-based control of a thermal regenerator. Part 1: dynamic model*, Computers and Chemical Engineering, 24, 2519–2531, (2000).

[11] Dufour, P., Michaud, D.J., Toure, Y., Dhurjati, P.S., *A partial differential equation model predictive control strategy, application to autoclave composite processing*, Computers and Chemical Engineering, 28(4), 545–556, (2004).

[12] Dufour, P., Toure, Y., Blanc, D., Laurent, P., *On nonlinear distributed parameter model predictive control strategy: On-line calculation time reduction and application to an experimental drying process*, Computers and Chemical Engineering, 27(11), 1533–1542, (2003).

[13] Cristea, V.M., Roman, R., Agachi, S.P., *Neural Networks Based Model Predictive Control of the Drying Process*, ESCAPE 13, Computer Aided Chemical Engineering, 14, 1–4, June, Lappeenranta, Finland, 389–394, (2003).

[14] Iancu, M., Cristea, M.V., Agachi, P.S., *Retrofit design of heat exchanger network of a fluid catalytic cracking plant and control based on MPC*, Computers and Chemical Engineering, 49, 205–216, 2013.

[15] Van overloop, P.J., Weijs, S., Dijkstra, S., *Multiple model predictive control on a drainage canal system*, Control Engineering Practice, 16(5), 531–540, 2008.

[16] Gomez, M., Rodellar, J., Mantecon, J.A., *Predictive control method for decentralized operation of irrigation canals*, Applied Mathematical Modelling, 26(11), 1039–1056, 2002.

[17] Shen, W., Chen, X., Corrioub, J.P., *Application of model predictive control to the BSM1 benchmark of wastewater treatment process*, Computers and Chemical Engineering, 32, 2849–2856, 2008.

[18] Ostace, G.S., Cristea, V.M., Agachi, P.S., *Cost reduction of the wastewater treatment plant operation by MPC based on modified ASM1 with two-step nitrification/denitrification model*, Computers and Chemical Engineering, 15(11), 2469–2479, 2011.

[19] Garcia-Gabin, W., Zambrano, D., Camacho, E.F., *Sliding mode predictive control of a solar air conditioning plant*, Control Engineering Practice, 17(6), 652–663, 2009.

[20] Roca, L., Guzman, J.L., Normey-Rico, J.E., Berenguel, M., Yebra, L., *Robust constrained predictive feedback linearization controller in a solar desalination plant collector field*, Control Engineering Practice, 17(9), 1076–1088, 2009.

[21] Asadi, B., Vahidi, A., *Predictive cruise control: utilizing upcoming traffic signal information for improving fuel economy and reducing trip time*, IEEE Transactions on Control Systems Technology, 19(3), 707–714, 2011.

[22] Keviczky, T., Balas, G.J., *Receding horizon control of an F-16 aircraft: a comparative study*, Control Engineering Practice, 14(9), 1023–1033, 2006.

[23] Alexis, K., Nikolakopoulos, G., Tzes, A., *Switching model predictive attitude control for a quadrotor helicopter subject to atmospheric disturbances*, Control Engineering Practice, 19(10), 1195–1207, 2011.

[24] From, P.J., Gravdahl, J.T., Lillehagen, T., Abbeel, P., *Motion planning and control of robotic manipulators on seaborne platforms*, Control Engineering Practice, 19(8), 809–819, 2011.

[25] Percival, M.W., Wang, Y., Grosman, B., Dassau, E., Zisser, H., Jovanovic, L., Doyle, F.J., III., *Development of a multiparametric model predictive control algorithm for insulin delivery in type 1 diabetes mellitus using clinical parameters*, Journal of Process Control, 21(3), 391–404, 2011.

[26] Xi, Y.-G., Li, D.-W., Lin, S., *Model predictive control – status and challenges*, Acta Automatica Sinica, 39(3), 222–236, 2013.

[27] Kalman, R.E., *Contributions to the theory of optimal control*, Bulletin de la Societe Mathematique de Mexicana, 5, 102–119, 1960.

[28] Kalman, R.E., *A new approach to linear filtering and prediction problems*, Transactions of ASME, Journal of Basic Engineering, 87, 35–45, 1960.

[29] Zadeh, L.A., Whalen, B.H., *On optimal control and linear programming*, IRE Transactions on Autonomic Control, 7(4), 45, 1962.

[30] Propoi, A.I., *Use of LP methods for synthesizing sampled-data automatic systems*, Autonomic Remote Control, 24, 837, 1963.

[31] Richalet, J.A., Rault, A., Testud, J.L., Papon, J., *Model predictive heuristic control: applications to an industrial process*, Automatica, 14, 413–428, 1978.

[32] Prett, D.M., Gillette, R.D., *Optimization and Constrained Multivariable Control of a Catalytic Cracking Unit*, in *AIChE National Meeting*, Houston, TX, 1979.

[33] Cutler, C., Morshedi, A., Haydel, J., *An industrial perspective on advanced control*. in *AICHE Annual Meeting*, Washington, DC, October 1983.

[34] Garcıa, C.E., Morshedi, A.M., *Quadratic programming solution of dynamic matrix control (QDMC)*, Chemical Engineering Communications, 46, 73–87, 1986.

[35] Grosdidier, P., Froisy, B., Hammann, M., *The IDCOM-M Controller*, p. 31–36 in McAvoy, T.J., Arkun, Y., Zafiriou, E., (Eds.), *Proceedings of the 1988 IFAC Workshop On Model Based Process Control*, Pergamon Press, Oxford, 1988.

[36] Froisy, J.B., Matsko, T., *IDCOM-M application to the Shell fundamental control problem*, AICHE Annual Meeting, November 1990.

[37] Clarke, D.W., Mohtadi, C., Tuffs, P.S., *Generalized predictive control – part 1. The basic algorithm*, Automatica, 23(2), 137–148, 1987.

[38] Clarke, D.W., Mohtadi, C., P.s, T., *Generalized predictive control – part 2 Extensions and interpretations*, Automatica, 23(2), 149–160, 1987.

[39] Van den boom, T.J.J., *Model Based Predictive Control*, LernModul 6. Swiss Society for Automatic Control, 1997.

[40] Soeterboek, A.R.M., *Predictive Control-A Unified Approach*, Prentice Hall, New York, 1992.

[41] Seborg, D.E., Edgar, T.F., Mellichamp, D.A., *Process Dynamic and Control*, 649–669, John Wiley & Sons, 1989.

[42] Munske, K.R., Rawlings, J.B., *Model predictive control with linear models*, AIChE Journal, 39(2), 262–287, 1993.

[43] Lee, J.H., Morari, M., Garcia, C.E., *State-space interpretation of model predictive control*, Automatica, 30(4), 707–717, 1994.

[44] Balchen, J.G., Ljungquist, D., Strand, S., *State space predictive control*, Chemical Engineering Science, 47(4), 787–807, 1992.

[45] Garcia, C.E., Prett, D.M., Morari, M., *Model predictive control: theory and practice – a survey*, Automatica, 25(3), 335–348, 1989.

[46] Brosilow, C., Joseph, B., *Techniques of Model-Based Control*, Prentice Hall, New York, 2002.

[47] Clarke, D.W., Mohtadi, C., *Properties of Generalized*, Automatica, 25(6), 859–875, 1989.

[48] Zafiriou, E., Morari, M., *Design of robust digital controllers and sampling-time selection for SISO systems*, International Journal of Control, 44(3), 711–735, 1986.

[49] Kalra, L., Georgakis, C., *Effect of process nonlinearity on the performance of linear model predictive controllers for the environmentally safe operation of a fluid catalytic cracking unit*, Industrial and Engineering Chemistry Research, 33, 3063–3069, 1994.

[50] Cristea, M.V., Agachi, S.P., *Model predictive control of inferred variables and dynamic sensitivity analysis applied to MPC tuning*, Buletinul Universitatii "Petrol-Gaze" Ploiesti, 52(1), 52–57, 2000.

[51] Mayne, D.Q., Rawlings, J.B., Rao, C.V., Scokaert, P.O.M., *Constrained model predictive control: stability and optimality*, Automatica, 36(6), 789–814, 2000.

[52] Zheng, A., Morari, M., *Global Stabilization of Linear Discrete-Time Systems with Bounded Controls – A Model Predictive Control Approach*, ACC, 1994.

[53] Mayne, D.Q., Rawlings, J.B., Rao, C.V., Scokaert, P.O.M., *Constrained model predictive control: stability and optimality*, Automatica, 36, 789–814, 2000.

[54] Mayne, D.Q., *Model predictive control: recent developments and future promise*, Automatica, 50, 2967–2986, 2014.

[55] Qin, S.J., Badgwell, T.A., Allgower, F., Zheng, A., *An Overview of Nonlinear Model Predictive Control Applications*, (Eds.), Birkhauser, Switzerland, 2000.

[56] Bauer, M., Craig, I.K., *Economic assessment of advanced process control – a survey and framework*, Journal of Process Control, 18, 2–18, 2008.

[57] Findeisen, R., Allgöwer, F., L.t, B., *Assessment and Future Directions of Nonlinear Model Predictive Control*, Springer, Berlin, 2007.

[58] Magni, L., Raimondo, D.M., Allgöwer, F., *Nonlinear Model Predictive Control: Towards New Challenging Applications*, Springer-Verlag, Berlin, 2009.

[59] Klatt, K.U., Marquadt, W., *Perspectives for process systems engineering | personal views from academia and industry*, Computers and Chemical Engineering, 33(3), 536–550, 2009.

[60] Manenti, F., *Considerations on nonlinear model predictive control techniques*, Computers and Chemical Engineering, 35, 2491–2509, 2011.

[61] Allgöwer, F., Findeisen, R., Nagy, Z.K., *Nonlinear model predictive control: From theory to application*, Journal of the Chinese Institute of Chemical Engineers, 35, 299–315, 2004.

[62] Zavala, V.M., Biegler, L.T., *The advanced-step NMPC controller: optimality, stability and robustness*, Automatica, 45, 86–93, 2009.

[63] Jang, S., Joseph, B., Mukai, H., *Control of constrained multivariable nonlinear processes using a two-stage approach*, Industrial and Engineering Chemistry Research, 26, 2106–2114, 1987.

[64] Bequette, B.W., *Nonlinear predictive control using multi-rate sampling*, Canadian Journal of Chemical Engineering, 69, 136–143, 1991.

[65] Patwardhan, A.A., Rawlings, J.B., Edgar, T.F., *Nonlinear model predictive control*, Chemical Engineering Communications, 87, 123–141, 1990.

3 Fuzzy control

3.1 Introduction

The control system is aimed to generate and send its control decisions to the controlled process such that its steady-state and dynamic behavior conforms to desired process performance. It is obvious that process characteristics govern the design of the controller. Consequently, control performance is directly related to the capability of describing the process behavior and to the way process information is embedded in the controller. Neither the model creation nor the controller design tasks is trivial. Among the important reasons hindering these endeavors are, process complexity, incomplete knowledge, uncertainty of the description, or stochastic behavior of the process. The knowledge on the process behavior may have different forms, i.e., models built on analytical (first principle) or statistical basis. The controller design, based on analytical models, is preferred, but as complexity increases, the approach may become stiff, consume more computing resources, become very sensitive to disturbances, become less robust, or accumulate errors. Statistical models may be an alternative for such cases because they may be built on heuristic methods. Accounting for the tolerance to incomplete determination and uncertainty may lead to the design of a controller able to fulfil the control task in a way humans do, i.e., not necessarily on the basis of very precise (mathematical) knowledge or representation (e.g., on ordinary or partially differential equations), but on a less-precise evaluation that involves the use of some acting rules. Finally, this approximate approach may result in an efficient control, able to overcome the problems mentioned before, and partially render the controller design, a human way of acting. The fuzzy controller is a representative of this class. The fuzzy approach, for both modeling and controller design, is based on the fuzzy logic. Its main elements are presented in the next sections.

3.2 Fuzzy sets

In classical (Boolean) set theory, a large(r) set X having a subset F, $F \subset X$, has elements $x \in X$ that either belong to the set F, $x \in F$, or do not belong to it, $x \notin F$. The classical set F is denoted as a crisp set. As opposed to the crisp set F, a fuzzy set F has elements that may belong to the set F, in a more or less extent. The membership degree of an element to a fuzzy set is defined by a real value in the interval $[0, 1]$ and shows how much truth is in the statement "x belongs to the set F". The membership value of 1 denotes total belonging and the membership value of 0 shows not at all belonging to the set F. It was in 1965 when Lotfi Zadeh introduced the notion of fuzzy set and opened up the new horizon of the fuzzy logic research and applications [1].

https://doi.org/10.1515/9783110789737-004

i **Example 3.1**

Consider a set of metal balls that have been coated with white, black, and white-black combinations of paints. This is the large set X, of both classical and fuzzy approach. Consider the set F to be the set of gray balls.

The *classical (crisp) set* of gray balls consists of balls (elements) having been coated with a combination of 50% white paint and 50% black paint. All other balls, coated with other combinations of white-black paints or exclusively coated white or black paint, do not belong to the crisp set F. This approach of building the set F is a very accurate one, as it considers only balls coated with exactly equal white-black paint mixture. Other crisp sets may be defined on the large set X of balls, such as white set of balls F_W, black set of balls F_B, etc.

The *fuzzy set* of gray balls F consists of balls (elements) having been coated not only with a combination of 50% white paint and 50% black paint but also with different percentages of black-and-white paint. According to this approach, balls having been coated with varying proportions of white-black paint (e.g., combination of 80% white paint and 20% black paint or combination of 20% white paint and 80% black paint), also belong to the fuzzy set F. Furthermore, it may be stated that balls from this fuzzy set belong to the set F more or less, according to how much the combination of their white-black paints comes close to the 50% white paint and 50% black paint combination. For example,

for a ball coated with a combination of 50% white paint and 50% black paint, it may be assigned total membership to the fuzzy set F (with associated membership value of 1),

for a ball coated with a combination of 75% white paint and 25% black paint, it may be assigned partial membership to the fuzzy set F (with associated membership value of 0.5),

for a ball coated with a combination of 25% white paint and 75% black paint, it may be assigned partial membership to the fuzzy set F (with associated membership value of 0.5),

for a ball exclusively coated with white paint (100%) or black paint (100%), it may be assigned membership to the fuzzy set F with associated membership value of 0.

Other fuzzy sets may be defined on the large set X of balls, such as white set of balls F_W or black set of balls F_B. Either of them contains balls coated with different combinations of white-black paints but have associated different membership values, proportional to the closeness of their color to the white or black paint.

Definition: A fuzzy set F (on a large set X, named universe of discourse) is represented by the set of pairs [2]:

$$F = \{(x,\ m_F(x))|x \in X\} \tag{3.1}$$

where m_F is a characteristic function, named membership function of the fuzzy set F, defined by

$$m_F{:}X \rightarrow [0,\ 1] \tag{3.2}$$

As a result of the fuzzy set definition, the membership function assigns to every element x from F, a real value $m_F(x)$ that belongs to the interval [0, 1]. The value $m_F(x)$ is denoted as the membership degree of the element x from the fuzzy set F and shows how much the element x belongs to the set F. The value of 0 denotes no membership (false) and the value of 1 shows total membership (true). Even though the fuzzy set is a time-invariant structure, it may be successfully used for the design of the fuzzy controller.

Note that for a crisp set F, such a characteristic function could be also associated but it would take values only from the set {0, 1}, i.e., if the element x belongs to the classical set F, it will take the value of 1 and if the element x does not belong to the classical set F, it will take the value of 0.

3.3 Typical membership functions of the fuzzy sets

For the description of the membership functions, a set of typical function forms are commonly used in applications, although there is no restriction about the form of their choice [3, 5].

Figure 3.1 shows the triangular membership function.

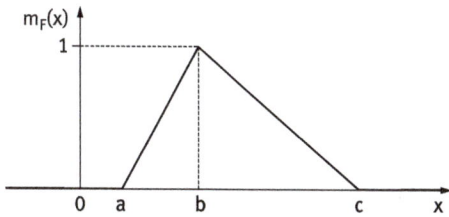

Fig. 3.1: Triangular membership function.

Its analytical form is

$$m_F(x) = \begin{cases} 0, & x < a \\ \dfrac{x-a}{b-a}, & a \le x \le b \\ \dfrac{c-x}{c-b}, & b \le x \le c \\ 0, & x > c \end{cases} \tag{3.3}$$

Figure 3.2 shows the trapezoidal membership function.

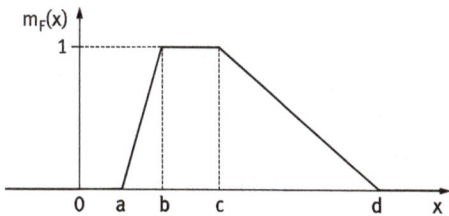

Fig. 3.2: Trapezoidal membership function.

Its analytical form is

$$
m_F(x) = \begin{cases}
0, & x < a \\
\dfrac{x-a}{b-a}, & a \le x \le b \\
1, & b \le x \le c \\
\dfrac{d-x}{d-c}, & c \le x \le d \\
0, & x > d
\end{cases} \tag{3.4}
$$

Figure 3.3 shows the saturation membership functions:

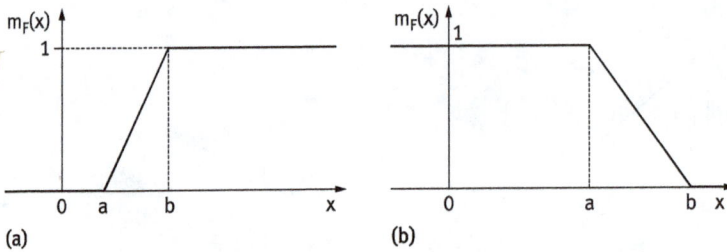

Fig. 3.3: High (a) and low (b) saturation membership functions.

Their analytical forms are

$$
m_F(x) = \begin{cases}
0, & x < a \\
\dfrac{x-a}{b-a}, & a \le x \le b \\
1, & x > b
\end{cases} \tag{3.5}
$$

$$
m_F(x) = \begin{cases}
1, & x < a \\
1 - \dfrac{x-a}{b-a}, & a \le x \le b \\
0, & x > b
\end{cases} \tag{3.6}
$$

For the triangular, trapezoidal, and saturation membership functions, the parameters *a, b, c,* and *d* are constants, but they may be adjusted according to the needs. These membership functions are piecewise continuous functions.

A particular membership function form is the singleton membership function. Its shape, presented in Fig. 3.4, emerges from a Dirac impulse function, with a finite width ε.

Smooth continuous membership functions are also used, such as sigmoidal saturation functions (including their difference) or Gauss bell-shaped function. The latter is presented in Fig. 3.5.

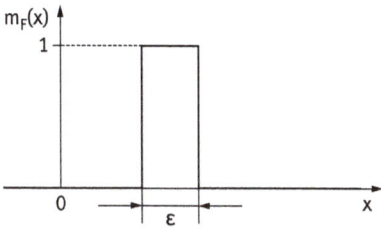

Fig. 3.4: Singleton membership function.

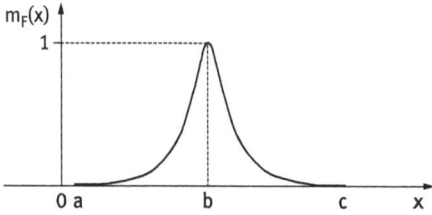

Fig. 3.5: Bell-shaped membership function.

Its equation is of the form

$$m_F(x) = e^{-\frac{(x-b)^2}{2(c-a)^2}} \qquad (3.7)$$

for which, the restriction to the interval $[a, c]$ is usually considered.

Example 3.2

Consider the system of a mixing tank in which two inlet flows are entering, one of cold water and the other of hot water. The tank outlet flow of mixed water feeds the downstream process with water at the desired temperature of $T° = 25$ °C. The temperature in the tank is controlled either by the inlet flow of hot water or by the inlet flow of cold water. The process variable of first importance is the temperature in the tank. It may be considered that the tank temperature can be classified into three categories (ranges): *cold water* with temperature in the interval [18, 23] °C, *warm water* with temperature in the interval [23, 27] °C, and *hot water* with temperature in the interval [27, 32] °C. It is possible to associate, both crisp and fuzzy sets to the three water temperature categories. The cold, warm, and hot temperature crisp sets are presented in Fig. 3.6.

According to the crisp set approach, a water temperature of $T° = 23.3$ °C belongs to the warm water set and does not belong to the cold water set or to the hot water set.

One possible choice of fuzzy sets associated with the three water temperature categories is presented in Fig. 3.7.

The fuzzy sets – cold and hot water – have saturation forms (low, respectively high), while the warm water fuzzy set is of triangular form. All of them have as universe of discourse, the temperature interval [18, 32] °C. The membership functions associated with the three fuzzy sets $m_C(T°)$, $m_W(T°)$, and $m_H(T°)$ show a gradual transition for the membership of the elements belonging to adjacent sets. For example, the temperature element of sets $T° = 23.3$ °C belongs to cold water fuzzy set to the extent (trueness) described by

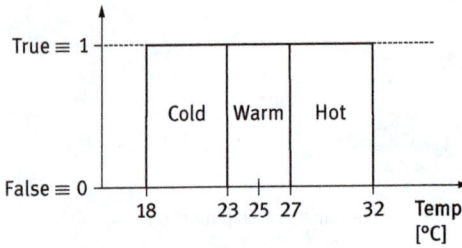

Fig. 3.6: Crisp water temperature sets.

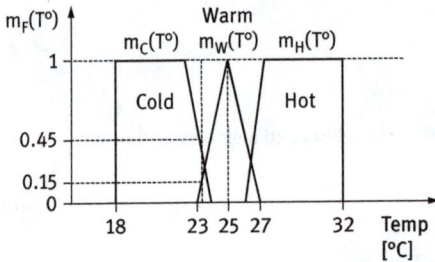

Fig. 3.7: Cold, warm, and hot water fuzzy water temperature sets.

the membership value $m_C(T°) = 0.45$, while the same element belongs to warm water fuzzy set to the extent (trueness) described by the membership value $m_W(T°) = 0.15$. This shows that the water temperature $T° = 23.3$ °C is considered to be cold in a larger extent (trueness) than warm. According to the way the fuzzy sets are defined, the same temperature element has a membership value of $m_H(T°) = 0$, showing that it may not be considered a hot temperature.

The triangular form of the warm water fuzzy set was selected for revealing a gradual measure of proximity to the temperature value of $T° = 25$ °C, which is of highest importance for the considered mixing process and its associated control system task of keeping the temperature to this setpoint value.

The control philosophy is simple and may be concentrated in the following linguistic statements: if the water temperature is cold, then increase the hot inlet flow; if the water temperature is hot, then decrease the hot inlet flow; if the water temperature is warm, then do not change the hot inlet flow (or gently increase of decrease the hot inlet flow, as the water temperature is slightly below or slightly above the desired value of $T° = 25$ °C).

As noticed from this example, for the temperature variable, three fuzzy sets were defined. First, they are described by their names, denoted by linguistic variables: *cold*, *warm*, and *hot*. Second, the membership functions associated with these sets show the extent a certain temperature (element) may be considered to belong to each of the sets.

3.4 Operations with fuzzy sets

Operations with classical sets may be extended to the fuzzy sets. The latter will involve the membership functions associated with the fuzzy sets when operations with fuzzy sets are performed.

For two fuzzy sets M and N, defined on the universe of discourse F and having the membership functions $m_M(x)$ and $m_N(x)$, the union $M \cup N$, intersection $M \cap N$, and the complement C_M operations for $x \in F$ are described by [3, 4]

$$m_{M \cup N}(x) = m_M(x) \vee m_N(x) = \max(m_M(x), \ m_N(x)) \tag{3.8}$$

$$m_{M \cap N}(x) = m_M(x) \wedge m_N(x) = \min(m_M(x), \ m_N(x)) \tag{3.9}$$

$$m_{C_M}(x) = 1 - m_M(x) \tag{3.10}$$

Other properties of the fuzzy set operations may also be developed. They are:
The empty set $\varnothing \subseteq F$, which has

$$m_\varnothing(x) = 0 \tag{3.11}$$

The total set, which is characterized by

$$m_F(x) = 1 \tag{3.12}$$

The two fuzzy sets are equal, $M = N$, if and only if their membership functions are identical:

$$m_M(x) = m_N(x) \tag{3.13}$$

Most of the operations with the classical sets are also met for fuzzy sets, except the excluded middle and the law of contraction, as follows:

$$M \cup C_M \neq F \tag{3.14}$$

$$M \cap C_M \neq \varnothing \tag{3.15}$$

On the other side, two fuzzy-specific operations may be performed:
the algebraic product of two fuzzy sets, $M \cdot N$, where the membership function of the product set is defined by

$$m_{M \cdot N}(x) = m_M(x) \cdot m_N(x) \tag{3.16}$$

the algebraic sum of two fuzzy sets, $M + N$, where the membership function of the sum set is:

$$m_{M+N}(x) = m_M(x) + m_N(x) - m_M(x) \cdot m_N(x) \tag{3.17}$$

The union, intersection, and the complement operations with the temperature fuzzy set for cold water C, and warm water W, from Example 3.2, are presented in Figs. 3.8–3.10.

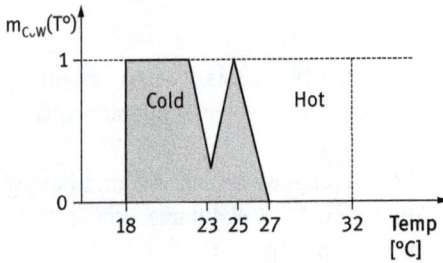

Fig. 3.8: Fuzzy union operation between cold and warm water temperature fuzzy sets.

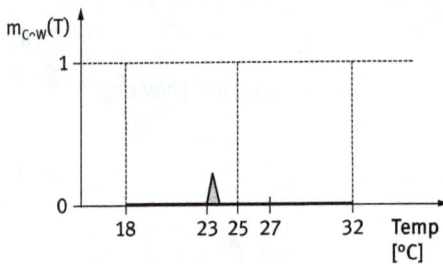

Fig. 3.9: Fuzzy intersection operation between cold and warm water temperature fuzzy sets.

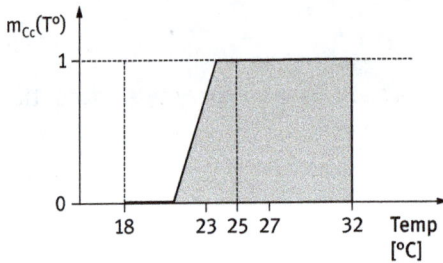

Fig. 3.10: Fuzzy complement operation of the cold water temperature fuzzy set.

Note that the results of the operations with fuzzy sets are fuzzy sets too.

3.5 Fuzzy logic

Having its origin in the human way of reasoning, fuzzy logic is actually its formaliza-
tion. Fuzzy logic is a generalization of the Boolean logic where the discrete approach
with the only two true-false (1 and 0) values are replaced by a continuous approach.
Its mathematical fundamentals may be found in the polyvalent logic of Lukasiewicz.

A linguistic statement, i.e., a proposition Pp, may only have the *True* (1) or the *False* (0) evaluation (result) when classical predicate logic is considered [3, 4]. In fuzzy logic, a proposition Pp may have intermediate evaluations between *True* and *False*, having assigned a fuzzy set. Its membership function reveals the degree of truth of the proposition by a numerical value in the interval [0, 1].

For one of the propositions mentioned in Example 3.2 related to the control philosophy, i.e., "if the water temperature is *cold* then . . .", the degree of trueness may be well described by the fuzzy set *cold water*. A water tank with a temperature of $T° = 19$ °C may very well be considered as cold water (membership value of $m_C(T°) = 1$), while a water tank with a temperature of $T° = 23$ °C may be only "partially" considered as cold water (membership value of $m_C(T°) \approx 0.5$).

As presented in the previous paragraphs, we associate to deterministic variables their fuzzy correspondents, i.e., the fuzzy variables. These fuzzy variables are linguistic variables. A linguistic variable uses words to characterize its values, as it is formulated in natural language. These words are described by fuzzy sets on the universe of discourse. Each word of the linguistic variable has an associated fuzzy set to show the membership measure of one *deterministic value* (element), from the universe of discourse, of belonging to the category described by that particular word. Equivalently, the fuzzy set associated with that word shows the degree of trueness of the linguistic statement, formulated by the words of the linguistic variable. To conclude, the deterministic variable has as values scalars (or vectors) expressed by numbers, while the fuzzy variable has as values different linguistic attributes expressed by words (formalized as fuzzy sets).

Usually, the linguistic variable uses two words to describe it. The first one (primary term) has the same name with the deterministic variable it originates from and the second one (linguistic class) denotes the category of the primary term, using adjectives or adverbs.

For Example 3.2, the deterministic variable "water temperature" has scalar values in the interval [18, 32] °C. To this deterministic variable, the fuzzy variable "water temperature" was associated. The latter has the following values, described in words: "cold water temperature", "warm water temperature", and "hot water temperature". The adjectives "cold", "warm", and "hot" denote the categories of the primary term "temperature". Each of the linguistic variables "cold water temperature", "warm water temperature", and "hot water temperature" have associated fuzzy sets. The linguistic statement "the water temperature is cold" is formalized as a fuzzy set. For a particular deterministic value of the variable "water temperature", e.g., for $T° = 23.3$ °C, the previously mentioned statement is true with a degree of trueness described by the membership value of the associated fuzzy set, "cold water", i.e., $m_C(23.3 °C) = 0.45$.

The transformation of a deterministic variable into its corresponding fuzzy variable (the latter having associated linguistic values) is denoted by *fuzzification*. The number of the linguistic (fuzzy) variables associated with a deterministic variable depends on the particular problem, but usually, a number of five linguistic values are sufficient. Typically, they are very small, small, mean, large, and very large.

For a fuzzy control system, the error-deterministic variable (difference between the setpoint and the controlled variable) may have the corresponding fuzzy linguistic variables: negative large (NL), negative small (NS), zero (Z), positive small (PS), and positive large (PL), as shown in Fig. 3.11.

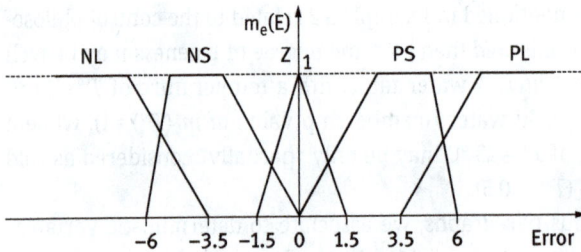

Fig. 3.11: Five fuzzy sets defined for the error variable of the fuzzy controller.

The human knowledge, represented in a linguistic way, is usually formulated in the form of *if-then* rules. It is in fact a transformation of the knowledge into a nonlinear (cause-effect) mapping. Fuzzy logic is based on information represented by fuzzy sets and performs fuzzy *if-then rules* on these linguistic variables. The fuzzy *if-then rule* is a conditional statement involving two propositions:

$$\textbf{if } \{\text{fuzzy proposition } Pa\} \textbf{ then } \{\text{fuzzy proposition } Qc\} \tag{3.18}$$

The *Pa* proposition is denoted as *antecedent* and the *Pc* proposition as *consequent*. The linguistic-formulated *if-then rule* makes a correlation between the premise/cause *Pa* and the consequent/effect *Pc*, building a relationship between them. *According to the degree of trueness of the antecedent statement, the degree of trueness of the consequent statement will be satisfied.* The mathematical correspondent for building this relationship is the *implication*. Although not similar, the fuzzy implication corresponds to the crisp composition of functions.

The fuzzy implication starts from the trueness assessment of one element x to be member of the fuzzy set P, and as a consequence, it follows by establishing the trueness of the elements y (of Q) to belong to the fuzzy set Q. The result of a fuzzy implication is also a fuzzy set Q^*. The fuzzy set Q^* has the same linguistic attributes as the fuzzy variable Q. The fuzzy implication is usually referred as *fuzzy inference* [3, 4, 6].

Recalling the control philosophy presented in Example 3.2, the following fuzzy sets may be considered:

P: The water temperature is *cold,*
Q: Hot inlet flow is *increased.*

which are involved in the control rule "if the water temperature is *cold* then *increase* the hot inlet flow". In this example, there are three fuzzy sets for the fuzzy variable, "Water temperature", according to the linguistic attributes: *Cold, Warm,* and *Hot.* Additionally, there are three fuzzy sets for the fuzzy variable, "Hot inlet flow", according to the linguistic attributes: *Increase, Decrease,* and *No change.* In this particular example, *x* is the *water temperature* and *y* is the *hot inlet flow.*

As a result, the fuzzy implication $P{\rightarrow}Q$, described by the rule "if the water temperature is *cold* then *increase* the hot inlet flow", is a new fuzzy set Q^* showing how much "hot inlet flow is *increased*".

The membership function of the fuzzy set Q^* (fuzzy inference) $m_{P{\rightarrow}Q}(x, y)$ is computed on the basis of the membership functions of the fuzzy sets P, $m_P(x)$, and Q, $m_Q(y)$. There are different ways of defining this fuzzy implication [3, 4]:

1. Mamdani inference:

$$m_{P\rightarrow Q}(x,y) = \min\left(m_P(x), m_Q(y)\right) \tag{3.19}$$

2. Tagaki-Sugeno-Kang (TSK) inference:

$$m_{P\rightarrow Q}(x,y) = g(x) \tag{3.20}$$

where $g(x)$ is a function of the crisp input variable x (or variables). For the case of the *first-order Sugeno inference*, the function $g(x_1, x_2, \ldots, x_n)$ has a linear dependence on the input variables $x_1, x_2, \ldots, x_n$:

$$m_{P\rightarrow Q}(x,y) = g(x) = p_0 + p_1 x_1 + p_2 x_2 + \cdots + p_n x_n \tag{3.21}$$

Boolean inference:

$$m_{P\rightarrow Q}(x,y) = \max\left(1 - m_P(x), m_Q(y)\right) \tag{3.22}$$

Zadeh I inference:

$$m_{P\rightarrow Q}(x,y) = \min\left(1, 1 - m_P(x) + m_Q(y)\right) \tag{3.23}$$

Zadeh II inference:

$$m_{P\rightarrow Q}(x,y) = \min\left[\max\left(m_P(x), m_Q(y)\right), 1 - m_P(x)\right] \tag{3.24}$$

In applications, the most used are the Mamdani and Tagaki-Sugeno-Kang fuzzy logic inference methods.

The antecedent part of the fuzzy rule may have a single statement or a compound statement that merges several statements (premises). Each statement involves a different fuzzy variable. They are connected to the antecedent proposition by the words *and* and *or*. The trueness of the antecedent is evaluated on the basis of the operations with fuzzy sets corresponding to the individual statements, i.e., union and/or intersection operators (as presented in the 3.4 paragraph).

To illustrate the combination of fuzzy variables, consider for Example 3.2, as additional input, the *rate of change* of the temperature in the tank. This crisp variable has a corresponding fuzzy variable, *temperature rate of change*, with three linguistic attributes:

Positive, Zero, and *Negative.* The following *if-then rule* may be added for the control strategy of the temperature in the tank:

if{{the water temperature is Cold} <u>and</u> {the water temperature change is Positive}}

    then{the hot inlet flow is No change}

(3.25)

The antecedent of this fuzzy rule implies the computation of the membership function as a compound statement, i.e., the intersection (*and* operator) of the fuzzy sets "the water temperature is cold" and "the water temperature change is positive". This means the truth of the antecedent is computed as the *min* value of the involved fuzzy sets' membership functions:

$$m_{C \cap \Delta+}(x) = m_C(x) \wedge m_{\Delta+}(x) = \min(m_C(x),\ m_{\Delta+}(x))$$

(3.26)

where the membership functions for the fuzzy sets, "the water temperature is cold" and "the water temperature change is positive" have been denoted by $m_C(x)$ and $m_{\Delta+}(x)$, respectively.

Every fuzzy system consists of a set of rules. As more than one fuzzy rule becomes active for a given set of antecedents, an *aggregation* process for the active rules is needed. There are two types of aggregation mechanisms. The first one is denoted by "and" aggregation and is determined by intersection of the individual rule consequents. The second one is named "or" aggregation and is determined by union of the individual rule consequents.

The last component of the fuzzy logic technique is the *defuzzifier.* This element transforms the result of the implications (aggregated consequent of the active set of rules), which are also fuzzy sets, into crisp (representative) values for the fuzzy variables (sets). This transformation is usually based on the *computation of the centroid C* (center-of-gravity) of the aggregated fuzzy set *F*, according to the following formula [4]:

$$C = \frac{\int_F x m_F(x)\,dx}{\int_F m_F(x)\,dx}$$

(3.27)

where the integral is computed over the universe of discourse.

It may be noticed that fuzzy logic may be successfully used for modeling applications, especially where first-principle models are inoperable or difficult to develop due to process complexity, but for which human expertise or human knowledge is available in the form of linguistic assessments. Furthermore, fuzzy logic may be used to design control systems able to better cope with process uncertainties or incomplete descriptions, compared to the analytical (equation-based) approach, and be successful in processing information affected by noise.

The general structure of the fuzzy system that uses the fuzzy logic is presented in Fig. 3.12.

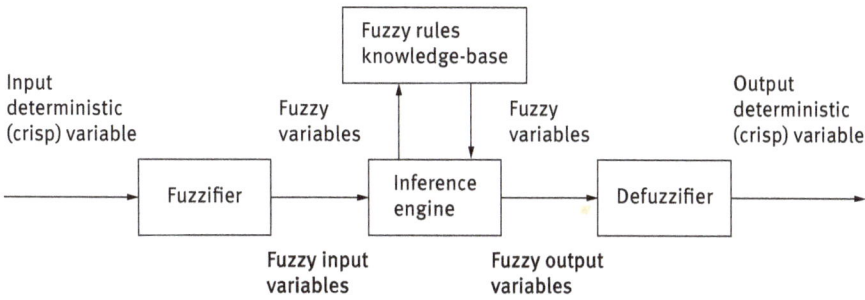

Fig. 3.12: General structure of a fuzzy system.

The general steps for developing a fuzzy system (model or controller) are [3]:
1. description of the knowledge base of the problem,
2. selection of the input-output (state) variables,
3. fuzzyfication of the crisp input and output variables,
4. building the fuzzy inference rules and setting the mechanism of aggregation of the rules,
5. defuzzyfication of the fuzzy output variables,
6. setting mechanisms for making a fuzzy system adaptive.

Example 3.3

Consider the example of a fuzzy controller designed to maintain the temperature in the mixing tank at the setpoint temperature value of $T° = 25 °C$, as presented in Example 3.2. The controller has two crisp inputs: the error e (difference between the setpoint and the current temperature) and the differentiated error de/dt. The controller output is the mixing tank inlet flow rate of cold water.

The associated fuzzy variables for the considered crisp inputs and the fuzzy membership functions are presented in Fig. 3.13(a) and (b).

For the error variable, three linguistic variables have been considered: *N*-negative (high water temperature), *Z*-zero (water temperature close to the setpoint value), and *P*-positive (low water temperature). For the differentiated error variable (error rate) three linguistic variables have also been considered: *Nė*, negative (rising tank water temperature), *Zė*, zero (lack of tank water temperature change), and *Pė*, positive (falling tank water temperature).

The associated fuzzy variable for the considered crisp output is presented in Fig. 3.14.

For the controller output variable, i.e., the mixing tank inlet flow rate of *cold water*, five linguistic fuzzy variables have been considered: *CF*, close fast; *CS*, close slow; *NC*, no change; *OS*, open slow; *OF*, open fast.

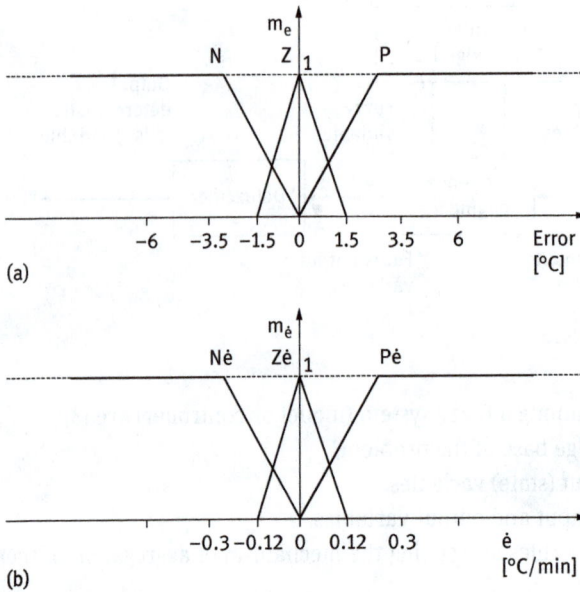

Fig. 3.13: Input fuzzy membership functions of the fuzzy controller (a) error and (b) differentiated error.

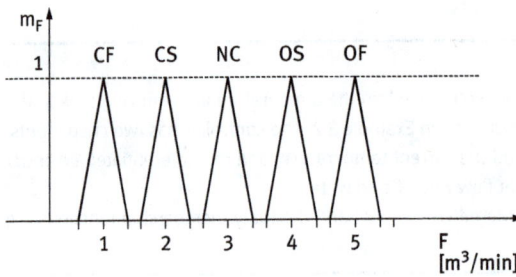

Fig. 3.14: Output fuzzy membership functions of the fuzzy controller.

Five fuzzy inference rules have been designed for the operation of the controller. They are
1. if (error e is negative N) then (cold water flow rate F is open fast OF),
2. if (error e is positive P) then (cold water flow rate F is close fast CF),
3. if (error e is zero Z) and (error rate $\dot{e}$ is positive $P\dot{e}$) then (cold water flow rate F is close slow CS),
4. if (error e is zero Z) and (error rate $\dot{e}$ is negative $N\dot{e}$) then (cold water flow rate F is open slow OS),
5. if (error e is zero Z) and (error rate $\dot{e}$ is zero $Z\dot{e}$) then (cold water flow rate F is no change NC).

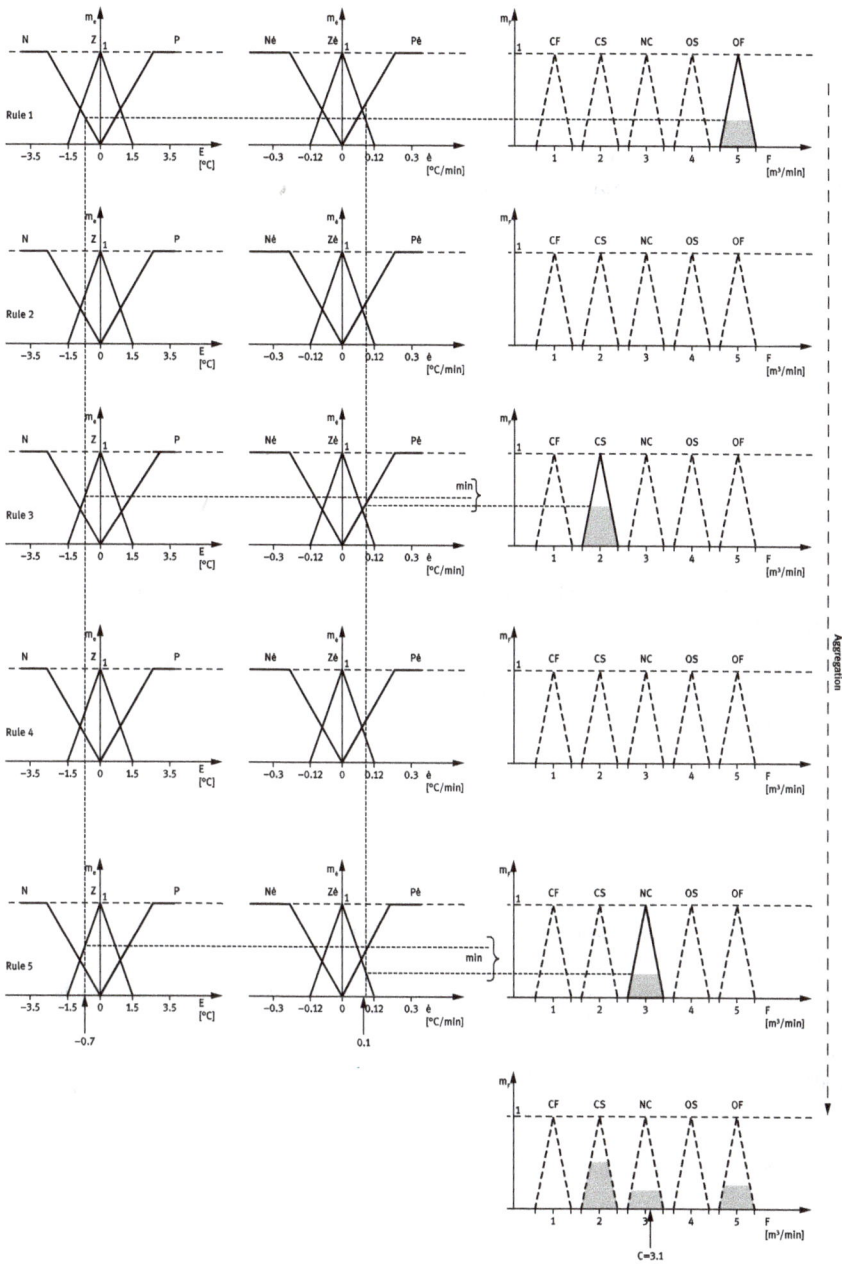

Fig. 3.15: PD fuzzy controller design.

Mamdani inference mechanism is chosen for the fuzzy implication operation. A graphical representation for the computation of the fuzzy controller output is presented in Fig. 3.15. The inlet values are the error, with a value of $e = -0.7$ °C, and the differentiated error, with a value of $\dot{e} = 0.10$ °C/min. For rules no. 3, 4, and 5, the minimum *min* function has been used, as the antecedent parts are joined by the *and* operator. For each rule, the Mamdani inference is used to compute the membership function of the consequent. Furthermore, the resulting membership functions of the rules are aggregated using the *or* rule. As a result, the aggregated membership function is obtained $m_{Faggregated}$.

Defuzzification is performed using the following centroid method:

$$C = \frac{\sum\limits_{i=1}^{n} x_i m_{Fi}(x)}{\sum\limits_{i=1}^{n} m_{Fi}(x)}, \tag{3.28}$$

where n is the number of segments (areas) in which the aggregated area has been divided in, x_i is the position of each segment, and $m_{Fi}(x)$ is the membership value (area) corresponding to each segment. The centroid is situated at the value of $C = 3.1$ m³/min.

The design of the fuzzy controller presented in Example 3.3 shows the procedure that resulted in a PD fuzzy controller. The design procedure may be used for building other types of fuzzy controllers.

References

[1] Zadeh, L.A., *Fuzzy sets*, Information and Control, 8, 338–353, 1965.
[2] Palm, R., Driankov D., Hellendorn H., *Model Based Fuzzy Control*, Springer-Verlag, Berlin, New York, Heidelberg, 1997.
[3] Sofron, E., Bizon, N., Ionita, S., Raducu, R., *Sisteme de Contro Fuzzy, Modelare Si Proiectare Asistate de Calculator*, ALL Educational, Bucuresti, 1998.
[4] Sablani, S.S., Rahman, M.S., Datta, A.K., Mujumdar, A.S. (Eds.), *Handbook of Food and Bioprocess Modeling Techniques*, Taylor and Francis Group, CRC Press, 2007.
[5] Klirk, J.G., Yuan, B., *Fuzzy Sets and Fuzzy Logic – Theory and Applications*, Prentice-Hall PTR, New Jersey, 1995.
[6] Siler, W., Buckley, J.J., *Fuzzy Expert Systems and Fuzzy Reasoning*, John Wiley & Sons, New Jersey, Wiley-Interscience, 2005.

4 Optimal control systems

It seems that optimal control was not very much in attention in recent years since model-based control techniques, which have embedded an optimization procedure (minimization of the error), were applied extensively in industry (see Chapter 2 of this book). It remains an important issue, since the minimization of costs via control became more and more important.

Optimal control of a process can refer to three distinct aspects:
- Optimal control in steady state, based on a scope or objective function (benefit, yield, pollution index, etc.) that has to be optimized, and thus, the optimal set-points of the controlled parameters are calculated [1]
- Optimal control in dynamic state of the batch processes, this being referred mainly at the minimization of the duration of the batch, or maximization of some characteristics as reaction conversion, respecting the constraints of quality [2–4]
- Optimal control in dynamic state of the continuous processes [5, 6], where the problem is to optimize a control quality criterion (IAE, ISE, ITAE, ITSE [7, 8], etc.)

Of course, in approaching optimal control, optimization techniques, which are not the subject of the present text, have to be known.

4.1 Steady-state optimal control

This type of control implies the existence of a steady-state model of the controlled process, having as an output an economic or technical variable and as input or state variables the key parameters of the process. The scope function is thus realized and subjected to optimization [9].

In the case of optimized control of the sequential processes, methods of identifying optimal policies are used [10].

4.1.1 Pontryagin's maximum principle

The method "maximum principle" was elaborated in 1956 by Pontryagin, a renowned Russian mathematician. This method is an extension of the variational calculus (or calculus of variations) upon the optimization problems described by ordinary differential equations. In the case of this type of problems, one single variable is considered, such as time or space. The condition imposed to the variable is that it has to have a fixed value at the beginning and the end of the evolution of the process. Thus, a value $t_0 = 0$ and another one t_f are imposed corresponding to the time domain on which the optimization

https://doi.org/10.1515/9783110789737-005

is done, or a $z_0 = 0$ and another $z_f = l$ corresponding to the space, where l is the length of the plug flow reactor, for example. Thus, considering a dynamic system (Fig. 4.1):

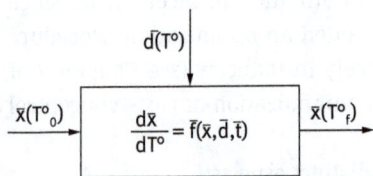

Fig. 4.1: The fundamental problem.

whose state is characterized by the state vector $\bar{x}(t)$, evolution in time is characterized by a set of decision variables that are also functions of time and that form the control/decision vector $\bar{d}(t)$; these decision variables can vary in an admissible domain D; the model of the system can be represented by a system of differential equation (4.1) such as

$$\frac{dx_1}{dt} = f_1(x_1, x_2, \ldots x_n, d_1, d_2, \ldots, d_m, t)$$

$$\frac{dx_2}{dt} = f_1(x_1, x_2, \ldots x_n, d_1, d_2, \ldots, d_m, t) \tag{4.1}$$

$$\ldots \ldots$$

$$\frac{dx_n}{dt} = f_1(x_1, x_2, \ldots x_n, d_1, d_2, \ldots, d_m, t)$$

The maximum principle proposes finding the decision vector $\bar{d} * (t)$, that is, the optimal solution among the admissible decision vectors $\bar{d}(t)$, which maximizes the objective/scope function:

$$f_{ob} = \int_{t_0}^{t} F(\bar{x}(t), \bar{d}(t)) dt \overset{\Delta}{=} \max \tag{4.2}$$

It is subject to local constraints of the form:

$$\frac{dx_i}{dt} = f_i(\bar{x}(t), \bar{d}(t)) \text{ for } i = 1, 2, \ldots, n \tag{4.3}$$

with the initial conditions $x_i = x_{i,0}$ at $t = t_0$ and $i = 1, 2, \ldots, n$.

According to the maximum principle, for the function (4.2) to be maximum, the decision vectors $\bar{d}(t)$ have to be chosen in such a way that the Hamiltonian,

$$H = F(\bar{x}(t), \bar{d}(t)) + \sum_{i=1}^{n} \lambda_i(t) f_i(\bar{x}(t), \bar{d}(t)) \tag{4.4}$$

is also at maximum; $\lambda_i(t)$ are the Lagrange multipliers.

A way to solve the problem is to determine the extremum of the Hamiltonian as follows:

1. Values for the functions $\bar{d}(t)$ are proposed for the whole domain of existence of $t \in [t_0, t_f]$.
2. The differential system (4.3) is solved numerically, where the functions $\bar{d}(t)$ have the previously established values and thus functions $\bar{x}(t)$ are obtained.
3. The equation:

$$\frac{d\lambda_i(t)}{dt} = -\sum_{i=1}^{n} \frac{\partial f_i}{\partial x_i} \lambda_i(t) - \frac{\partial F}{\partial x_i} \text{ for } i = 1, 2, \ldots, n \qquad (4.5)$$

is solved numerically for $\lambda_i = 0$ at $t = t_f$.
4. The approximation done in the first stage is improved by optimizing the Hamiltonian (eq. (4.4)) by using the gradient method:

$$d_i(t)^{(k+1)} = d_i(t)^{(k)} + s_g \frac{\partial H^{(k)}}{\partial d_i(t)} \qquad (4.6)$$

The cycle is retaken with operation 2. The optimization is finished when functions $\bar{d}(t)$ do not change anymore.

s_g is the step done on the Hamiltonian direction.

Example 4.1

[i]

Optimal control of the methanol process using Pontryagin's maximum principle: The method has been applied to control the temperature profile at the methanol plant at C.C. Victoria, Romania [11]. The calculated increase in production was 33.8%, but in reality it reached 27%.

Methanol mass production makes optimal control possible and tempting because a fairly small 1% increase of the production brings important economic benefits. The case study is for a methanol reactor with a production of 210,000 t/year. The process to which we refer is based on the low-pressure technology, the methanol synthesis being carried out in a multilayer reactor on a Co-ZnO-Al$_2$O$_3$ catalyst with external cooling after each layer (Fig. 4.2).

The reactor design led to the characteristics presented in Tab. 4.1. ξ_1 and ξ_2 are the chemical conversions referring to the following reactions:

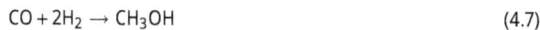

$$CO + 2H_2 \rightarrow CH_3OH \qquad (4.7)$$

and

$$CO_2 + H_2 \rightarrow CO + H_2O \qquad (4.8)$$

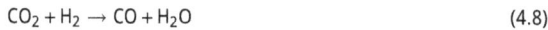

It is to be mentioned that a major design hypothesis was that of an adiabatic operation with the layer input temperature of 523 K.

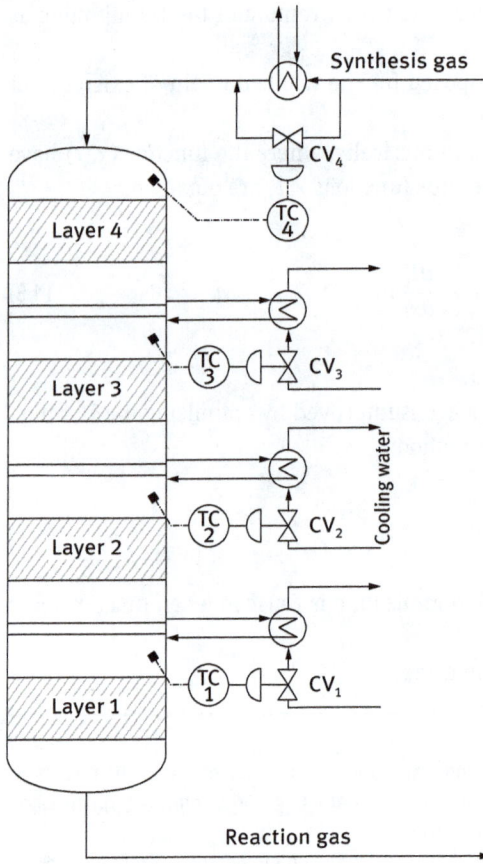

Fig. 4.2: Methanol synthesis reactor discussed in Example 4.1.

Tab. 4.1: Characteristics of the methanol synthesis reactor in Example 4.1.

Layer	Layer thickness (m)	Temperature (K)		Output conversion	
		Input	Output	ξ_1	ξ_2
1	0.3750	523	543.109	0.12421	0.000445
2	0.7050	523	543.136	0.25632	0.115788
3	1.0875	523	543.039	0.39172	0.201761
4	0.5925	523	543.983	0.44079	0.221330

4.1.2 Thermodynamics of the process

Thermodynamic data referring to reactions (4.7) and (4.8) are presented in Tab. 4.2.

Tab. 4.2: Thermodynamic data of the methanol synthesis process described in Example 4.1.

Reaction	$\Delta_r H^0_{298}$ (kcal/mol)	$\Delta_r S^0_{298}$ (cal/(mol*K))	$\Delta_r G^0_{298}$ (kcal/mol)
$CO + 2 H_2 \rightarrow CH_3OH$	−21.68	−52.31	−6.0916
$CO_2 + H_2 \rightarrow CO + H_2O$	8.64	10.02	5.8540

The equilibrium constants of both reactions are given by eqs. (4.9) and (4.10), respectively:

$$K_{p1} = \frac{K_{f1}}{K_{\gamma 1}} = \frac{p_{CH_3OH}}{p_{CO}p^2_{H_2}} \tag{4.9}$$

$$K_{p2} = \frac{K_{f2}}{K_{\gamma 2}} = \frac{p_{CO_2}p_{H_2O}}{p_{CO}p_{H_2}} \tag{4.10}$$

$K_{\gamma i}$ are the fugacity constants, p_{CH_3OH} ... are the partial pressures of the components of both reactions, and

$$K_{fi} = e^{-\frac{\Delta G^0_{T,i}}{RT^0}}, \quad i = 1, 2 \tag{4.11}$$

By means of a computer program, the equilibrium constants, the fugacity and K_{fi} constants were determined within the temperature range of 0–300 °C as well as pressure range of 10–100 bar.

4.1.3 Kinetics of the process

The catalytic process develops in the catalyst's pores (Fig. 4.3) by following the sequence:
1. reactant transport through the limit diffusion layer to the surface of the catalyst granule;
2. diffusion of the reactant in the catalyst's pores;
3. activated adsorption of the reactant R on the catalyst surface (chemisorption);
4. reaction on the catalyst centers on its surface, when reactant R transforms into product P;
5. desorption of product P from the catalyst;
6. diffusion of product through the catalyst's pores;
7. diffusion of the product away from the catalyst granule.

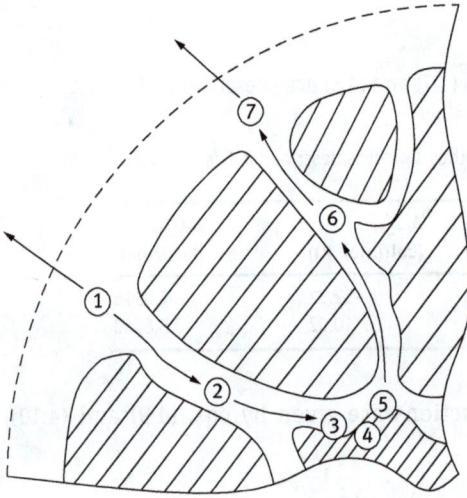

Fig. 4.3: The stages of the catalytic process described in Example 4.1.

Natta et al. [12] arrived at the conclusion that the rate-determining step is the chemical reaction between the specia adsorbed on the catalyst, with activation energy of 17.1 kcal/mol.

In Tab. 4.3, the different reported expressions for the reaction rate of the process are given.

Tab. 4.3: Rate of reaction forms proposed in the literature.

Authors	Reaction rate expression
Atroscenko, Zasorin [32]	$r_1 = k_1 p_{H_2} \sqrt[4]{\dfrac{p_{CO}}{p_{CH_3OH}}} - k_2 \sqrt[4]{\dfrac{p_{CH_3OH}}{p_{CO}}}$
Natta, Pasquon [43]	$r_1 = \dfrac{f_{CO}p_{CO}f_{H_2}^2\, p_{H_2}^2 - f_{CH_3OH}\dfrac{p_{CH_3OH}}{K_{cg}}}{A\left(1 - Bf_{CO}p_{CO} + Cf_{H_2}^2\, p_{H_2}^2 + Df_{CH_3OH}p_{CH_3OH} + Ef_{CO_2}p_{CO_2}\right)}$
Uchida, Ogino [58]	$r_1 = k_1\left[\sqrt[4]{p_{CO}p_{H_2}^2} - \dfrac{p_{CH_3OH}}{k_{p1}\sqrt[3]{p_{CO}p_{H_2}^2}}\right]$
Cappelli [11]	$r_1 = k_1\left(\dfrac{p_{H_2}\sqrt{p_{CO}}}{\sqrt[3]{p_{CH_3OH}^2}} - \dfrac{1}{k_{p1}p_{H_2}}\dfrac{\sqrt[3]{p_{CH_3OH}}}{\sqrt{p_{CO}}}\right)$

Tab. 4.3 (continued)

Authors	Reaction rate expression
Siminiceanu [48]	$r_2 = \dfrac{k_2\left(p_{CO_2}p_{H_2}k_{P_2} - p_{CO}p_{H_2O}\right)}{k'p_{CO} + p_{CO_2}}$
Experimentally determined on ICI catalyst	$r_1 = k_1\left(\dfrac{p_{H_2}\sqrt{p_{CO}}}{\sqrt[3]{p^2_{CH_3OH}}} - \dfrac{1}{k_{p1}}\dfrac{\sqrt[3]{p_{CH_3OH}}}{p_{H_2}\sqrt{p_{CO}}}\right)$ $k_1 = 9.24 \times 10^6 e^{-11,423.1/T}$

The terms in Tab. 4.3 are detailed in [11].

4.1.4 Mathematical model of the process

A quasi-homogeneous, unidimensional model was proposed by taking into consideration the following realities and simplifying assumptions:
- There are few data about the structural and textural characteristics of the catalyst.
- Assumption of plug-flow behavior of the gaseous phase is supported by the uniformity of its velocity profile in the cross-sectional area of the catalyst.
- Important temperature gradients can be neglected due to adiabatic operation.
- Pressure drop on the reactor is neglected in comparison with the operating pressure.
- Variation of gas velocity depending on temperature is considered.
- Variation of gas velocity due to the mole number change is considered.
- Average molar heat of the mixture is recalculated depending on composition and temperature.

Consequently, the mathematical model is composed of the relations of conversion and temperature:

$$\frac{dp_{CO}}{dt} = -r_1 + r_2 \tag{4.12}$$

where r_1 and r_2 are rates of consumption and formation of CO in reactions (4.7) and (4.8), respectively. They are presented in Tab. 4.3 (Siminiceanu and ICI catalyst [11]). The rates can be expressed through:

$$-r_1 = \frac{dp_{CO}}{d\xi_1} \cdot \frac{d\xi_1}{dt} \tag{4.13}$$

$$r_2 = \frac{dp_{CO}}{d\xi_2} \cdot \frac{d\xi_2}{dt} \tag{4.14}$$

Note: With v_0 the initial fictive gas mixture velocity in the reactor and with v the current one, one may pass to the space derivatives:

$$\frac{d}{dt} = \frac{d}{dz} \cdot \frac{dz}{dt} \quad \text{where} \quad \frac{dz}{dt} = v = v_0(1+\varepsilon\xi_1)\frac{T^\circ}{T_0^\circ} \quad (4.15)$$

Equations (4.14) and (4.15) become

$$-r_1 = v_0(1+\varepsilon\xi_1)\frac{T^\circ}{T_0^\circ}\frac{dp_{CO}}{d\xi_1} \cdot \frac{d\xi_1}{dz} \quad (4.16)$$

and

$$r_2 = v_0(1+\varepsilon\xi_1)\frac{T^\circ}{T_0^\circ}\frac{dp_{CO}}{d\xi_2} \cdot \frac{d\xi_2}{dz} \quad (4.17)$$

Considering that:

$$p_{CO} = x_{CO}p$$

and

$$x_{CO} = x_{CO}^0\frac{1-\xi_1+\beta\xi_2}{1+\varepsilon\xi_1}$$

equations (4.16) and (4.17) become [4, 5]:

$$-r_1 = x_{CO}pv_0(1+\varepsilon\xi_1)\frac{T^\circ}{T_0^\circ}\frac{1-\xi_1+\beta\xi_2}{1+\varepsilon\xi_1} \cdot \frac{d\xi_1}{dz} \quad (4.18)$$

$$r_2 = \beta pv_0x_{CO}\frac{T^\circ}{T_0^\circ} \cdot \frac{d\xi_2}{dz} \quad (4.19)$$

and

$$x_{H_2} = x_{CO}^0\frac{a-2\xi_1-\beta\xi_2}{1+\varepsilon\xi_1}$$
$$x_{CH_3OH} = x_{CO}^0\frac{\gamma+\xi_1}{1+\varepsilon\xi_1}$$
$$x_{H_2O} = x_{CO_2}^0\frac{1-\xi_2}{1+\varepsilon\xi_1} \quad (4.20)$$
$$x_{CH_3OH} = x_{CO}^0\frac{\delta+\xi_2}{1+\varepsilon\xi_1}$$

where

$$a = \frac{x_{H_2}^0}{x_{CO}^0}; \quad \beta = \frac{x_{CO_2}^0}{x_{CO}^0}; \quad \gamma = \frac{x_{CH_3OH}^0}{x_{CO}^0}; \quad \delta = \frac{x_{H_2O}^0}{x_{CO}^0}; \quad \varepsilon = -2x_{CO}^0$$

When replacing the molar fractions (4.20), together with expressions (4.18) and (4.19) in the rate expressions [4, 5], the following is obtained:

$$r_1 = k_1 \left(\frac{px_{CO}^0}{1+\varepsilon\xi_1}\right)^{0.84} \left[\frac{(1-\xi_1+\beta\xi_2)^{0.8}(a-2\xi_1-\beta\xi_2)}{(\gamma+\xi_1)^{0.66}} - \frac{1}{K_{p1}(px_{CO})^2(a-2\varepsilon_1-\beta\varepsilon_2)} \frac{(\gamma+\xi_1)^{0.34}(1+\varepsilon\xi_1)^2}{\sqrt{(1-\xi_1+\beta\xi_2)}}\right]$$

(4.21)

$$r_2 = k_2 \frac{px_{CO}^0}{1+\varepsilon\xi_1} \frac{K_{p2}(a-2\xi_1-\beta\xi_2)(1-\xi_2)-(1-\xi_1+\beta\xi_2)(\delta+\varepsilon_2)}{k'(1-\xi_1+\beta\xi_2)+\beta(1-\xi_2)}$$

(4.22)

Finally, from eqs. (4.18), (4.19), (4.21), and (4.22), conversion-space function is obtained:

$$\frac{d\xi_1}{dz} = \frac{k_1 T_0^\circ}{v_0 T^\circ} \frac{1}{1+\varepsilon+\beta\varepsilon\xi_1} \left[\frac{(a-2\xi_1-\beta\xi_2)\sqrt{1-\xi_1+\beta\xi_2}}{(\gamma+\xi_1)^{0.66}} - \frac{1}{(px_{CO}^0)^2 K_{p1}(a-2\varepsilon_1-\beta\varepsilon_2)} \frac{(\gamma+\xi_1)^{0.34}(1+\varepsilon\xi_1)^2}{\sqrt{1-\xi_1+\beta\xi_2}}\right]$$

$$\left(\frac{1+\varepsilon\xi_1}{px_{CO}^0}\right)^{0.16}$$

(4.23)

$$\frac{d\xi_2}{dz} = \frac{k_2 T_0^\circ K_{p2}(a-2\xi_1-\beta\xi_2)(1-\xi_2)-(1-\xi_1+\beta\xi_2)(\delta+\varepsilon_2)}{v_0 T^\circ \qquad k'(1-\xi_1+\beta\xi_2)+\beta(1-\xi_2)}$$

(4.24)

Since both rate and equilibrium constants depend on temperature, the equation of heat balance (since the reactor operates under adiabatic regime) has to complete the model:

$$\frac{dT^\circ}{dz} = \frac{1}{1+\varepsilon\xi_1}\left(x_{CO}^0 \frac{\Delta_r H_1}{c_p}\frac{d\xi_1}{dz} - x_{CO_2}^0 \frac{\Delta_r H_2}{c_p}\frac{d\xi_2}{dz}\right)$$

(4.25)

4.1.5 Optimal temperature profile in the methanol synthesis reactor

The maximization of the methanol synthesis is targeted and the control policy strives for a temperature profile along the reactor, $T^\circ(z)$.

The objective function is

$$f_{ob} = \xi_1 = \int_0^L \frac{d\xi_1}{dz} dz$$

(4.26)

By comparing (4.27) with (4.2), one may observe that $F = \frac{d\xi_1}{dz}$ and that ξ_1 and ξ_2 are corresponding to the dependent variables $\overline{x}(z)$.

The local constraints (eq. (4.3)) are

$$\frac{d\xi_1}{dz} = f_1(\xi_1, \xi_2, T^\circ)$$
$$\frac{d\xi_2}{dz} = f_2(\xi_1, \xi_2, T^\circ)$$

(4.27)

while the initial conditions are

$$\xi_1 = 0, \ \xi_2 = 0 \quad \text{at} \quad z = z_0 = 0$$

Equation (4.5) can be written under the condition $F = \frac{d\xi_1}{dz} = f_1$, as

$$\frac{d\lambda_1}{dz} = -\frac{\partial f_1}{\partial \xi_1}\lambda_1 - \frac{\partial f_2}{\partial \xi_1}\lambda_2 - \frac{\partial F}{\partial \xi_1} = -(1+\lambda_1)\frac{\partial f_1}{\partial \xi_1} - \lambda_2\frac{\partial f_2}{\partial \xi_1}$$

$$\frac{d\lambda_2}{dz} = -\frac{\partial f_1}{\partial \xi_2}\lambda_1 - \frac{\partial f_2}{\partial \xi_2}\lambda_2 - \frac{\partial F}{\partial \xi_2} = -(1+\lambda_1)\frac{\partial f_1}{\partial \xi_2} - \lambda_2\frac{\partial f_2}{\partial \xi_2}$$

(4.28)

with the final conditions $\lambda_1 = 0, \ \lambda_2 = 0$ at $z = L$
The Hamiltonian H becomes

$$H = F + \lambda_1 f_1 + \lambda_2 f_2 = (1+\lambda_1)f_1 + \lambda_2 f_2$$

(4.29)

and its derivative related to the control policy is

$$\frac{\partial H}{\partial T^\circ} = (1+\lambda_1)\frac{\partial f_1}{\partial T^\circ} + \lambda_2\frac{\partial f_2}{\partial T^\circ}$$

(4.30)

Equation (4.30) is used to correct the temperature profile:

$$T^\circ(z)^{(k+1)} = T^\circ(z)^k + S_g\left(\frac{\partial H}{\partial T^\circ}\right)^{(k)}$$

(4.31)

The computing program for calculating the optimum temperature profile has the following sequence:

1. Initially, an isothermal regime for the whole reactor is supposed, with the input temperature T_0°; thus, for $k = 1$, $T_i^{(1)} = T_0^\circ$, $i = 1, 2, \ldots n$.
2. The equation system (4.25) and (4.27) is integrated by using the Euler method, and by considering the conditions from (4.27).
3. System (4.28) is integrated from $z = L$ to $z = 0$ with the final conditions $\xi_{1,n} = 0$, $\xi_{2,n} = 0$:

$$\lambda_{i,j} = \lambda_{i,j+1} + \Delta z\left[-(1+\lambda_{i,j+1})\left(\frac{\partial f_1}{\partial \xi_1}\right)_{j+1} - (\lambda_{2,j+1})\left(\frac{\partial f_2}{\partial \xi_1}\right)_{j+1}\right]$$

where
$i = 1, 2$ and $j = n - 1, n - 2, \ldots 1$.

4. Temperatures are recalculated as

$$T_i^{\circ(k+1)} = T_i^{\circ(k)} + S_g\left(\frac{\partial H}{\partial T_i^\circ}\right)^{(k)}$$

where $i = 1, 2, \ldots, n$.
One passes to a new iteration from point 2.

Tab. 4.4: Optimal theoretic temperature profile and theoretical optimal performances in the methanol reactor of Example 4.1.

z (m)	ξ_1	ξ_2	λ_1	λ_2	T (K)
0.000	0.000	0.000	1.338	-1.338	573.660
0.276	0.517	0.487	1.001	-0.820	546.150
0.552	0.546	0.447	0.775	-0.554	542.940
0.828	0.570	0.430	0.605	-0.386	540.990
1.104	0.591	0.420	0.469	-0.271	539.480
1.380	0.610	0.414	0.357	-0.189	538.170
1.656	0.628	0.410	0.263	-0.129	536.980
1.932	0.644	0.408	0.183	-0.084	535.860
2.208	0.658	0.406	0.114	-0.049	534.810
2.484	0.672	0.405	0.053	-0.022	533.820
2.760	0.685	0.404	0.000	0.000	532.910

The results of the iterations and the optimum temperature profile that ensure $\xi_1 = \max$ are given in Tab. 4.4.

The optimum temperature profile from Tab. 4.4 cannot be practically obtained because its tuning can be carried out only between the catalyst layers by using the external heat exchangers (see Fig. 4.2). The input temperature in each next layer has to be controlled in such a way that all input and output temperatures have values around the optimal profile (see Tab. 4.5 and Fig. 4.4).

Tab. 4.5: Optimal real temperature profile and the consequent performances in the methanol reactor of Example 4.1.

Layer	Layer thickness (m)	Temperature (K)		Output conversion	
		Input	Output	ξ_1	ξ_2
1	0.3750	563	586.36	0.2562	0.4161
2	0.7050	543	571.71	0.4115	0.4160
3	1.0875	538	561.51	0.5369	0.4139
4	0.5925	538	548.40	0.5893	0.4035

It is important to observe that the real performance ($\xi_1 = 0.589$) is inferior to the theoretical performance obtained after applying the optimization procedure ($\xi_1 = 0.685$), but with 33.8% higher than the performance obtained in the traditional adiabatic operation, $\xi_1 = 0.440$ (Tab. 4.1). The sawtooth temperature profile is obtained by fixing each controller to the appropriate setpoint value: $T^{\circ}_{1set} = 563$ K, $T^{\circ}_{2set} = 543$ K, $T^{\circ}_{3set} = 538$ K, $T^{\circ}_{4set} = 538$ K.

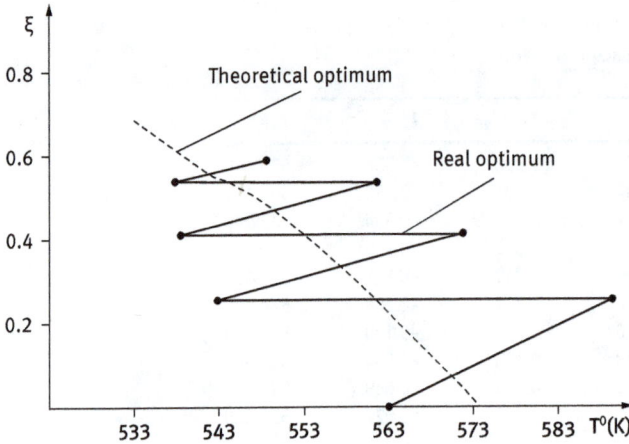

Fig. 4.4: The quasi-optimal sawtooth profile of temperatures in the methanol reactor.

4.2 Dynamic optimal control of batch processes

In the given situation, there are two ways of approaching the problem:
- To obtain a *maximum* of quantity Q, in a given time period T
- To obtain a certain product quantity Q, in a *minimum* operational time

In the first of these situations, a maximization of a function of type:

$$I = \int_0^T Q dt \text{ is subjected to the constraint} \tag{4.32}$$

$\bar{y} = \dfrac{\int_0^T y Q dt}{\int_0^T Q dt} = y*$, where $\bar{y}$ is the average specification of the quality of the product (it can be a mass or a molar concentration or fraction).

The second situation requires minimization of a time function:

$$t_T^* = \min \int_0^{t_T} dt, \text{ or the dividing of the process in } N \text{ discrete stages,} \tag{4.33}$$

$$t_T^* = \min \sum_{i=1}^{N} t_i, \text{ where } t_i \text{ is the duration of one stage.}$$

Example 4.2
One has to determine the optimal operation policy of a batch distillation by manipulating the reflux flow to obtain a certain given quantity in minimum time using Dynamic Programming [13].

A technique of dynamic programming was applied [10].

The process of batch distillation of one fraction resides in the following operations:
1. Loading the batch to be separated into the bottom of the column
2. Bringing the temperature to the boiling point
3. Separation of the lightest fraction based on the difference of volatility, until the composition reaches a certain minimum value

Corresponding to the continuous decrease of the light component molar fraction in the bottom, x_B, the molar fraction of the light component in the distillate, x_D, decreases as well (when the reflux ratio is kept constant).

After writing the mass balance for the column, the continuous decrease of the batch quantity in the bottom, M_B, is expressed through eqs. (4.34) and (4.35):

$$\frac{dM_B(t)}{dt} = -Q_D(t) \tag{4.34}$$

$$\frac{dx_B(t)}{dt} = \frac{Q_D(t)}{M_B(t)} \{x_B(t) - x_D[x_B(t), Q_D(t)]\} \tag{4.35}$$

where $Q_D(t)$ and $M_B(t)$ are the quantities separated at time t at the top and the bottom of the load, respectively.

If t_T is the total batch distillation time and $\overline{x}_D$ the average desired molar fraction of the collected distillate, the mass balance below describes the process:

$$\overline{x}_D \int_0^{t_T} Q_D(t)dt = \int_0^{t_T} Q_D(t) \cdot x_D[x_B(t), Q_D(t)]dt \tag{4.36}$$

The vapor flow produced in the column reboiler is considered constant, a fact which ensures an adequate thermal regime for the bottom of the column.

As observed in Fig. 4.5, the column can be operated with either a constant reflux ratio (the molar fraction at the top is decreasing in time) or with a progressive reflux ratio increase (with maintaining an average $\overline{x}_D$ constant; this means the continuous decrease of the distillate flow rate D). This happens simultaneously with the decrease of the light component concentration in the bottom.

The goal of control action is to minimize the duration of the distillation process of the prescribed distillate quantity, by also achieving its specified purity. In this situa-

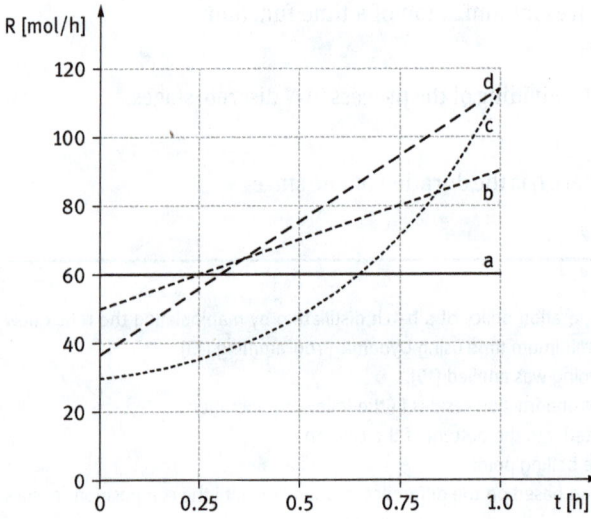

(a)

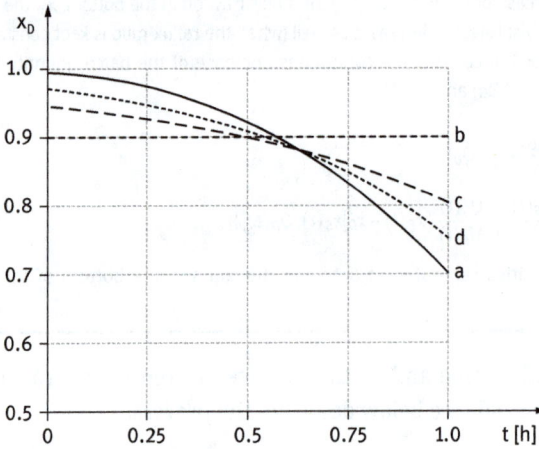

(b)

Fig. 4.5: Batch distillation operation: operation with constant reflux ratio (a); operation with constant molar fraction (b); optimal operation with maximum distillate quantity in a given time (c); and optimal operation with a certain quantity Q in a minimum of time (d).

tion, the optimization of the batch distillation process becomes a problem of allocation [10, 13]. It is solved in the following sequence:

1. The considered function

$$f(x_1, x_2, \ldots, x_i, \ldots, x_N) = g_1(x_1) + g_2(x_2) + \cdots + g_i(x_i) + \cdots + g_N(x_N) \qquad (4.37)$$

is subjected to the constraints:

$$\sum_{i=1}^{N} x_i = y, \ x_i \geq 0$$

where y is the total quantity of resources and the functions $g_i(x_i)$ can have any form. The variables x_i are considered resources to be allocated optimally to N activities, each activity being characterized through its own objective function $g_i(x_i)$. The objective function of the system is a sum of functions that characterizes each separate activity. The problem has a finite optimum only in the case of the limited quantity of resources. Thus, the optimal value of the function (4.37) is

$$f^*(y) = \mathrm{opt}\, f(x_1, \ x_2, \ \ldots, \ x_N) \tag{4.38}$$

2. The properties to be considered are:
a. $g_i(0) = 0$ for any $i = \overline{1, N}$ when $f_i^*(0) = 0$
b. $f_i^*(y) = g_i(y), y \geq 0$; thus, a recurrence equation between f_i^* and f_{i-1}^* can be found:

$$f_i(y) = g_i(x_i) + f_{i-1}(y - x_i)$$

c. The optimum choice of x_i is the one optimizing $f_i(y)$,

$$f_i^*(y) = \mathrm{opt}_{0 \leq x_i \leq y}\left[g_i(x_i) + f_{i-1}^*(y - x_i)\right] \text{ for } i = 2, \ 3, \ \ldots, \ N, y \geq 0 \tag{4.39}$$

Consequently, by using eq. (4.33), the value of t_i, which is the duration of one distillation stage, can be expressed. It is defined as the interval in which the bottom molar fraction decreases from $x_{B,i-1}$ to $x_{B,i}$.

By taking into account the known quantity and quality of the load in the bottom of the column at the beginning of the batch distillation process, $M_{B,0}$ and $x_{B,0}$, the final (after N stages) molar fraction of the volatile component in the bottom, $x_{B,N}$, as well as the average molar fraction of the collected distillate, $\overline{x_D}$, the following equations (4.40–4.43) can be written:

$$M_{B,0} - M_{B,N} = P_N \tag{4.40}$$

$$M_{B,0}x_{B,0} - M_{B,N}x_{B,N} = P_N \overline{x_D} \tag{4.41}$$

where the total quantity of distillate is:

$$P_N = \sum_{i=1}^{N} D_i, \text{ with } D_i \text{ as the quantity extracted at each stage} \tag{4.42}$$

The problem can be thus seen as a problem of allocation where the activities are the stages 1, 2, . . ., N, characterized by the variation Δx_w of the concentration of the volatile component in the bottom:

$$\Delta x_B = \frac{x_{B,0} - x_{B,N}}{N} \tag{4.43}$$

The resources are the quantities of distillate D_i. The functions g_i from the allocation problems are the time periods t_i.

To determine the optimal function, the mass balance equations for stage i are written as follows:

$$M_{B,i-1} - M_{B,i} = D_i \tag{4.44}$$

$$M_{B,i-1}x_{B,i-1} - M_{B,i}x_{B,i} = D_i x_{D,i} \tag{4.45}$$

But:

$$M_{B,i} = M_{B,0} - \sum_{j=1}^{i} D_j \tag{4.46}$$

and

$$x_{B,i-1} = x_{B,0} - (i-1)\Delta x_B \tag{4.47}$$

or

$$x_{B,i} = x_{B,0} - i\Delta x_B \tag{4.48}$$

From eqs. (4.44) to (4.48):

$$x_{D,i} = x_{B,0} + \left(\frac{M_{B,0} - \sum_{j=1}^{i} D_j}{D_i} - i + 1 \right) \Delta x_B \tag{4.49}$$

Thus,

$$x_{D,i} = x_{D,i}\left(D_i, \sum_{j=1}^{i} D_j \right) \tag{4.50}$$

and

$$x_{B,iav} = x_{B,i-1} - \frac{\Delta x_B}{2} \tag{4.51}$$

When having the average concentrations in the bottom and at the top, the average value of the reflux in stage i can be determined. To reduce the volume of calculations, the Fenske-Underwood-Gilliland method [14] and not the iterative one (tray-to-tray) is used. With this method, coefficients

$$A = \frac{R - R_{\min}}{R + 1} \text{ and } B = \frac{S - S_{\min}}{S + 1} \tag{4.52}$$

are calculated.

The minimum reflux R_{min} and minimum number of theoretical transfer units S_{min} are calculated with this method, where the feed molar fraction x_F is replaced with $x_{B,i\ av}$. To determine the reflux R corresponding to a given S, the function $A = f(B)$ is obtained through the least square method:

$$A = 0.96 - 2,09B + 0,21B^2 + 2.82B^3 - 2.90B^4$$

$$\text{for} \quad 0 \leq B \leq 0.51 \text{ and}$$

$$A = 0.86 - 0.95B - 2.48B^2 + 3.00B^3$$

$$\text{for} \quad B > 0.51$$

If $S < S_{min}$ and $B < 0$, respectively, the fractionation is impossible even at infinite reflux. Finally, $R_i(x_{Di})$ is calculated as

$$R_i = R_i\left(D_i, \sum_{j=1}^{i} D_j\right) \tag{4.53}$$

The mass balance equation for the vapors produced at the bottom of the column during one stage is

$$Q_V t_i = D_i(R_i + 1) \tag{4.54}$$

where Q_V is the constant, known vapor flow rate produced in the reboiler.
From previous relations (4.38) and (4.39),

$$t_i = g_i\left(D_i, \sum_{j=1}^{i} D_j\right) \tag{4.55}$$

and the desired optimal value [7] is

$$t_T^* = \min_{D_1, D_2, \ldots D_N}[g_1(D_1) + g_1(D_2, \ D_1 + D_2) + \cdots + g_N(D_N, \ D_1 + D_2 + \cdots + D_N)] \tag{4.56}$$

To calculate the optimal policy $f_i^*(P)$, one has to consider the allocations D_i.
For $x_{Di} < 1$,

$$D_i > \frac{(M_{B,0} - P)\Delta x_B}{1 - x_{B,0} + (i-1)\Delta x_B} > 0 \text{ and for } i = 1, D_i = P, \text{ and thus}$$

$$D_i > \frac{M_{B,0}\Delta x_B}{1 - x_{B,0}} > 0$$

Imposing $D_1 > 0$, $D_2 > 0$, and $D_{i-1} > 0$, it results that $D_i < P$. Let us apply this principle in the case of batch fractionation:

$$f_i^*(P) = \min_{D_{i\ min} < D_i < P}[g_i(D_i, P) + f_{i-1}^*(P - D_i)] \tag{4.57}$$

with $i = 2, 3, \ldots, N$ and $f_i^*(P) = g_i(D_i)$

As the objective function for each stage i depends on the allocation D_i and the total quantity of resources P, the solution will be modified, and the functions g_i can not be separately calculated (they depend on P). Dividing the domain $[0, P_N]$ in N intervals of dimension ΔP, and because D_i takes discrete values in the nodes of the network, the results of the minimization procedure (eq. (4.57)) need $1 + 2 + \cdots + (M - 1) = \frac{(M-1)M}{2}$ evaluations of the function $f_i(P)$ in each stage, instead of $N + 1$ as in the usual procedure. As in any stage i, $D_{i+1}, D_{i+2}, \ldots, D_N$ cannot be zero but at most equal to ΔP, the highest rational value for P is $[M - (N - i)]\Delta P$. Consequently, for a series of values $j(1 \le j \le N)$, a fake, very high value will be allocated to the functions $f_i^*(j\Delta P)$ and similarly to the conventional value of the corresponding decision $D_i^*(j\Delta P)$.

Finally, the optimal fractionation policy will be obtained by swiping the *optimal decisions matrix*, starting from the element in the row $j = M$, corresponding to P_N and the column $i = N$, corresponding to $x_{B,N}$.

The matrix resolution procedure is as follows:

1. The domain $0 \le y \le y_N$ in which the resources y can take values is divided in M intervals of magnitude Δ. This way, only the discrete values of the network R will be considered: $R = R\{0, \Delta, 2\Delta, \ldots, j\Delta, \ldots, M\Delta\}$; it will be admitted that each term of the sequence of functions $f_i^*(y)$ will be evaluated only in the nodes of this network and all allocation variables will be discretized in these points to avoid interpolation calculations.

2. For $i = 1, f_i^*(y) = g_1(y)$ will be determined immediately. Further,

$$f_i^*(j\Delta) = g_1(j\Delta) \tag{4.58}$$

where $j = 0, 1, 2, \ldots, M$, and the total resources quantity takes the discrete values of the R network.

3. For $i = 2$, eq. (4.39) becomes:

$$f_2^*(y) = \text{opt}_{0 \le x_i \le y}\left[g_2(x_2) + f_1^*(y - x_2)\right], \text{ or, in the discrete form,}$$

$$f_2^*(j\Delta) = \text{opt}_{0 \le k \le j}\left\{g_2(k\Delta) + f_1^*[(j - k)\Delta]\right\}, j = 0, 1, 2, \ldots, M \tag{4.59}$$

The optimization process evolves through direct evaluation (exhaustive search):

- the string of values $g_2(k\Delta)$ for $k = 0, 1, 2, \ldots, M$ is calculated.
- because $f_1^*[(j - k)\Delta]$ has been previously calculated, for each of the $M + 1$ data sets out of the $j + 1$ values of $f_2^*(j\Delta)$, the optimal (maximum or minimum) value is chosen (eq. (4.59)).

4. The procedure at point 3 is repeated for the following activities for $i = 3, 4, \ldots, N$. Finally, the optimal decisions matrix $k_i^*(j)$ summarized in Tab. 4.6 is obtained.

Tab. 4.6: The matrix of optimal decisions for Example 4.2.

Stage State	1	2	–	i	–	N – 1	N
0		$k_1^*(0)$	–	$k_i^*(0)$	–	$k_{N-1}^*(0)$	$k_N^*(0)$
1		$k_2^*(1)$	–	$k_i^*(1)$	–	$k_{N-1}^*(1)$	$k_N^*(1)$
2		$k_2^*(2)$	–	$k_i^*(2)$	–	$k_{N-1}^*(2)$	$k_N^*(2)$
–		–	–	–	–	–	–
–		–	–	–	–	–	–
j		$k_2^*(j)$	–	$k_i^*(j)$	–	$k_{N-1}^*(j)$	$k_N^*(j)$
–		–	–	–	–	–	–
–	–	–	–	–	–	–	–
M – 1	–	$k_2^*(M-1)$	–	$k^*(M-1)$	–	$k_{N-1}^*(M-1)$	$k_N^*(M-1)$
M		$k_2^*(M)$	–	$k_i^*(M)$	–	$k_{N-1}^*(M)$	$k_N^*(M)$

5. To obtain the optimal allocation policy for a process with a total number of i activities ($i \leq N$) and $y = j\Delta$ ($y \leq y_M$, $j \leq M$), the following procedure is used:
   –   in column i and row j from the optimal decisions matrix, the solution $k_i^*(j)$ is found, corresponding to $x_i^*(y)$;
   –   this value is subtracted from the total quantity of resources, the available resources for the other $i - 1$ activities, being obtained.
   –   on the line corresponding to row $j - k_i^*(j)$ and column $i - 1$ of the optimal decisions matrix, the optimal allocation for the activity $i - 1$, $k_{i-1}^*[j - k_i^*(j)]$ is found.
   –   the procedure is repeated for the other decreasing values of i until all values k_i^* of $\Delta = x_i^*$ are found. This is the desired optimal allocation policy.

The advantage of the formulation of the optimal fractionation policy as a resource allocation policy is that it allows the establishment of the final optimal bottom composition and determines its degree of separation that is economically appropriate to operate.

The optimal decisions matrix allows solving any allocation problem with maximum M activities and maximum quantity of resources y_M. In the present situation, this means that for any final concentration of the residue larger or equal to $x_{B,N}$ and for any quantity less or equal to P_N, the distillation does not advance until the final concentration $x_{B,N}$ and stops at $x_{B,i}(x_{B,i} \geq x_{B,N})$.

By keeping in mind the distillate's desired purity, the following equations can be written:

$$M_{B,0} - M_{B,i} = P_i, \text{ the mass balance equation} \tag{4.60}$$

$$M_{B,0}x_{B,0} - M_{B,i}x_{B,i} = P_i\bar{x}_D, \text{ the component balance equation} \tag{4.61}$$

from where

$$P_i = M_{B,0}\frac{x_{B,0} - x_{B,i}}{\bar{x}_D - x_{B,i}} \tag{4.62}$$

with $P_i \le P_N$ because $x_{B,i} \ge x_{B,N}$ and $\bar{x}_D \ge x_{B,0}$.

The optimal policy will be obtained by starting from the element of the optimal decisions matrix from column i and row j where $j = \frac{P_i}{\Delta P}$. Since for any $x_{B,i} \ge x_{B,N}$ an optimal distillation policy can be obtained, the one corresponding to the most convenient value $x_{B,i}^*$ will be chosen. Finally, the following can be determined:

- the optimal policy D_i^*, $i = 1, 2, \ldots, i^*$ (i^* corresponds to $x_{w,i}$)
- R_i^*, the corresponding optimal policy of the reflux

The numerical application is run with the following initial data:

$$M_{B,0} = 500 \text{ mol}; \ x_{B,0} = 0.2; \ x_{B,N} = x_{min} = 0 \cdot 05; \ x_{B,max} = 0.12; \ \bar{x}_D = 0.8; \ a = 2.5;$$
$$Q_V = 1000 \text{ mol/h}; \ M = 15; \ N = 200; \ \Delta P = \frac{P_N}{N} = \frac{100}{200} = 0.5 \text{ mol}; \ \Delta x_B = 0 \cdot 01.$$

A code for calculating the optimal policy was written and the result is presented in Tab. 4.7.

Tab. 4.7: The optimal operation policy of the batch distillation in Example 4.2.

Stage no. (i)	Total time (h)	Composition in the bottom (x_B, mol%)	Total distilled quantity (P, mol)	Values on interval				
				Duration of the interval (h)	Distilled quantity (P_i, mol)	Distillate flow rate (D, mol/h)	Reflux rate	Distillate composition (x_D, mol%)
1	0.030	0.1200	7.500	0.030	7.5	251.890	2.97	0.8567
2	0.060	0.1800	15.000	0.030	7.5	250.978	2.98	0.8367
3	0.091	0.1700	22.000	0.031	7.0	223.159	3.43	0.8625
4	0.123	0.1600	29.000	0.032	7.0	221.111	3.52	0.8229
5	0.155	0.1500	36.000	0.032	7.0	217.711	3.59	0.8029
6	0.188	0.1400	43.000	0.033	7.0	213.059	3.69	0.8931
7	0.223	0.1300	49.500	0.035	6.5	184.912	4.41	0.8131
8	0.259	0.1200	56.000	0.036	6.5	179.078	4.58	0.7931
9	0.297	0.1100	62.500	0.036	6.5	171.953	4.82	0.7731
10	0.337	0.1000	69.000	0.040	6.5	163.561	5.11	0.7531
11	0.379	0.0900	75.500	0.042	6.5	153.896	5.50	0.7875
12	0.426	0.0800	81.500	0.047	6.0	126.877	6.88	0.7675
13	0.478	0.0700	87.500	0.052	6.0	115.988	7.62	0.7475
14	0.536	0.0600	93.500	0.058	6.0	103.775	8.64	0.7044
15	0.600	0.0500	100.000	0.064	6.5	101.705	8.83	0.6754

The optimal operation shows a nonlinear decrease of the distillate flow D, 2.5 times from the beginning, and a 3-fold nonlinear increase of the reflux rate. The result is a quasi-constant composition x_D. The reduction of the batch time, as compared to the classical operation (linear decrease of D), is by 5%. In the volatile essential oils industry, a batch can be of 24-h duration, and 5% means 1.2 h per batch with the corresponding steam consumption. The results are process intensification and reduction of energy consumption.

4.3 Dynamic optimal control of continuous processes

This optimal control procedure targets to stabilize the behavior of an automatic control system (ACS), of minimizing the steady-state error, as well as the transient time, with the ultimate goal of increasing the economic efficiency of the control action.

The approach is based on the minimization of the *quadratic performance index* as J_{IAE} or J_{ITAE}. The system is described through the state equation:

$$\dot{\bar{x}} = A\bar{x} + B\bar{u} + C\bar{\bar{d}} \tag{4.63}$$

where $\bar{x}$ is the *state vector* (dimension $nx1$), $\bar{u}$ is the *control vector* (dimension $mx1$), and $\bar{d}$ is the *disturbance vector* (dimension $px1$).

Such a system is derived from the mathematical model of a process [15], but with the condition that the equations are either linear or linearized.

The control vector that minimizes the performance index is

$$J = \int_0^\infty P(\bar{x}, \bar{u}) dt \tag{4.64}$$

where $P(\bar{x}, \bar{u})$ is a quadratic function of $\bar{x}$ and $\bar{u}$.

It can be shown [16] that for such a performance index, where the limits are 0 and ∞, the resulting control law is a function of the state vector $\bar{x}$:

$$\bar{u}(t) = K\bar{x}(t) \tag{4.65}$$

where K is a matrix of dimensions $m \times n$.

The performance index to be minimized is quadratic,

$$J = \int_0^\infty (\bar{x}^T Q\bar{x} + \bar{u}^T R\bar{u}) dt \tag{4.66}$$

where Q and R are symmetric matrices, positive-definite, called *weight* matrices. These matrices determine the importance of the states of the system and the energy penalties via control actions, in the cost function. The vector $\bar{u}$ is not supposed to be subjected to constraints.

If matrix K can be determined in such a way that the index J from (4.66) is minimized, then the vector $\bar{u}(t)$ is optimal for any initial state $\bar{x}(0)$.

By substituting (4.65) in (4.63) and considering the disturbance as being 0,

$$\dot{\bar{x}} = (A - BK)\bar{x} \tag{4.67}$$

and then inserting eq. (4.65) in eq. (4.66), the performance index becomes

$$J = \int_0^\infty \bar{x}^T (Q + K^T RK)\bar{x}dt \tag{4.68}$$

By considering that:

$$\bar{x}^T (Q + K^T RK) = -\frac{d}{dt}\bar{x}^T S\bar{x} \tag{4.69}$$

where S is a positive-definite symmetrical matrix, the following can be written:

$$\bar{x}^T (Q + K^T RK)\bar{x} = -\bar{x}^T S\bar{x} - \bar{x}^T S\dot{\bar{x}} = \bar{x}^T \left[(A - BK)^T S + S(A - BK) \right]\bar{x} \tag{4.70}$$

Since this equation has to be true for any $\bar{x}$,

$$(A - BK)^T S + S(A - BK) = -(Q + K^T RK) \tag{4.71}$$

If $(A - BK)$ is a stable matrix, there is a positive-definite matrix so that S satisfies the eq. (4.71). As all *eigenvalues* of the matrix $(A - BK)$ have real negative parts,

$$\bar{x}(\infty) \rightarrow 0 \text{ and } J = \bar{x}^T S\bar{x}(0).$$

R is a symmetric positive-definite matrix, $R = T^T T$, where T is a nonsingular matrix, so that eq. (4.70) can be written in another way as:

$$(A^T - K^T B)S + S(A - BK) + Q + K^T T^T TK = 0$$

or

$$A^T S + SA + [TK - (T^T)^{-1}B^T S]^T [TK - (T^T)^{-1}B^T S] - SBR^{-1}B^T S + Q = 0 \tag{4.72}$$

Minimization of J relative to K requires the minimization of

$$\bar{x}^T [TK - (T^T)^{-1}B^T S]^T [TK - (T^T)^{-1}B^T S]\bar{x} \text{ relative to } K. \tag{4.73}$$

As this value is nonnegative, the minimum is located where the expression (4.73) is equal to 0 and $TK = (T^T)^{-1}B^T S$.

The compensating matrix is

$$K = T^{-1}(T^T)^{-1}B^T S = R^1 B^T S \tag{4.74}$$

Matrix S from (4.74) has to satisfy eq. (4.71) or the reduced Riccati equation:

$$A^T S + SA - SBR^{-1}B^T S + Q = 0 \tag{4.75}$$

This way, the synthesis procedure of the optimal control is reduced to:
- finding the solution of the reduced matrix Riccati equation relative to S
- substitution of the matrix S in eq. (4.75).

Example 4.3
Considering the system described through

$$\dot{\bar{x}} = \begin{bmatrix} 0 & 1 \\ 0 & 0 \end{bmatrix}\bar{x} + \begin{bmatrix} 0 \\ 1 \end{bmatrix}\bar{u}$$ and using control law (4.66)

$$\bar{u}(t) = -K\bar{x}(t)$$

K has to be determined to minimize the performance index:

$$J = \int_0^\infty (\bar{x}^T Q\bar{x} + \bar{u}^T\bar{u})dt \tag{4.76}$$

Q has the form:

$$Q = \begin{bmatrix} 1 & 0 \\ 0 & a \end{bmatrix} a \geq 0 \text{ and } R = [1] \tag{4.77}$$

The Riccati equation that has to be solved is (4.76) and thus:

$$\begin{bmatrix} 1 & 0 \\ 0 & 0 \end{bmatrix}\begin{bmatrix} S_{11} & S_{12} \\ S_{21} & S_{22} \end{bmatrix} + \begin{bmatrix} S_{11} & S_{12} \\ S_{21} & S_{22} \end{bmatrix}\begin{bmatrix} 0 & 1 \\ 0 & 0 \end{bmatrix} - \begin{bmatrix} S_{11} & S_{12} \\ S_{21} & S_{22} \end{bmatrix}\begin{bmatrix} 0 \\ 1 \end{bmatrix}[1][0\ 1]\begin{bmatrix} S_{11} & S_{12} \\ S_{21} & S_{22} \end{bmatrix} + \begin{bmatrix} 1 & 0 \\ 0 & a \end{bmatrix}$$

$$= \begin{bmatrix} 0 & 0 \\ 0 & 0 \end{bmatrix}$$ and further $$\tag{4.78}$$

$$\begin{bmatrix} 0 & 0 \\ S_{11} & S_{12} \end{bmatrix} + \begin{bmatrix} 0 & S_{11} \\ 0 & S_{21} \end{bmatrix} - \begin{bmatrix} S_{12}^2 & S_{12}S_{22} \\ S_{12}S_{22} & S_{22}^2 \end{bmatrix} + \begin{bmatrix} 1 & 0 \\ 0 & a \end{bmatrix} = \begin{bmatrix} 0 & 0 \\ 0 & 0 \end{bmatrix}$$

resulting in the equations (S is symmetric):

$$1 - S_{12}^2 = 0$$
$$S_{11} - S_{12}S_{22} = 0 \tag{4.79}$$
$$a + 2S_{12}S_{22}^2 = 0$$

From here,

$$S = \begin{bmatrix} S_{11} & S_{12} \\ S_{21} & S_{22} \end{bmatrix} = \begin{bmatrix} \sqrt{a+2} & 1 \\ 1 & \sqrt{a+2} \end{bmatrix} \text{ and the optimal control matrix}$$

$$K = R^{-1}B^{T}S = \begin{bmatrix} 1\sqrt{a+2} \end{bmatrix} \text{ and the optimal control law:}$$

$$u = -K\bar{x} = -x_1 - \sqrt{a+2}x_2$$

(4.80)

Example 4.4 ([17])

Optimal control of a binary distillation column (Fig. 4.6). The column has the control configuration given in the figure below. An optimal control law of the distillate composition has to be written by using the distillate flow (D) as control variable when the feed flow (F) suffers disturbances.

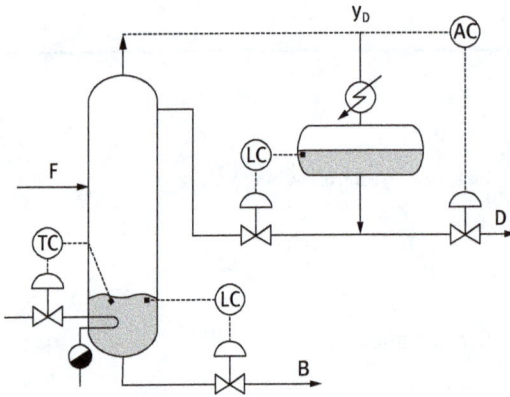

Fig. 4.6: Binary continuous distillation column with the corresponding control loops.

The block scheme of the distillation process is presented in Fig. 4.7.

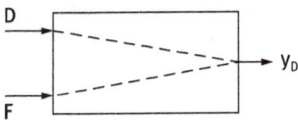

Fig. 4.7: The block scheme of the distillation process having input D (distillate flow) as control action, F (feed flow) as disturbance, and y_D (distillate concentration) as output.

The influence of the control action (D) and the disturbance (F) on the distillate composition, y_D, is given by the *transfer function*:

$$Y_D(s) = G_1(s)D(s) + G_2(s)F(s) = D_e(s) + F_e(s)$$

(4.81)

where $G_1(s)$ and $G_2(s)$ are the transfer functions corresponding to the transfer paths $D \rightarrow y_D$ and $F \rightarrow y_D$ and D_e and F_e are equivalent transfer functions.

For G_1, the expression found in [17] is

$$G_1(s) = \frac{K_1}{T_1 s + 1} = -\frac{0.031}{0.55 s + 1}. \tag{4.82}$$

Hence,

$$Y_D(s) = -\frac{0.031}{0.55 s + 1} D(s) + F_e(s) \text{ and when passing to time domain then} \tag{4.83}$$

$$0.55 \frac{dy_D(t)}{dt} + y_D(t) = -0.031 D(t) + d_1(t) \tag{4.84}$$

where $d_1(t)$ is the expression in time of the effect of the feed disturbance on the distillate composition.

When trying to put the equation in the form of the state eq. (4.63), the result is

$$\frac{dy_D(t)}{dt} = -\frac{1}{0.55} y_D(t) - \frac{0.031}{0.55} D(t) - \frac{1}{0.55} d_1(t) \tag{4.85}$$

By noting the variables $x_1 = y_D - y_{D,\text{ref}}$, $u = D - D_{\text{ref}}$, $d = d_1 - d_{1,\text{ref}}$, the above equation becomes the state equation (4.63):

$$\frac{dx_1}{dt} = ax_1(t) + bu(t) + cd(t) \quad \text{with } a = -\frac{1}{0.55}; \ b = 0.031a; \ c = a$$

At $t = 0$, the feed disturbance is considered to occur (step disturbance $d(t) = d_0$), the previous deviations of the variables being considered 0: $x_1(0) = 0$ and $d(0) = 0$.

We note with x_2 the state variable, $x_2 = bu(t) + cd_0$ and $u_1 = u(t)$.

The previous state equation thus becomes a system:

$$\frac{dx_1}{dt} = ax_1(t) + x_2(t)$$

$$\frac{dx_2(t)}{dt} = bu_1(t) \tag{4.86}$$

The proposed performance index is

$$J = \frac{1}{2} \int_0^{t_f} \left(x_1^2 + ru^2 \right) dt = \frac{1}{2} \int_0^{t_f} \left[\left(y_D - y_{D,\text{ref}} \right)^2 + rD^2 \right] dt \tag{4.87}$$

in which the first term represents the deviation of the distillate composition from the steady-state value while the second stands for the cost of the distillate flow change. r is the penalty for the change and has to be ≤ 1.

u_1 is sought after. By applying the optimal control law (4.65) and relationship (4.74):

$$\bar{u}_1(t) = K\bar{x}_1(t) = -R^{-1}B^T S\bar{x}_1(t) \tag{4.88}$$

where $[R] = [r]$; $[B]^T = \begin{bmatrix} 0 & -\frac{0.031}{0.55} \end{bmatrix}$; $[S] = \begin{bmatrix} S_{11} & S_{12} \\ S_{21} & S_{22} \end{bmatrix}$ where S_{ij} are calculated from the Riccati equation (4.75):

$$\begin{bmatrix} -\frac{1}{0.55} & 0 \\ 1 & 0 \end{bmatrix} \begin{bmatrix} S_{11} & S_{12} \\ S_{21} & S_{22} \end{bmatrix} + \begin{bmatrix} S_{11} & S_{12} \\ S_{21} & S_{22} \end{bmatrix} \begin{bmatrix} -\frac{1}{0.55} & 1 \\ 0 & 0 \end{bmatrix} -$$

$$\begin{bmatrix} S_{11} & S_{12} \\ S_{21} & S_{22} \end{bmatrix} \begin{bmatrix} 0 \\ -\frac{0.031}{0.55} \end{bmatrix} [r] \begin{bmatrix} 0 & -\frac{0.031}{0.55} \end{bmatrix} \begin{bmatrix} S_{11} & S_{12} \\ S_{21} & S_{22} \end{bmatrix} + \begin{bmatrix} 1 & 0 \\ 0 & 0 \end{bmatrix} = \begin{bmatrix} 0 & 0 \\ 0 & 0 \end{bmatrix} \text{ equivalent with}$$

$$-\frac{2}{0.55} S_{11} + \frac{0.031^2}{0.55^2} S_{11} S_{12} \ r = -1$$

$$S_{11} - \frac{1}{0.55} S_{12} - \frac{0.031^2}{0.55^2} S_{11} S_{21} \ r = 0$$

$$S_{11} - \frac{1}{0.55} S_{21} - \frac{0.031^2}{0.55^2} S_{21} S_{22} \ r = 0 \tag{4.89}$$

$$S_{21} + S_{22} - \frac{0.031^2}{0.55^2} S_{22}^2 \ r = 0$$

The optimal solution is then:

$$u_1(t) = \begin{bmatrix} -\frac{1}{r} \end{bmatrix} \begin{bmatrix} 0 & -\frac{0.031}{0.55} \end{bmatrix} \begin{bmatrix} S_{11} & S_{12} \\ S_{21} & S_{22} \end{bmatrix} \begin{bmatrix} x_1 \\ x_2 \end{bmatrix} = \frac{0.031}{0.55} (S_{21}x_1 + S_{22}x_2) \text{ or}$$

$$u(t) = \frac{0.056}{r} \left[S_{21}x_1 + S_{22} \left(x_1 + \frac{1}{0.55} x_1 \right) \right] = \frac{0.056}{r} \left(S_{21} + \frac{1}{0.55} S_{22} \right) x_1 + \frac{0.056}{r} S_{22}x_1 \tag{4.90}$$

Integrating,

$$u(t) = \frac{0.056}{r} S_{22}x_1(t) + \frac{0.056}{r} \left(S_{21} + \frac{1}{0.55} S_{22} \right) \int_0^{t_f} x_1(t)dt + u_0 \tag{4.91}$$

we find S_{22} by solving the equation:

$$4q^4 S_{22}^6 - 4(pq^3 + 2q^3)S_{22}^5 - (p^2q^2 - 8pq^2 + 4q^2 - 4pq^3)S_{22}^4 - (2p^2q + 4pq - 4p^2q^2 + 8pq^2)$$

$$\times S_{22}^3 + (p^2 - 4p^2q - p^3q)S_{22}^2 - (p^3 - q)S_{22} - \frac{1}{2}p = 0, \tag{4.92}$$

S_{21} is calculated from equation:

$$S_{21} = -\frac{1 + pS_{11}}{qS_{11}} \tag{4.93}$$

and S_{11} from:

$$S_{11} = S_{22}(qS_{22} - 1)(2S_{22}q - p) \tag{4.94}$$

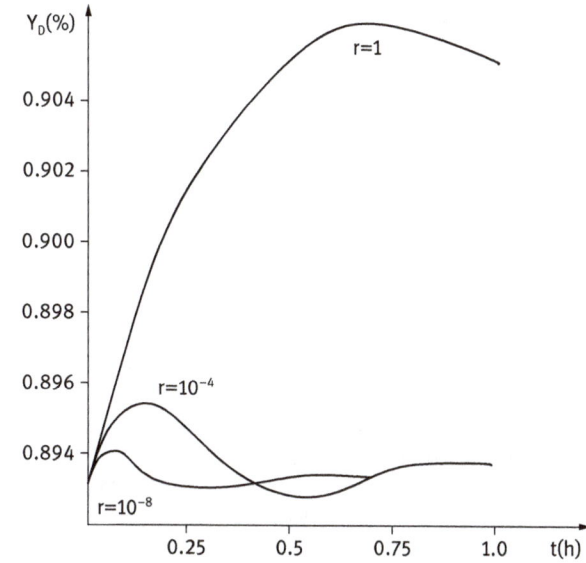

(a)

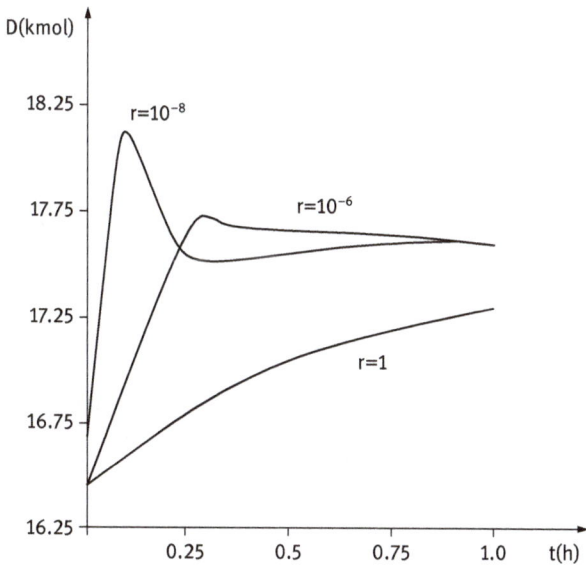

(b)

Fig. 4.8: The optimal control of the distillate purity (y_D): (a) through the distillate flow (D) and (b) depends on the penalty term.

where $p = -\frac{2}{0.55}$ and $q = \frac{0.031^2}{0.55^2}\,r$.

Consequently, from eqs. (4.91) to (4.94), the optimal control law is

$$D(t) = \frac{0.056}{r} - S_{22}(y_D - y_{D,\text{ref}}) + \frac{0.056}{r}\left(S_{21} + \frac{1}{0.55}S_{22}\right)\int_0^{t_f}(y_D - y_{D,\text{ref}})\,dt + D_0 \qquad (4.95)$$

where D_0 is the distillate flow at $y_D = y_{D,\text{ref}}$.

It can be observed that a PI-type control function is obtained but which is depending on the penalty imposed to the second term (Fig. 4.8).

References

[1] Kirk, D.E., *Optimal Control Theory – An Introduction*, p. 4,Prentice Hall, 2004.
[2] Zhang, G.P., Rohani, S., *On line optimal control of a seeded batch cooling crystallizer*, Chemical Engineering Science, 58(9), 1887, 2003.
[3] Sheikhzadeh, M., Trifkovic, M., Rohani, S., *Real-time optimal control of an anti-solvent isothermal crystallization process*, Chemical Engineering Science, 63(3), 829, 2008.
[4] Smets, I., Claes, J., November, E., Bastin, G., Van Impe, J., *Optimal adaptive control of (bio)chemical reactors: past, present and future*, Journal of Process Control, 14, 795, 2004.
[5] Manon, P., Valentin-Roubinet, C., Gilles, G., *Optimal control of hybrid dynamical systems: application in process engineering*, Control Engineering Practice, 10(2), 133, 2002.
[6] Engell, S., *Feedback control for optimal process operation*, Journal of Process Control, 17(3), 203, 2007.
[7] Agachi, P.S., Cristea, V.M., *Basic Process Engineering Control*, p. 297, Walter de Gruyter, Berlin/ Boston, 2014.
[8] Yi, C., *Learning PID Tuning III: Performance Index Optimization, a Tool and Tutorial to Perform Optimal PID Tuning*, File Exchange, Matlab Central, http://www.mathworks.com/matlab-central/fileexchange/18674-learning-pid-tuning-iii–performance-index-optimization/content/html/optimalpidtuning.html.
[9] Imre Lucaci, A., Agachi, P.Ş., *Optimizarea proceselor din industria chimică (Optimization of the Chemical Industry Processes)*, Editura Tehnică, Bucureşti, 5, 2002.
[10] Imre Lucaci, A., Agachi, P.Ş., *Optimizarea proceselor din industria chimică (Optimization of the Chemical Industry Processes)*, Editura Tehnică, Bucureşti, 7, 2002.
[11] Agachi, S.P., Vass, E., *Optimization of a methanol reactor using Pontryagin's maximum principle*, Revista De Chimie, 6, 513–521, 1995.
[12] Natta, G., Pino, P., Mazzanti, G., Pasquon, J., *Kinetics of methanol synthesis reaction*, Chimie et Industrie(Milano), 35(6), 705, 1953.
[13] Agachi, S., *Automatizarea proceselor chimice (Chemical Process Control)*, p. 337, Ed. Casa Cărţii de Ştiinţă, Cluj, 1994.
[14] Seader, J.D., *Perry's Chemical Engineer's Handbook*, 6th Edition., McGraw-Hill, New York, NY, 13–37–13-39, 1984.
[15] Agachi, P.S., Cristea, V.M., *Basic Process Engineering Control*, p. 63, Walter de Gruyter, Berlin/Boston, 2014.
[16] Singh, M., *Applied Industrial Control: An Introduction (International Series of Systems and Control)*, Elsevier, prg.5.6, 1980.
[17] Bozga, G., *Conducerea cu calculatoare a proceselor chimice (Chemical process computer control)*, Institutul Politehnic Bucureşti, 224, 1989.

5 Multivariable control

5.1 Introduction

Usually, the control of a plant aims to keep more than one of its outputs at desired setpoints, either constant or changing, to counteract the action of disturbances, to provide both safe and efficient operation, and to conform to environmental requirements. The single variable control is in the large majority of cases not sufficient to fulfil the multiple requests demanded by the control tasks and the need for the control of multiple output variables is obvious. Multivariable control is the solution for multiple-input-multiple-output (MIMO) processes to fulfil their associated multiple control objectives. Multivariable control involves multiple manipulated variables and the design of the control system becomes complex, sometimes asking for the support of advanced control methodologies.

Although the general control concepts of the single loop control are also valid for the multivariable control, there are some additional and particular aspects derived from the multiple-variables characteristic. Some of these aspects are presented in the following.

The starting point of designing any multivariable control system is to clearly define the control objectives. From the control objectives arises the need for specifying which of the process outputs needs to be controlled. Choosing the controlled outputs must rely first on the analysis of the possibility to either directly measure the outputs intended to be controlled or to infer their values from other available measured outputs.

Following the selection of the controlled outputs, the selection of the manipulated variables (inputs) is of first importance. This category of process input variables are the handles used by the controllers to act on the controlled outputs. Additionally, the analysis must be carried on identifying other relevant process inputs affecting the controlled outputs, i.e., the disturbances. Having both qualitative and quantitative knowledge on the disturbances means offering valuable information for the control system design. Particularly, the identification of the measured disturbances offers the opportunity to perform feedforward control and efficiently counteract part of the undesired effects of the disturbances. Comprehensive information on the disturbances is useful for the multivariable control system design as it may reveal both the source of disturbances and the amplitude of their effect on the different controlled variables.

The major problems for the multivariable control design originate in the existence of the interaction among inputs and outputs of the MIMO systems, i.e., a manipulated variable may influence several controlled variables. This feature, not considered in the case of the single variable control design, may affect not only the multivariable control performance but even the stability of the controlled process. The interaction among the process variables that are controlled may produce upsets in all the other controlled (or noncontrolled) variables. Design methods for the multivariable control should be developed to either reduce the effects of the interactions or eliminate them.

https://doi.org/10.1515/9783110789737-006

There are two fundamental approaches to multivariable control. They are the *multiloop control*, also known as *decentralized control*, and the *centralized* (coordinated) *control* [1]. The multiloop control uses multiple single variable controllers where each manipulated variable is assigned to an output variable (and establishes a control loop), with the aim of controlling the output. The centralized control considers all measured outputs and simultaneously generates all manipulated variables in such a way that process interacting effects are counteracted when the manipulated variables computation is performed.

After the selection of the sets of controlled and manipulated variables, the multiloop control design implies the specification of the control loops configuration. This task consists in deciding the *pairing* of the controlled variables and the manipulated variables. This pairing is of most importance for reducing the interaction effects among the multiple control loops and, as a result, for the attainable control performance desired for the multivariable control system [1–3].

For the processes where the number of controlled and the number of manipulated variables are not the same, the multivariable control design has to cope with this circumstance and provide specific control methods.

5.2 Multiloop control

The most commonly applied approach of the multivariable control is the multiloop control strategy, although from theoretical point of view, the centralized control presents obvious incentives. However, the latter too may have to comply with possible implementation impediments [1].

The multiloop control uses several single-loop controllers, each of them devoted to control one process output variable. Besides historical reasons of precedency, the incentives of this control approach rely on its simplicity, both with respect to the use of unsophisticated architecture and to the exploitation of common control laws, such as the PID algorithm. Furthermore, the multiloop control is easily understood and accepted by the operators especially when their direct intervention is needed for coping with instrumentation malfunction or significant process upsets.

5.2.1 Interaction among control loops

The key aspect in designing the multiloop control system is the interaction among the manipulated and controlled variables, i.e., the way each input (manipulated) variable influences all of the output variables of the process. As a result, any single-loop controller can no longer be considered decoupled with respect to the other single-loop controllers. The presence of the process interaction induces interaction among the individual control loops. Example 5.1 presents a simple but intuitive process with interaction among two control loops.

Example 5.1

Consider a mixing process of two liquid streams presented in Fig. 5.1.

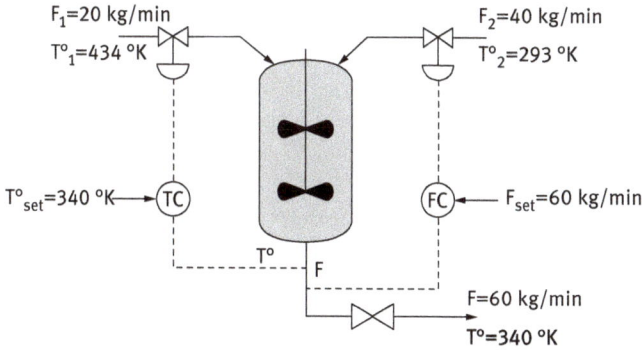

Fig. 5.1: Interacting flow and temperature control loops in the mixing tank.

The streams entering the mixing tank have the inlet mass flows F_1 and F_2, $(F_1 < F_2)$, and the inlet temperature T_1^o and T_2^o, $(T_1^o > T_2^o)$. The outlet mass flow F has the outlet temperature T^o. It is assumed that mixing is perfect, and the densities of the input flows are equal. The specific heats of the inlet flows are equal, too. The mixing process goal is to keep both the outlet mass flow and outlet temperature at desired setpoints to provide smooth operation for the downstream unit. This is performed by the outlet flow FC and outlet temperature TC controllers presented in Fig. 5.1 [3].

Consider the system to be initially in steady state. Suppose that a change in the setpoint F_{set} of the mass flow controller is made to increase the desired outlet flow. This task is accomplished by the flow controller FC by increasing the inlet flow F_2. As the inlet flow F_2 is changed (increases) by FC, the temperature in the mixing tank T^o changes also (decreases). This change is due to the different (lower) temperature of the F_2 stream, compared to the temperature of the F_1 stream. Consequently, as the temperature in the mixing tank is changed from its initial value (equal to the setpoint value), the temperature controller starts acting by changing (increasing) the inlet flow F_1. In conclusion, the setpoint change in the flow control loop produces the need for control intervention provided by the temperature control loop. The same interacting behavior may be observed if a change of the outlet temperature setpoint T_{set}^o is initiated, resulting in the need for control action in the flow control loop. The two control loops also show interacting effects if a disturbance is acting on any of the inlet flows or temperatures.

The interaction among the control loops may destabilize the process and, as the number of control loops increases, this undesired effect may be intensified. Therefore, a new task comes to the control design. This task is the loop pairing and consists in specifying for each controlled output the corresponding manipulated variable to be used for its control, in such a way that the loop interaction is minimal [1, 2, 8].

In the following, the case of an interacting control system for a multivariable process with two manipulated variables and two controlled outputs will be considered. The 2 × 2 open-loop system is presented in Fig. 5.2 [1, 2].

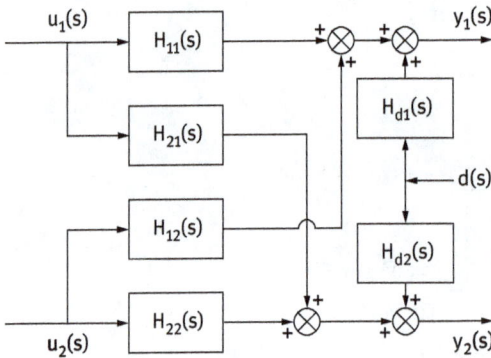

Fig. 5.2: Diagram of the 2 × 2 open-loop system.

The transfer matrix of the 2 × 2 open-loop system may be described by:

$$\begin{bmatrix} y_1(s) \\ y_2(s) \end{bmatrix} = \begin{bmatrix} H_{11}(s) & H_{12}(s) \\ H_{21}(s) & H_{22}(s) \end{bmatrix} \begin{bmatrix} u_1(s) \\ u_2(s) \end{bmatrix} + \begin{bmatrix} H_{d1}(s) \\ H_{d2}(s) \end{bmatrix} d(s), \tag{5.1}$$

where $H_{ij}(s)$ is the transfer function relating output $y_i(s)$ to input $u_j(s)$ and $H_{di}(s)$ is the transfer function relating output $y_i(s)$ to disturbance $d(s)$.

For the multivariable process to be steady-state controllable, i.e., for the controlled outputs to be brought at steady state to the desired setpoints in the presence of disturbances, the inverse of the steady-state gain matrix should exist. This is equivalent to the condition that the determinant of the steady-state gain matrix H_s is nonzero [1].

Example 5.2

The transfer matrix for the mixing process described in Example 5.1 can be obtained from the mass and heat balance equations:

$$\frac{dm}{dt} = F_1 + F_2 - F \tag{5.2}$$

$$\frac{d}{dt}(mc_pT^\circ) = F_1c_pT_1^\circ + F_2c_pT_2^\circ - Fc_pT^\circ \tag{5.3}$$

Considering that the mass of liquid in the mixing tank ($m = 600$ kg) does not change, $dm/dt = 0$, eqs. (5.2) and (5.3) become:

$$F = F_1 + F_2 \tag{5.4}$$

$$m\frac{dT^\circ}{dt} = F_1 T_1^\circ + F_2 T_2^\circ - FT^\circ \tag{5.5}$$

Linearization of eq. (5.5) around the steady-state values of variables presented in Fig. 5.1 leads to the following linear forms, using the deviation variables $\Delta T^\circ = T^\circ - T_s^\circ$, $\Delta F_1 = F_1 - F_{1s}$, and $\Delta F_2 = F_2 - F_{2s}$ (subscript s has been used for *steady state*).

$$m\frac{d\Delta T^\circ}{dt} = T_1^\circ \Delta F_1 + T_2^\circ \Delta F_2 - T_s^\circ \Delta F_1 - T_s^\circ \Delta F_2 - (F_{1s} + F_{2s})\Delta T^\circ \tag{5.6}$$

and

$$\frac{m}{(F_{1s} + F_{2s})}\frac{d\Delta T^\circ}{dt} + \Delta T^\circ = \left(\frac{F_{2s}(T_1^\circ - T_2^\circ)}{(F_{1s} + F_{2s})^2}\right)\Delta F_1 + \left(\frac{F_{1s}(T_2^\circ - T_1^\circ)}{(F_{1s} + F_{2s})}\right)\Delta F_2 \tag{5.7}$$

The matrix transfer function of the system becomes

$$H(s) = \begin{bmatrix} H_{11}(s) & H_{12}(s) \\ H_{21}(s) & H_{22}(s) \end{bmatrix} = \begin{bmatrix} \dfrac{1}{\dfrac{F_{2s}(T_1^\circ - T_2^\circ)}{(F_{1s}+F_{2s})^2}} & \dfrac{1}{\dfrac{F_{1s}(T_2^\circ - T_1^\circ)}{(F_{1s}+F_{2s})^2}} \\ \dfrac{m}{(F_{1s}+F_{2s})}s+1 & \dfrac{m}{(F_{1s}+F_{2s})}s+1 \end{bmatrix} = \begin{bmatrix} \dfrac{1}{1567} & \dfrac{1}{-0.783} \\ \dfrac{10s+1}{} & \dfrac{}{10s+1} \end{bmatrix}, \tag{5.8}$$

revealing a time constant of $T_p = 10$ [min] for the temperature change.

Finally, the static (steady-state) gain matrix of the open-loop process has the following form:

$$H_s = \begin{bmatrix} 1 & 1 \\ 1567 & -0.783 \end{bmatrix} \tag{5.9}$$

As $\det(H_s) = -2.349 \neq 0$, it may be concluded that the process is controllable with the chosen pairing of the manipulated and controlled variables.

The selection of the manipulated and controlled variables for building the multiloop control configuration aims to find the multivariable control structure that presents the weakest interaction among the individual control loops. For n-manipulated and n-controlled variables, the number of possible pairing is equal to $n!$ alternatives [1, 2].

The multiloop control system must cope both with circumstances when all loops work in control mode and with situations when only part of the individual loops are operating in automatic mode and the others are in manual mode.

The interaction effect on the control performance is investigated by considering two cases. The first one assesses the closed-loop operation and the transfer function when only one loop is closed (in automatic mode) and all the other loops are in open loop (in manual mode). The second evaluation is performed for the case when all control loops are closed. For simplicity of presentation, this investigation is done for the 2×2 loop control system presented in Fig. 5.3, but the extension to a larger dimension of the MIMO control is straightforward. For the same reason, the transfer function of the final control elements and measuring devices are considered equal to unity.

Fig. 5.3 shows the first case when *one loop is closed* and the other is open.

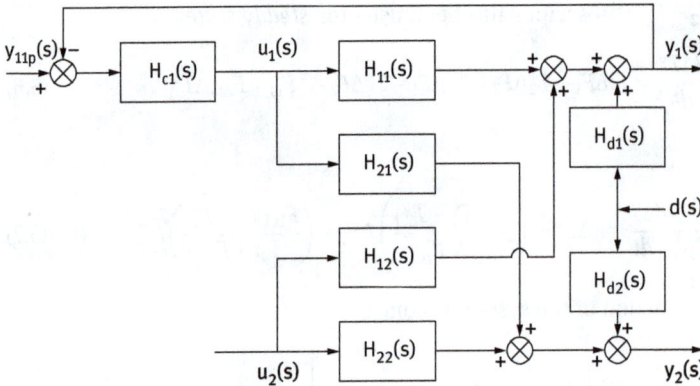

Fig. 5.3: Interaction assessment for the case when only one loop is closed.

It may be noticed that both outputs are influenced by the setpoint of the closed loop, according to the following transfer functions:

$$y_1(s) = \frac{H_{c1}(s)H_{11}(s)}{1 + H_{c1}(s)H_{11}(s)} y_{1sp}(s) + H_{d1}(s)d(s) \tag{5.10}$$

$$y_2(s) = \frac{H_{c1}(s)H_{21}(s)}{1 + H_{c1}(s)H_{11}(s)} y_{1sp}(s) + H_{d2}(s)d(s) \tag{5.11}$$

This is due to the transfer path provided by $H_{21}(s)$.

The second case, when *all control loops are closed*, is presented in Fig. 5.4.

When both loops are closed, the change of the loop 1 setpoint y_{1sp} is producing a change of the controlled output y_1 due to the action of the loop 1 control system. This is the direct effect of the manipulated variable u_1 change on output y_1 and it is represented in Fig. 5.4 by the dash-dotted line. At the same time, the change of the manipulated variable u_1 produces a change on output y_2 due to the H_{21} transfer path. As a result, y_2 deviates from its desired setpoint y_{2sp} and loop 2 starts to operate for counteracting the developed offset. Meanwhile, the action of the manipulated variable u_2 of loop 2 produces a second, indirect effect on output y_1, due to the H_{12} transfer path. This indirect effect is represented in Fig. 5.4 by the dashed line. Moreover, due to this new change of output y_1, the control action is again activated in loop 1 and this produces a new change of y_2. This interaction between the control loops induces oscillating behavior on both controlled variables, producing an extended settling time or, in the extreme case, instability [7]. Similar behavior may be initiated by the change of the loop 2 setpoint or by the action of disturbances.

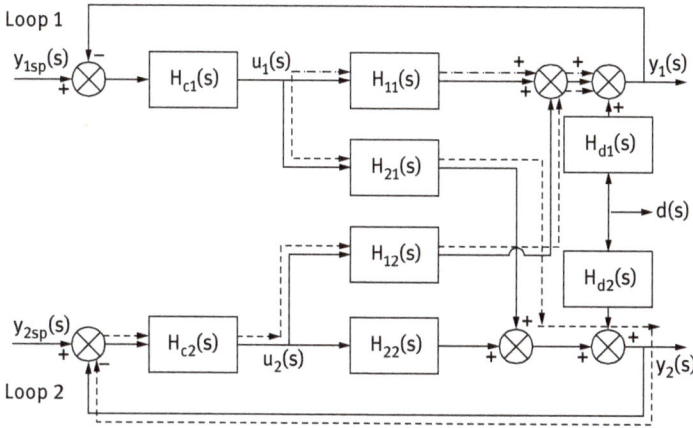

Fig. 5.4: Interaction assessment for the case when all loops are closed.

The closed-loop transfer functions of the controlled system, when both control loops are closed, may be described by:

$$y_1(s) = G_{11}(s)y_{1sp}(s) + G_{12}(s)y_{2sp}(s) + H_{d1}(s)d(s) \tag{5.12}$$

$$y_2(s) = G_{21}(s)y_{1sp}(s) + G_{22}(s)y_{2sp}(s) + H_{d2}(s)d(s) \tag{5.13}$$

with

$$G_{11}(s) = \frac{H_{11}H_{c1} + (H_{11}H_{22} - H_{12}H_{21})H_{c1}H_{c2}}{(1 + H_{11}H_{c1})(1 + H_{22}H_{c2}) - H_{12}H_{21}H_{c1}H_{c2}} \tag{5.14}$$

$$G_{12}(s) = \frac{H_{12}H_{c2}}{(1 + H_{11}H_{c1})(1 + H_{22}H_{c2}) - H_{12}H_{21}H_{c1}H_{c2}} \tag{5.15}$$

$$G_{21}(s) = \frac{H_{21}H_{c1}}{(1 + H_{11}H_{c1})(1 + H_{22}H_{c2}) - H_{12}H_{21}H_{c1}H_{c2}} \tag{5.16}$$

$$G_{22}(s) = \frac{H_{22}H_{c2} + (H_{11}H_{22} - H_{12}H_{21})H_{c1}H_{c2}}{(1 + H_{11}H_{c1})(1 + H_{22}H_{c2}) - H_{12}H_{21}H_{c1}H_{c2}} \tag{5.17}$$

The transfer functions developed for the two investigated cases show that to ensure stability for both operating scenarios, i.e., when either of the loops is closed and the other one is an open loop, and when both loops are closed, it is necessary that simultaneously the roots of the following characteristic equations should have negative real parts:

$$1 + H_{c1}(s)H_{11}(s) = 0,$$

$$1 + H_{c2}(s)H_{22}(s) = 0,$$

$$(1 + H_{11}H_{c1})(1 + H_{22}H_{c2}) - H_{12}H_{21}H_{c1}H_{c2} = 0$$

(5.18)

The conclusion is that due to the interacting effects the controllers should be tuned simultaneously to provide the desired control performance and stability.

5.2.2 Pairing the control loops

The interacting effects of the multiloop control system is dependent on the form of the MIMO system transfer matrix $H(s)$. A diagonal form of $H(s)$ reveals the lack of interaction. The selection of the pairing among manipulated variables and controlled variables for finding the multiloop control structure with minimum interaction between the loops is based on the Relative Gain Array (RGA). RGA, also called Bristol array, was first introduced by Bristol [5], and it measures the interaction in the process [6]. The elements of the RGA matrix, denoted by Λ, consist in the ratio of the open loop to the closed-loop gain for each input-output path of the process. The generic element of the RGA matrix λ_{ij}, associated to the transfer path starting from input u_j and finishing to output y_i is described by either of the following ratio forms:

$$\lambda_{ij} = \frac{\left(\dfrac{\Delta y_i}{\Delta u_j}\right)_{u_l = \text{const.},l \neq j}}{\left(\dfrac{\Delta y_i}{\Delta u_j}\right)_{y_p = \text{const.},p \neq i}} \quad \text{or} \quad \lambda_{ij} = \frac{\left(\dfrac{\partial y_i}{\partial u_j}\right)_{u_l = \text{const.},l \neq j}}{\left(\dfrac{\partial y_i}{\partial u_j}\right)_{y_p = \text{const.},p \neq i}}$$

(5.19)

The numerator of λ_{ij}, i.e., the open-loop gain, is the corresponding element of the static gain matrix H_s and shows the gain between input u_j and output y_i when only one loop (the one involving input u_j and output y_i) is closed and the others are open. The denominator of λ_{ij}, i.e., the closed-loop gain, is the steady-state gain between input u_j and output y_i when all control loops are closed [1, 6].

The RGA matrix can be determined either by experiments or by computation methods. The first methodology implies two experiments for obtaining the numerator and denominator of each λ_{ij}, and the second one computes the Hadamard product between the static gain matrix H_s and its inverse transposed matrix $\left(H_s^{-1}\right)^T$:

$$\Lambda = H_s^* [H_s^{-1}]^T$$

(5.20)

where the symbol * denotes the element-by-element matrix multiplication [3].

The RGA matrix is scale-independent and the sums of its elements on each row or column are equal to unity.

The values of the RGA matrix may be interpreted for the selection of the multiple-loop control pairing, according to the following remarks [1]:

1. When $\lambda_{ij} = 0$, there is no direct steady-state open-loop relationship between input u_j and output y_i. It means this pairing is not desirable and in most cases it is not acceptable. Nevertheless, when all loops are closed the controller paired with the value of $\lambda_{ij} = 0$ will operate and in very particular cases this pairing might be acceptable (due to its indirect effect).

2. When $\lambda_{ij} = 1$, the change of input u_j directly influences only output y_i and not the other indirect paths (of the other closed loops), i.e., the loop paired according to $\lambda_{ij} = 1$ is decoupled with respect to the other loops. It means this pairing is the most desirable to be applied for the control of output y_i. Nevertheless, it should be noticed that changes in input (manipulated) u_j used in this loop may also influence other controlled outputs.

3. When $\lambda_{ij} < 0$, the direct effect of input u_j on output y_i has a different sign compared to the indirect effect generated by the manipulated variables of the other closed loops. This may change the feedback of the control loop from negative to positive, with undesired consequences on the stability. This pairing is not acceptable (unless special measures for ensuring stability are introduced).

4. When $0 < \lambda_{ij} < 1$, the direct effect of input u_j on output y_i is smaller compared to the sum of direct and indirect effect generated by the manipulated variables of the other closed loops. As λ_{ij} approaches zero, this indirect effect is larger and the interaction effect is larger too. This pairing is acceptable especially when λ_{ij} approaches 1.

5. When $\lambda_{ij} > 1$, the direct effect of input u_j on output y_i is larger compared to the sum of direct and indirect effects generated by the manipulated variables of the other closed loops. As λ_{ij} approaches 1, the indirect effect is smaller and the interaction effect is smaller too. This pairing is acceptable especially when λ_{ij} approaches 1.

6. When $\lambda_{ij} = \infty$, the sum of the direct and indirect effects of input u_j on output y_i has a zero value and this means output y_i may not be controlled by input u_j. This pairing is not acceptable.

Example 5.3

The computation of the RGA matrix for the mixing process described in Example 5.1 results in the following form of Λ:

$$\Lambda = H_s^* [H_s^{-1}]^T = \begin{bmatrix} 1 & 1 \\ 1567 & -0.783 \end{bmatrix} * \begin{bmatrix} 0.333 & 0.667 \\ 0.425 & -0.425 \end{bmatrix} = \begin{matrix} & u_1 = F_1 \ \ u_2 = F_2 \\ \begin{bmatrix} 0.333 & 0.667 \\ 0.667 & 0.333 \end{bmatrix} & \begin{matrix} y_1 = F \\ y_2 = T^\circ \end{matrix} \end{matrix} \qquad (5.21)$$

According to the elements of Λ from eq. (5.21) the most favorable pairing is the following. The mixer output mass flow F should be controlled by cold inlet flow stream $F_2(\lambda_{12} = 0.667)$ and the mixer output temperature T° should be controlled by hot inlet flow stream $F_1(\lambda_{21} = 0.667)$. This pairing is already presented in Fig. 5.1. The appropriateness of the pairing suggested by the RGA matrix may be also confirmed by the phenomenological and practical assessment relying on the observation that the high value of the cold inlet flow stream $F_2 = 40$ [kg/min] is able to influence output flow F to a greater extent than the cold inlet flow $F_1 = 20$ [kg/min] would.

RGA is a measure of interaction for steady state, and this limitation should be considered as the transient regime is not revealed by it. Furthermore, the analytical computation of the RGA matrix is based on linearized models of the process. The accuracy of the linearized models may be dramatically affected when the operating point of the process changes significantly or the nonlinearity is pronounced. During the last decades, important research effort has been devoted to developing a theoretical frame for using the frequency-dependent RGA in interaction assessment, but the results do not yet have the power of insight and generality characteristic to the steady-state RGA [9].

Measuring the interaction in a multiloop control system may be performed on the basis of the singular value decomposition of the static gain matrix H_s [1, 14]. This decomposition reveals the directions in the controlled variables space that make the manipulated variables have the largest and the smallest change. The *condition number* is the ratio between the largest and the smallest changes in these identified directions, i.e., ratio between the maximum and the minimum singular values. As the condition number has a high value, it shows high interaction being present in the multiloop control system and it offers a measure for pairing the control loops. Correlations between RGA and condition number may be performed when analyzing pairing [10, 12].

One of the problems that emerged from the practice of multiloop control is to cope with situations when part of the control loops are in automatic mode while the others are not in closed-loop mode due to operation in manual mode or are out of service because of failures. Pairing of the control loops should consider and ground its design on the assessment of the control performance for such operation situations. Maintaining the integrity of the system means preserving the stability of the multiloop control system in case of operating with several control loops open and without the need for changing the signs of the feedback controllers [1, 13]. Assessment of the integral stabilizability may be performed by the *Niederlinski index* (NI) test, assuming that multiloop controllers are fitted with integral action. The NI can be computed by the following formula:

$$NI = \frac{\det\left(H_s^*\right)}{\prod_{k=1}^{n} \left(H_s^*\right)_{kk}} \tag{5.22}$$

where H_s^* is the static gain matrix of the process that emerges from the H_s gain matrix by arranging the loop pairing on the main diagonal of the matrix.

If the NI value is negative, the loop pairing design should not be accepted. Designs with positive NI may be considered as candidates for pairing the control loops.

5.2.3 Tuning the multiloop controllers

Tuning the multiloop controllers does not have a very well-defined theoretical approach and clearly formulated methodology. However, there are general guidelines that may be used in association with the interaction analysis and the assessment of the disturbance effects on the multiloop control performance [11].

First, the tuning should take into consideration the objectives of the control problem and, directly emerged from that, generate a hierarchy of importance among the controlled variables. To the most important of them, the multiloop control pairing and tuning of the associated control loops should provide tighter control performance (setpoint tracking, disturbance rejection, and fast response), while to the others, the performance might be less strict [1].

A first choice of tuning can be the experimental tuning made on the real process by repeated adjustments of the tuning parameters of the controllers and by assessment of the control performance or made on a dynamic simulator of the process. Results obtained from the latter greatly depend on the model accuracy and might be less precise but may be suitable when producing experiments on the process is not convenient. Usually, for this tuning approach the starting values of the tuning parameters are those specific to the single-loop tunings (without considering the interaction) but adjusted for obtaining stability. In most of the cases, the tuning may proceed by detuning these values of initial tuning parameters for coping with the interaction between the control loops.

The best tuning parameters may be also obtained by minimizing one or several of the control performance indices (e.g., ISE, IAE, ITAE), on the basis of a dynamic simulator and by the use of efficient optimization algorithms. Again, the obtained best tuning parameters are dependent on the involved dynamic model.

The multiloop controllers should include the integral mode (effect) to eliminate the steady-state offset. RGA provides the basic information for the pairing design and is the main instrument used for this purpose. NI and condition number complement the RGA information. The tuning is performed after the loop-pairing step. Controlled variables having higher assigned importance will be paired with manipulated variables demonstrating fast effect. Additionally, an analysis may be performed to assess the effect of disturbances on the controlled variables. This may be done by the relative disturbance gain [8].

5.2.4 Decoupling interaction for multiloop control

The RGA offers information for the design of the loop pairing such as that the interaction between the multiloop control loops is minimal. However, it is possible that even for this best multiloop design and proper tuning, the control performance is still poor. RGA proposes a "palliative" solution to treat the interacting problem, but a "surgical" one still exists. This is called decoupling and consists in the design of special elements, called decouplers, whose role is to reduce or even eliminate loops interaction by transforming the closed-loop transfer matrix in a new one, but with a diagonal form [1]. As a result, the control loops would operate such that they are independent and noninteracting. This solution introduces new control paths from each manipulated variable to all other controlled outputs, such as the added effect of loops interaction and the one generated by the decoupler should compensate each other. The process inherent interactions, described by $H_{ij}(s)$, cannot be removed but the decouplers introduce new effects (paths) of the manipulated variables u_j on the controlled outputs $y_i(i \neq j)$ in such a way to produce compensation for this interaction. As a result, the control loops will not interact anymore. The decoupler is designed according to the following relationship:

$$D_{ij}(s) = -\frac{H_{ij}(s)}{H_{ii}(s)} \tag{5.23}$$

The 2×2 system with the associated multiloop controllers presented in Fig. 5.4 will be further used for showing the decoupler design. Consider a disturbance or setpoint change in loop 1. The manipulated variable u_1 will change to bring the controlled output to the setpoint y_{1sp}. This change of u_1 will produce a change on controlled output y_2, due to process interaction. To keep y_2 unchanged, the decoupler $D_{21}(s)$ will be introduced such that the process effect of u_1 to y_2 is eliminated. The transfer function of this decoupler is

$$D_{21}(s) = -\frac{H_{21}(s)}{H_{22}(s)} \tag{5.24}$$

The control loops with D_{21} decoupler is presented in Fig. 5.5.

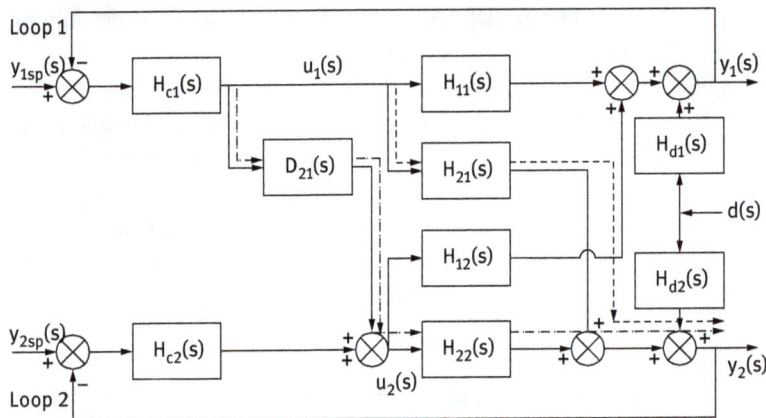

Fig. 5.5: Decoupling effect of loop 1 on loop 2.

As presented in Fig. 5.5, the effect of u_1 on y_2 (dotted line) is compensated by the effect introduced by the decoupler D_{21} path (dash-dotted line). The transfer function of the decoupler $D_{21}(s)$ is obvious. This decoupler prevents the transmission of any change from loop 1 to loop 2 (one-way decoupling).

The same approach may be used to make the decoupling of loop 2 to loop 1, by designing the decoupler D_{12}:

$$D_{12}(s) = -\frac{H_{12}(s)}{H_{11}(s)} \tag{5.25}$$

The 2×2 system with the associated multiloop controllers and decouplers is presented in Fig. 5.6.

It may be observed that the decoupling effect is very similar to the feedforward control and the decouplers are feedforward elements. The decoupler acts immediately as the manipulated variable changes and produces at every moment of time an effect that is equal but with opposed sign to the process interacting effect.

Depending on the time constants and dead time of the process transfer functions $H_{ij}(s)$, the design of the decoupler may be physically unrealizable. Partial decoupling or steady-state decoupling may be used in such cases.

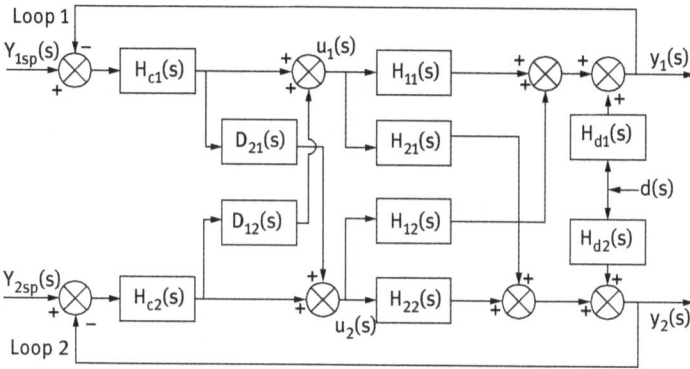

Fig. 5.6: Complete decoupling of the multiloop control for the 2×2 system.

The multiloop control system with complete decoupling has the following closed-loop transfer function dependence on the setpoints and disturbance:

$$y_1(s) = \frac{\left(H_{11} - \dfrac{H_{12}H_{21}}{H_{22}}\right)H_{c1}}{1 + \left(H_{11} - \dfrac{H_{12}H_{21}}{H_{22}}\right)H_{c1}} y_{1sp}(s) + H_{d1}(s)d(s) \tag{5.26}$$

$$y_2(s) = \frac{\left(H_{22} - \dfrac{H_{12}H_{21}}{H_{11}}\right)H_{c2}}{1 + \left(H_{22} - \dfrac{H_{12}H_{21}}{H_{11}}\right)H_{c2}} y_{2sp}(s) + H_{d2}(s)d(s) \tag{5.27}$$

The block diagram representation of the decoupled closed loops is presented in Fig. 5.7.

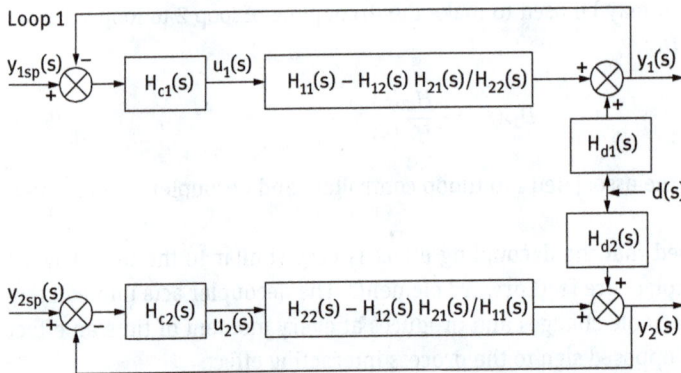

Fig. 5.7: Equivalent block diagram showing the complete decoupling of the multiloop control system for the 2 × 2 system.

Although the dynamics of the process with decouplers is different, compared to the original process, tuning of the decupled control loops may be performed in an independent way and both control performance and stability may be straightforwardly achieved.

The complete decoupling is very sensitive to the accuracy of describing the process and transfer functions of decouplers. Due to this reason complete decoupling is almost impossible for the processes featuring strong nonlinearity and without considering an adaptive strategy to cope with changes in the linearized model or in the nonstationary parameters of the process. This situation is especially cumbersome for the processes where the pairing was made according to large relative gain values ($\lambda_{ij} > 1$).

The sensitivity of the decoupling to the accuracy of the model description is very much reduced when only part of the decouplers are implemented in the multiloop control structure. Partial decoupling, also denoted as one-way decoupling, may bring important benefits especially if it is applied for the most important controlled variables.

Even though complete dynamic decoupling is desired from the theoretical perspective of effectiveness, the steady-state decoupling based on the static gains of the decouplers transfer functions may bring benefits compared to the lack of any decoupling.

Example 5.4

The computation of the decouplers for the multiloop control of the mixing process described in Example 5.1 results in the following form of the decouplers:

$$D_{12}(s) = -\frac{H_{12}(s)}{H_{11}(s)} = -\frac{1}{1} = -1 \tag{5.28}$$

$$D_{21}(s) = -\frac{H_{21}(s)}{H_{22}(s)} = -\frac{1.567}{-0.783} = -2 \tag{5.29}$$

It may be noticed that, in this particular case, the dynamic and static decouplers have the same form. This is because both manipulated variables have equal dynamic effects on each of the controlled outputs (in fact, the first controlled variable is instantaneously affected by both manipulated variables and its dynamics is lacking).

The form of the equivalent transfer functions for the decoupled control loops are:

$$y_1(s) = -\frac{3H_{c1}}{1+3H_{c1}} y_{1sp}(s)$$ (5.30)

$$y_2(s) = -\frac{2.35\ H_{c2}}{10s+1-2.35\ H_{c2}} y_{2sp}(s)$$ (5.31)

5.3 Multivariable centralized control

Multivariable centralized control aims to develop the control law for the action of all manipulated variables of a MIMO system based on the inputs and outputs of the process and considering all process interactions. The emerged control law considers the system as a whole and not divided in decentralized control loops. The way the multivariable controller outputs are computed considers, at each instance of time, the effect of all interacting effects of the manipulated variables on all controlled outputs. The multivariable controller is described by a transfer matrix $H_c(s)$ having a form that, in the general case, is not diagonal.

The most fundamental concept of *perfect control* for a MIMO process, i.e., the design of a controller that should be able to keep the process outputs equal to the desired input values (setpoints) at any moment of time (featuring no time lag or time delay), would imply a controller whose transfer matrix is equal to the inverse of the process transfer matrix. But such a MIMO controller should also operate for removing the effect of the measured and unmeasured disturbances. Due to the action of unmeasured disturbances and due to the process transfer matrix lack of accuracy of representing the real process, the controller should have to rely on feedback information on the outputs deviation from their desired setpoints. Based on feedback, the controller should generate the manipulated variables with the aim of keeping outputs to the desired setpoints. Unfortunately, although it is mathematically possible to design such a perfect controller, its transfer matrix form would not be physically realizable due to the fundamental lack of conforming to the cause-effect principle or equivalently, due to the need of a pure anticipative behavior of the controller. However, it is still possible to design a multivariable controller to have only the physically realizable part of the process inverse transfer matrix (e.g., at least a controller equal to the inverse of the process gain matrix), but the control performance will be degraded accordingly.

The design of such a multivariable controller does not yet have a general, very well defined, and practical methodology (although for the SISO case the internal model control has developed a design framework). Nevertheless, there are some methodologies that design such a multivariable centralized controller having an explicit or an implicit form of the emerged control law.

The most renowned and validated practice implementations of centralized multivariable control are the model predictive control (MPC)-based designs [15]. MPC is founded on the computation of its control algorithm using different forms of the multivariable process model that are used for predicting the future behavior of the controlled variables. Taking into consideration the model-based predictions, a control performance index is optimized with respect to the manipulated variables while considering the future prediction horizon. Based on measured information describing the process outputs, feedback is introduced by the receding horizon approach and by making corrections to the process predictions at every sampling moment. MPC has straightforward means for distributing both the control importance among the controlled variables and the control effort among the manipulated variables. They are performed by means of the weighting matrices within the optimization index. The success of the MPC applications owes not only to its capability of coping with constraints in a systematic way but also to its multivariable centralized approach. The MPC principle is described in Chapter 2 of the book.

References

[1]   Marlin, T.E., *Process Control – Designing Processes and Control Systems for Dynamic Performance*, McGraw-Hill, Inc., 2000.
[2]   Stephanopoulos, G., *Chemical Process Control – An Introduction to Theory and Practice*, Prentice-Hall, Inc., Englewood Cliffs, New Jersey, 1984.
[3]   Agachi, S.P., Cristea M.V., *Basic Process Engineering Control*, Walter de Gruyter GmbH, Berling/Boston, 2014.
[4]   Romagnoli, J.A., Palazoglu, A., *Introduction in Process Control*, CRC Press, Taylor and Francis Group, Boca Raton, FL, 2006.
[5]   Bristol, E., *On a new measure of interaction for multivariable process control*, IEEE Transaction on Automatic Control, AC-11, 133–134, 1996.
[6]   Shinskey, F.G., *Process Control Systems*, McGraw-Hill, New York, 1988.
[7]   McAvoy, T., *Connection between relative gain and control loop stability and design*, AIChE Journal, 27(4), 613–619, 1981.
[8]   McAvoy, T., *Interaction Analysis*, Instrument Society of America, Research Triangle Park, NC, 1983.
[9]   Skogestad, S., Lundstrom, P., Morari M., *Selecting the best distillation control configuration*, AIChE Journal, 36, 753–764, 1990.
[10] Ortega, J.M., *Matrix Theory, A Second Course*, Plenum Press, New York, 1987.
[11] Skogestad, S., Morari M., *Effect of disturbance directions on closed-loop performance*, IEC Research, 26, 2029–2035, 1987.
[12] Grosdidier, P., Morari, M., Holt, B., *Closed loop properties from steady state gain information*, Industrial and Engineering Chemistry Fundamentals, 24, 221–235, 1985.
[13] Chiu, M., Arkun, Y., *Decentralized control structure selection based on integrity considerations*, IEC Research, 29, 369–373, 1990.
[14] Campo, P., Morari, M., *Achievable closed-loop properties of systems under decentralized control: conditions involving steady-state gain*, IEEE Transaction Autonomic Control, 39, 932–943, 1994.
[15] Agachi, P.S., Nagy Z.K., Cristea, V.M., Imre-Lucaci A., *Model Based Control – Case Studies in Process Engineering*, Wiley-VCH, Weinheim, 2006.

6 Plantwide control

6.1 Introduction

Process engineering concepts of automatic control are applied in a wide range of industries such as production of chemicals, pharmaceuticals, petroleum, plastics, pulp and paper, automotive, food, textiles, glass, metals, building materials, electricity, and electric and electronic devices. The challenge of the continuously changing market demands and conditions, associated with the need to spare natural resources, and the requirements to save energy while preventing environmental pollution are asking for safe, flexible, economically efficient, and environmentally friendly solutions provided by automatic control. The control systems represent efficient tools to accomplish such demanding tasks, and as intelligent control solutions are conceived, they may also represent a cost-effective and environmentally friendly way of responding to these challenges.

The generic structure of a comprehensive control system consists in a hierarchical architecture of layers, as presented in Fig. 6.1 [1].

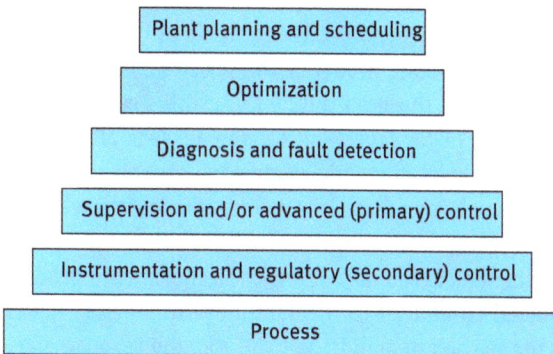

Fig. 6.1: Hierarchical structure of the whole plant control system.

The *instrumentation and regulatory control layer* performs not only the measurement and data acquisition from the plant but also the regulatory control at the most basic level of the control hierarchy. The layer interacts with the higher layers by sending and receiving data. This is the core of a distributed control system (DCS). The control systems at this level usually have single loops, with typically PID structure, and are intended to control the secondary variables of the process such as inventory, level, pressure, flow, and temperature. Its modularity allows the quick and prompt intervention for maintaining the secondary variables at their desired setpoints despite the action of disturbances.

https://doi.org/10.1515/9783110789737-007

The *supervision and/or advanced layer* computes the reference values for the regulatory layer or directly controls process variables on the basis of a multivariable model of the process. Primary controlled variables may be identified at this level, and they are subject to control, in association with or by the means of the regulatory layer. Besides fulfillment of control of performance-based objectives, at this layer, economic objective functions might be considered for the computation of the reference values that are further sent to the underlying regulatory layer. Model predictive control is the most applied advanced control algorithm for this layer, although it may be used in other layers, too.

The *diagnosis and fault detection layer* aims to make the operation of the process, as much as possible, independent of the faults and hence enhances the safe and reliable work of the subsystems belonging to all other layers. Models of the process may be also used for the identification of the causes producing malfunction of either process or instrumentation equipment, to offer early information and to manually or automatically mitigate their future negative consequences. This layer is not always present in the comprehensive control hierarchy, but its beneficial outcomes are highly appreciated.

The *optimization layer* uses a model of the process to compute the best manipulated variables or references for the supervisory/advanced lower layer. The computation is performed by optimization of an economic objective function or of the process throughput. Usually, at this layer, a steady-state model of the process is involved in the computation of the optimal values for the manipulated variables representing references for the supervisory/advanced lower layer and for the degrees of freedom remained from the lower levels. At this level, dynamic models may be also used for optimization, but their sampling time is larger than for the ones associated with supervisory/ advanced and regulatory layers.

The objective of the *plant planning and scheduling layer* is to generate the production planning for an extended period compared to the sampling time of the lower layers. It takes into consideration the market demands, sets the way the resources are allocated, and schedules the plan of operation. The economic objective type is dominating at this level and the decisions are mainly made for maximizing the profit. Optimization algorithms can be also used at this level to find the optimal solutions to be sent as targets to the lower layer.

6.2 Premises of plantwide control

Plantwide control consists of the control system aimed to operate the whole plant with the task of achieving the desired objectives, despite the action of disturbances and conforming to imposed constraints. Plantwide control may also imply the specification and placement of the measured variables, specification of the manipulated variables, and the decomposition of the control task into smaller problems to be solved based on the overall control approach [2].

The systematic design of the plantwide control system is becoming more and more important due to the increasing complexity of the processes to be controlled and the ever-growing economic, energetic, and environmental constraints to be fulfilled. Despite the stress exerted by these diving forces, the plantwide control in a large number of industrial applications is very much based on heuristic approaches, making use of experts' knowledge. The research performed during the last two decades in this field has investigated several approaches [3]. They are considering either heuristic- or optimization-based methods and struggle to propose an algorithmic approach for the selection of the most favorable plantwide control structure [4–9].

Traditional plantwide control practice starts from control solutions specific to different unit operations intended to make the plant operation reliable, while getting good understanding and acceptance by the operators [10]. This approach usually begins with fixing the production throughput at the plant input and is followed by the sequential design of the control system along the unit operations contributing to the successive processing stages. But this approach does not guarantee the attainment of the optimal control structure or the best operation of the plantwide control system. Nevertheless, some systematic approaches have been proposed during the last decades, and they have been successfully tested by simulation and industrial applications [3, 8].

It is obvious that the fundamental objective of any plantwide control system is to maintain the mass, energy, and momentum balance of the whole process [3]. Generally, it may be considered that momentum balance can be obtained with moderate effort, as flow and pressure measurements are available and control valves, pumps, and compressor are efficient manipulated variables. Energy balance may also be achieved by the support of the low-cost temperature measurements and by the manipulation of the cooling or heating utilities. The most difficult task is to maintain the balance of the material entering and leaving the whole system. This not only considers the overall mass balance but also the mass balance for almost every component. As concentration measurements are usually not available, due to cost and complexity of the instruments and as the independent manipulation of each chemical species flow is also problematic, the mission of preserving the mass balance of the components is the most challenging task. This balance should prevent the accumulation or consumption of components in the process and becomes an important objective for the control system. Such duty of the control system is aggravated by the existence of the recycle streams that may negatively affect the process controllability, and in the extreme case, losing stability. It is also well recognized that for processes containing reduced inventory units, such as buffer tanks, the sequential plantwide design of the control system methodology based on unit operations may show poor control performance.

It is noteworthy to mention that plantwide control design may have to meet two different requirements. The first one is specific to the case of the newly designed process and has the objective of maximizing the efficiency of the operation for the nominal throughput of the plant. This situation may also be encountered in cases of upgrading the old control system. This case is driven by the economic target of minimizing the utility, energy, raw material, and environment protection costs.

The second requirement is characterized by the objective of maximizing the throughput of the process. It may be driven by the market conditions of favorable prices for the products and asks for obtaining profit by maximizing the throughput. The maximized throughput approach implies the appearance of a bottleneck on the process production flow. The localization of the throughput manipulator may affect the structure of the control system. It may be also concluded that control system design must be performed such that it will show a robust response to most of the possible modifications in the operating conditions.

Operators' acceptance of the plantwide control strategy is an aspect worthy to consider, as very complex control systems require trained personnel to operate in the challenging industrial environment. Changing the traditional way of operation to the plantwide control solutions implies the formation of qualified human resources by special operators training programs and the use of operating training systems.

6.3 Designing the plantwide control strategy

It is obvious that plantwide control design should be directly related to the process design. Traditionally, the control design was performed after the process design was completed. The result of this sequencing may affect the controllability of the process and reduce the achievable control performance. One major problem of this historical approach for the control system design may come out from the fact that process design is performed almost exclusively on the basis of the steady-state description of the process. Unfortunately, this does not reveal possible problems arising during transient operating regimes. Specifically, the process design implies the setting of the nominal steady-state operating point that might be not attainable in a simple manner during the dynamic operating mode. Additionally, the effect of the disturbances, of new imposed constraints, or of process load changes may affect the controllability of the process and the efficiency of the control system design may be degraded.

As a result of these circumstances, it becomes obvious that both process and control system design should be done by an integrated approach. This implies a joint design approach that relies on merging the process knowledge with the control theory competency and possibly associated with understanding economic indicators and their assessment methods.

The plantwide control system design, in its most favorable theoretical approach, should have the capability of bringing the process operation at the optimum operating point (or in its neighborhood). The optimal control, for a centralized multivariable approach, may emerge from running plantwide real-time optimization to obtain the best control solution and by using comprehensive models [11].

Meanwhile, the plantwide control system could be designed to make it simple, possibly with single-loop control structure, and have the benefit of the self-regulating or self-optimizing property of the process subsystems that facilitates the accomplishment of the optimal operation in a natural way.

Although a comprehensive solution to the plantwide control design problem is not yet maturated, one systematic approach for designing the control system is proposed in [4, 8]. This plantwide control design approach is considered to provide answers to the two main groups of design tasks: design of the regulatory control layer and design of the economics-driven control layer.

The first task objective is to identify the suitable secondary controlled variables and their associated control variables (pairing), including control of the mass and energy inventories. The proposed design of the regulatory control layer is performed such that the remaining degrees of freedom (primary control variables) are straightforwardly used for the design of the plantwide economic optimization strategy.

The second design task objective aims to integrate the economic optimization outcomes in the plantwide control design while identifying the primary controlled variables. This part of the design is centered on economic objectives, with their associated constraints (such as the environmental or safety requirements), and is based on steady-state models. The main challenges for the economic optimization integration in the plantwide control system design are the specification of the primary controlled variables and the best localization of the throughput manipulator.

The control structure design methodology consists of two parts with seven steps [4]. The *top-down part* deals with economic optimization group of tasks and the *bottom-up part* with the regulatory group of tasks, to which are added the supervision and the dynamic optimization steps.

The *top-down part* of the control system design consists of four steps. They are [4]:

1. Defining the economic objectives and their associated constraints. Based on the steady-state models of the process an optimization economic index I is defined. This economic optimization objective consists in either maximizing the profit or maximizing the production rate (throughput). Typical disturbances are included in the steady models such as feed flows and composition, kinetic parameters, or potential changes of market parameters.

2. Selecting the set of control variables (degrees of freedom) and solving the economic optimization problem. Optimization is done with respect to the control variables, u, and in the presence of typical disturbances, d. Steps 1 and 2 analyze the optimal operation of the process and are prerequisites for the selection of the most important variables to control or to maintain constant, from an economic point of view.

3. Choosing the primary controlled variables. This is the most difficult part of the top-down part, and it has to specify which variables should be controlled to get an optimal (or at least close-to-optimal) economic operation. The most fundamental method of selecting the primary controlled variables is to choose them equal to the gradient of the optimization index function, i.e., $I_u = dI/du$. This approach is based on the observation that by keeping $I_u = 0$, independent of disturbances, it fulfills the economic optimization criterion (as necessary condition for optimality). However, it may be difficult to develop an analytical expression for I or to have available process measurements for all variables of I_u. The selection of the primary controlled variables may be done based on process insight. Two categories of variables are important to be considered as good candidates for primary controlled variables. They are the so-called *active constraints variables* and the *self-optimizing variables*.

The *active constraints variables* are characterized by the fact that the optimum is very much dependent on these variables and therefore they should be strictly maintained at desired setpoints. A small deviation of the active constraints variables from their optimal values has large effect on the deviation from optimality. Their desired values are usually maximum or minimum values of process variables such as composition, temperature, and flow. Very good control performance for the control of active constraints variables should be accomplished.

The *self-optimizing variables* are characterized by the fact that when they are maintained constant, they drive the process close to the optimal operation, despite the action of the disturbances. Small deviations of the self-optimizing variables from their optimal values do not have large effect on the deviation from optimality. Therefore, in most cases, a very good control performance for the control of self-optimizing variables is not necessary. To be chosen as primary controlled variables, the self-optimizing variables must be measurable variables or at least to have the possibility of being inferred from other measured variables [7]. The self-optimizing variables must be considered in relation to remaining unconstrained degrees of freedom (of the regulatory or economic optimization layers).

Besides the physical insight approach, a set of four systematic methods have been developed for discovering appropriate self-optimizing variables [4]:

(a) The first method assesses different sets of candidates for the controlled (self-optimizing) variables, each of them considered to be maintained at constant values. An evaluation is made for each candidate set with respect to the deviation of the optimization index function, computed with the constant values of the self-optimizing variables, from the optimal value of the performance function, in the presence of disturbances. The set presenting the most reduced deviation is selected.

(b) Another method selects the self-optimizing variables to be equal to the gradient of the optimization index with respect to the degrees of freedom (as already mentioned).

(c) If the previous method is not feasible due to lack of available measurements of the variables involved in I_u, a null space method may be employed to find optimal measurement combinations to be used in I_u [4, 12, 13].

(d) The last method selects the self-optimizing variables on the basis of the variables presenting large-scale gain, computed as the ratio between the process gain and the span of the controlled variable [4, 14].

4. Setting the location of the throughput manipulator. Setting the production rate and the position of the throughput manipulator has direct impact on the maximization of the economic objective. At the same time, the structure of the inventory control loops and their pairing are affected by the throughput manipulator placement. Although the traditional location of the throughput manipulator is set at the inlet feed position, placing it in other positions may bring benefits, such as the location at the bottleneck unit of the process.

The *bottom-up part* of the control system design consists of the following three steps [4]:

5. Selection of the *secondary controlled variables* used in the regulatory control layer, whose intended task is to stabilize the process. Pairing of the secondary controlled variables with control (manipulated) variables should be also accomplished at this step. Inventory control will be addressed, too. The main task of the regulatory layer is to reduce the deviation (by providing short transient time) of the secondary controlled variables from their associated setpoints.

A systematic procedure for the selection of the secondary controlled variables is presented in [15]. This approach involves three stages in the design, aiming to select the secondary controlled variables in such a way as to provide efficient indirect control of the primary controlled variables, considering both the servo and the regulatory control performance. The design approach is based on minimizing the scaled integral absolute error (IAE) of the primary controlled variables. IAE is used as a measure for indicating the most favorable secondary controlled variables selection intended to indirectly control the primary controlled variables. The three stages of the selection methodology are [15]:

(a) The first stage generates an initial candidate set of secondary controlled variables, identifies disturbances, and formulates a subset selection constraint to find if a control variable may be used for control of a candidate-controlled variable. Additionally, the input to secondary and primary controlled output models and disturbance models are developed during this stage. The IAEs of the cascaded primary control loops are computed for all possible pairings of the triplet controlled-secondary controlled-primary controlled variables. IAEs are scaled to account for the economic importance of the primary controlled variable. The generation of the candidate set of secondary controlled variables can be done either by considering a larger set of candidate variables or by reducing the set using process insight. The selection of relevant disturbances can be done only on process insight.

The subset selection constraint formulates a matrix of only logical 1 and logical 0 values that shows if the pairing of each candidate control variable with each controlled variable is acceptable from the servo and regulatory performance. The subset selection constraint is then used for pruning (of super nodes) in the branch and bound algorithm.

(b) The second stage consists in the selection of the set of secondary controlled variables based on the summed and scaled IAEs of the primary controlled variables, considering a cascaded structure where the secondary controlled variables are used as control variables for the control of the primary ones. Constrained minimization of the summed and scaled IAEs of the primary control loops is used for the selection of the optimal set of secondary controlled variables. The constraints of this mixed integer optimization problem take into account the interactions between control loops, at the regulatory and supervisory control layers, and the servo control performance at the primary control loop. RGAs are computed for interaction assessment for both the regulatory control layer and for the supervisory control layer [16, 17].

(c) The third stage of the methodology consists in the performance evaluation of the control system that emerges from the selection of the secondary controlled variables in case of off-design operating conditions. As the design is based on linear models and nominal operating conditions, the evaluation of the secondary controlled variables control performance has to be tested on the nonlinear model or the process. Particularly, gain switching may occur due to operation changes in the process and they should be avoided for the selected pairing. The final selection of the secondary controlled variables is made based on this last stage.

6. Design of the supervisory control layer. Economic considerations may be considered at this control layer, such as whether good control performance at this level can comply with economic objective targets. The design must decide which of the controlled variables of the secondary control layer have to be controlled by the supervisory layer by means of imposed setpoints. The control variables of the supervisory control layer are usually the secondary controlled variables of the regulatory layer. At this control layer, both decentralized and centralized control may be used.

7. Design of the optimization layer. At this control layer the design is dominated by economic objectives accomplishment. It is based on a centralized control approach and it uses steady-state models or dynamic models with large sampling time, compared to the one of the subordinated layers.

Plantwide control design may be needed both for new plants and for upgrading older ones. The latter is driven by the economic context of the market, desired specifications of the products, and the constraints on the production feed.

References

[1] Zhu, Y., *Multivariable System Identification for Process Control*, Pergamon, Elsevier, Oxford, UK, 2001.
[2] Dimian, A.C., Bildea, S., Kiss, A., *Integrated Design and Simulation of Chemical Processes*, in *Computer Aided Chemical Engineering*, Vol. 35, 2nd Edition, Elsevier, Amsterdam, 2014.
[3] Larsson, T., Skogestad, S., *Plantwide control: A review and a new design procedure. Modeling, Identification and Control*, 21, 209–240, 2000.
[4] Downs, J.J., Skogestad, S., *An industrial and academic perspective on plantwide control*, Annual Reviews in Control, 35, 99–110, 2011.
[5] Kookos, I.K., Perkins, J.D., *An algorithmic method for the selection of multivariable process control structures*, Journal of Process Control, 12, 85–99, 2002.
[6] Luyben, W.L., Tyreus, B.D., Luyben, M.L., *Plantwide process control*, McGraw-Hill, New York, 1998.
[7] Skogestad, S., *Plantwide control: The search for the self-optimizing control structure*, Journal of Process Control, 10, 487–507, 2000.
[8] Skogestad, S., *Control structure design for complete chemical plants*, Computers and Chemical Engineering, 28(1–2), 219–234, 2004.
[9] Zheng, A., Mahajanam, R.V., Douglas, J.M., *Hierarchical procedure for plantwide control system synthesis*, AIChE Journal, 45(6), 1255–1265, 1999.
[10] Agachi, S.P., Cristea, M.V., *Basic Process Engineering Control*, Walter de Gruyter, Berlin/Boston, 2014.
[11] Agachi, P.S., Nagy, Z.K., Cristea, V.M., Imre-Lucaci, A., *Model Based Control – Case Studies in Process Engineering*, Wiley-VCH Verlag, Weinheim, 2006.
[12] Alstad, V., Skogestad, S., *Null space method for selecting optimal measurement combinations as controlled variables*, Industrial and Engineering Chemistry Research, 46(3), 846–853, 2007.
[13] Alstad, V., Skogestad, S., Hori, E.S., *Optimal measurement combinations as controlled variables*, Journal of Process Control, 19, 138–148, 2009.
[14] Halvorsen, I.J., Skogestad, S., Morud, J.C., Alstad, V., *Optimal selection of controlled variables*, Industrial and Engineering Chemistry Research, 42(14), 3273–3284, 2003.
[15] Dustin Jones, D., Bhattacharyyaa, D., Turtona, R., Zitney, S.E., *Plant-wide control system design: Secondary controlled variable selection*, Computers and Chemical Engineering, 71, 253–262, 2014.
[16] Marlin, T.E., *Process Control – Designing Processes and Control Systems for Dynamic Performance*, McGraw-Hill, New York, 2000.
[17] Stephanopoulos, G., *Chemical Process Control – An Introduction to Theory and Practice*, Prentice-Hall, Englewood Cliffs, NJ, 1984.

7 Linear discrete systems and the Z transform

7.1 Introduction

Continuous time systems deal with systems having continuous (with respect to time) input $u(t)$, output $y(t)$, and state $x(t)$ variables, and having mathematical representations based on functions with time as the independent variable [1]. These independent variables have a continuous set of values for time, $t \in R$. Unlike continuous systems, discrete systems (also named sampled systems) have the input $u(n \cdot T)$, output $y(n \cdot T)$, and state $x(n \cdot T)$ variables represented by discrete functions with respect to time, functions that are only defined at particular moments of time, $t \in Z$. These time moments are usually chosen equally spaced, $t = n \cdot T$, $n = 0, 1, 2, \ldots$, and, consequently, are multiples of the time interval T, named the *sampling time*. A discrete time function $f(t)$ is defined as [1]

$$f(t) = \begin{cases} \text{exist}, & \forall\, t = nT, \quad n \in N \\ 0 \quad \text{or} \quad \text{undefined}, & \forall\, t \neq nT \end{cases} \tag{7.1}$$

Although the need for describing discrete systems appeared before the spectacular development of the digital control systems using the computer, it may be stated that the perspective of these applications represented the main motivation for the progress of the discrete system theory. The explanation is that every digital system is working on the basis of a sequential mode of operation. This sequential mode puts in a successive arrangement, all operations that have to be performed [2–4]. Thus, a digital control system that is based on a computing system is not able to generate a control action at each moment of time. The arithmetic and logic unit of the computer has to perform several operations that may not be simultaneous, such as reading data from the process, analogue-to-digital (A/D) conversion, processing the acquired data on the basis of a computation algorithm, generation of the control variable, and digital-to-analogue conversion (D/A) [3]. A digital control system using a computer is presented in Fig. 7.1 [1].

The output of the process is a continuous signal. This continuous signal is transformed in a digital form using the A/D converter. The A/D conversion is performed at discrete moments of time $t = n \cdot T$, $n = 0, 1, 2, \ldots$, also named sampling moments. The clock of the system sets the sampling moments. The sampling is generally defined as the operation of extracting a small part, named sample, from the whole. In the context of the computer control, sampling is the transformation of a continuous time signal into a succession of numeric values, representing the values of the continuous signal at the time moments $y(n \cdot T)$. The sampled signals are discrete signals obtained from the continuous signals by the sampling operation. The succession of numeric values obtained after the sampling operation is processed by the computing algorithm that generates a new succession of numeric values, $u(n \cdot T)$, representing the control action (commands). These commands are determined only at discrete moments of time and

https://doi.org/10.1515/9783110789737-008

sent to the D/A converter. The D/A converter transforms the succession of the commands' values into a continuous signal $u(t)$ that is further sent to the controlled process. Usually, this transformation into the continuous signal consists in keeping the
computed value at a sample moment, constant in time, until the next one. All operations, A/D conversion, computation of the command, and D/A conversion, happen at
discrete and successive moments of time, marked by the clock of the computing system. Consequently, the computing system operates with discrete signals, with values
changing only at the sampling moments of time.

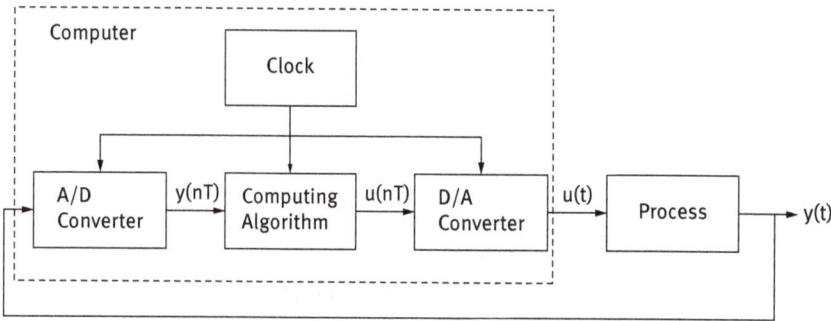

Fig. 7.1: Schematic representation of the process controlled by a computer system.

In the control system using the computer, both continuous and discrete signals are
present, leading to the following consequences:
-   before being "read" by the computer, it is necessary to transform the continuous
    signals $y(t)$ into discrete signals $y(n \cdot T)$,
-   in order to have an efficient action upon the process, it is necessary to transform
    the discrete signals produced by the computer $u(n \cdot T)$ into continuous signals $u(t)$,
-   the mathematical instruments used for describing the behavior of the continuous
    linear systems (differential equations or transfer functions) are not convenient
    for describing the dynamic behavior of the control system based on the computer.
    Therefore, it is necessary to find specific ways for describing the discrete systems.

Beside the computer-based systems, there are also other systems being implicitly discrete. Radar localization is an example of such a system that processes information
on the spatial position of an object only after the electromagnetic reflected wave is
again generated and processed (after a full rotation of the electromagnetic wave
beam). As a result, it produces a signal with new values only at equally spaced and
discrete intervals of time. An important category of analytical instruments, such as
the mass spectrometers or the gas chromatographs, are implicit discrete systems because they work sequentially. Usually, these instruments require disjoint (successive)
time intervals for extracting the sample, performing the analysis, and preparing the
new analysis [1].

7.2 Discrete systems described by an input-output relationship

7.2.1 Sampling the continuous signals

Consider the way that a continuous signal may be transformed into a discrete signal, an operation commonly named as sampling the continuous signal [1, 4, 5, 16]. Consider a continuous signal $y(t)$ transmitted over a transmission line that is interrupted by a switch, named sampling element (sampler), as schematically presented in Fig. 7.2.

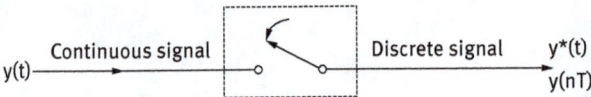

Fig. 7.2: Basic representation of the sampling element.

The sampling element (switch) is closing every T seconds and remains closed for a very short (approaching zero) time interval.

Figure 7.3(a) presents the continuous signal $y(t)$ and Fig. 7.3(b) the discrete signal obtained after the sampling operation, denoted by $y^*(t)$.

The sampled signal consists of a series of (point-wise defined) values that are equal to the values of the continuous signal at each moment of time, the latter being a multiple of the sampling period $t = n \cdot T$, $n = 0, 1, 2, \ldots$. Figure 7.3(c) and (d) presents the signals obtained after sampling the same continuous signal with double and triple sampling period (compared with the reference sampling period, presented in Fig. 7.3(b)).

On this basis, the following remarks may be formulated [1]:

i) As the sampling period gets smaller and approaches zero $T \to 0$, the sampled signal y^* approaches the continuous signal $y(t)$, but this situation requires a very large number of sampled values.

ii) As the sampling time gets larger, less sampled values are produced but the sampled signal deteriorates, meaning that the reconstruction of the original signal from the discrete values may become imperfect or even impossible.

Therefore, there is a need to analyze the conditions for which the reconstruction of continuous signals is possible from the sampled values and the way the sampling period is chosen. To answer these questions, the following particular cases are considered.

Consider the case of a continuous first-order system having a step response presented in Fig. 7.4 [1].

Analyzing the response, it may be stated that the sampling operation has to be performed with a sampling period smaller than the time constant of the system (process), $T < T_p$, to capture the dynamics of this signal. Practical considerations suggest the selection of the sampling time between $T = 0.1 \cdot T_p$ and $T = 0.2 \cdot T_p$ to obtain sampled values

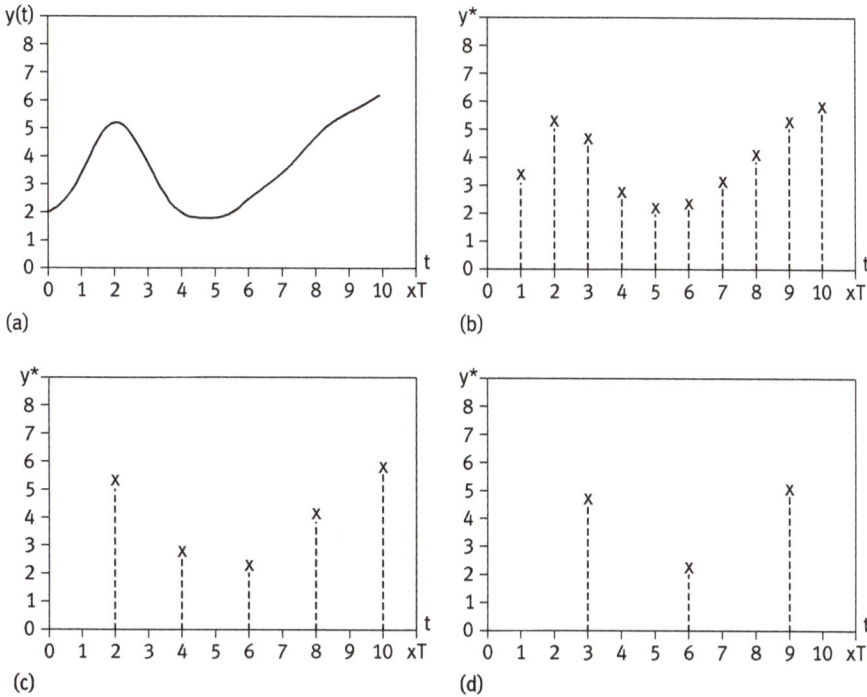

Fig. 7.3: Representation of (a) continuous signal, (b) sampled signal with the reference sampling period, (c) sampled signal with double sampling period, (d) sampled signal with triple sampling period.

that are able to reflect, in a convenient way, the output variable dynamics of the first-order system.

For the case of the first-order system with dead time τ_m, having the step response presented in Fig. 7.5, the selection of the sampling time is performed as follows [1]:

– if the dead time is of the same order of magnitude as the time constant of the process $\tau_m \approx T_p$, the sampling time has to be about one-tenth of the time constant $T = 0.1 \cdot T_p$ or of the dead time $T = 0.1 \cdot \tau_m$, whichever is smaller,

– if the dead time is much larger than the time constant of the system (process) $\tau_m \gg T_p$, the sampling time is chosen to be about one-tenth of the dead time value, $T = 0.1 \cdot \tau_m$,

– if the dead time is much smaller than the time constant of the system $\tau_m \ll T_p$, the sampling time is chosen to be about one-tenth of the time constant of the system $T = 0.1 \cdot T_p$.

For the case of the overdamped second-order system, the selection of the sampling period may be reduced to the case of the equivalent first-order system having an equivalent time constant T_{pe} and an equivalent dead time τ_{me}, as presented in Fig. 7.6 [1, 6, 15].

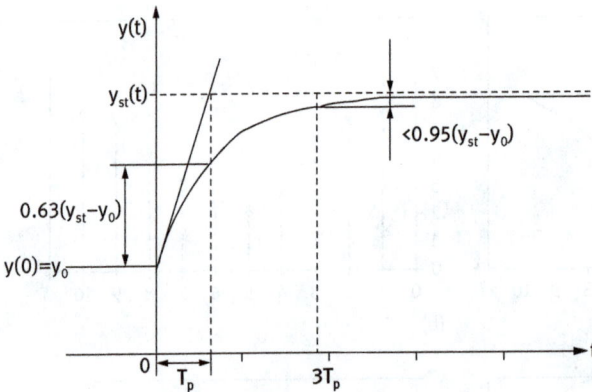

Fig. 7.4: Step response of the first-order system.

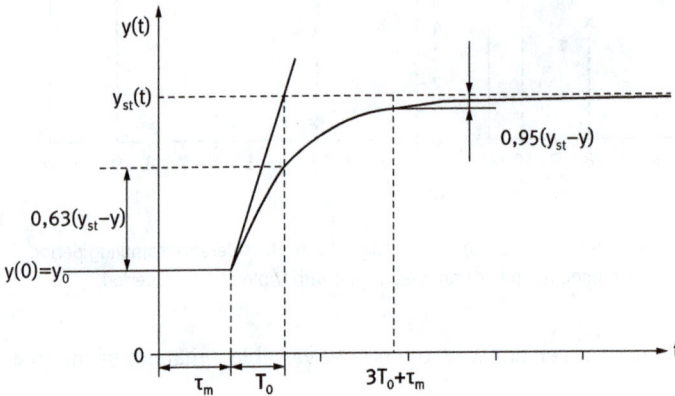

Fig. 7.5: Step response of the first-order system with dead time.

The equivalence is accomplished graphically by drawing the tangent line, at the inflexion point I, to the step-response plot of the second-order system and obtaining the segments $OM = \tau_{me}$ and $MN = T_{pe}$. These segments are determined by the projections M and N, on the abscissa axis, of the tangent line intersection points A and B with the asymptotes of the initial and final steady-state values. Fig. 7.6(b) presents the step response of the first-order system with dead time equivalent to the step response of the second-order system presented in Fig. 7.6(a).

For the case of the underdamped second-order system or for the case of the linear systems subject to the sinusoidal-shaped input signal, the sampling period is chosen to be less than the half of the period corresponding to the sine signal. The plots presented in Fig. 7.7 suggest this choice [1, 4].

From Fig. 7.7, it may be noticed that for the case of the sampling operation performed with a sampling period equal to the period of the sinusoidal signal $T = 2\pi/\omega$,

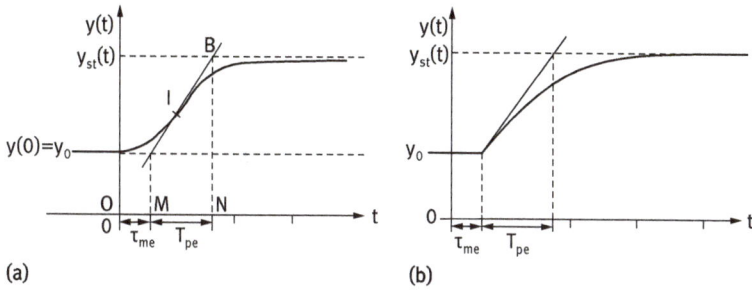

(a)

(b)

Fig. 7.6: (a) Step response of the second-order system. (b) Equivalent (first order with dead time) step response of the second-order system.

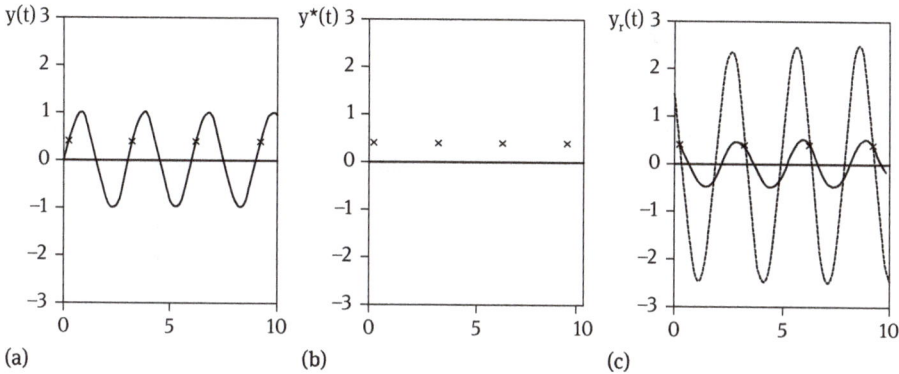

(a)

(b)

(c)

Fig. 7.7: (a) Sinusoidal signal subject to the sampling operation, with the sampling period T equal to the period of the signal, (b) Values of the sampled signal $y^*(t)$, (c) Reconstructed signals from the sampled values.

the sampled signal $y^*(t)$ from Fig. 7.7(b) is obtained. This signal has values that do not reflect the shape of the continuous sinusoidal signal $y(t)$ that was subject to the sampling operation. If the reconstruction of the continuous signal is desired from the values of the sampled signal, different reconstructed signals $y_r(t)$ may be obtained, as presented in Fig. 7.7(c). Obviously, they are different than the original signal subjected to the sampling operation. It is concluded that the sampling period has to be chosen less than or equal to the period of the sinusoidal signal.

Figure 7.8 presents the case of the sampling with a period less than (with a ¾ factor) the sinusoidal signal period $T = 0.75(2\pi/\omega)$. For this case again, the reconstructed signal $y_r(t)$ from the values of the sampled is not satisfying the desired quality, as presented in Fig. 7.8(c).

The theorem of Shannon offers an exact quantitative measure for choosing the sampling period. It is based on the fact that a periodic signal may be represented as an infinite sum of periodic sinusoidal component functions (Fourier series) [1, 4].

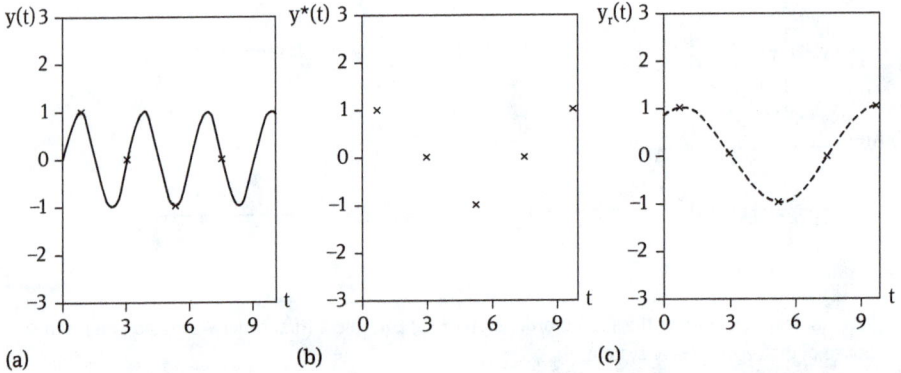

Fig. 7.8: (a) Sinusoidal signal sampled with the sampling period equal to three-quarters of the period of the continuous signal. (b) Values of the sampled signal $y^*(t)$. (c) Reconstructed signal from the sampled values.

Theorem of Shannon: The sampling period of a continuous signal has to be chosen less than or equal to half of the period corresponding to the highest frequency ω_{max} contained in the continuous signal, i.e., $T \leq \pi/\omega_{max}$.

In the following, a way for quantitatively describing the sampling operation will be investigated using a mathematically coherent model [1]. To accomplish this task, the starting point is the sampling element represented as a switch situated in the closed position for a very short time interval Δt, around the sampling moments $t = n \cdot T$, $n = 0$, 1, 2, ... The signal obtained as a result of this *physical sampling operation*, for a generic sampling moment $t = n \cdot T$, is an impulse of the form presented in Fig. 7.9(a). If the real impulse signal is considered as having ideal (sharp) flanks, the form of the impulse in the Fig. 7.9(b) is obtained.

To develop a compact mathematical description, it is assumed that the ideal sampling element acts instantly, being in the closed position only for a very short time interval $\Delta t \rightarrow 0$. To keep the same "power" (area) of the impulse when $\Delta t \rightarrow 0$, it is necessary that the height of the impulse becomes infinite. This impulse having a duration approaching zero and a height approaching infinity has an *area equal to the amplitude of the continuous signal* at the corresponding sampling moment. But this impulse, having a duration approaching zero, an infinite amplitude, and a finite area of the impulse is the Dirac function. The representation of this Dirac impulse, for the generic time moment $t = n \cdot T$, is presented in Fig. 7.9(c) [1, 4].

The mathematical form of the impulse presented in Fig. 7.9(c) is

$$y^*(n \cdot T) = y(n \cdot T) \cdot \delta(t - n \cdot T), \tag{7.2}$$

where the notation $\delta(t\text{-}n \cdot T)$ has been used for the shifted Dirac impulse at the moment $t = n \cdot T$.

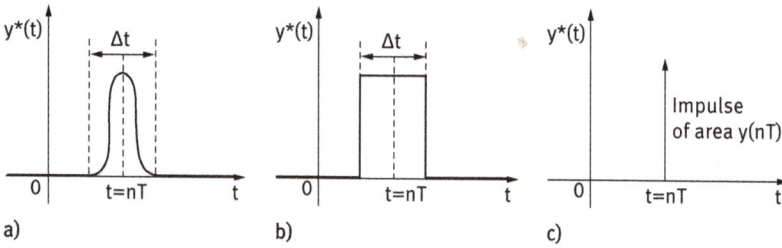

Fig. 7.9: (a) Real impulse signal (b) Real impulse signal having ideal flanks (c) Ideal impulse signal.

Consequently, the *ideal sampling element* provides a series of Dirac impulses at the sampling moments $t = n \cdot T$, $n = 0, 1, 2, \ldots$, with each Dirac impulse having an area equal to the value of the continuous signal at the corresponding moment of time. Therefore, the sampled signal $y^*(t)$ may be mathematically represented as a sum of Dirac impulses weighted by the value of the continuous signal from the sampling moments:

$$y^*(T) = y^*(0) + y^*(T) + y^*(2T) + \cdots =$$
$$= y(0)\ \delta(t) + y(T)\ \delta(t - T) + y(2T)\ \delta(t - 2T) + \cdots \qquad (7.3)$$

or

$$y^*(t) = \sum_{n=0}^{\infty} y(nT)\ \delta(t - nT) \qquad (7.4)$$

This ideal way of representing the sampling signal, obtained from a continuous signal, is based on the assumption that at each sampling moment, the area ("power") of the impulse is equal to the amplitude of the continuous signal. The notation "*", in super-script, is usually used for denoting the sampled signal. Between the sampling moments, the value ("power") of the sampled signal is equal to zero.

For the series of discrete values, the discrete signal (sampled signal) preserves the character of the *function with respect to time* (with continuous argument). Following this fact, the Laplace transform may be applied for both members of eq. (7.4). The Laplace transform of the sampled signal $Y^*(s)$ may be described by [1, 4]:

$$Y^*(s) = L(y^*(t)) = L\left(\sum_{n=0}^{\infty} y(nT)\delta(t - nT) \right)$$
$$= \sum_{n=0}^{\infty} y(nT)L(\delta(t - nT)) = \sum_{n=0}^{\infty} y(nT)e^{-nTs}L(\delta(t))$$
$$= \sum_{n=0}^{\infty} y(nT)e^{-nTs} \qquad (7.5)$$

7.2.2 Reconstruction of the continuous signals from their discrete values

The reconstruction of the continuous signals from the discrete signals is important because the large majority of the real processes are continuous and their input variables are also continuous [1]. For the case of the process controlled by the computer, the commands generated by the computer are provided periodically and have discrete-time values. They are represented by a succession of nonzero values just for certain moments of time, marked by the clock of the computer system (between the sampling moments, the values are equal to zero or, more generally, indefinite). This succession of impulses is not able to act efficiently in the process because such command would make the control valve to open for a very short time interval and then to close until the next sampling moment, in a periodic operating mode. It is therefore necessary to reconstruct a continuous signal from the discrete time values [4, 5].

Consider a discrete time signal produced by a discrete system (e.g., by a computer) consisting of a succession of values represented by a series of impulses:

$$u^*(0) = u(0)\ \delta(t),\ \ u^*(T) = u(T)\ \delta(t-T), u^*(2T) = u(2T)\ \delta(t-2T),\ \ldots \tag{7.6}$$

This discrete signal is represented in Fig. 7.10(a) [1]:

The simplest way for reconstructing the continuous signal is to maintain constant the discrete value from a sampling moment up to the next sampling moment:

$$u(t) = u(nT)\ \ \text{for}\ \ nT \le t < (n+1)\ T,\ n = 0,\ 1,\ 2,\ \ldots \tag{7.7}$$

This way of reconstructing the continuous signal, presented in Fig. 7.10(b), may be assigned to a system that transforms the discrete values in continuous variables, according to eq. (7.7). This system is denoted as the *zero-order hold* (ZOH) element.

There are also other ways of reconstructing the continuous signal from the discrete values. If two successive discrete values are considered, for example, $u[(n-1)T]$ and $u(nT)$, a linear extrapolation based on these two values may be considered to determine the values of the signal over the next sampling interval. The following continuous signal is obtained:

$$u(t) = u(nT) + \frac{u(nT) - u[(n-1)T]}{T}(t - nT) \tag{7.8}$$
$$\text{for}\ nT \le t < (n+1)T,\ \ n = 1, 2, \ldots$$

This way of reconstructing the continuous signal, presented in Fig. 7.10(c), may be assigned to a system that transforms the discrete values in continuous variables, according to eq. (7.8). This system is denoted as the *first-order hold* (FOH) element.

The mathematical basis for determining the equation for the hold element, irrespective of its order, is the Taylor series development of the function $u(t)$ [1, 2, 4]. The development is performed around the discrete value $u(nT)$:

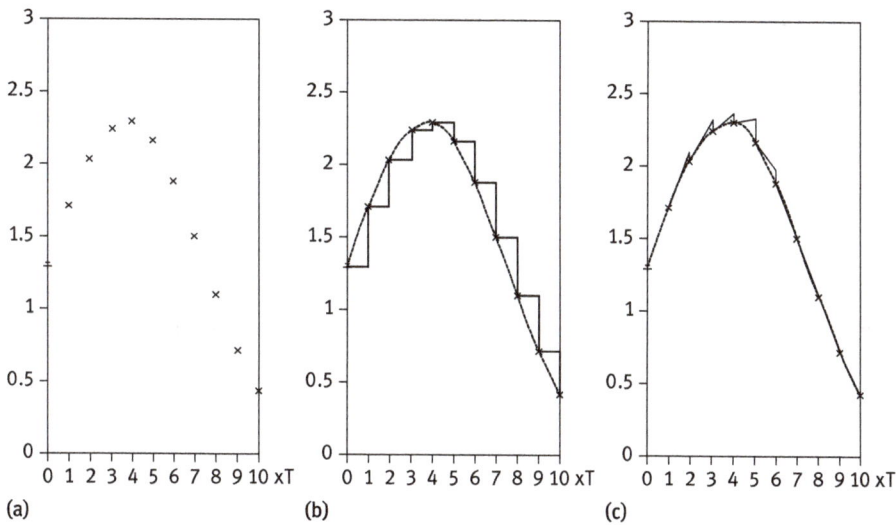

(a)   (b)   (c)

Fig. 7.10: (a) Discrete signal. (b) Reconstructed signal using the zero-order hold (ZOH) element.
(c) Reconstructed signal using the first-order hold (FOH) element.

$$u(t) = u(nT) + \left(\frac{du}{dt}\right)_{t=nT}(t-nT) + \frac{1}{2!}\left(\frac{d^2u}{dt^2}\right)_{t=nT}(t-nT)^2 + \cdots + \tag{7.9}$$

If only the first term from the Taylor series is taken into consideration, the zero-order hold element is obtained. If two terms of the Taylor series are considered, the first-order hold element is obtained.

The first-order hold element needs at least two values for beginning the reconstruction of the continuous signal, but the zero-order hold element needs just one value for the reconstruction. The second-, third-, and higher-order hold elements may be developed for reconstructing the continuous signal, but they need three, four, or more values for beginning it. As the order of the hold element increases, the computation effort grows, bringing poor improvement in the precision of the continuous signal reconstruction. If the sampling period is small, the differences between the hold elements are relatively insignificant. These are the reasons for using, with very good results, the zero-order hold element in practical applications. Increasing the precision of the reconstructed signal is merely performed by decreasing the sampling period than by increasing the order of the hold element.

In general, it may be considered that when the sampling period is too large, the reconstruction of the continuous signal from its discrete values with a first-order element is better for the case of slow varying signals. The reconstruction of the continuous signal from its discrete values is unsatisfactory, both with a first- or zero-order hold element, for rapid varying signals. For the case of rapid varying signals, the reconstruction of the continuous signal using the first-hold element may lead to peaks and may determine excessive large values of the command.

The output of the zero-order hold element may be considered as a succession of rectangular impulses having the duration equal to the sampling period. Each of these impulses may be considered as being formed from the difference of two-step signals shifted with a sampling period of time:

$$u(t) = u(nT)[u_0(t - nT) - u_0(t - (n+1)T)].$$ (7.10)

where $u_0(t)$ is the unit step signal. This impulse signal, for the generic time moment $t = nT$, obtained from the difference of the two step signals is presented in Fig. 7.11 [1, 2, 4].

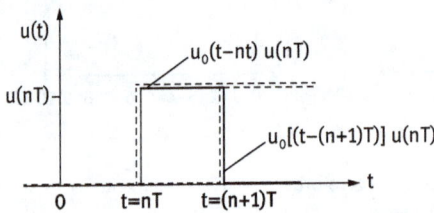

Fig. 7.11: Impulse signal obtained from the difference of two step signals.

Applying the Laplace transform for both members of eq. (7.10) leads to the transfer function of the zero-order hold element $H_0(s)$ [1, 2, 4]:

$$U(s) = u(nT)\left[e^{-snT}\frac{1}{s} - e^{-s(n+1)T}\frac{1}{s}\right]$$

$$U(s) = u(nT)e^{-snT}\left[\frac{1 - e^{-sT}}{s}\right]$$ (7.11)

$$U(s) = L(u(nT)\delta(t - nT))\frac{1 - e^{-sT}}{s}$$

$$\frac{U(s)}{L(u(nT)\delta(t - nT))} = \frac{1 - e^{-sT}}{s} = H_0(s)$$

In the above equations, it has been considered that the input of the zero-order hold element is the impulse $u(nT) \cdot \delta(t - nT)$, having the Laplace transform $u(nT) \cdot e^{-nTs}$, and the output is the signal $u(t)$ with the Laplace transform $U(s)$.

7.2.3 Analytical description of the discrete systems

The discrete system has the input and output variables represented by sequences of numeric values for the input $u(n \cdot T)$ and the output $y(n \cdot T)$, respectively, where $n = 0$, $1, 2, \ldots$. The discrete system is characterized by the relationship between these sequences of numeric values. The mathematical form of this relationship is represented by the *difference equations*, also named recurrent equations. For a discrete system, the difference equation stating the input and output relationships has the general form [1]:

$$F(y(n), y(n+1), \ldots, y(n+N), u(n), u(n+1), \ldots, u(n+M), n) = 0, \qquad (7.12)$$

where $F: C^{N+M+2} \times T \to C$ is, in general, a time-dependent nonlinear function. The relationship (7.12) is true for every integer $n \in T = \{n_0, n_0 + 1, n_0 + 2, \ldots\}$ or $n \in \mathbf{Z}$. M and N are integer and positive constants.

In the following, the *linear and discrete time-invariant* systems will be investigated. They are described by [1]

$$q_N y(n+N) + q_{N-1} y(n+N-1) + \cdots + q_1 y(n+1) + q_0 y(n) =$$

$$= p_M u(n+M) + p_{M-1} u(n+M-1) + \cdots + p_1 u(n+1) + p_0 u(n) \qquad (7.13)$$

where q_i, $i = 0, 1, \ldots, N$ and p_j, $j = 0, 1, \ldots, M$ are real or complex constant coefficients. For non-anticipating discrete systems, the condition $N > M$ is fulfilled.

Example 7.1

An example of a discrete system is a digital computer system that receives as input a sequence of values of a physical variable $u = \{u(0), u(1), u(2), \ldots\}$ and transforms it into a sequence of numeric values $y = \{y(0), y(1), y(2), \ldots\}$, according to the recurrent algorithm [1, 4]:

$$y(n+1) = ay(n) + (1-a)\, u(n+1), \qquad (7.14)$$

where $n = 0, 1, 2, \ldots$, and a is a positive constant, $0 < a < 1$.

This digital system is named *exponential smoothing filter*. It may be noticed that the output $y(n + 1)$ at any moment of time $(n + 1)T$ is determined as a weighted mean of the current input $u(n + 1)$ and of the output $y(n)$ from the previous moment of time nT. As the constant a approaches the value 1, the output from the previous moment is given higher importance and the output from the current moment of time is smoother.

If the equation describing the discrete system is repeatedly applied, the following equation is obtained:

$$y(n) = a^n y(0) + (1-a) \sum_{k=0}^{n-1} a^k u(n-k). \qquad (7.15)$$

This equation presents the fact that the output $y(n)$ is equal with the weighted sum of all inputs from the previous moments of time, starting with $u(1)$ and of the initial output $y(0)$. For $|a| < 1$, the effect on the output of the initial value $y(0)$ decreases asymptotically as n increases. The weighting coefficients a^k exponentially decrease as they are weighting values situated further in the past.

The question is how can the continuous system description (model) be transformed into a discrete description (model)? It is known that algebraic, differential, or integral

equations can represent continuous systems. Operations equivalent to the basic operations met by the description of the continuous systems may be also developed [1, 2, 14].

1. The *algebraic* relations of the continuous models, between values of the input
   and output at the same moment of time, are transformed into relationships of the
   same form for discrete models. For example, the simple relationship $y(t) = k \cdot u(t)$
   of a continuous time gain element model is transformed into the discrete equivalent relationship $y(nT) = k \cdot u(nT)$.
2. The *integration* operation, from the continuous models, is transformed into a *sum*
   of the discrete models. The integral $\int u(t)dt$ has the significance of the area under
   the plot of the function $u(t)$ and may be approximated with a sum of areas corresponding to the rectangles obtained by sampling the time. These rectangles have
   one side equal to the sampling period and the other side equal to the value of the
   function at the corresponding sampling moment:

$$\int_0^t u(t)dt \cong T \sum_{k=0}^n u(kT), \tag{7.16}$$

These rectangles are represented in Fig. 7.12 [1].

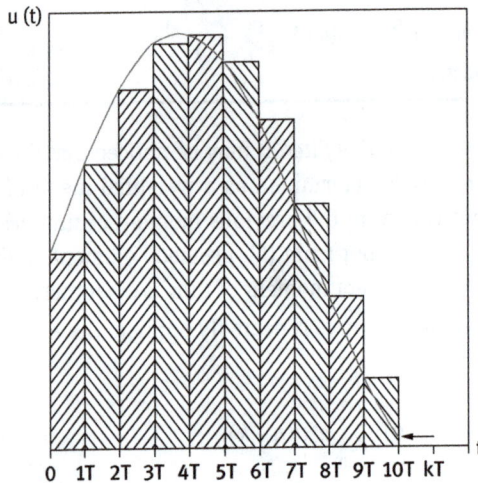

Fig. 7.12: Approximation of the integral from the continuous models with the sum of the discrete
models.

The way of transforming the integral of a function from the continuous case into
its discrete correspondent is not unique, depending on the different numerical
methods for approximating the integral: rectangle, trapezoid, etc.

3. The *derivative du/dt* from the continuous models is transformed into the differ-ence equations. The simplest numerical method for approximating the derivative is given by

$$\frac{du}{dt} \cong \frac{u(nT) - u[(n-1)\ T]}{T} \tag{7.17}$$

The slope of the straight line AB from Fig. 7.13 is the approximation of the derivative [1].

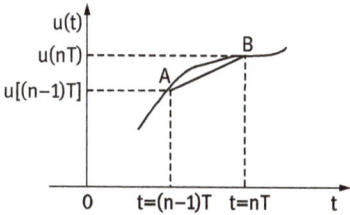

Fig. 7.13: Approximation of the derivative from the continuous models with the difference equation for the discrete models.

The transformation of the derivative into its discrete model correspondent is not unique, depending on the different numeric methods for approximating the deriva-tive: Euler, Runge-Kutta, etc. There are methods for the numeric approximation of the derivative, based on the right (forward) difference $dy/dt = (1/T)(y[(n+1)T] - y(nT))$ or on the left (backward) difference $dy/dt = (1/T)(y(nT) - y[(n-1)T])$. High-order de-rivatives are approximated by high-order difference equations. Usually, for terms of the form $y(nT)$, the notation y_n is used.

For obtaining the discrete time model, starting from the continuous time model, the following steps have to be performed [1, 2, 14]:
1. Start with the continuous model described by algebraic, differential, and integral equations,
2. Approximate the derivatives of any order by finite differences,
3. Approximate the integrals by a method of numeric integration,
4. Introduce the values of the algebraic terms directly with their values from the corresponding sampling moments.

Example 7.2

The discrete model of the continuous first-order system is determined as follows [1, 2, 14]. The nonlinear first-order continuous system is described by

$$\frac{dy}{dt} = f(u,y), \tag{7.18}$$

where f is a nonlinear function.

Approximating the derivative by the first-order forward difference $dy/dt = (1/T) \cdot (y_{n+1} - y_n)$, the following difference equation is obtained:

$$y_{n+1} = y_n + Tf(u_n, y_n), \tag{7.19}$$

showing the way that output at the next moment of time y_{n+1} depends on the values of the input u_n and output y_n from the current moment.

For the case of the linear system, as is the process of liquid accumulation in a tank, the continuous linear system is described by the differential equation:

$$T_p \frac{dy}{dt} + y = ku, \tag{7.20}$$

and the discrete time form of the model, based on the same first-order forward difference, becomes

$$y_{n+1} = \left(1 - \frac{T}{T_p} y_n\right) + \frac{kT}{T_p} u_n. \tag{7.21}$$

---

i Example 7.3

The discrete time model of the linear second-order system, consisting of the process of liquid accumulation in a series of two tanks, will be determined [1, 2, 14]. The second-order system is described by the second-order differential equation:

$$T_1 T_2 \frac{d^2 y}{dt^2} + (T_1 + T_2) \frac{dy}{dt} + y = ku \tag{7.22}$$

The second-order derivative is approximated by the second-order forward difference:

$$\frac{d^2 y}{dt^2} = \frac{d}{dt}\left(\frac{dy}{dt}\right) \approx \frac{d}{dt}\left[\frac{y_{n+1} - y_n}{T}\right] \approx \frac{1}{T}\left[\frac{y_{n+2} - y_{n+1}}{T} - \frac{y_{n+1} - y_n}{T}\right]$$

$$= \frac{1}{T^2}(y_{n+2} - 2y_{n+1} + y_n) \tag{7.23}$$

Replacing the first- and second-order derivatives with their approximations, the following discrete time form is obtained:

$$T_1 T_2 \left[\frac{1}{T^2}(y_{n+2} - 2y_{n+1} + y_n)\right] + (T_1 + T_2)\frac{1}{T}(y_{n+1} - y_n) + y_n = ku_n$$

$$y_{n+2} = \frac{1}{T_1 T_2}\left\{[2T_1 T_2 - (T_1 + T_2)T]y_{n+1} + [(T_1 + T_2)T - T_1 T_2 - T^2]y_n + kT^2 u_n\right\} \tag{7.24}$$

Determination of the derivatives on the basis of numeric approximations, for the case of continuous physical (real) signal, may bring practical difficulties due to the noise superposed on the pure signal. The process noise has low amplitude but high frequency (rapid change in the time unit) and may generate components with unacceptable high amplitude when the derivation operation is used. To overcome this inconvenience, digital filters (Example 7.1) are used to smooth the measured signal [1, 2, 14].

The transformation of the continuous model, presenting dead time, into its discrete correspondent may be performed directly if the dead time is approximated with a multiple (integer) of the sampling period $\tau_m = k \cdot T$. Consider a first-order system with dead time described by the differential equation:

$$T_p \frac{dy}{dt} + y = k_p u(t - \tau_m). \tag{7.25}$$

Its equivalent discrete form is

$$y_{n+1} = \left(1 - \frac{T}{T_p}\right) y_n + \frac{k_p T}{T_p} u_{n-k}. \tag{7.26}$$

7.2.4 Z transform

According to the concepts presented in the previous paragraphs, it may be concluded that difference equations represent a simple and natural way for the mathematical description of the discrete time systems behavior. This is similar to the description of continuous time systems by differential equations [1, 4, 12]. The *Z transform* offers an elegant and convenient method for the analysis and synthesis of the linear discrete systems in the same way as Laplace transform is used for continuous systems [6]. The incentives brought by the use of the Z transform consist in the simplicity of representation and directly offering an instrument for qualitative and quantitative analysis of linear discrete systems [7].

The following row of values is obtained by the sampling of a continuous signal $y(t)$ using the sampling time T:

$$y(0), y(T), y(2T), y(3T), \ldots \tag{7.27}$$

Definition: The Z transform of the row of sampled values is defined by the sum (series):

$$Z(y(0), y(T), y(2T), \ldots) = \sum_{n=0}^{\infty} y(nT) z^{-n} \tag{7.28}$$

The usual appellation for Z transform of the row of sampled values is of *Z transform of the function* $y(t)$. It is denoted by $Y(z)$, and its notation is

$$Z(y(t)) = Y(z) = \sum_{n=0}^{\infty} y(nT) z^{-n} \tag{7.29}$$

The Z transform exists only for those values of the independent variable z for which the infinite series, given by eq. (7.28) or (7.29), has finite values (the series is convergent) [1, 4, 12].

This way of defining the Z transform makes the Z transform a sequence of numeric values originating from different continuous functions, but having the same values at the sampling moments, to have the same Z transform.

The Z transform associates a signal (function) from the so-called Z domain (having the z independent variable) to a discrete time signal (function).

The Z transform of a time function depends on the value chosen for the sampling time T.

According to the concepts presented, when the sampling of the continuous signal has been investigated for a sequence of sampled values, it may be associated with a time function $y^*(t)$ described by a sum of the shifted Dirac impulses multiplied by the values of the sampled function corresponding to the sampling moments. The integral (area) of each of the terms from the sum is equal to the value of the function at the corresponding sampling moment:

$$y^*(t) = \sum_{n=0}^{\infty} y(nT)\ \delta(t - nT). \tag{7.30}$$

Applying the Laplace transform for this continuous time function leads to

$$Y^*(s) = \sum_{n=0}^{\infty} y(nT)\ e^{-nTs}. \tag{7.31}$$

If the notation $z = e^{Ts}$ is used, the Z transform of the sequence (7.27) is

$$Y^*(s)|_{z = e^{Ts}} = \sum_{n=0}^{\infty} y(nT)z^{-n} = Y(z). \tag{7.32}$$

It may be concluded that the Z transform of the row of sampled values is a particular form of the Laplace transform. The variables of the two transforms, s and z, are linked by the relationship $z = e^{Ts}$.

7.2.5 Z transform of several simple functions

The Z transform of the following simple functions is important for analyzing the behavior and the synthesis of discrete systems [1, 4, 12]. They are presented in Tab. 7.1.

It may be noticed that the Z transform of the presented functions has the form of a ratio between two polynomials with the independent variable z^{-1} or z.

7.2.6 Inverse of the Z transform

The determination of the inverse Z transform is similar, as a methodology, with the determination of the inverse Laplace transform. In the case of the inverse Z transform

Tab. 7.1: Z transform of usual functions [4].

Time signal	Laplace transform	Z transform
$\delta(t)$	1	1
$u_0(t)$	$\dfrac{1}{s}$	$\dfrac{1}{1-z^{-1}}$
$a \cdot t$	$\dfrac{a}{s^2}$	$\dfrac{a \cdot T \cdot z^{-1}}{(1-z^{-1})^2}$
e^{-at}	$\dfrac{1}{s+a}$	$\dfrac{1}{1-e^{-aT} \cdot 1-z^{-1}}$
$t \cdot e^{-at}$	$\dfrac{1}{(s+a)^2}$	$\dfrac{T \cdot e^{-aT} \cdot z^{-1}}{(1-e^{-aT} \cdot z^{-1})^2}$
$\sin(\omega t)$	$\dfrac{\omega}{(s^2+\omega^2)}$	$\dfrac{z^{-1} \cdot \sin \omega T}{1-2 \cdot z^{-1} \cos \omega T + z^{-2}}$
$\cos(\omega t)$	$\dfrac{s}{(s^2+\omega^2)}$	$\dfrac{1-z^{-1} \cdot \cos \omega T}{1-2 \cdot z^{-1} \cdot \cos \omega T + z^{-2}}$
$1-e^{-at}$	$\dfrac{a}{s(s+a)}$	$\dfrac{(1-e^{-aT}) \cdot z^{-1}}{(1-z^{-1})(1-e^{-aT} \cdot z^{-1})}$
$e^{-at} \cdot \sin(\omega t)$	$\dfrac{\omega}{(s+a)^2+\omega^2}$	$\dfrac{z^{-1} \cdot e^{-aT} \cdot \sin \omega T}{1-2 \cdot z^{-1} \cdot e^{-aT} \cdot \cos \omega T + e^{-2aT} \cdot z^{-2}}$
$e^{-at} \cdot \cos(\omega t)$	$\dfrac{s+a}{(s+a)^2+\omega^2}$	$\dfrac{1-z^{-1} \cdot e^{-aT} \cdot \cos \omega T}{1-2 \cdot z^{-1} \cdot e^{-aT} \cdot \cos \omega T + e^{-2aT} \cdot z^{-2}}$

of a function $G(z)$, the values of the function $g(t)$ at the multiples of the sampling time, $t = n \cdot T$, $n = 0, 1, 2, \ldots$, are computed [1, 4, 12]:

$$Z^{-1}(G(z)) = \{g(0),\ g(T),\ g(2T),\ \ldots\}. \tag{7.33}$$

The original (continuous time) function $g(t)$ may not be recovered from the image $G(z)$ by the inverse Z transform, but only the values $g(nT)$, $n = 0, 1, 2, \ldots$. This fact is natural because it is not expected to get more information from the inverse Z transform than was embedded by the direct Z transform. As a consequence, the sequence of values obtained by the inverse Z transform may correspond to different (continuous time) functions but which share the same values at the sampling moments. It is not possible to compute the sampling time T by the inverse Z transform.

From practical reasons, finding the inverse Z transform of a function having the form of a ratio between two polynomials in the z independent variable is desired:

$$G(z) = \frac{p_n z^n + p_{n-1} z^{n-1} + \cdots + p_0}{z^n + q_{n-1} z^{n-1} + \cdots + q_0} = \frac{P(z)}{Q(z)} \tag{7.34}$$

A first way to determine the inverse Z transform of the function $G(z)$ is the direct division of the polynomials $P(z)$ and $Q(z)$. The order of the polynomial $Q(z)$ has to be always higher than or equal to the order of the polynomial $P(z)$, for the case of causal systems. If the following form is obtained after the direct division:

$$G(z) = \frac{P(z)}{Q(z)} = g_0 + g_1 z^{-1} + g_2 z^{-2} + \cdots \tag{7.35}$$

and considering the definition of the Z transform, it may be concluded that

$$g(0) = g_0, \ \ g(T) = g_1, \ \ g(2T) = g_2, \ \ldots \tag{7.36}$$

This means that the row of the successive quotients obtained by the direct division of the polynomials $P(z)$ and $Q(z)$ represent the values of the sequence and they form the inverse Z transform of the function $G(z)$.

Example 7.4

Consider the Z transform of a sequence of values with the following form [1]:

$$G(z) = \frac{z}{2z^2 - 3z + 1}. \tag{7.37}$$

The inverse Z transform may be determined by the division of the polynomials $P(z) = z$ and $Q(z) = 2z^2 - 3z + 1$:

$$
\begin{array}{ll}
z & \left| \, 2z^2 - 3z + 1 \right. \\
\underline{-z + 1.5 - 0.5z^{-1}} & \left| \, 0.5z^{-1} + 0.75z^{-2} + 0.875z^{-3} + 0.9375z^{-4} + \cdots \right. \\
\quad 1.5 - 0.5z^{-1} & \\
\quad \underline{-1.5 + 2.25z^{-1} - 0.75z^{-2}} & \\
\qquad 1.75z^{-1} - 0.75z^{-2} & \\
\qquad \underline{-1.75z^{-1} + 2.625z^{-2} - 0.875z^{-3}} & \\
\qquad\quad 1.875z^{-2} - 0.875z^{-3} & \\
\qquad\quad \underline{-1.875z^{-2} + 2.8125z^{-3} - 0.9375z^{-4}} & \\
\qquad\qquad 1.9375z^{-3} - 0.9375z^{-4} &
\end{array}
$$

etc.

From this division, the form of the function $G(z)$ is

$$G(z) = 0 + 0.5z^{-1} + 0.75z^{-2} + 0.875z^{-3} + 0.9375z^{-4} + \cdots \tag{7.38}$$

The sequence of values at the sampling moments is

$$g(0) = 0, \ \ g(T) = 0.5, \ \ g(2T) = 0.75, \ \ g(3T) = 0.875, \ \ g(4T) = 0.9375, \ \ldots \tag{7.39}$$

The second way to determine the inverse Z transform is the partial fraction expansion of the function $G(z)$ in simple fractions having well-known inverse Z transform (accord-

ing to the direct Z transform of the set of functions presented in Tab. 7.1). This procedure is similar to the one used for determining the inverse Laplace transform [1].

There are two equivalent ways for obtaining the partial fraction expansion:

a. Taking into account that all Z transforms of the functions in Tab. 7.1 have the z factor at the numerator, it is simple to determine first the partial fraction expansion of
the function $G(z)/z$. Subsequently, for the partial fraction expansion of the function
$G(z)$, the inverse Z transform is applied (distributed for each term of the expansion).

b. The function $G(z)$ is transformed into a ratio of two polynomials $P'(z^{-1})$ and $Q'(z^{-1})$
having z^{-1} independent variable:

$$G(z) = \frac{P(z)}{Q(z)} = \frac{p_n + p_{n-1}z^{-1} + p_{n-2}z^{-2} + \cdots + p_0 z^{-n}}{1 + q_{n-1}z^{-1} + q_{n-2}z^{-2} + \cdots + q_0 z^{-n}} = \frac{P'(z^{-1})}{Q'(z^{-1})} \tag{7.40}$$

The partial fraction expansion of this ratio of polynomials is performed, obtaining
simple fractions with well-known (from Tab. 7.1) inverse Z transforms:

$$G(z) = \frac{P'(z^{-1})}{Q'(z^{-1})} = \frac{c_1}{v_1(z^{-1})} + \frac{c_2}{v_2(z^{-1})} + \cdots + \frac{c_n}{v_n(z^{-1})}. \tag{7.41}$$

Coefficients $c_1, c_2, \ldots, c_n$ may be determined according to a simple methodology. The
polynomials $v_1(z^{-1}), v_2(z^{-1}), \ldots, v_n(z^{-1})$ are usually of low order and have known inverse Z transform.

Applying the inverse Z transform to the members of eq. (7.41), the function $f(t)$ is
determined as a sum of elementary time functions, $r_1(t), r_2(t), \ldots, r_n(t)$:

$$g(t) = Z^{-1}(G(z)) = Z^{-1}\left(\frac{c_1}{v_1(z^{-1})}\right) + Z^{-1}\left(\frac{c_2}{v_2(z^{-2})}\right) + \cdots + Z^{-1}\left(\frac{c_n}{v_n(z^{-n})}\right)$$

$$= r_1(t) + r_2(t) + \cdots + r_n(t) \tag{7.42}$$

It has to be mentioned that the values of the function $g(t)$, from expression (7.42), are
only meaningful for the sampling moments:

$$g(0) = r_1(0) + r_2(0) + \cdots + r_n(0)$$
$$g(T) = r_1(T) + r_2(T) + \cdots + r_n(T)$$
$$g(2T) = r_1(2T) + r_2(2T) + \cdots + r_n(2T) \tag{7.43}$$

$$\vdots$$

Example 7.5

Consider the Z transform of a sequence of values, with the form [1]:

$$G(z) = \frac{z}{2z^2 - 3z + 1} = \frac{z^{-1}}{z^{-2} - 3z^{-1} + 2} \tag{7.44}$$

The roots of the polynomial $Q'(z^{-1})$, from the denominator of the function $G(z^{-1})$, are equal to $z^{-1} = 1$ and $z^{-1} = 2$ The partial fraction expansion of the function $G(z^{-1})$ is

$$G(z) = \frac{z^{-1}}{z^{-2} - 3z^{-1} + 2} = \frac{1}{1 - z^{-1}} - \frac{1}{1 - \frac{1}{2}z^{-1}} \tag{7.45}$$

Applying the inverse Z transform, the following are obtained:
- for the term $1/(1 - z^{-1})$, the unit step function $u_0(t)$ is obtained;
- for the term $-1/(1 - 0.5z^{-1})$, the original (discrete time) function is an exponential, having for the sampling moments, the form $- e^{n \cdot \ln(0.5)}$, $n = 0, 1, 2, \ldots$, it has been accounted that $e^{aT} = 0.5$ and equivalently $aT = \ln(0.5)$.

Consequently, the sequence of values $g(nT)$ at the sampling moments, for the function $g(t)$, may be determined by

$$g(nT) = u_0(nT) - e^{n \ln(0.5)} = 1 - e^{n \ln(0.5)}, \quad n = 0, 1, 2, \ldots \tag{7.46}$$

and the sequence of values is

$$g(0) = 1 - 1 = 0$$

$$g(T) = 1 - e^{\ln(0.5)} = 1 - 0.5 = 0.5$$

$$g(2T) = 1 - e^{2\ln(0.5)} = 1 - (0.5)^2 = 0.75 \tag{7.47}$$

$$g(3T) = 1 - e^{3\ln(0.5)} = 1 - (0.5)^3 = 0.875$$

$$\vdots$$

It may be noticed that irrespective of the applied methods, the result of the inverse Z transform from Examples 7.4 and 7.5 are identical.

7.2.7 Z transfer function

Resuming the difference equation (7.13) describing the discrete system behavior. Taking into account the property of the causal systems $N \geq M$, it is considered that $N = M = n$ and, $q_{N=n} = 1$, the difference equation may be reformulated with the following equivalent forms [1]:

$$y(k+n) + q_{n-1}y(k+n-1) + \cdots + q_0 y(k) =$$
$$= p_n u(k+n) + p_{n-1}u(k+n-1) + \cdots + p_0 u(k) \tag{7.48}$$

$$y(k) + q_{n-1}y(k-1) + \cdots + q_0 y(k-n) =$$
$$= p_n u(k) + p_{n-1}u(k-1) + \cdots + p_0 u(k-n) \tag{7.49}$$

Applying the Z transform for eq. (7.49) and taking into account the Z transform of a shifted function, the following forms are obtained [8]:

$$
\begin{aligned}
Y(z) + q_{n-1}z^{-1}Y(z) + \cdots + q_0 z^{-n}Y(z) &= \\
= p_n U(z) + p_{n-1}z^{-1}U(z) + \cdots + p_0 z^{-n}U(z) & \\
Y(z) = \frac{p_n + p_{n-1}z^{-1} + \cdots + p_0 z^{-n}}{1 + q_{n-1}z^{-1} + \cdots + q_0 z^{-n}} U(z) &= H(z)U(z)
\end{aligned}
\tag{7.50}
$$

or, in a more compact form:

$$
Y(z) = H(z)U(z) \tag{7.51}
$$

The notation $H(z)$ has been used for the Z *transfer function* of the discrete system, defined by the expression:

$$
H(z) = \frac{p_n + p_{n-1}z^{-1} + \cdots + p_0 z^{-n}}{1 + q_{n-1}z^{-1} + \cdots + q_0 z^{-n}} = \frac{p_n z^n + p_{n-1}z^{n-1} + \cdots + p_0}{z^n + q_{n-1}z^{n-1} + \cdots + q_0} = \frac{Y(z)}{U(z)} \tag{7.52}
$$

The Z transfer function $H(z)$ describes the relationship between the input and output of a discrete system in the z domain. Zero value is assumed for the input and output at time moments previous to the present (initial) moment [9].

Similar to the continuous case, the roots of the polynomial $P(z)$ from the numerator of the Z transfer function are denoted as *zeros* of the system and the roots of the polynomial $Q(z)$ from the denominator of the transfer function are denoted as *poles* of the system. To specify the poles and the zeros, it is preferable to use the representation based on the z variable instead of the z^{-1} variable. The same poles are specified by both the representations, but only the first clearly indicates the existence of a pole at the origin.

Example 7.6

To sustain the aforementioned statement, it is considered, for example, the Z transfer function $H(z) = z/(z-1)$, clearly indicating the presence of a zero at the origin [1, 4, 12]. The equivalent form $H(z) = 1/(1-z^{-1})$ does not directly reveal this zero.

Similar to the case of the linear continuous systems, the effect superposition principle and the cause-effect proportionality principle have direct consequences for the way of determining the discrete response of a linear discrete system for any given discrete input. The response (output) of a linear discrete system to a certain discrete input may be determined by the summation of the discrete Dirac impulse responses, weighted by the input function values from the sampling moments (the discrete input function may be approximated by this sum of Dirac impulses weighted by the input function values at the sampling moments). Knowing the response to an input represented by the discrete Dirac impulse, the response to any given input may be determined by the help of the *discrete convolution* (having the form of a sum):

$$
y(k) = \sum_{i=0}^{k} h(i)u(k-i) \tag{7.53}
$$

Discrete Dirac impulse has been denoted by the succession of values 1, 0, 0, 0, . . ., from the sampling moments 0, T, 2T, 3T, ... The discrete Dirac impulse response, having the notation h(iT) or simply h(i), is represented by the succession of values h(0), h(T), h(2T), ...

Analyzing the expressions from eqs. (7.51) and (7.53), it may be noticed that, similar to the continuous systems case, for the discrete systems, there is also a direct correspondence between the *multiplication operation of the Z transform images* (from the z domain) and the *discrete convolution operation* of the sequences of values of the sampled signals (in the discrete time domain). The discrete Dirac impulse response h(iT) i = 1, 2, 3, . . . is the inverse Z transform of the Z transfer function H(z). In the following, the discrete Dirac impulse will be simply denoted as Dirac impulse and the discrete Dirac impulse response as Dirac impulse response.

7.2.8 Z transfer function of the sampled system

For the digital (discrete) control systems, it is often like the case when a continuous system is controlled by a digital computing system. For a rigorous analysis, it is necessary to describe the sampled system, presented in Fig. 7.14(a), by the z domain representation, according to Fig. 7.14(b) [1, 9].

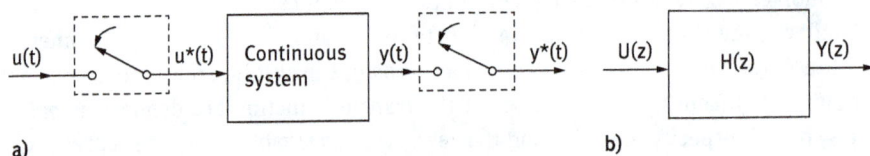

Fig. 7.14: (a) Sampled continuous system and (b) Sampled continuous system using Z transform representation.

The input function u(t) of the continuous system is sampled $u^*(t) = \sum(u(kT) \cdot \delta(t-kT))$ but the output is continuous.

If the differential equation describing the continuous system is known, the Z transfer function H(z) may be determined by the help of the Z transform. Knowing the transfer function of the continuous system H(s), the Z transfer function may be determined by the help of the ζ *transform*, defined by [9]

$$H(z) = Z\left[L^{-1}(H(s))|_{t=kT}\right] = \zeta(H(s)) \tag{7.54}$$

It may be mentioned that the inverse ζ transform $H(s) = \zeta^{-1}(H(z))$ is not unique due to the same reason as the inverse Z transform is not unique.

For practical applications, the control of a continuous system using a sequence of Dirac impulses $u^*(t)$ is not possible. The control action must have a "continuous" characteristic, usually given by the zero-order hold element. Its place is between the sampling element and the process, as presented in Fig. 7.15(a) [1, 9].

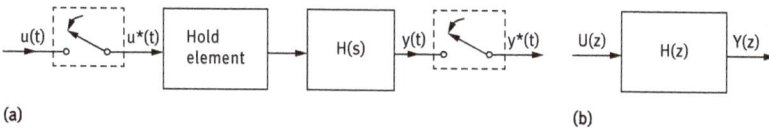

(a)   (b)

Fig. 7.15: (a) Discretized continuous system with hold element. (b) Discretized continuous system represented using the Z transform.

Knowing the transfer function of the zero-order hold element, $H_0(s) = (1 - e^{-sT})/s$, and taking into account the translation property of the Laplace transform, the determination of the Z transfer function for the system presented in Fig. 7.15(b) leads to

$$H(z) = \zeta\left(\frac{1 - e^{-sT}}{s} H(s)\right) = Z\left(L^{-1}\left((1 - e^{-sT})\frac{H(s)}{s}\right)\Big|_{t=kT}\right)$$

$$= Z(g(kT) - g(kT - T)) \tag{7.55}$$

$$H(z) = (1 - z^{-1})\zeta\left(\frac{H(s)}{s}\right)$$

The notation $g(t)$ has been used for the inverse Laplace transform of the function $H(s)/s$, with $g(t) = \zeta^{-1}(H(s)/s)$.

7.2.9 Z transfer function of the interconnected systems

A few simple cases of discrete control systems having sampling elements will be presented in the following [1, 9]. The results may be obtained using the convolution property. It may be noticed that the ζ transform presents the property $\zeta(H_1(z) \cdot H_2(z)) \neq \zeta(H_1(z)) \cdot \zeta(H_2(z))$.

1. Series interconnected discrete systems, Fig. 7.16:

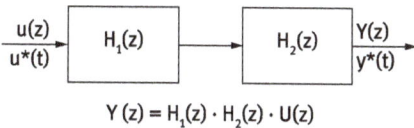

$$Y(z) = H_1(z) \cdot H_2(z) \cdot U(z)$$

Fig. 7.16: Series interconnected discrete systems.

2. Series interconnected continuous systems without intermediate sampling element, Fig. 7.17 [1]:

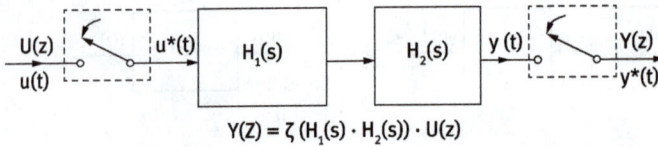

$$Y(Z) = \zeta\,(H_1(s) \cdot H_2(s)) \cdot U(z)$$

Fig. 7.17: Series interconnected continuous systems without intermediate sampling element.

3. Series interconnected continuous systems with intermediate sampling element, Fig. 7.18 [1, 9]:

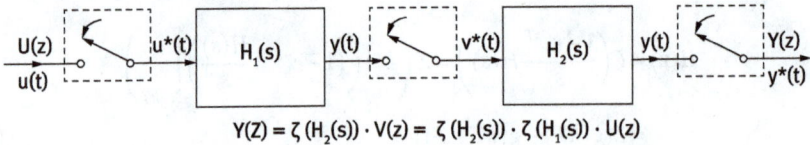

$$Y(Z) = \zeta\,(H_2(s)) \cdot V(z) = \zeta\,(H_2(s)) \cdot \zeta\,(H_1(s)) \cdot U(z)$$

Fig. 7.18: Series interconnected continuous systems with intermediate sampling element.

4. Continuous system before the sampling element, Fig. 7.19 [1, 9]:

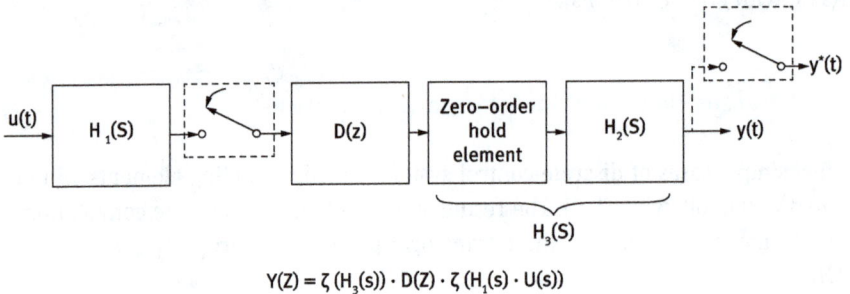

$$Y(Z) = \zeta\,(H_3(s)) \cdot D(Z) \cdot \zeta\,(H_1(s) \cdot U(s))$$

Fig. 7.19: Series interconnected systems with continuous system placed before the sampling element.

5. Discrete feedback control system, Fig. 7.20 [1, 9]:

7.3 Discrete PID controller

The discrete PID controller algorithm emerges from the discretization of the continuous form of the PID control law. As it works on a discrete time bases, the control variable computation has, as input, the sampled value of the error (or controlled variable), and the computed control variable for a sampled time moment is kept constant until the next sampling time by the zero-order hold element. Generally, the discrete PID tuning and the desired performance may be assessed and achieved based on similar consider-

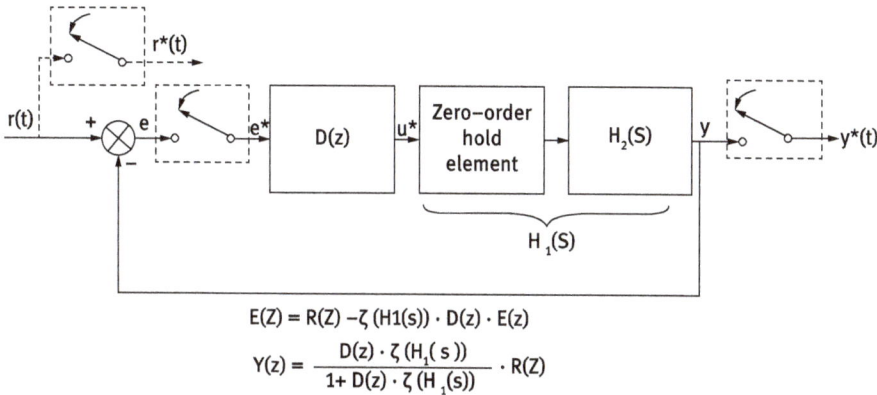

$$E(Z) = R(Z) - \zeta\,(H1(s)) \cdot D(z) \cdot E(z)$$

$$Y(z) = \frac{D(z) \cdot \zeta\,(H_1(\,s\,))}{1 + D(z) \cdot \zeta\,(H_{\,1}(s))} \cdot R(Z)$$

Fig. 7.20: Discrete feedback control system.

ations used for its continuous correspondent [10]. The discrete PID control law may be described by [4, 11, 12]:

$$u(kT) = K_c \left[e(kT) + \frac{T}{T_I} \sum_{i=0}^{k} e(iT) + \frac{T_D}{T} \left(e(kT) - e((k-1)\,T) \right) \right] + u_0 \qquad (7.56)$$

where the discrete integral control mode has been obtained by the rectangular approximation of the continuous integral of the error and the discrete derivative control mode by the backward difference approximation of the continuous derivative of the error. The initialization term u_0 is the initial value of the manipulated variable when the error should be zero.

A commonly used form of the PID controller law does not take into account the derivative effect of the setpoint variable (inherent in the error) and changes to the following discrete PID control law [4, 11]:

$$u(kT) = K_c \left[e(kT) + \frac{T}{T_I} \sum_{i=0}^{k} e(iT) - \frac{T_D}{T} \left(y(kT) - y((k-1)\,T) \right) \right] + u_0 \qquad (7.57)$$

where $y(kT)$ is the sampled controlled variable. Forms (7.56) and (7.57) of the discrete PID are denoted as *position form* of the algorithm since they compute the full control variable for every sampling moment.

The summation implied by the integral mode does not have to be performed at every sampling moment as the sum of errors up to the previous sampling moment:

$$S((k-1)T) = \sum_{i=0}^{k-1} e(iT) \qquad (7.58)$$

may be saved (stored) and used in the following form of the discrete PID:

$$u(kT) = K_c\left[e(kT) + \frac{T}{T_I}\left(S((k-1)T + e(kT)) - \frac{T_D}{T}(y(kT) - y((k-1)\ T))\right)\right] + u_0 \quad (7.59)$$

In the *velocity form* of the discrete PID algorithm, the control variable change is computed, according to the following relationship [4, 11]:

$$\Delta u(kT) = u(kT) - u((k-1)T)$$

$$= K_c\left[e(kT) - e((k-1)T) + \frac{T}{T_I}e(kT) - \frac{T_D}{T}(y(kT) - 2y((k-1)T) + y((k-2)T))\right]$$

$$= K_c\left[\left(1 + \frac{T}{T_I}\right)e(kT) - e((k-1)T) - \frac{T_D}{T}(y(kT) - 2y((k-1)T) + y((k-2)T))\right]$$

$$(7.60)$$

The (7.60) form of the velocity algorithm can be obtained by subtracting the position forms corresponding to two consecutive sampling moments, i.e., $u(kT)$ and $u((k-1)T)$.

The velocity form of the discrete PID algorithm has appreciated incentives compared to the discrete position PID as it does not need initialization and does not show the integral windup undesired behavior.

Due to sampling, an additional time delay (dead time) appears in the control loop. This dead time is equal to the sampling time and has negative effect on the control performance. As a result, it may be expected that discrete controller's performance is not as good as its continuous counterpart. This aspect is important especially when the sampling time is large compared to the feedback dynamics. Some authors suggest the choosing of the sample time for the discrete controller such that it is less than 5% of the sum between the feedback dead time and the largest time constant [11].

Nevertheless, tuning the discrete PID controllers may be performed on the bases of the same methods used for tuning the continuous PID controllers (Ziegler-Nichols, Cohen-Coon, minimizing the integral square error or integral absolute error, etc.) [13]. Some authors propose that tuning of the PID discrete controller should be made in a similar way as the continuous PID tuning but considering an additional dead time equal to half the sampling time [12].

Stability of the discrete systems may be analyzed on the system's Z transfer function, which has the form of a ratio between polynomials in the z variable, $H(z) = P(z)/Q(z)$. The discrete system is stable if all roots of the denominator polynomial $Q(z)$, denoted as *poles* of the system, are situated inside the unit circle of the complex plane. If at least one of the roots of the polynomial $Q(z)$ is situated outside the unit circle, the system shows instability. Roots of $Q(z)$ that have real parts near the value -1 (but inside the unit circle) show a strongly oscillating behavior, denoted as *ringing* [4, 11].

The discrete controller may produce the ringing undesired effect if it is not properly designed or tuned. Controller ringing is produced by the changes of the controller output, with high amplitude and different sign (slowly damping oscillations of the control variable), producing degradation of the control system performance. Basically, the reduction of the controller gain and increasing the integral time can reduce the controller ringing of the discrete PID. For the general case of the discrete controllers, ringing may be avoided by designing the poles of the controller such that they are placed inside the positive region of the unit circle and have absolute values in the [0.4 0.6] interval [11].

Particular attention must be given to the integral mode of the PID controller, when the control valve reaches saturation, i.e., the valve is in a totally open or a totally closed position, and the error has not been yet eliminated. This situation makes the control system to be unable of further acting efficiently for obtaining zero offset. However, as time passes, the effect of the control variable on the controlled variable increases in time and the error is reduced. At some moment, this delayed reduction of the error determines the control valve to leave its completely open or completely closed position and the control signal will re-enter in the active domain, i.e., between the 0% and the 100% positions of the valve opening and finally, the error will be eliminated. The problem generated by the integral mode is that during the period of saturation of the control valve, as error is still present, the integral of the error produces a continuous increase or continuous decrease of the control variable value. As a result, the control variable reaches very large or very small values, a phenomenon called *integral* (or *reset*) *windup*. Bringing back the control variable value in the active region of the control valve may imply a long period of time and this causes the control system to be inefficient in the meanwhile.

Integral windup can be generated by improper design of the control system equipment or by unexpectedly large disturbances. Avoiding the integral windup can be achieved by appropriate equipment design or by special *anti-reset windup* adaptations brought to the traditional PID controller structure. The latter implies the use of a saturation element. Such an anti-reset windup structure, for a PI controller, is presented in Fig. 7.21 (for the continuous case) [11]. As long as the signal u does not reach the saturation limits, the PI controller works as a traditional one. But when saturation is reached, the controller output is limited to u_{lim}.

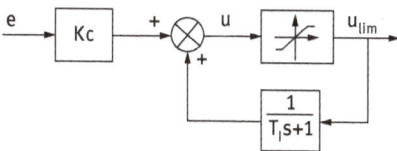

Fig. 7.21: PI controller with anti-reset windup.

The velocity form of the discrete PID controller does not accumulate the integral error and it avoids the integral windup by its design.

7.4 Other forms of the discrete controllers

The discrete approach of designing controllers allows the computation of new forms of the controllers that may differ from the classical discrete PID while satisfying the desired control performance [4, 11, 12].

Starting from the Z transfer function of the discrete feedback control loop presented in Fig. 7.20, it is possible to design the controller Z transfer function $D(z)$ such as for a given form of the Z transform of the setpoint change $R(z)$, a desired closed loop Z transform response $Y(z)$ is desired (imposed). The controller satisfying this desired behavior of the closed control loop is described by the following Z transfer function:

$$D(z) = \frac{\frac{Y(z)}{R(z)}}{\zeta(H_1(s))\left(1 - \frac{Y(z)}{R(z)}\right)} = \frac{\frac{Y(z)}{R(z)}}{H_1(z)\left(1 - \frac{Y(z)}{R(z)}\right)} = \frac{Y(z)}{H_1(z)(R(z) - Y(z))} \tag{7.61}$$

where $R(z)$, $Y(z)$, and $H_1(z)$ are known.

This newly designed discrete controller has the form of a ratio of two polynomials in z^{-1}:

$$D(z) = \frac{U(z)}{E(z)} = \frac{p_0 + p_1 z^{-1} + \cdots + p_n z^{-n}}{1 + q_1 z^{-1} + \cdots + q_m z^{-m}} \tag{7.62}$$

This is equivalent to

$$U(z)\left(1 + q_1 z^{-1} + \cdots + q_m z^{-m}\right) = E(z)\left(p_0 + p_1 z^{-1} + \cdots + p_n z^{-n}\right) \tag{7.63}$$

Making the inverse Z transform of eq. (7.63), the following algorithm form for computing the control variable $u(kT)$ at the sampling moment $t = kT$ may be developed:

$$u(kT) = p_0 e(kT) + p_1 e((k-1)T) + \cdots + p_n e((k-n)T) - \cdots$$
$$- q_1 u((k-1)T) - \cdots - q_m u((k-m)T) \tag{7.64}$$

Care should be taken that the discrete controller is physically realizable, i.e., the computation of the control variable $u(kT)$ at the sampling moment $t = kT$ depends only on past or utmost present values of the error and of the control variable. This is equivalent to the condition that powers of z in the polynomial $P(z)$ to be strictly negative.

Two of the most renowned results of such new discrete controller designs are the deadbeat and the Dahlin controllers [4].

The *deadbeat controller* is designed, requiring the controller (for the sampling moments) to respond to the step change of the setpoint by producing the same step change on the controlled variable but delayed by one sampling period. Usually, this design builds controllers that are very aggressive and show strongly oscillating behavior. This design may also be applied to processes having dead time, but physically realizable forms may be only obtained if the desired step response of the controlled variable cumulates one sampling time period, to which is added an integer number of sampling time periods exceeding the process dead time (in which are also included the dead time of the control element and measuring device).

The *Dahlin controller* is designed, requiring the controller (for the sampling moments) to respond to the step change of the setpoint by producing a first order with the dead time behavior of the controlled variable. This design makes the control to be less aggressive and the oscillatory behavior is diminished. Physically realizable controllers may only be obtained if the desired response of the controlled variable cumulates one sampling time period, to which is added an integer number of sampling time periods exceeding the process dead time (considering also included the dead time of the control element and measuring device).

References

[1] Cristea, M.V., Agachi, S.P., *Elemente de Teoria Sistemelor*, Editura Risoprint, Cluj-Napoca, 2002.
[2] Äström, K.J., Wittenmark, B., *Computer-Controlled Systems Theory and Design*, Prentice Hall, Englewood Cliffs, NJ, 1997.
[3] Miclea, L., *Fiabilitatea și diagnoza sistemelor digitale*, Universitatea Tehnică Press, Cluj-Napoca, 1998.
[4] Stephanopoulos, G., *Chemical Process Control An Introduction to Theory and Practice*, Prentice Hall, Englewood Cliffs, NJ, 1984.
[5] Edward, A.L., Varaiya, P., *Structure and Interpretation of Signals and Systems*, University of California, Berkeley, CA, 2000.
[6] Schroer, W., *Systementheorie. Vorlesungsmanuskript*, Ulm Universität, 1998.
[7] Corovei, I., Pop, V., *Transformări integrale, calcul operațional*, Universitatea Tehnică Press, Cluj-Napoca, 1993.
[8] Mihoc, D., Ceaprău, M., Iliescu, S.S., Borangiu, I., *Teoria și elementele sistemelor de reglare automata (Theory and elements of automatic control)*, Editura Didactică și Pedagogică, București, 1980.
[9] Kottmann, M., Kraus, F., Schaufelberger, W., *Signal und Systemtheorie II*, Skript zur Vorlesung, ETH Institut für Automatik, Zürich, 2001.
[10] Feștilă, C., *Regulatoare automate-îndrumător de proiectare (Automatic controllers – Design guide)*, Institutul Politehnic, Cluj-Napoca, 1990.
[11] Marlin, T.E., *Process Control-Designing Processes and Control Systems for Dynamic Performance*, McGraw-Hill, 2000.
[12] Franklin, G., Powel, J., Workman, M., *Digital Control of Dynamic Systems*, Addison-Wesley, Reading, MA, 1990.
[13] Seborg, D.E., Edgar, T.F., Mellichamp, D.A., Doyle III, J.F., *Process Dynamics and Control*, 3rd Edition, John Wiley & Sons, New York, 2003.
[14] Gerald, C.F., Whealtley, P.O., *Applied Numerical Analysis*, 7th Edition, Addison Wesley, New York, 2003.
[15] Agachi, P.S., Cristea, M.V., *Basic Process Engineering Control*, Walter de Gruyter, Berlin/Boston, 2014.
[16] Agachi, S., *Automatizarea Proceselor Chimice (Chemical Process Control)*, p. 320, Casa Cărții de Știință, Cluj Napoca, 1994.

Part II: **Applied Process Engineering Control**

The reason for the addition of this second part in the present book is that the control of multivariable processes is a complex enterprise, even if handled with independent control loops. Industrial processes can be grouped mainly as belonging either to the synthesis or the separation stage; each is complex and has multiple inputs and outputs. The synthesis stage occurs in reactors of various types (such as CSTR, BSTR, PFR, and a multitude of other kind of reactors). The separation stage is chosen in agreement with characteristics of the products to be separated. In addition, we would like to draw attention upon the importance of utilities' control (steam, water, electricity, etc.). These are extremely important from economical point of view.

This part, *Applied Process Engineering Control*, applies the theoretical concepts of the first part. The control of any process, either a synthesis or a separation, is discussed from the point of view of the intrinsic needs of the process. Thus, control strategy is proposed in agreement with these requirements.

Consequently, the principles of the control system design used in this chapter are the following:
1. Study of process to be controlled as well as of correlation between controlled and all possible manipulated variables;
2. Design of steady-state mathematical model;
3. Quantifying goal variables (e.g. chemical conversion, material humidity, waste composition) as well as input and state variables;
4. If necessary, sensitivity and relative gain array analysis of the process to eliminate the unimportant correlations between output and input/state variables and simplify the control scheme;
5. Design of control scheme.

A new chapter, "Case Studies" was added in this Edition approaching complex plants (e.g. Catalytic Cracking, or Cement Manufacturing)

https://doi.org/10.1515/9783110789737-009

8 Reaction unit control

8.1 Introduction

This chapter deals with basic control strategies of chemical synthesis units. Since temperature affects the rates of both physical and chemical processes, its efficient control has to be omnipresent. Consequently, this topic is given special attention, and thermal instability and ways to avoid it are described in detail. Yet, pressure and liquid level control as well as ways of coping with the logarithmic dependence of pH on the acid or base concentration are also discussed. Process end-point detection is coupled with aspects of product quality control. The principles of designing the control structure for some homogeneous and heterogeneous, continuous or batch units are presented, and electrochemical reactors are also mentioned within this context. The picture is completed with pertinent examples.

The *reactor* is the core of most chemical production facilities; other operations are related to its necessities and performance. Therefore, its smooth and stable operation under well-defined parameters ensures both *profitability* and *safety*. The goal of each control strategy is to minimize as much as possible, if not entirely eliminate, the effect of disturbing parameters on the desired quality of the process, in terms of product distribution, energy consumption, exploitation safety, etc. There is no general "recipe" to achieve all of these; smart solutions differ from one process to another but rely on the same basic principles. An appropriately optimized design for both batch and continuous chemical reactors can improve plant productivity up to 25% [1]. Such results may be achieved using adequate individual control loops and strategies to program the main parameters but also to provide sequencing and record-keeping functions.

Success or failure of a certain control system strongly relates to the reactor design. An incorrect layout could result in an unstable reactor, regardless of the employed control strategy and even in the absence of a significant disturbance. For example, strong oscillations in temperature or pressure could compromise product or catalyst, damage equipment, or injure personnel. Hence, the choice of adequate control should be in conjunction with the inherent characteristics of each reactor type.

8.2 Basic concepts of ideal continuous and batch units

The very basic unit types [1–4] dealt with in chemical reaction engineering and design are *continuous stirred tank reactors* (CSTRs), *tubular plug-flow reactors* (PFRs) and *batch stirred tank reactors* (BSTRs). In terms of flow and thermal conditions, they are referred to as *ideal* and may work in *dynamic* or *steady-state*. Descriptions of industrial units, either homogeneous or heterogeneous, are derived from these basic models.

https://doi.org/10.1515/9783110789737-010

Uniform inlet and outlet flows are assumed for continuous units – CSTR and PFR. The average amount of time spent by a molecule in the unit is called *residence time*. It is defined by V/F, the ratio between the reactor volume V (m³) and the total volumetric flow F (m³/s).

A CSTR assumes ideal mixing; therefore, parameters are uniform within the entire volume of the reaction mixture. Each point in space is characterized by the same value of residence time, temperature, and conversion. These are also the attributes of the outlet flow. A PFR assumes total radial but no axial (flow direction) mixing; the reaction mixture advances like a *plug* along the longitudinal axis of the tube. Thus, reactor parameters are distributed along it, and at each point of residence time, temperature and conversion have different values. In other words, such a reactor functions like a cascade of infinitely small stirred tank reactors.

A BSTR is operated in cycles of one load each. The time amount necessary for the load to reach the desired end-point of the chemical process is called *reaction time* $t_{reaction}$. Parameters will vary as it advances. This is the reason for their more difficult control (see Section 8.7). Yet, since ideal mixing in assumed in such units, the temperature and conversion values will be uniformly distributed at a certain moment through the entire volume of the reaction mass.

Two main types of *batch* units are operated in chemical industry: the pure-batch and the *fed-batch* (for example neutralizing processes). In a *pure-batch* unit, reactants are charged all at once at the beginning of the process. The fed-batch is loaded only partially at the beginning of a cycle and some reactants are added at a well-defined rate during the progress of the reaction.

Primordial performance criteria for a reactor are residence time V/F or reaction time $t_{reaction}$, *conversion* ξ, and *selectivity* Φ [1, 3–6]. Conversion is the only advance variable of a chemical process that is defined with respect to the limiting reactant (species A in equations below), whereas selectivity relates to the desired product (species P). The *limiting reactant* is the species present in the smallest stoichiometric amount in the mixture at reaction initiation, while all others are in *excess*.

Definitions (8.1) and (8.2) assign both ξ and Φ values between 0 and 1. n_{Ao} and n_{Po} stand for the influent amount (expressed in moles) of A and P, whereas n_A and n_P stand for the corresponding effluent values. If the total volume of the reaction mass does not vary significantly during the course of the reaction, the molar amounts can be replaced with the corresponding molar concentrations:

$$\xi = \frac{n_{Ao} - n_A}{n_{Ao}} \tag{8.1}$$

$$\Phi = \frac{n_P - n_{Po}}{n_{Ao} - n_A} \tag{8.2}$$

For an isolated first-order reaction $A \xrightarrow{k} P$, eqs. (8.3)–(8.5) describe conversion versus temperature $\xi = f(T^\circ)$ patterns for various unit types [1, 3–4]. k stands for the rate coefficient (s^{-1}) and, as a rule [5–6], depends exponentially on temperature – see eq. (8.6).

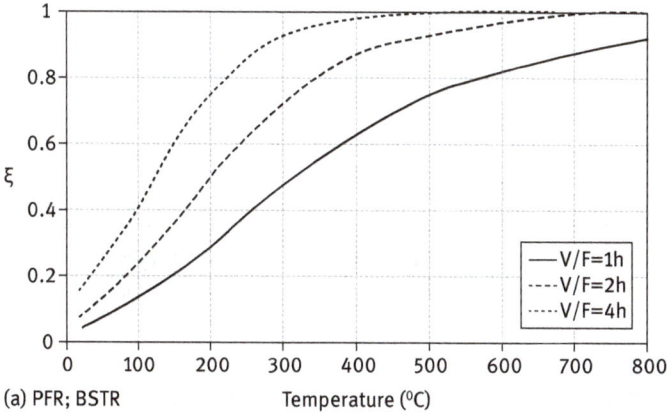

(a) PFR; BSTR

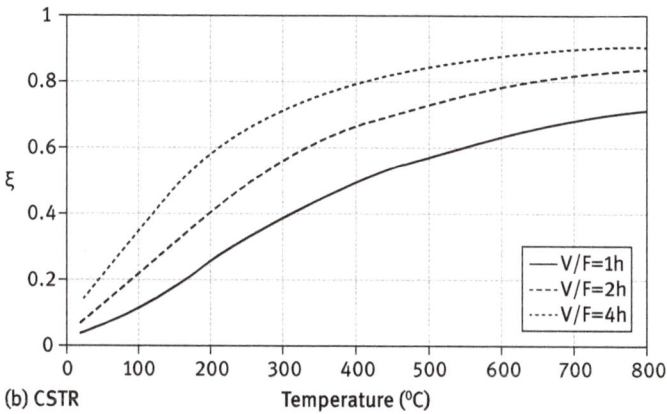

(b) CSTR

Fig. 8.1: Conversion versus temperature patterns for: (a) steady-state PFR and BSTR; (b) steady-state CSTR. Kinetic data are $k_0 = 11.87\ h^{-1}$, $E_a = 14{,}000\ J/mol$.

$$\xi_{CSTR} = \frac{k\frac{V}{F}}{1 + k\frac{V}{F}} \tag{8.3}$$

$$\xi_{PFR} = 1 - \exp\left(-k\frac{V}{F}\right) \tag{8.4}$$

$$\xi_{BSTR} = 1 - \exp(-k\ t_{reaction}) \tag{8.5}$$

$$k = k_0\ \exp\left(-\frac{E_a}{RT^\circ}\right) \tag{8.6}$$

where
- k_0 stands for the pre-exponential coefficient (s^{-1}, same units as k),
- E_a for the activation energy (J/mol),
- $R = 8.314$ J/mol·K is the universal gas constant,
- and $T°$ stands for the reaction temperature (K).

The above relationships as well as Fig. 8.1 demonstrate that these profiles are nonlinear and vary from one reactor design to the other. For the same process as well as for identical time and temperature values, PFR and BSTR (Fig. 8.1(a)) are kinetically superior to CSTR (Fig. 8.1(b)) since the corresponding ξ is higher.

Optimizing ξ and Φ ensures minimum raw material consumption and the best product distribution. However, other constraints and limitations also have to be considered: avoiding local overheating of the catalyst, ensuring the desired particle sizes or molecular weights for a product, batch-to-batch uniformity, etc. Under real operating conditions, residence times are not uniform and their distribution pattern also affects ξ and Φ values [3–4]. Catalyzed or uncatalyzed heterogeneous reaction systems that involve fluid diffusion through various media, fluid adsorption/desorption to a solid surface, or gas absorption/dissolution into a liquid are influenced, in addition, by the nature of flow regime, diffusion characteristics, or thermodynamics and kinetics of the interphase contact processes.

8.3 Temperature control

Reaction temperature is almost always one of the controlled variables. Some examples of extreme temperature-sensitive goals are: lowering the side product accumulation rates, keeping narrow polymer molecular weight distributions, or protecting the integrity of glass-lined reactors. Therefore, some systems demand high-performance temperature regulation – within 0.2–0.3 °C [1].

Because runaway *exothermic* reactions can destabilize an operating reactor, special attention is given to *thermal stability. Endothermic* reactions are self-regulating; if external heat-delivery suffers, the process slows down or stops. Heat transfer is also self-regulating; its rate increases by improvement of any parameter that describes heat exchange.

8.3.1 Into thermal instability

Thermal instability of a unit in function may occur only for exothermic processes when the slope of the heat-release Q_r curve exceeds that of the heat transfer Q_{ev} (see Fig. 8.2). Such a situation can start *runaway* behavior: an uncontrolled increase of temperature speeds up the process, which in return releases more heat and further elevates the temperature. Thus, the process responds to a temperature disturbance with a positive feed-back. A controller compensates it with a negative feedback by increasing the heat trans-fer rate. The stronger feedback decides whether the unit is thermally stable or not.

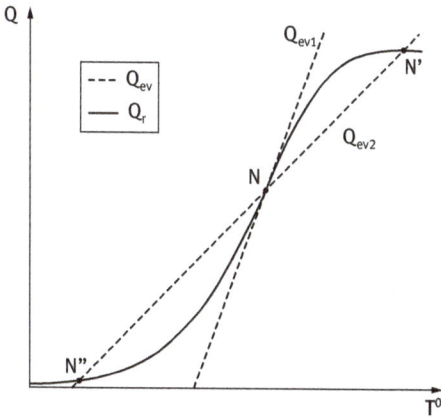

Fig. 8.2: Reaction generated Q_r and evacuated Q_{ev} heat rate versus temperature patterns. Qualitative representation (adapted after [3, 4]).

Thermal stability is ensured when:

$$\left(\frac{dQ_{ev}}{dT^\circ}\right) > \left(\frac{dQ_r}{dT^\circ}\right) \tag{8.7}$$

Figure 8.3 illustrates the dynamic variation of conversion from steady-state 1 to steady-state 2. It is obvious that the instantaneous variation $(d\xi/dT^\circ)$ of conversion with temperature is higher for the dynamic mode than that for the steady-state (since $\xi < 1$). The relationship between them is described in the literature [7–9]:

$$\left(\frac{d\xi}{dT^\circ}\right)_{dynamic} = \left[\frac{1}{1-\xi}\right]\left(\frac{d\xi}{dT^\circ}\right)_{steady} \tag{8.8}$$

A singular first-order reaction $A \xrightarrow{k} P$ is characterized by the rate law:

$$r = kC_A = k_0 \exp\left(-\frac{E_a}{RT^\circ}\right)C_A \tag{8.9}$$

where r stands for reaction rate (kmol/m^3·s), k for the first-order rate coefficient (s^{-1}), and C_A for the momentary molar concentration (kmol/m^3) of the reactant species A. The significance of E_a, R, and T° are the same as given for eq. (8.6). Using eqs. (8.1)–(8.6), the statement in eq. (8.8) will verify for both a CSTR and a PFR.

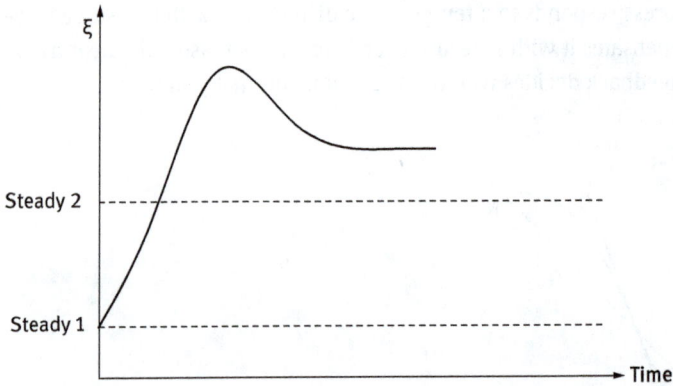

Fig. 8.3: Dynamic variation of conversion between two steady states. Qualitative representation (adapted after [8]).

- For a continuous stirred tank reactor:

$$\left(\frac{d\xi}{dT^\circ}\right)_{dynamic} = \left(\frac{E_a}{R(T^\circ)^2}\right) \xi > \left(\frac{d\xi}{dT^\circ}\right)_{steady} = \left(\frac{E_a}{R(T^\circ)^2}\right) \xi \, (1-\xi) \qquad (8.10)$$

- For a plug-flow reactor:

$$\left(\frac{d\xi}{dT^\circ}\right)_{dynamic} = \left(\frac{E_a}{R(T^\circ)^2}\right) \ln(1-\xi) > \left(\frac{d\xi}{dT^\circ}\right)_{steady} = \left(\frac{E_a}{R(T^\circ)^2}\right) (1-\xi) \ln(1-\xi) \qquad (8.11)$$

Depending on ξ, the value of $(\frac{d\xi}{dT})$ at the same temperature T° can be higher for the dynamic state. Thus, thermal stability condition (8.7) is not satisfied and the unit becomes unstable. As a consequence, correct design requires checking the validity of constraint (8.7) for both dynamic and steady states at the setpoint temperature. Otherwise, a reactor that is apparently thermally stable under steady-state becomes unstable when operated dynamically – see relationship (8.8).

8.3.2 Out of thermal instability

Thermal instability can be kept under control, which requires simultaneous solutions for both mass and energy balances. For the exothermic first-order reaction $A \xrightarrow{k} P$ occurring in a CSTR, the mass balance for reactant A is

$$VkC_A = F(C_{A0} - C_A) = F\xi C_{A0} \qquad (8.12)$$

The corresponding reaction-*generated heat rate* is

$$Q_r = -\Delta H_r V k C_A = -\Delta H_r F \xi C_{Ao} \qquad (8.13)$$

where C_{Ao} is the initial molar concentration of the reactant A (kmol/m^3) and ΔH_r is the exothermic reaction heat (kJ/kmol). The *evacuated heat rate* depends on the geometry of the reactor (by means of A_T) and the characteristics of the material from which it was built and coated (by means of K_T); ρ is the reaction mixture density (kg/m^3), C_P is the mean reaction mixture heat capacity (J/kg·K), A_T is the heat transfer surface (m^2), K_T is the mean heat transfer coefficient between the reaction mixture and the coolant (J/s·m^2·K), and T_{ag}° is the coolant's average temperature.

$$Q_{ev} = F\rho\, C_p\, T^\circ + K_T A_T \left(T^\circ - T_{ag}^\circ \right) \qquad (8.14)$$

Relationships (8.12)–(8.14) are written under the assumption that no volume variation occurs due to reaction advance. In other words, the density of the entire mixture remains constant, regardless of ξ (such a dependence is described by linear functions in the literature [5, 6]). Similarly, the value of C_p is considered to characterize the entire mixture as temperature-independent and therefore constant. Thus, Q_{ev} in eq. (8.14) depends linearly on T° (see Fig. 8.2).

Meanwhile, because of the nonlinear temperature dependence of terms k and ΔH_r [5–6], the generated heat Q_r exhibits nonlinearity when plotted against temperature. This fact is illustrated by Fig. 8.2. The thermal stability requirement (8.7) is fulfilled in setpoint **N** for the line indicating Q_{ev1}. If the temperature varies either way around its value in **N**, the process will slip back and recover to the parameters of **N**. Hence, the reactor is thermally stable. Meanwhile, the line corresponding to Q_{ev2} intersects the curve of Q_r at more than one point, among which **N** is unstable and both **N'** and **N''** are stable. For example, if the temperature rises over the setpoint value in **N** where $Q_r > Q_{ev}$, the disturbance will result in a positive feedback of the process and the temperature will further increase till point **N'**. Beyond it, $Q_r < Q_{ev}$, and the system recovers to the operating conditions of **N'**. Thus, it describes a thermally stable state. The same judgment can be applied to point **N''**, proving its stability as opposed to **N**. The instability of setpoint **N** can be eased by enhancing the heat transfer trough adjustment of A_T, K_T, and T_{ag}° values, respectively.

Analysis of Fig. 8.2 leads to the conclusion that the thermally stable unit has an adequate heat transfer surface. Therefore, it requires an unsophisticated temperature control loop, as the one exemplified in Fig. 8.4, for a CSTR. This may involve an ordinary negative response proportional-integral controller [1, 2, 7–9]. Meanwhile, the thermally unstable reactor demands a fast and strong enough negative feedback in the control loop to adequately compensate the positive response of the chemical process. Hence, the control system has to fulfill some *constraints* to stabilize this open-loop instability.

Quantitative assessment of thermal stability or instability of functioning units is based on checking the heat balance for the setpoint temperature $T°$. For the first-order reaction discussed above, thermal equilibrium means

$$F\rho \ C_p \ T_{in}° - \Delta H_r F \xi \ C_{Ao} = F\rho \ C_p \ T° + K_T A_T \left(T° - T_{ag}°\right) \tag{8.15}$$

The left side of eq. (8.15) stands for the heat-inlet rate, while the right side stands for the outlet rate. $T_{in}°$ stands for the influent reactant temperature. Thermal stability, that is the satisfaction of condition (8.7), translates into eq. (8.16), where $(d\xi/dT°)$ is assessed for both dynamic and steady states:

$$F\rho \ C_p + K_T A_T > \left| \Delta H_r F \ C_{Ao} \left(\frac{d\xi}{dT°}\right) \right| \tag{8.16}$$

Figure 8.1 shows that the conversion ξ depends nonlinearly on the temperature $T°$. Therefore, $(d\xi/dT°)$ is approximated by a linearization with respect to the setpoint described by $\bar{\xi}$ and $\overline{T°}$ (see Example 8.1 where calculus is carried out for both dynamic and steady states):

$$\bar{\xi} = \xi + \left(\frac{d\xi}{dT°}\right)_{\overline{T°}} \left(\overline{T°} - T°\right) \tag{8.17}$$

The above data serve the computation of time constant T_{rm} (seconds) and amplification factor K_{rm} (dimensionless) of the reaction mass [1, 7, 8]. When these values are positive, the reactor is thermally stable and its temperature control is conventional (see Fig. 8.4):

$$T_{rm} = \frac{V \ \rho \ C_p}{F \ \rho \ C_p + K_T A_T + \Delta H_r \ F \ C_{Ao} \left(\frac{d\xi}{dT°}\right)_{steady}} \tag{8.18}$$

$$K_{rm} = \frac{K_T A_T}{F \ \rho \ C_p + K_T A_T + \Delta H_r \ F \ C_{Ao} \left(\frac{d\xi}{dT°}\right)_{steady}} \tag{8.19}$$

If the reverse is true, and the total heat-outlet rate is lower than the inlet rate, the values of T_{rm} and K_{rm} become negative. As a result, the process amplification $(K_{rm}/\sqrt{1 + T_{rm}^2 \omega^2})$ is also negative. The phase shift φ_{rm} (rad) is a function of oscillating frequency ω (rad/s) and varies between $-180°$ and $90°$:

$$\varphi_{rm} = -180° - arctg(T_{rm}\omega) = -180° + arctg(|T_{rm}|\omega) \tag{8.20}$$

The term $(-180°)$ indicates a negative gain under steady state and the sign (+) in front of the arctg function implies a negative time constant.

A chemical reactor can function without reaction mass temperature regulation if the control of the coolant temperature is satisfactory [9] (see Fig. 8.4). However, there are two constraints to be imposed on a control system to achieve thermal stability of an unstable reactor:

– First constraint – Imposing lower proportional band PB values, in agreement with relationship (8.21); ω_{osc} stands for the oscillation frequency of the control system:

$$PB < 200 \left(\frac{K_{rm}}{\sqrt{1 + T_{rm}^2 \ \omega_{osc}^2}} \right) \tag{8.21}$$

– Second constraint – Exclusion of integral time T_i because of the negative phase shift it generates. Therefore, eq. (8.20) will modify because of regulator phase shift φ_R. The introduction of T_i brings the system closer to its stability limit, described by $\varphi_{total} = -180°$:

$$\varphi_{total} = -180° = \varphi_{rm} + \varphi_R = -180° + \text{arctg}(|T_{rm}|\omega) - \text{arctg}\left(\frac{1}{T_i\omega}\right) \tag{8.22}$$

Example 8.1

Let us consider the described first-order reaction $A \xrightarrow{k} P$ occurring in liquid phase in a CSTR for which the $\xi = f(T°)$ dependence corresponds to that in Fig. 8.1(b). The setpoint is given by the ratio $\frac{V}{F} = 2$ h and $\overline{T°} = 200$ °C. Inlet flow is preheated to $T_{in}° = \overline{T°}$. Other technical data are $K_T = 2.10^3 \text{kJ}/\text{m}^2 \cdot \text{h} \cdot °\text{C}$; $A_T = 4$ m^2; $F = 5$ m^3/h; $\Delta H_r = -15$ kJ/kmol; $C_{Ao} = 10$ kmol/m^3; $\rho = 10^3$ kg/m^3; $C_p = 1$ kJ/kg $\cdot$ °C.

Data in Fig. 8.1(b) indicate that the setpoint corresponds to $\overline{\xi} = 0.4$. The coolant temperature that ensures thermal stability can be calculated from the energy balance (8.15) by compelling that the heat-outlet flow has to be higher than that of the inlet. Replacement of data yields: $8 * (200 - T_{ag}°) \geq 300$ kJ/h, where $T_{ag}°$ is expressed in °C. The result is $T_{ag}° \leq 162.5$ °C.

A stable reactor, in either dynamic or steady state, has to check eq. (8.7). Fig. 8.1(b) provides data for the calculus of steady-state $(d\xi/dT°)$ by employing linearization (8.17) between the setpoint $\overline{T°} = 200$ °C; $\overline{\xi} = 0.4$ and point $\overline{T°} = 100$ °C; $\overline{\xi} = 0.2$. The dynamic value can be further obtained using eq. (8.8):

$$\left(\frac{d\xi}{dT°} \right)_{steady} = \left(\frac{\overline{\xi} - \xi}{\overline{T°} - T°} \right) = \frac{0.4 - 0.2}{200 - 100} \frac{1}{°\text{C}} = 0.0020 \frac{1}{°\text{C}}$$

$$\left(\frac{d\xi}{dT°} \right)_{dynamic} = \left[\frac{1}{1-\xi} \right] \left(\frac{d\xi}{dT°} \right)_{steady} = \frac{1}{1-0.4} \left(0.0020 \frac{1}{°\text{C}} \right) = 0.0033 \frac{1}{°\text{C}}$$

Both values verify thermal stability requirement (8.16): for steady state $13 > 1.5$ (kJ/h·°C) and for dynamic state $13 > 2.5$ (kJ/h·°C), respectively. Thus, both operation modes are stable under the specified conditions.

If the setpoint is switched to $(\bar{V}/F) = 4\ h$ at the same $T° = 200\ °C$, then according to Fig. 8.1(b), $\bar{\xi} = 0.575$. The same judgment leads to $T°_{ag} \leq 146.1\ °C$. Estimation of $(d\xi/dT°)$ is calculated with respect to $\bar{T}° = 100\ °C$ and $\bar{\xi} = 0.2$:

$$\left(\frac{d\xi}{dT°}\right)_{steady} = \frac{0.575 - 0.342}{200 - 100}\frac{1}{°C} = 0.0023\frac{1}{°C}$$

and

$$\left(\frac{d\xi}{dT°}\right)_{dynamic} = \frac{1}{1 - 0.575}\left(0.0023\frac{1}{°C}\right) = 0.0055\frac{1}{°C}$$

Again, both dynamic and steady-state are stable.

Yet, if the reaction heat increases eightfold ($\Delta H_r = -120\ kJ/kmol$), then at 200 °C, the first setpoint $\bar{V}/F = 2\ h$; $\bar{\xi} = 0.4$ is stable under steady-state operation but unstable under dynamic conditions. The second setpoint $(\bar{V}/F) = 4\ h$; $\bar{\xi} = 0.575$ is unstable in both operation modes. This example proves the importance of thermal stability check calculus in the reactor design.

---

i **Example 8.2**

Let us consider a process for which the time constant and the amplification factor are $T_{rm} = -2.5\ h$ and $K_{rm} = -4$, respectively [8]. The dead times of equipment [7] add up to $\tau_e = 0.1\ h$. The goal is to determine the parameters of a regulator able to thermally stabilize the unit. The total phase shift is written by applying (8.20) and (8.22):

$$\varphi_{total} = -180° = -180° + \text{arctg } |T_{rm}| - 57.3\ t_e\omega$$

It yields $\omega = 15.5\frac{rad}{h} = 2.46\frac{1}{h}$ and $\left(\frac{K_{rm}}{\sqrt{1 + T_{rm}^2\ \omega_{osc}^2}}\right) = -0.103$. The latter corresponds to the limit value of the proportional band of 64%. Yet, the real PB has to obey constraint (8.21) and is therefore 32%. The controller gain [7] is $K_C = \frac{100}{PB} = 1.56$.

8.3.3 Temperature control in practice – continuous units

Thermal stability can be accomplished in many ways. One option is to ensure a difference of approximately 30 °C between the coolant and the bulk reaction mixture to satisfy condition (8.7). Depending on the desired temperature or on whether the employment of a single media is either insufficient or uneconomical, the adequate heat transfer agent can be mono- or multiple-media: for example, either water or a mixture of steam and water (also see Section 8.3.4, Fig. 8.10).

Figure 8.4 depicts a simple control scheme based on manipulating the inlet flow rate of the reactor jacket as a function of the controlled reaction mixture temperature; for continuous units, the jacket inlet/outlet valve is open (reactor influent and effluent valves are also continuously open). Its disadvantages derive from the possible changes in coolant flow rate and temperature, leading to variations of the jacket dead-time. This influences the controller parameters by means of eqs. (8.18)–(8.19) and (8.20)–(8.22).

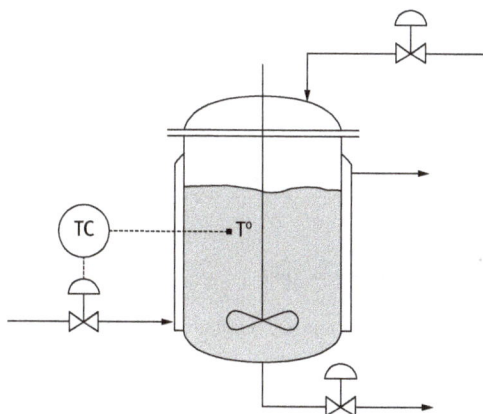

Fig. 8.4: Temperature regulation for a CSTR by manipulating the coolant inlet flow rate (adapted after [1, 8]).

Another handy option is to add to the reaction mixture, some inert ingredient that boils at the desired (setpoint) temperature. This way, the process will be kept almost isothermal whereas production costs might increase because of larger unit volumes, product separation, inert purification and recirculation, etc. A gas phase example is the hydrocracking unit, where the temperature can be controlled by manipulating the hydrogen feed rate because it both dilutes the reactants and cools the mixture (also see Fig. 8.9).

When the equivalent dead time τ_e of the controller does not exceed 33%–35% of the reactor's time constant T_e [7–9], the process can be thermally stabilized, whereas if $\tau_e/T_e > 0.35$, the ability of a single control loop to regulate the temperature in the unit decreases. If it approaches 1, the reactor is uncontrollable. As a result, the temperature oscillates and its profile will resemble the shape of saw teeth as depicted in Fig. 8.5.

In such cases, one suitable solution is a *cascade control* (see also Chapter 1). This includes *master* and *slave* loops. Because the controlled variable (reaction core temperature) has a slow response to changes in the manipulated variable (coolant flow), it is allowed to adjust the setpoint of a secondary loop whose response to coolant changes is rapid. Thus, the master reaction temperature controller varies the setpoint of the slave jacket temperature loop. The addition of the second loop reduces the oscillation period of the main one; in other words, it corrects for all outside disturbances without allowing them to affect the reaction temperature (compensates the disadvantages of control loop presented in Fig. 8.4). An example of such a configuration is depicted in Fig. 8.6.

Cascades are stable and function adequately only if the slave is faster than the master loop; the time constant of the slave should not exceed 10% of that of the master

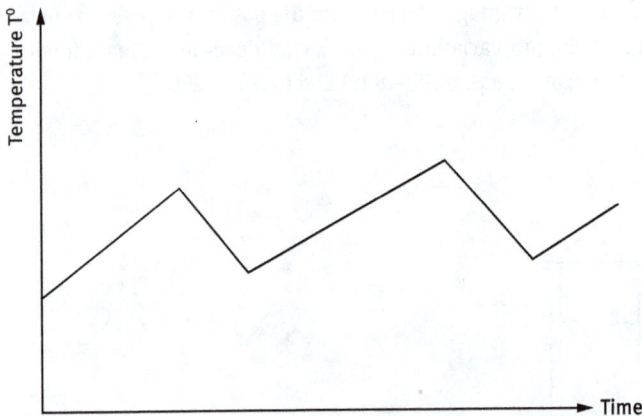

Fig. 8.5: Regulated temperature profile (saw teeth) for an unstable reactor. Qualitative representation (adapted after [8]).

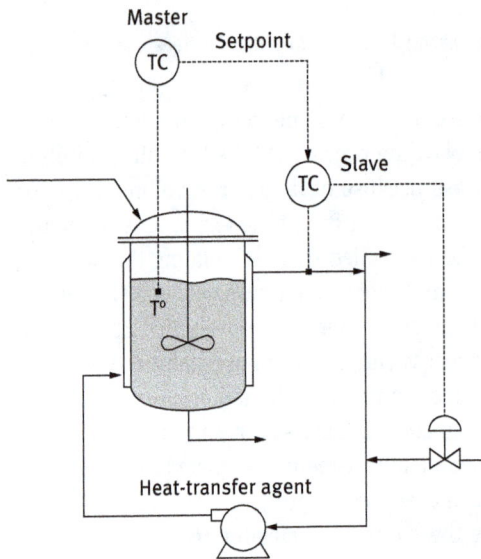

Fig. 8.6: Example of cascade temperature control (adapted after [1]).

[1–2, 9]. Otherwise, it is not able to respond in time to the signal variation detected by the master and the overall loop performance of the cascade will decay.

The slave control loop regulates either the *jacket outlet* or *inlet* temperature. The first option is usually preferred because the jacket time constant and the dynamic response as well as the nonlinearity between the jacket-outlet temperature and heat-transfer agent will be included in the slave loop. As a result, the main nonlinear term within the master loop is removed and the reaction temperature will depend linearly

on the jacket's outlet. If the master's loop oscillation period is cut in half when connected to a slave, the *PB* might change from 30% to 15% and the derivative and integral settings of an interacting controller could also be diminished to half. This would
mean a fourfold overall loop performance improvement [1–2, 9].

Chemical industry often involves processes for which the jacket temperature must
be limited and tightly controlled. Such cases are crystallization units or safety systems
for thermal shock protection of glass-lined synthesis units (because thermal expansion
coefficients for lining and reactor material differ). Thus, the slave control loop is set to
regulate the jacket *inlet* temperature. The disadvantage is that the nonlinear dynamics of
the jacket will be included in the master loop.

Model-based controllers for temperature are recommended when either the reaction
mechanisms are complex (because of autocatalysis or oscillations) or the process is highly
exothermic (danger of runaway behavior). The *predictor* block of such a controller estimates, for instance, the heat release rate, and feeds forward this value to the slave loop
to bias its setpoint. The total effect will be corrected by the *corrector* block and applied
on the jacket temperature setpoint. The disadvantage is related to the complexity and accuracy of the model because it has to contain thermodynamic, kinetic, mass, energy, and
momentum transfer models of both chemical processes and employed equipment.

According to various descriptions above, a cascade usually controls the reaction
temperature by regulating the coolant temperature to match the setpoint. However,
in fast and stable processes, a reversed configuration could perform better [9].

The performance of a conventional temperature control loop must be enhanced for
chemical processes in which the reaction temperature is correlated with reactor pressure.
A good example is a reaction for which pressure is mainly due to the vapor pressure of a
reactant or a product. In such cases, pressure changes represent almost instantaneous
responses to temperature changes or vice versa (see Example 8.10). Another example is a
gas-phase reaction occurring at both high temperature and pressure (oxidation, hydrogenation, etc.) or a high-pressure polymerization. In such cases, partial pressures and
temperature are interconnected within the rate law as well. For these kinds of processes
(both temperature- and pressure-sensitive), a better temperature regulation is carried out
using *pressure-compensated temperature control* loops [10].

In tubular plug-flow reactors, the parameters are distributed on the longitudinal
flow direction; residence time, conversion, and temperature vary as the reaction mixtures advance on the length L of the tube. Fig. 8.7 shows the qualitative temperature
profiles versus axial position OZ for *adiabatic* and *isotherm* chemical processes in profiles (a) and (c), respectively. Profile (b) describes the heat transfer agent that ensures
quasi-isotherm conditions of (c). The terms "adiabatic" and "isotherm" refer to reactor
operation modes, whereas the process itself is exothermic.

To satisfy the demands of an isotherm operation mode, the energy balance in eq.
(8.15) is written for infinitely small volumes dV, which correspond to dZ advances of the
plug flow and have $dA_{T,Z}$ heat transfer surfaces. The reaction mixture enters these micro
units having $T_{in,z}^{\circ}$ temperature. It operates and leaves at T_Z° by reaching conversion ξ_Z°.

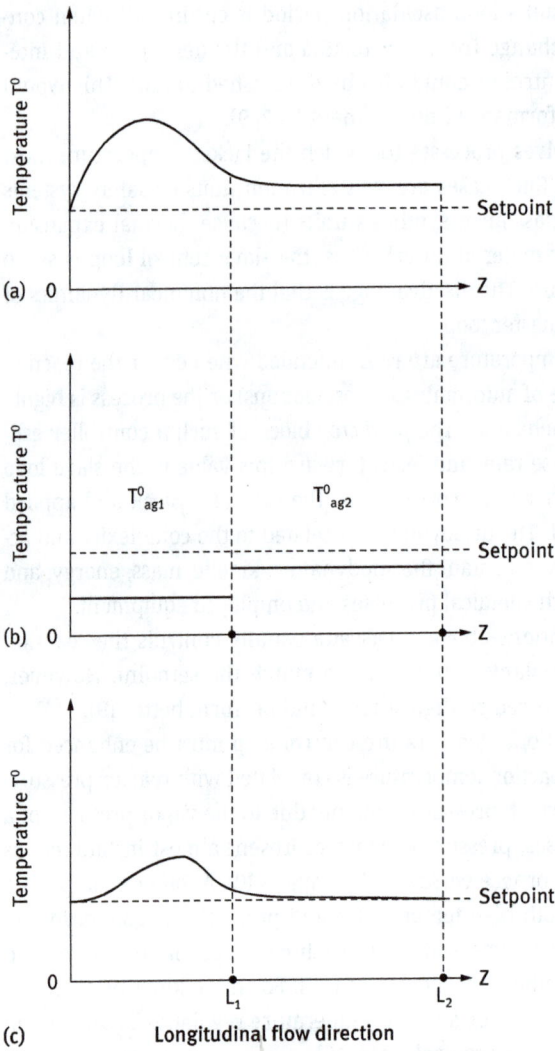

Fig. 8.7: Temperature profiles versus axial position OZ for (a) adiabatic process; (b) quasi-isothermal process; (c) heat transfer agent ensuring conditions in (b). Qualitative representation (adapted after [8]).

$$F \rho C_P (T^{\circ}_{in,Z} - T^{\circ}_Z) - (\Delta H_r)_Z F \, \xi_Z \, C_{Ao} = K_T \, dA_{T,Z} (T^{\circ}_Z - T^{\circ}_{ag}) \tag{8.23}$$

Integration between process-imposed limits (tube length L and V/F residence time) yields $T^{\circ}_{opt} = f(Z)$ profiles like those in Fig. 8.7 (that is optimum T°_Z values at various reactor Z coordinates). From (8.23), values of T°_{ag} are calculated for a desired number of tube portions. As depicted by Fig. 8.8, a quasi-isotherm operation mode such as that in Fig. 8.7(c) requires two distinctly operating jackets, each at another T°_{ag} (see Fig. 8.7(b)).

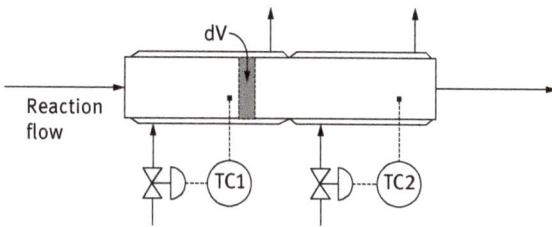

Fig. 8.8: Example of temperature control with separately operating jackets on different portions of a PFR (adapted after [8]).

Heterogeneous multibed reactors are often modeled as PFR reactors. Practical ways to accomplish the desired temperature profiles are *interstage heat exchange* (see Fig. 4.2) or *cold stream injection* (see Fig. 8.9) [1, 2, 8, 11].

Interstage cooling or heating does not dilute the flow; hence, does not require additional beds. Cooling is used for exothermic (sulfur dioxide oxidation, some hydrodealkylation reactions), whereas heating is used for endothermic equilibrium reactions (dehydrogenation of ethylbenzene to styrene).

Inter-bed injection may be carried out using either a cold gas/liquid phase inert or a cooled reaction mixture. The first option dilutes the main stream and the depth of the bed has to be adapted accordingly to maintain the same conversion. The second option lowers the conversion and additional beds are required to reach its desired value at the reactor outlet.

In the industrial ammonia and methanol synthesis (see Example 8.9), hydrogenation of benzene or for some hydrocracking reactions, a fresh and cooled gas is injected between the individual catalyst beds. Its flow rate is designed to satisfy the energy balance in the injection point. Figure 8.9(a) illustrates such a control strategy. The disadvantage of a costly multiloop configuration is compensated by the possibility of ensuring the *optimum temperature profile* on the longitude of the reactor, a profile that may be neither isotherm nor adiabatic. By using the optimizing principle of Pontryagin [12–13], the setpoint of each individual control loop in Fig. 8.9(a) is chosen to regulate the injected flow in a manner that ensures certain optimum T_{opt}° for the main gas flow after each catalyst bed. In other words, the effluent temperature T_1° after the first bed is combined with a fresh and cool stream (T_{in}°) in order to yield the influent T_{opt1}° in the second bed (see also Fig. 8.9(b)), and so on.

The individual values T_{opt1}° (after the first bed), T_{opt2}° (after the second bed), etc., should ideally lay on the optimum profile $T_{opt}^\circ = f(Z) = f\left(\frac{V}{F}\right)_Z$, depicted in Fig. 8.9(b). The points Z_i along the axis of the PFR correspond to various residence time values. The real (achieved) regulated temperature profile is not as smooth as the theoretical one (see Fig. 8.9(b)); it has a saw-teeth shape because the temperature cannot be controlled within the catalyst beds. However, it is obvious that it follows the evolution of the theoretical curve.

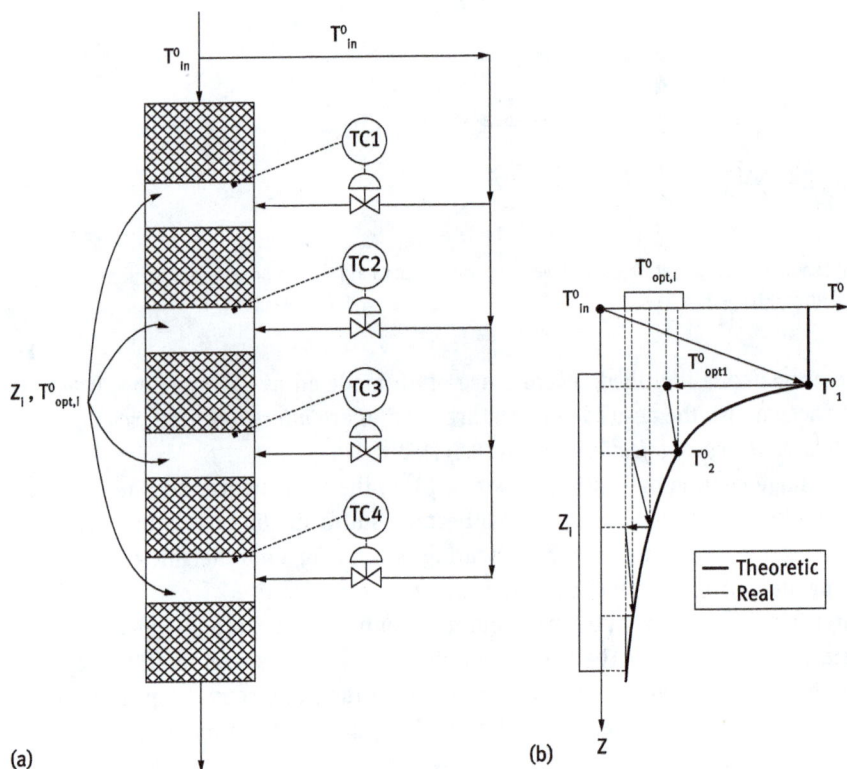

Fig. 8.9: Heterogeneous multibed reactor – temperature control strategy by inter-stage injection of cooled reaction mixture. (a) configuration of control loops; (b) optimal temperature profile on reactor length. Qualitative representation (adapted after [1, 8, 11]).

A $T_{opt}^{\circ} = f(Z)$ shape, like that in Fig. 8.9(b), favors reversible exothermic reactions. With the advance of the plug-shaped reaction mixture, the released heat constrains the process into regeneration of reactants. Hence, the final conversion at the reactor outlet drops. Meanwhile, if the temperature is lowered with process advance, the reverse reaction will lose significance and conversion gains of up to 20–40% can be achieved – see also Chapter 2.

Lowering the overall costs can be achieved by preheating the reactant flow at the expense of the products. It implies regulating the PFR inlet temperature as a function of the outlet value by means of a three-position valve (see Example 8.9 and Fig. 8.26). This is also a necessary safety measure; evacuation of gases has to be possible when pressure builds up because of overheated reactor outlet flows.

If the chemical process imposes extra constraints on temperature regulation and large heat transfer surfaces are required, a multitubular reactor [11] could be a suitable solution. Yet, it limits the throughput because of the smaller available cross sections for the fluid flow.

8.3.4 Temperature control in practice – batch units

Batch units operate in *cycles* that start with charging the solvent and reactants into the reactor, and then mixing and (usually) heating up to the reaction temperature. Initiation of the chemical process can occur during heating (reaction rates speed up to a significant value due to temperature increase) or after the addition of a catalyst or an initiator (reaction mechanism switches to a less energy-demanding path and occurs with a measurable rate). During the course of the reaction, the desired temperature profile can be kept linear or nonlinear. After reaction completion or achievement of the required conversion (process *end-point*), the unit is, as a rule, cooled off and emptied. The time amount necessary to complete one cycle (t_{load}) is obtained by adding to the actual reaction time, the durations of all other necessary operations (heat-up, cool-off, reactor emptying and cleaning, etc.).

For such a succession of operations, the process cannot be described by a steady state, neither can the temperature controller be tuned to a single setpoint. An adequate control system should permit rapid heat-up to the reaction initializing temperature by, meanwhile, minimizing its overshoot. It should also adjust rapidly and accurately the load temperature to the desired (optimal) profile.

This is a complex task since the dynamics of the batch unit varies in time. Thus, process variables as well as gains and all time constants vary during the reaction time. In the case of exothermic chemical reactions, the positive feedback of the process has also to be compensated by suitable self-regulating cooling controllers. In addition, batch-to-batch uniformity must also be maintained. The latter depends on many factors, starting with the purity of chemicals and ending with the repeatability of measurements and control equipment.

An example of synchronized temperature versus reaction time $T° = f(t_{reaction})$ profiles for both the reaction mixture (a) and the corresponding heat transfer medium (b) are illustrated in Fig. 8.10. The following operating modes can be distinguished within one batch cycle: I – heat-up, II – maintain/chemical reaction mode, and III – cool-down.

The control loop depicted in Fig. 8.4 regulates the load temperature also in a BTSR by manipulating the coolant inlet valve. For batch units, the jacket outlet valve could occasionally be closed during the duration of a complete cycle, whereas the other valves are operated as follows: during unit load, the effluent valve is closed, and during unit discharge, the influent valve is usually closed. The temperature of the heat transfer agent can be adjusted by mixing, for example, steam and water inlet flows in various ratios (see Fig. 8.10(b)).

A three-mode conventional PID master temperature controller could be employed by tuning it to the setpoint of the reaction mode II of a batch cycle (see Fig. 8.10(a)).

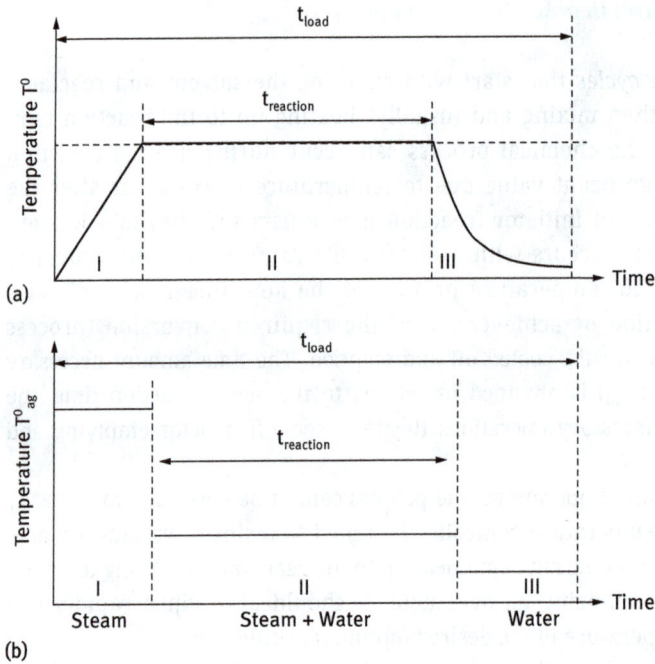

Fig. 8.10: Synchronized temperature versus reaction time profile for: (a) reaction mixture; (b) corresponding heat transfer medium. Operating modes within a batch cycle: I – heat-up, II – maintain/chemical reaction mode, III – cool-down. Qualitative representation (adapted after [8]).

A regulator with a PB of 30% and a 5 min setting for both integral and derivative times would generate a significant overshoot if kept on automatic during heat-up. Therefore, regular PID controllers require some added features that minimize this inconvenience. Depending on the required proportional band, the addition of a so-called *batch unit* feature (for PB < 50%) (see Fig. 8.11) or a *dual-mode unit* feature (for PB > 50%) to the controller will ensure smarter start-up characteristics [1–2, 7–9].

The common goal during heat-up is to raise the temperature within the unit from its initial to the elevated setpoint value as quickly as possible, but by avoiding *overshoot* and oscillations. The integral term of a conventional PID regulator will cause overshoot because it stays at the high output limit until the setpoint is reached – see top curve in Fig. 8.11. If this integral term is reduced to match the expected load, then the controller output is imposed to hold the variable at setpoint in the new steady state. Thus, overshoot is eliminated – see middle curve in Fig. 8.11. Without preloading, the controller will *undershoot* – see bottom curve in Fig. 8.11. If a derivative is not used, integration commences as soon as the output leaves its limit before crossing the setpoint. Hence, the preload setting has to be reduced somewhat below the expected load.

A dual-mode temperature controller also minimizes overshoot in a rapid heat-up. It couples an *on/off switch* controller and a preloaded PID controller. The on/off switch initiates either heating or cooling till a close value of the setpoint is reached. After this time-delay, the PID takes over and controls within a narrow temperature range. Such control is successfully applied to polystyrene synthesis reactors.

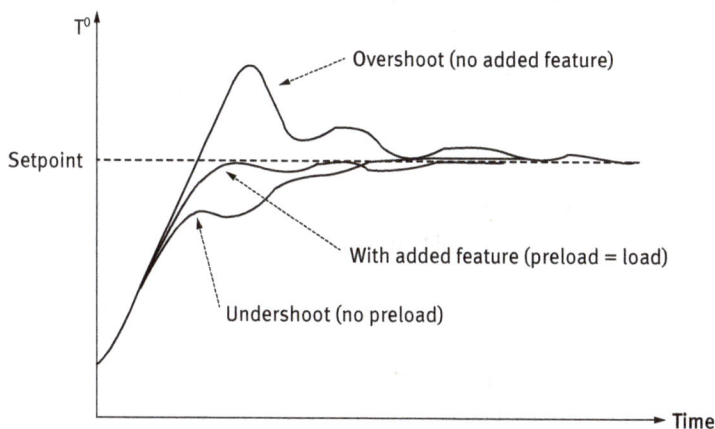

Fig. 8.11: Response of a preloaded "batch unit" PID temperature controller during heat-up. Qualitative representation (adapted after [1, 2]).

While each batch cycle benefits from the well-performing heat-up control, employment of either of the above described added feature to PID regulators has some drawback; settings need to be adapted from one batch to the other (even for the same unit) since heat transfer coefficients usually worsen. Yet, a preloaded PID controller having an external reset feedback from a secondary temperature is much more robust to disturbances [1, 2].

Chemical processes that are endangered to runaway behavior can be protected by setting a permissible rate of temperature rise during heat-up. In this case, the setpoint of the master loop depends on the heat-up rate. As the reaction mode temperature II is approached near the end of the heat-up mode I (see Fig. 8.10(a)), the value of the permissible rise rate is reset to lower values than at the beginning of the cycle. A model-based heat-up controller, for example, would use, in this case, the heat input rate as the variable, manipulated by the predictor block. Such an approach prevents building up of thermal inertia and keeps the process out of the potentially unsafe temperature domains.

A smart temperature control strategy ensures an *optimal* (*linear* or *nonlinear*) load temperature during the reaction course. Hence, the end-point is reached in a minimum of reaction time and the product distribution (described by the values of conversion and selectivity) is favorable. The optimum $T° = f(t_{reaction})$ profile is to be obtained by solving eq. (8.24), since the reaction rate r depends on both temperature $T°$ and conversion ξ, and the latter on the reaction time $t_{reaction}$ (see Example 8.3):

$$\frac{dr}{dT^\circ} = 0 \qquad (8.24)$$

A *nonlinear* setting is desirable for exothermic reversible (equilibrium) reactions (see also Section 8.3.3 and Fig. 8.9(b)). At process beginning, the accumulation of heat accelerates both reactions but favors mostly the direct step (because of high reactant concentrations), whereas when the products accumulate, the rate of the reverse reaction gains significance. Moreover, temperature elevation will also push the mass balance toward the reactants. Therefore, a decreasing temperature profile during the course of the reaction is an adequate strategy.

Meanwhile, irreversible processes that are only slightly exothermic could benefit from an *adiabatic* operation mode. Accumulated heat and the resulting temperature increase can compensate for the decay of the reaction rate due to the consumption of reactant species, but only as long as thermal safety requirements are met. As a result, reaction end-point can be reached in smaller volume units.

Jacket and heat transfer media temperature regulation can also benefit from non-linear settings for runaway reactions. Cooling must be very effective at the beginning of the reaction mode II, but as the process advances, the reaction rates decrease and hence the process-generated heat rate also decays. Thus, cooling rates have to be adjusted accordingly.

Let us consider the example of a reversible first-order reaction $A \Leftrightarrow P$ in both directions and occurring in liquid phase. Its rate is given by eq. (8.25). Terms k_1 and k_2 (both expressed in s^{-1}) stand for the first-order rate coefficients of the direct and reverse reactions, respectively, whereas C_A and C_P stand for the momentary molar concentration (kmol/m^3) of reactant A and product P, respectively:

$$r = k_1\ C_A - k_2\ C_P = k_{01} \exp\left(-\frac{E_{a1}}{RT^\circ}\right) C_A - k_{02} \exp\left(-\frac{E_{a2}}{RT^\circ}\right) C_P \qquad (8.25)$$

If the initial concentration of the product is zero ($C_{Po} = 0$), then

$$C_A = C_{Ao}(1 - \xi); \quad C_P = C_{Ao}\xi \qquad (8.26)$$

and eq. (8.25) can be written as a function of ξ as follows:

$$r = k_{01} \exp\left(-\frac{E_{a1}}{RT^\circ}\right) C_{Ao}(1 - \xi) - k_{02} \exp\left(-\frac{E_{a2}}{RT^\circ}\right) C_{Ao}\xi \qquad (8.27)$$

According to eq. (8.24), it further becomes

$$\frac{dr}{dT^\circ} = \frac{C_{Ao}}{R(T^\circ)^2}\left[(1 - \xi)\ k_{01}\ E_{a1} \exp\left(-\frac{E_{a1}}{RT^\circ}\right)\right] - \xi\ k_{02}\ E_{a2} \exp\left(-\frac{E_{a2}}{RT^\circ}\right) = 0 \qquad (8.28)$$

and the optimum temperature value T°_{opt} in relationship (8.30) can be calculated from it. Because the equilibrium composition is related to the rate coefficient ratio k_1/k_2 by means of equilibrium constant K (see (8.29), where ξ_{eq} stands for the equilibrium value of ξ), T°_{opt} is controlled by both kinetic and thermodynamic data [14]:

$$K = \frac{k_1}{k_2} = \frac{(C_A)_{eq}}{(C_B)_{eq}} = \frac{\xi_{eq}}{1-\xi_{eq}} \tag{8.29}$$

$$T^{\circ}_{opt} = \frac{E_{a1} - E_{a2}}{R \ln\left[\left(\frac{E_{a1}\, k_{01}}{E_{a2}\, k_{02}}\right)\left(\frac{1-\xi}{\xi}\right)\right]} \tag{8.30}$$

The resulting $T^{\circ}_{opt} = f(\xi)$ in eq. (8.30) and its correlation with $\xi = f(t_{reaction})$ will generate the optimal $T^{\circ}_{opt} = f(t_{reaction})$ profile (see Fig. 8.12(a) and Example 8.3).

Conversion is always defined with respect to the limiting reactant – see eq. (8.1). Hence, if there are two or more reactants involved, the concentration of species that affects the rate should take into account the *excess ratio* y of other species over the limiting one.

For the reversible second-order reaction $A + B \Leftrightarrow P$ on the direct reaction path and of first on the reverse, y is defined as

$$y = \frac{C_{Bo}}{C_{Ao}} \tag{8.31}$$

where A is the limiting and B is the *excess reactant* [5]. According to this definition, $y > 1$. A value $y = 1$ means that the reactants are mixed in stoichiometric ratios. T°_{opt} – see result in eq. (8.34) – can be calculated by using the reaction rate (8.33), and C_A and C_P from eq. (8.26), together with the expression of C_B in eq. (8.32):

$$C_B = C_{Ao}(y - \xi) \tag{8.32}$$

$$r = k_1\, C_A\, C_B - k_2\, C_P = k_1\, (C_{Ao})^2(1-\xi)\,(y-\xi) - k_2\, C_{Ao}\, \xi \tag{8.33}$$

$$T^{\circ}_{opt} = \frac{E_{a1} - E_{a2}}{R \ln\left[\left(\frac{E_{a1}\, k_{01}}{E_{a2}\, k_{02}}\right)\left(\frac{C_{Ao}(1-\xi)\,(y-\xi)}{\xi}\right)\right]} \tag{8.34}$$

Terms k_1 and k_2 in eqs. (8.23)–(8.34) signify the rate coefficients of the direct and reverse reactions: k_1 is of the second (m^3/kmol · s) and k_2 of the first order (s^{-1}), respectively.

Example 8.3
The desired conversion for the above discussed first-order reversible reaction $A \Leftrightarrow P$ is $\xi = 0.8$. The maximum permitted temperature by the equipment is $T^{\circ}_{max} = 900$ K. Kinetic data are $k_{01} = 30$ min^{-1}, $k_{02} = 10^6$ min^{-1}, $E_{a1}/R = 6000$ K, and $E_{a2}/R = 16,000$ K. Heat-up starts from 300 K and occurs at a rate of 10 K/min.

Advancing ξ values from 0 to 0.8 yield T°_{opt} values by the relationship (8.30). These result in k_1 and k_2, which further lead to ξ_{eq} and $t_{reaction}$ [5], respectively – see eqs. (8.35)–(8.36). At each temperature, ξ_{eq} is calculated and the reversibility requirement $\xi \le \xi_{eq} \le 1$ has to be fulfilled:

$$\xi_{eq} = \frac{k_1}{k_1 + k_2} \tag{8.35}$$

(a)

(b)

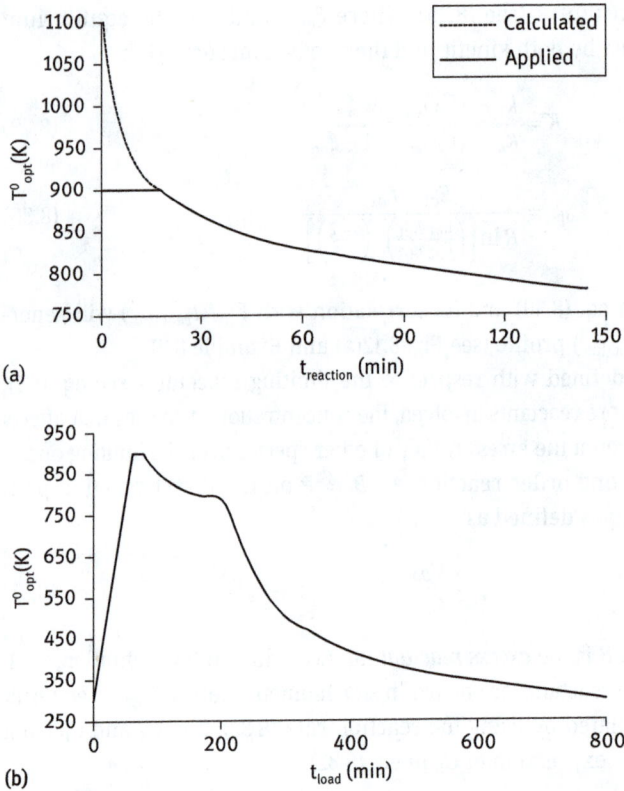

Fig. 8.12: Optimum temperature versus time profiles for the batch unit in Example 8.3: (a) duration of chemical process and (b) duration of a batch cycle (I – heat-up mode, II – chemical reaction mode, III – cool-down mode).

$$t_{reaction} = \frac{k_1}{k_1 + k_2} \ln\left(\frac{\xi_{eq}}{\xi_{eq} - \xi}\right) \qquad (8.36)$$

The results, correlated into a nonlinear $T_{opt}^o = f(t_{reaction})$ profile for the duration of the chemical process (operation mode II in Fig. 8.10(a)), are presented in Fig. 8.12(a). The dotted line stays for the calculated (theoretical) optimum temperature values. Since the unit is limited to T_{max}^o of 900 K, the setpoint temperature profile corresponding to the continuous line settles at exactly this value in the region for which the computed T_{opt}^o exceeds T_{max}^o. Figure 8.12(b) illustrates the entire temperature profile of a cycle for this batch unit. It may be observed that the linear setting of the chemical operation mode II in Fig. 8.10(a) is replaced by an optimized nonlinear one.

8.4 Pressure control

Section 8.3.3 mentions some examples of chemical processes that are, besides temperature-, also pressure-sensitive. Figure 8.13 presents examples of pressure control loops for batch units where a gas-phase species is either (a) the reactant (chemical reaction is preceded by the absorption of gas into liquid) or (b) the reaction product (for example, carbon dioxide) or both the reactant and the product, respectively. In both cases, the bulk liquid phase concentration C (kmol/m^3) of the species in question is related to its partial pressure in the vapor space p_{gas} (atm) by Henry's law [6, 14]. In relationship (8.37), k_{Henry} (kmol/m$^3 \cdot$ atm) is a temperature-dependent coefficient.

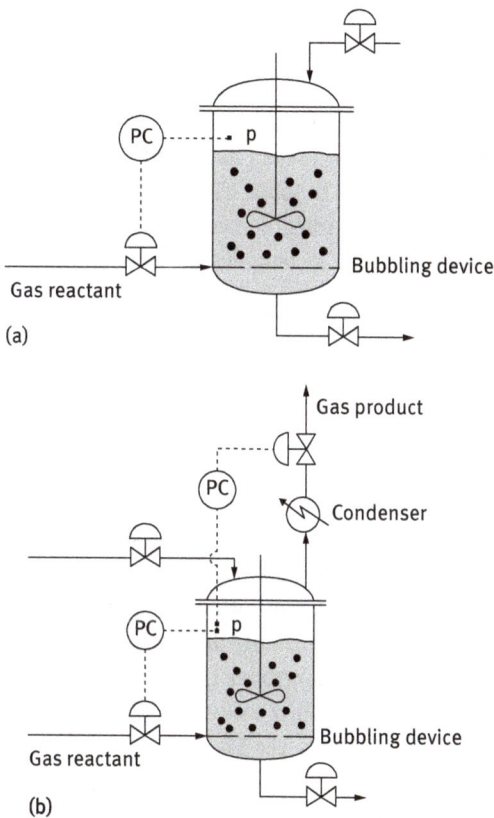

Fig. 8.13: Pressure control in batch units: (a) gas is reactant; (b) gas is a product or both a reactant and a product (adapted after [1]).

$$C = k_{Henry} p_{gas} \qquad (8.37)$$

Hence, pressure changes are linked to reaction advance (by means of conversion ξ) as well as to temperature changes.

Both pressure regulators depicted in Fig. 8.13 are fast and easily controlled. Depending on the measured vapor phase total pressure, they manipulate either an inlet or an outlet gas flow by means of a valve. A gas vent is usually employed for the loop in Fig. 8.13(b). Moreover, a condenser is often placed in front of the vent to minimize liquid product loss. If the chemical process occurs with both gas consumption as well as generation, the gas inlet pipe in Fig. 8.13(b) may benefit from a flow controller, also manipulated by reactor pressure. This configuration is part of the necessary safety strategy (it may shut down gas inlet) for the undesired situation of pressure building up in the unit and the outlet vent being not able to properly reduce it.

In continuous units operating both on liquid and gas streams, placement of at least one of the flow controllers on ratio (see Chapter 1) is useful because it ensures the desired setpoint initial mixture of the reactants.

For processes requiring vacuum, its source is often a steam jet-type ejector. Such constant-capacity sources are commonly matched to the variable-capacity reactor by wasting the excess capacity of the ejector [1]. A special situation is that of vacuum stripping in polymerization units: solvent and unreacted monomers have to be removed as rapidly as possible without causing the reaction mass to foam. The load temperature depends on the balance between heating (due to steam inlet) and cooling (due to monomer vaporization). The vacuum level is tightly linked to temperature, therefore also controlled by it.

8.5 Liquid level control

Liquid level in an operating unit is an indicator of the reaction mass volume. Tank reactors are filled only up to 70%–80% of their total volume. This fraction depends on process characteristics; if reaction-caused foaming occurs, then it will be smaller.

Well-controlled residence time for a continuous unit translates into a tight V/F ratio between the reaction mixture volume V and the reactor outflow F. It also ensures both desired conversion – see eqs. (8.3) and (8.4) – and product distribution – see eq. (8.2). This is commonly accomplished in practice by a controller that approximates V through a level measurement and manipulates the outflow valve. Such a loop is depicted in Fig. 8.14(a). It operates similarly in a batch unit – see Fig. 8.14(b) – where the desired volume, and thus dilution, conversion, and selectivity – see eqs. (8.1), (8.2), and (8.5) – is regulated by means of inlet flows. When the fluid level in the reactor reaches the setpoint value, inlets are shut down.

Volume V (liquid level) in Fig. 8.14(a) is maintained constant by a proportional regulator that modifies the effluent rate F (controller output) according to:

$$F = K_V(V - V_{min}) \tag{8.38}$$

The controlled variable V_{min} stands for the minimum value of V of the setpoint. For a batch unit, such as in Fig. 8.14(b), V_{min} corresponds to $F = 0$. The proportional gain K_V (s^{-1}) is a coefficient that depends on both the regulator and the valve characteristics.

8.6 pH control

8.6.1 pH and titration curves

The pH is a numeric scale used in chemistry to specify the acidity or basicity of an aqueous solution. It is defined as the negative decimal logarithm of the *hydrogen ion activity*. This differs from the concentration $[H^+]$ by a factor called *activity coefficient*. The value of the latter depends on the *ionic strength* (load) of the aqueous media and lies within 0 and 1. Yet, for diluted solutions, such as wastewater, it is approximated to 1. Thus, pH can be written as

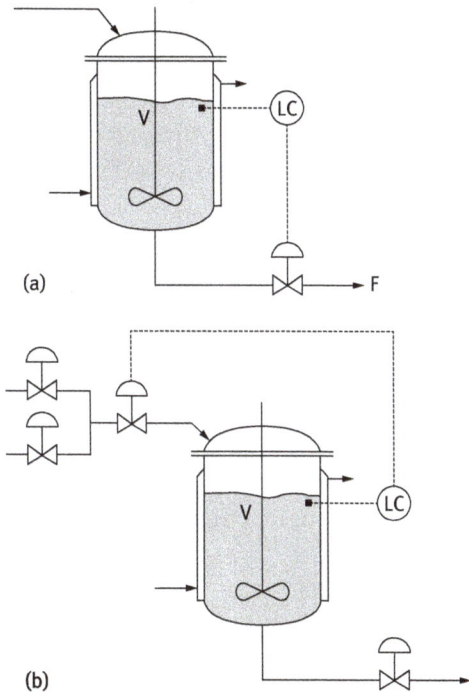

Fig. 8.14: Liquid level control for (a) a continuous unit (residence time control); (b) a batch unit (volume control). Simplified scheme for isotherm operation (adapted after [8]).

$$pH = -\log[H^+] \text{ or } [H^+] = 10^{-pH} \tag{8.39}$$

At 25 °C, the neutral value of pH is 7, where $[H^+] = [OH^-] = 10^{-7}N$. The brackets indicate *Normal* concentrations, designated as N and expressed in g-ions/dm³. At 25 °C,

$$[H^+] * [OH^-] = 10^{-14} = K_{Water} \tag{8.40}$$

K_{Water} is the dissociation equilibrium constant of water. It varies between 10^{-15} at 0 °C and 10^{-12} at 100 °C [6]. When $K_{Water} = 10^{-14}$, the *acidic* range is covered by pH of 0–7, whereas the *basic* is covered by 7–14 units. It is obvious that a tenfold variation of either $[H^+]$ or $[OH^-]$ concentrations yields only a 1 unit change in pH. This strong non-linear (logarithmic) relationship – see eq. (8.39) – makes pH so difficult to control.

In practice, pH control is required mainly for neutralization of industrial waste-waters. These contain *strong* or *weak* and *acid* or *basic* species. *Buffering* chemicals may also be present. Among the latter category are carbonates, silicates, phosphates, sulfonates, organic origin citric acid, or basic amines, and metal ions such as iron, copper, lead, chromium, cobalt, nickel, tin, zinc, cadmium, vanadium, manganese [15], and others. Buffers and weak agents, either acid or base, behave alike.

Some of the common strong acids are hydrochloric, nitric, or sulfuric acid. The strong bases include sodium or potassium hydroxide. They dissociate completely into ionic species when dissolved in water, whereas the weak agents dissociate (ionize) only partially and in more than one stage. These elementary steps are characterized by ionization constants: K_{acid} for acids, K_{base} for bases. Values of pK_{acid} and pK_{base} are commonly listed in databases; these stand for the negative decimal logarithm values of K_{acid} or K_{base}, respectively, and represent the pH value that neutralizes half of the amount in an aqueous solution.

When strong hydrochloric acid or sodium hydroxide base is added to the solution, it reaches concentrations C_{acid} and C_{base}, respectively. Negative and positive charges must balance [16], so that

$$C_{acid} + [H^+] = C_{base} + [OH^-] \tag{8.41}$$

Coupling eqs. (8.40) and (8.41) yield eq. (8.42), the titration curve equation of these two strong agents [16]. It is plotted in Fig. 8.15(a) in $pH = f(C_{base} - C_{acid})$ coordinates:

$$C_{base} - C_{acid} = 10^{(pH-14)} - 10^{-pH} \tag{8.42}$$

If weak acetic acid is neutralized with the same base, eq. (8.42) and its plot modify accordingly (see Fig. 8.15(b)) [16]. Depending on the number of dissociation steps and the K_{acid} values, the ratio's value in eq. (8.43) will change. The same is true for bases:

$$C_{base} - C_{acid} = 10^{(pH-14)} - 10^{-pH} + \frac{C_{Acetic\,acid}}{1 + 10^{(pK_{Acetic\,acid}-pH)}} \tag{8.43}$$

In Fig. 8.15(a), point 0 on the abscissa means that the acid and base characters of the aqueous solution are in balance $(C_{base} - C_{acid} = 0)$ and neutralization is complete. Negative abscise values indicate an acidic character, whereas positive values, a basic character. The reverse titration of sodium hydroxide with hydrochloric acid yields the mirror image of this curve.

The shape of curve in Fig. 8.15.a illustrates clearly the difficulties of a proper pH control. Bringing the pH in the vicinity of neutral 7 translates in crossing the huge gap, on a concentration scale of a few orders of magnitude, between the actual value of the controlled variable and the setpoint. Thus, a linear regulator is not suitable.

Example 8.4

Let us consider an ion exchanger wastewater entering treatment procedure at acidic pH 3. The goal is to raise its pH to 6. This requires matching the inlet concentration of 10^{-3} N with an accuracy of $\pm 10^{-6}$ N; in other words, the amount of caustic reagent added has to match the load in ratios of 1 part to 1000 [16]. Only 1 unit less in the pH of the inlet flow makes the goal more difficult; since precision has to be con-

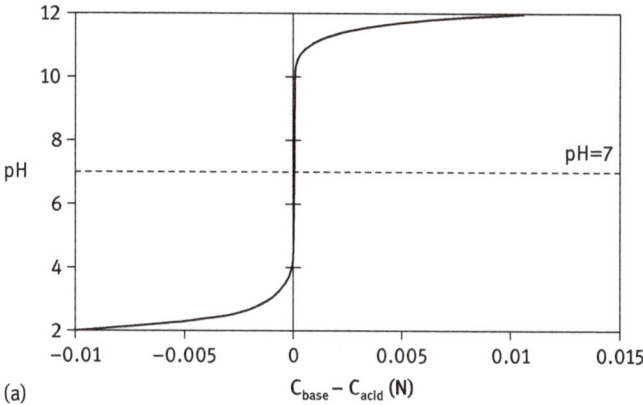

(a)

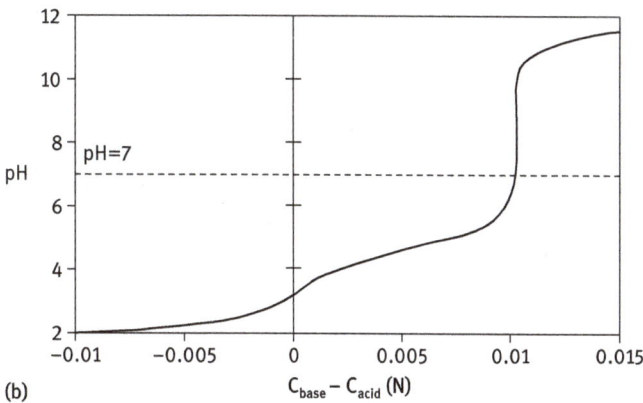

(b)

Fig. 8.15: Titration curve at 25 °C: sodium hydroxide neutralizes an aqueous solution of (a) 10^{-2} N hydrochloric acid; (b) 10^{-2} N acetic acid (adapted after [16]).

served, the amount to control is in the ratio of 1 part to 10,000. For the example of increasing pH 3 to 6, 1 m^3 of 10^{-3} N hydrochloric acid is neutralized by 1 m^3 of 10^{-3} N sodium hydroxide (see titration curve in Fig. 8.15(a)).

Meanwhile, the shape of curve in Fig. 8.15(b) shows that weaker acids or buffers are less sensitive to added reagent amounts; thus, pH control becomes somewhat easier. Here, neutralization is complete at $C_{base} - C_{acid} = 0.01$, the exact concentration of acetic acid.

A comparison between the curves in Fig. 8.15 leads to the apparent conclusion that any weak or buffering species present, besides the strong agent in an aqueous solution to be neutralized, makes the pH control task easier. Yet, real wastewaters contain a wide variety of species in large concentration ranges. Some of these are only partially ionized but all must be neutralized. The problem lies in the fact that the pH analyzer senses all and delivers an overall value, regardless of the species generating it. This again makes pH control difficult since different amounts of reagent have to be added if the same pH is generated by a strong or a weak agent. Yet, the demand for an extremely precise material balance remains.

---

i **Example 8.5**

Let us consider the same wastewater (pH of 3) as in Example 8.4. If the inlet value is due to acetic acid instead of hydrochloric acid, 1 m^3 of $\approx 5.6 \times 10^{-2}$ N sodium hydroxide is required to neutralize 1 m^3 of wastewater (56-fold more concentrated than for 10^{-3} N hydrochloric acid, also yielding pH 3). A similar calculation shows that a 10^{-3} N acetic acid solution has a pH of ≈ 3.6; this is seemingly closer to the neutral pH 7. Yet, 1 m^3 of it needs 1 m^3 of base of $\approx 9.5 \times 10^{-3}$ N caustic agent, almost a 10-fold higher amount than expected [16].

It may be concluded that feedforward signals of pH from the inlet flow are not useful in regulating the neutralizing reagent flow because overdosing or underdosing might occur. The only exception could be the situation of known composition inlet flows.

8.6.2 pH regulator characteristics

pH sensors have to fulfill constraints described generally for analyzers in Section 8.7. The electrical impedance of a widely used glass sensor is in the range of 10^2 MΩ. It contains a galvanic cell with two electrodes: a standard hydrogen electrode opposed to a reference, usually calomel or silver chloride electrode. The electromotive force of the cell depends linearly on the potential E_{H^+/H_2} (V) of the hydrogen ion-selective electrode, which also depends linearly on the pH. Ideally, it obeys the Nernst equation [6, 14]:

$$E_{H^+/H_2} = E^{\circ}_{H^+/H_2} - \frac{2.303 \; R \; T^{\circ}}{F_{araday}} pH \qquad (8.44)$$

In eq. (8.44), the standard electrode potential is $E^{\circ}_{H^+/H_2} = 0$, and the value of the slope is of 59.1 mV (also called the *Nernstian slope*) when: $R = 8.314$ J/mol · K, $T^{\circ} = 298$ K and

$F_{araday} = 96485\,C\,/\mathrm{mol}$. It also points out that pH is temperature-dependent, so that both calibration as well as measured readings must be temperature-compensated.

Because of the unique shape of titration curves, PID regulators employed in pH control need special nonlinear characterizers to linearize the loop. Therefore, there is no universal pH characterizer. However, three main types [16] are commonly in use since rigorous control has to be applied mainly around the setpoint and not over the entire curve (see Fig. 8.16).

One option is to *match* the titration curve (curve a in Fig. 8.16). This might be applicable if the influent to be neutralized is well described in terms of chemical species and their concentrations. However, these situations are rare and a great variety of titration curves overlap into a *real* one, which might even differ for the same influent. For example, municipal wastewater has different composition when household or storm water is looked at. In these cases, a very simple *three-piece* characterizer may be used (curve c in Fig. 8.16). It consists of three lines, each one applicable for a certain pH range. The *error-square* characterizer (curve b in Fig. 8.16) relies on a simple parabolic transfer function of the deviation [16], $f(e) = e|e|$. All are qualitatively illustrated in Fig. 8.16, where the matched curve is considered to superpose the real titration curve. Other characterizers are also feasible, but the above-described types have found the most extensive use.

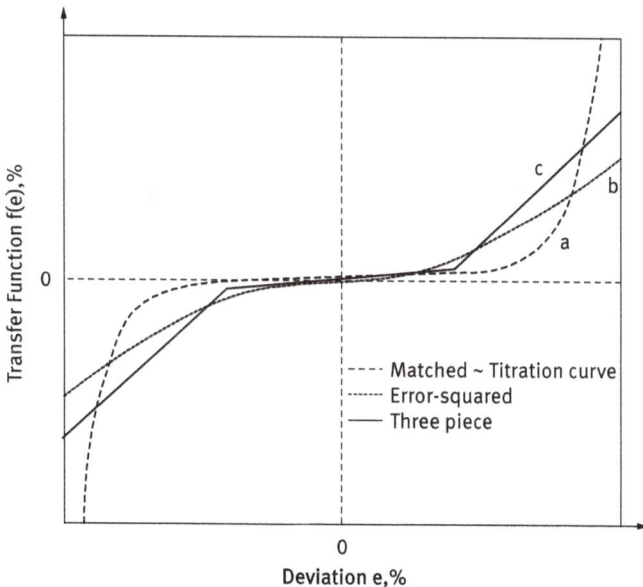

Fig. 8.16: Most extensively employed pH characterizers as compared to a real titration curve: (a) matched; (b) three-piece; (c) error-square (adapted after [16]).

Among the three, the matched curve provides the best response. However, recovery time from significant load changes is still longer than the oscillation period around the setpoint. Thus, even if matching is performed by some hundreds of points, the control loop will still not behave like a linear one. The three-piece characterizer is a very crude description of the titration curve, but still performs well; it has an adjustable minimum gain, whereas the error-squared curve has a zero gain at zero deviation. Moreover, the domain for which the gain of the three-piece description is valid is also adjustable [16].

Because of the logarithmic feature of titration curves, the PID regulator's proportional band and gain have to be chosen adequately to slightly damp the loop in the vicinity of setpoint. This is to avoid cycling. When cycling occurs for a controller that operates both acid and base agent valves in sequence, the amplitude could fall out of the specification range. Hence, the controller will command alternative addition of each reagent even if there is no real influent load change; only the added agents will neutralize each other (waste of chemicals) [1–2]. Limit-cycling may be acceptable, for example, in the case of wastewater where the target specification range is large (between pH 6–9) and the cycling amplitude falls within it.

The integral time of the employed PID should be at least twofold higher as compared to the corresponding optimum of a linear loop; yet it is limited to the window necessary for the pH to recover from a significant change. The fastest part of the cycle (near setpoint) yields the optimum derivative time. It is comparable to that of a corresponding linear loop.

Commercially available *self-tuning PID* controllers are usefully employed when process parameters change frequently and substantially [16]. These are tuned for the worst-case scenario and let self-tuning do the rest: observe loop behavior under upset conditions and estimate the settings that improve performance. The drawback is that no distinction is made among pH readings due to load changes, dirty, decalibrated, or malfunctioning sensors; thus, the PID may readjust to unreal parameters.

8.6.3 Aspects of pH control in practice

According to Hoyle et al. [16], neutralization is a process that *must* be designed for controllability. There are many aspects to be considered and only a few, generally valid, recommendations are mentioned here. Preferred continuous units should be stirred tanks with influent entering at the top and effluent evacuating at the bottom. The unit diameter should equal the liquid column height. The residence time should be of at least 5 min and the dead-time maximum, 1/20 of it. The mixer has to ensure a vessel turnover of maximum 1 min and a pumping rate of a minimum of 20-fold the throughput flow.

In batch units, a liquid is placed in the vessel and its pH is adjusted to the desired value by adding a certain reagent. Since no flow is continuously entering or leaving the unit, the load is zero. Thus, the controller output bias should be fixed at zero when setpoint pH is reached so that the valve dosing the neutralizing agent closes at setpoint. This involves elimination of integral action and, consequently, either a simple proportional or a proportional-plus-derivative regulator is recommended. The absence of a proportional controller can be replaced by building into the calculation block a (8.37)-like relationship, where K_{pH} stands for the proportional gain:

$$\text{Controller output} = K_{pH}\left(pH_{setpoint} - pH_{controlled}\right) \tag{8.45}$$

In other words, the logarithmic (nonlinear) feature of titration curves is compensated with equal-percentage valves for the neutralizing agents. The valve does not have to match the entire curve because controlled pH reaches the setpoint only once, at the end of the batch cycle, when the valve is closed. The dynamic pH profile over a cycle time will show a slight undershoot because the derivative component of the regulator closes the neutralizing reagent valve just shortly before the setpoint pH is reached. Even though such course of actions might slightly increase the duration of a cycle, it is preferred over a permanent overshoot (see also Section 8.7.1). Figure 8.17 illustrates such a pH control loop.

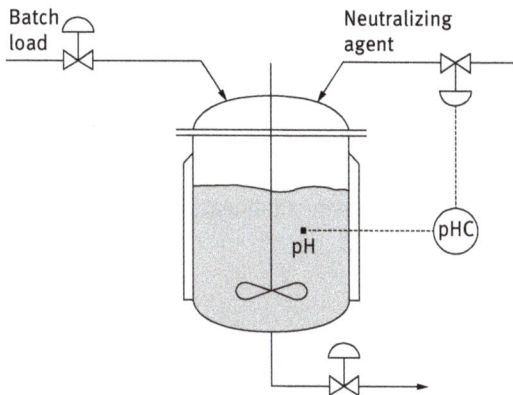

Fig. 8.17: A pH control loop for a batch unit.

Figure 8.18 illustrates a series of three consecutive CSTRs. Such a configuration alleviates the effort of adjusting pH over more than 2 units, which is over more than 100-fold concentration range. Manipulating valves and flows within a precision of ±1% is thus manageable. In a single neutralizing stage, pH would most probably cycle excessively because of valve imprecision.

Units I, II, and III in Fig. 8.18 cover approximately two pH units each, starting from either strong acidic or basic media in I (pH of 1 or 13) and closing up stepwise (through II with pH of 3 or 11) to the desired pH in III (pH of 6 or 9) [16]. The successive steps units are served by smaller valves; thus, each setpoint has to be very well defined. Otherwise, a failure upstream passes unmanageable disturbances (overloads) downstream.

The target specification of the final outlet flow is the setpoint of the last stage, which is responsible for the fine-tuning of pH. Each unit has its own independent pH control loop. Yet, these are still able to pass along undesired cyclings to the next loop. To avoid a cyclic disturbance upstream that matches the exact period of a controller downstream, the loops should have distinct/different oscillation periods.

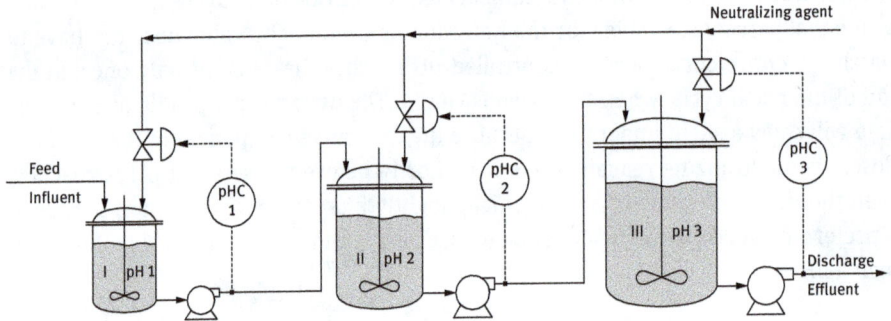

Fig. 8.18: Series of successive continuous units to adjust to a wide-range pH change (adapted after [16]).

This is the reason behind sizing the units progressively larger from the feed to discharge (from I to III).

Each of the three continuous units in Fig. 8.18 may also operate separately. PIDs with settings as described in Section 8.6.2 are recommended; linear PI controllers can cause large overshoots in such units – see Section 8.6.1.

8.7 End-point detection and product-quality control

Composition assessment by measuring various physical and chemical proprieties of a reaction mixture has many benefits. Analysis procedures inserted in either continuous or batch unit technological flows contribute to:
–   early detection of deviations from setpoint values (ensures desired quality products, avoids loss of chemicals);
–   proper detection of equipment malfunctions (contributes to plant safety, avoids economic loss);
–   correct identification of process end-points (reduces batch cycle time, increases productivity, avoids undesired side reactions).

Even though analytical instruments need adequate maintenance/calibration and often raise reliability questions, their use (if available and economically feasible) is recommended for end-point detection. However, other methods are also commonly in use for completion assessment in batch units. For example, the reactant flow can be terminated based on the total load weight or a decline in heat release is noticed when the limiting reactant is totally consumed. For polymerization reactions, the disappearance of the monomers is accompanied by a pressure drop.

8.7.1 Some analyzer types

Analyzers are used to determine some physical or chemical property of the reaction mixture. This has to be relevant for the conversion and vary enough to be detected properly if the system is disturbed. Therefore, instruments should be calibrated in a manner that senses even slight changes in the measured property. Besides pH (see Section 8.6), these could be resistivity, conductivity, density, viscosity, absorbance, reflectance, transmittance, etc. [7]. Commercially available automatic on-stream detectors also involve infrared (IR) beam attenuation, gas chromatography (GC), high-performance liquid chromatography (HPLC), or mass spectrometry (MS) and even coupling of some of the mentioned techniques [7].

The sensors can be placed on-stream for continuous units, either in the reactor itself or in the outlet pipe. Reaction vessel placement is chosen for batch units. Many variations are encountered in practice: retractable or fixed, inserted sidewise or immersed from the lid of the vessel, with sample extraction or in situ detection, etc. The analyzers must perform accurately and rapidly since control variables are often in their grasp (for example, temperature, pressure, or pH control). Thus, automatic cleaning and rinsing is to be considered.

Some analytical procedures require grab-sampling, followed by bench analysis. Here, sampling and sample preparation difficulties have to be overcome. These are more time consuming, hence intervention in the process is delayed.

In continuous units, the measured (controlled) variable has a single setpoint value. However, the analytical property in a batch reactor is away from its setpoint for almost the entire reaction duration. Therefore, integral action must be avoided or else it leads to irrecoverable overshoot (see pH control in Section 8.6, or heat-up mode in Section 8.3.4). In continuous operation, overshoot is temporary; the outflow will eventually purge away its effects.

8.7.2 End-point detection reliability issues

Reliable analyzers are necessary for the proper operation of reaction vessels as well as for overall plant safety. Reliable *readings* are conditioned by an adequate placement of the sensor into a representative and uniformly mixed zone of the reaction mass. Malfunction or failure that might drive the unit into an unsafe state has to be avoided; therefore data provided by the analyzers should be *temperature-compensated* and the sensing devices well cleaned, maintained, and calibrated. Deposits of tars, precipitates, or biological growth on the sensor cause variable bias readings; they raise the response dead-time and alter the signal value. In the case of electrodes or pH measurements, a fluid velocity of approximately 2–3 m/s [16] keeps them clean and causes no significant abrasion.

Extra care is to be granted to pH sensors, since their ideal response to 1 unit change of pH away from 7 is of only 59.1 mV [6, 14]. In glass bulb pH sensors, electrolyte fillings of both measuring as well as reference electrodes are buffered to pH 7. Hence, neutral pH yields a zero reading. A dead-short also yields 0 mV, indicating a false-neutral value.

In practice, such shortcomings are alleviated by the use of a *multiple sensor configuration* (redundant, voting or median selector). The example in Fig. 8.19 illustrates *redundant* readings by two equal-performance sensors inserted in the outflow pipe and linked to a high signal selector. If the controlled variable is, for example, a product concentration, the manipulated variable (inlet flow rate) is protected, regardless of which analyzer fails. If detector 1 deteriorates, then upstream detector 2 will take over automatically. If detector 2 malfunctions, the selector will command automatic shut down of the inlet. Thus, the reactor is safe until breakdown is remediated.

If no pauses in the production are permitted, a *voting* configuration (see Fig. 8.20) of at least three equivalent sensors is much safer. All are placed in the same media. Reliability is gained as any reading that disagrees with the majority is disregarded by the system. Data offered by the majority is further employed in closed-loop control. Another option is also presented in Fig. 8.20 and relies on a *median selector* in junction with the three analyzers. This rejects both the highest and the lowest of the three readings and thus transmits the median to the control loop. Reliability is enhanced because the selector also filters noise and transients that are uncommon to two of the signals, and protects the reactor against consequences of sensor failure.

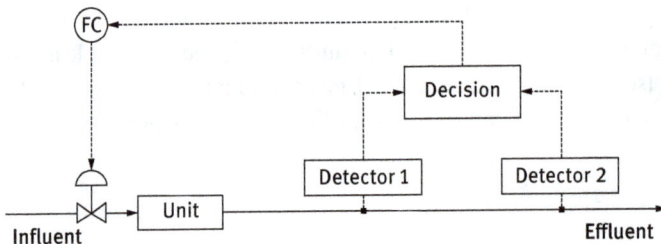

Fig. 8.19: Example of redundant analyzer configuration for adequate end-point detection and inlet flow rate regulation (adapted after [1]).

Execution

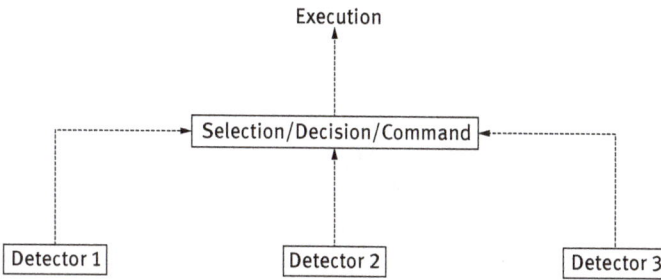

Fig. 8.20: A voting/median selector configuration of three analyzers (adapted after [1]).

In some cases, the value of the controlled variable (measured property) is not evenly distributed within the reaction mass. Thus, multiple sensor readings are necessary to decide on the value to be transmitted to the control loop. For example, the temperature in a heterogeneous fixed-bed reactor depends on the position of preferential flows, flow rates or catalyst age, etc. Consequently, temperature control is based on the highest reading of sensors aligned in the catalyst bed.

8.8 Control structure design for reaction units

8.8.1 Principles of control structure design

Designing the *control structure* of a chemical reactor means optimum selection of the controlled and manipulated variable pairs as well as adequate choice of controller loops and regulator configurations. Architecting control structures starts with an exhaustive characterization of the chemical process as well as of the unit that houses it, followed by a well-defined sequence of steps [17–23] – see also Chapters 5 and 6. For the first task, a *generic steady-state model* is usually sufficient. It consists of an array of equations describing mass, energy and momentum balances, description of physical and chemical phenomena, characterization of various raw and construction materials, etc. *Dynamic models* are mainly needed for batch units or continuous systems that face frequent and/or significant changes [17–18, 20]. Both model types are completed with *case-specific equations* when the controller parameters are defined and tuned.

Step 1. Definition of operational and control objectives
Control structure design begins with the definition of process *operational objectives* in terms of:
- product quality specifications (desired product distribution/purity, desired molecular weight distribution for polymers, etc.);
- production specifications (desired flow rate, desired residence time, etc.);

- operational constraints (avoid overheating reaction mass, avoid overflowing the unit, avoid drying out unit, maintain certain precision and accuracy of measurements, etc.);
- environmental and economic issues (avoid pollutant overload in wastewater and emissions, save material and energy, minimize operating costs, etc.).

The *control objectives* are derived from lists as above. After their qualitative descriptions are derived, these have to be quantified usually in terms of process output variables. For example, a continuous unit requiring thermal stability during steady-state operation translates quantitatively into core temperature that does not vary more than ±5 °C away from setpoint. A batch unit may require minimized cycle duration, that is, optimized reaction time. It is demonstrated in Example 8.3 that this translates into respecting a well-defined temperature versus time profile.

A list of possible *disturbance sources* that might upset the desired objectives is helpful in the choice of adequate measured, controlled, and manipulated variables. Some examples of disturbance generators are flow rate, composition, and temperature of influent stream; flow rate and temperature of coolant; pressure and purity of inlet gases; etc.

Step 2. Identification, classification, and selection of variables

The next stage of control design deals with identification, classification, and selection of variables. *Input* and *output* are rather general terms to define chemical process variables; they can describe characteristics of the feed and discharge streams of either a reactor or its jacket, but can also refer to the control system. For example, a reactor exit flow rate is an output for the chemical unit, but is also the input of a liquid level controller.

Thus, when targeting a control architecture, it is more straightforward to refer to *controlled* and *manipulated* variables; either may appear among process inlets or outlets. Depending on the accessibility of their values, variables may be *measurable* (onstream, in situ, or by grab-sampling) or *unmeasurable*. The latter are *estimated* by means of measured values of some other parameters and the controller's built-in mathematical models. A typical example is the liquid level measurement that yields calculated volume and residence time. Special attention is granted to various kinds of *disturbances* [7, 18, 19] (step, ramp, impulse, or sinusoidal shaped); these are (often measurable) input variables that upset an operating unit.

The list of *possible* controlled (usually, process outputs) as well as manipulated (usually, process inputs) variables is put together by following some rules and recommendations [17–20].

The selected controlled variable:
- is not self-regulating (such as the liquid level for a unit that has a pump in the discharge pipe);

 – may exceed equipment and operating constraints (such as temperature, pressure, concentration, pH, etc.);
 – reflects the desired quality (such as viscosity for polymerization or polycondensation reactions, pH for neutralizing processes, a certain chromatographic profile in organic synthesis, etc.);
 – is directly related to and affects significantly operational objectives (such as temperature affecting the selectivity and product distribution by means of side reaction rates);
 – can interact seriously with other controlled variables (such as upstream pH in a series of neutralizing units – see Fig. 8.18);
 – has favorable dynamic and static characteristics; it responds rapidly to adjustments imposed by the manipulated variable (such as inter-stage gas temperature for multibed heterogeneous reactors – see Figs. 8.9(a) and 8.26).

The manipulated variable:
 – has significant effect on the controlled variable;
 – results in rapid response of the controlled variable; in other words, affects it preferably directly (dead-time associated with manipulated variable should be relatively small as compared to process time constant);
 – does not recycle disturbances (such as an inlet stream that could propagate forward a disturbance, or a recycle flow that could recycle it back to the process).

The controlled variable has to be measured or estimated. However, if possible and economically feasible, it is useful to also measure the manipulated variables as well as the disturbances. Any measured value has to be reliable, accurate, and precise, recorded with minimum of time delay and from portions of the reaction mass, flows, etc. that significantly represent the changes of the parameter in question (see Section 8.7.2). Within this context, the choice of the measured variables from the list of controlled ones has to be carefully considered.

Step 3. Freedom degree analysis

The *freedom degree* $N_{Freedom}$ of a reactor [17–23] equals the number of variables that have to be specified to describe its operation. In other words, it is the difference between the number of independent variables $N_{Variables}$ and independent equations $N_{Equations}$ that interconnect these – see eq. (8.46). Any control loop raises the value of $N_{Equations}$ (because of control laws), thus lowering $N_{Freedom}$. Because in the dynamic balance equations, the accumulation terms differ from zero (and are equal to zero for steady state), $(N_{Freedom})_{Dynamic} \geq (N_{Freedom})_{Steady}$:

$$N_{Freedom} = N_{Variables} - N_{Equations} \tag{8.46}$$

However, for an underspecified process $N_{\text{Freedom}} > 0$. Some of the freedom degrees are imposed by the process environment (such as the flow rate F of an upstream unit feeding the reactor in question). Hence, the number of controlled variables N is at maximum equal to N_{Freedom}:

$$N = N_{\text{Controlled variables}} = N_{\text{Freedom}} - N_{\text{Fixed variables}} = N_{\text{Manipulated variables}} \qquad (8.47)$$

The number of independent manipulated variables has to be equal to the number of controlled ones (N), although it is possible to design some control systems that can regulate more variables than the number derived from eq. (8.47).

Step 4. Possible control configurations
After identification and selection in a correct number of possible controlled and manipulated variables, they have to be *paired* in control loops. When each manipulated variable is coupled to a single controlled variable (such as conventional feedback control), the chemical unit has a *multiloop* control configuration. On the other hand, when one manipulated variable depends on two or more controlled variables, a *multivariable* control configuration has been employed (such as model predictive control) – see also Chapters 2 and 5.

An operating unit with N controlled and N manipulated variables has $N!$ theoretically possible different control configurations. As N increases, the task of finding the best one is more difficult, because for $N = 2$, $N! = 1 \times 2 = 2$; for $N = 3$, $N! = 1 \times 2 \times 3 = 6$; for $N = 4$, $N! = 1 \times 2 \times 3 \times 4 = 24$; for $N = 5$, $N! = 1 \times 2 \times 3 \times 4 \times 5 = 120$ and so on. The value of N also reflects the evolution from SISO (single-input, single-output) to MIMO (multiple-input, multiple-output) control systems.

Figure 8.21 illustrates the two possible feedback multiloop configurations for a $N \times N = 2 \times 2$ chemical process. *Controlled V_1* and *Controlled V_2* stay for the 2 controlled variables, whereas *Manipulated V_1* and *Manipulated V_2* for the 2 corresponding manipulated ones.

As it is probably clear to the reader by now, the most widely employed control configurations are the feedback, inferential, and feedforward pairings [7–9, 17, 18] (see also Chapter 1). The *feedback loop* exploits a directly measured *controlled variable* – see Fig. 8.22(a), whereas the *feedforward loop* a measurement of the *disturbance* – see Fig. 8.22(b). The *inferential loop*, in either feedback or feedforward configuration, employs an *estimator block* instead of a direct (primary) measurement for the controlled variable; its value is computed from the results of *secondary* measurements (also see Step 2). The estimator is placed on the information stream between the measurement points and the controller (see Fig. 8.22).

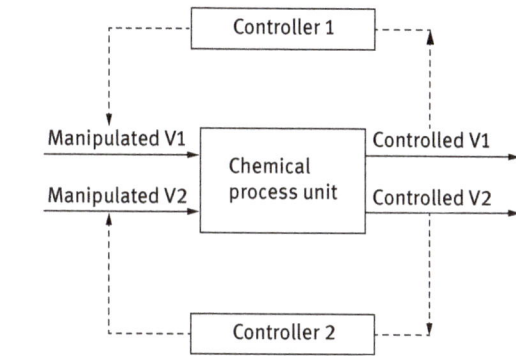

(a) First Configuration

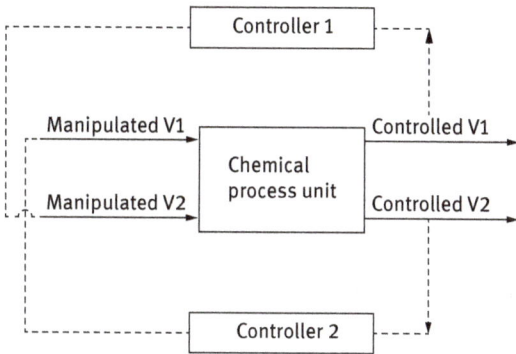

(b) Second Configuration

Fig. 8.21: Possible feedback multiloop configurations for a 2 × 2 chemical process.

Step 5. Control loop interaction analysis

Although pairing of controlled and manipulated variables follows common sense rules (see also Step 2), the "ultimate" criteria to decide upon the best control system configuration is minimization of the cross-interaction among the individual loops (see also Chapters 4 and 5).

Bristol [24] introduced in 1966 a very simple an effective interaction assessment, the *relative gain array* (RGA). RGA calculus and pairing rules are described in detail in Chapter 5. Because it relies on process steady-state gains, it is used in conjunction with the *Niederlinski index* NI to ensure system stability [19]. According to the RGA–NI rules [17–20], suitable controlled and manipulated variable pairs have positive, and as close to one as possible, RGA elements. Any pairing resulting in negative NI is unacceptable.

Even though the RGA–NI approach is commonly and widely used in the industry, there is scarce systematic information about how to effectively treat control

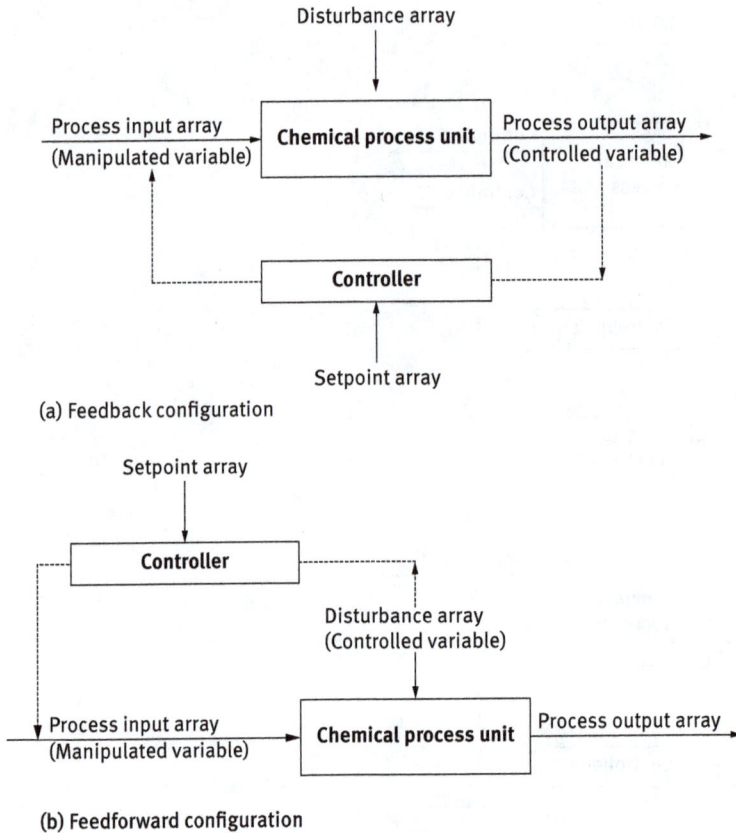

Disturbance array

Process input array | Chemical process unit | Process output array
(Manipulated variable) | | (Controlled variable)

Controller

Setpoint array

(a) Feedback configuration

Setpoint array

Controller

Disturbance array
(Controlled variable)

Process input array | Chemical process unit | Process output array
(Manipulated variable)

(b) Feedforward configuration

Fig. 8.22: Loop information flow for: (a) feedback control; (b) feedforward control.

configurations of high-dimensional processes. The RGA–NI serves well for 2 × 2 processes, since negative NI values just state loop instability for $N > 2$, but do not prove stability for positive values; it only indicates that the system is not definitely unstable and should be tested *via* dynamic simulation before implementation. Moreover, situations with RGA elements of equal values for individually different, yet feasible loops, are unsolved. Therefore, improved approaches have been proposed [25], such as the dynamic RGA (DRGA) [26–28], the block relative gain (BRG) [29], the relative normalized gain array (RNGA) [30], the improved controller-dependent DRGA [31], the effective gain array (ERGA) [32], or the effective relative energy array (EREA) [33].

Step 6. Controller design
The controller is a key element in any control loop. It is the active component that receives the information regarding the value of the controlled variable (either measured

or estimated), and by means of a smart transfer function, gives the command information to the manipulated variable for the process to rapidly regain its desired setpoint state.

Thus, they are chosen, depending both on the identity of the controlled variable (temperature, pressure, pH, conductance, color, etc.) and the manipulated variable (flow rate, flow ratios, temperatures of heat transfer agents, etc.). Typical regulator designs are widely described [7–9, 17, 18] and the user may choose from a variety of appliances that have:

- linear or nonlinear (such as "on-off switchers") relationships among the manipulated variable and the applied command;
- continuous or discrete actions;
- various conventional transfer functions (such as P, PI, PID), some special added features to the latter (such as a preloaded temperature PID controller used to heat-up a batch unit – see Section 8.3.4 and Fig. 8.11) or combinations of these (such as the dual-mode controller also described in Section 8.3.4).

8.8.2 Control structure design for homogeneous ideal units

To demonstrate how the abovementioned principles are employed, some simple chemical processes will be considered to occur in homogeneous phase in ideal chemical reaction units.

For the first-order exothermic chemical process $A \xrightarrow{k} P$, reaction rate r is described by relationship (8.9). Notations and significance of the terms are given in Section 8.3. The influent is characterized by flow rate F_{in}, temperature T_{in}° and concentration C_{Ao}, whereas the effluent by F, T°, and C_A, respectively. The reaction mass does not suffer density and heat capacity variations during the reaction course. Therefore, the values of ρ and C_P, respectively, are constant. Reaction heat is ΔH_r and reaction mass volume is V. The inlet heat transfer agent has a flow rate of $F_{ag,in}$ and a temperature of $T_{ag,in}^\circ$. They are mixed perfectly (the jacket behaves like a CSTR) and the mixture leaves the jacket with F_{ag} and T_{ag}°, respectively. The agent density and heat capacity are also constant at ρ_{ag} and $C_{P,ag}$, respectively. K_T and A_T refer to heat transfer characteristics of the jacket (see also Section 8.3). The reactor is considered to have no thermal inertia.

Example 8.6

The CSTR unit.

The above reaction takes place in a continuously stirred tank reactor operating in steady state as illustrated in Fig. 8.23. The generic process model consists of six independent equations: mass balances (8.48)–(8.50) (expressed in kmol/s), the heat balances (8.51) and (8.52) (expressed in J/s), and level control law (8.38). The accumulation terms of heat and mass balances are equal to zero (steady-state model). These relationships are completed with rate law (8.9), when necessary, with C_P, $C_{P,ag}$, and ΔH_r, temperature dependencies [6, 14], $A_T - V$ connection, and other relationships that characterize the system:

$$F_{in} - F = 0 \tag{8.48}$$

$$F_{ag,in} - F_{ag} = 0 \tag{8.49}$$

$$F_{in} \, C_{Ao} - F \, C_A - V \, r = 0 \tag{8.50}$$

$$F_{in} \, \rho \, C_P \, T_{in}^\circ - F \, \rho \, C_P \, T^\circ - \Delta H_r \, V \, r - K_T A_T \left(T - T_{ag}^\circ \right) = 0 \tag{8.51}$$

$$F_{ag,in} \, \rho_{ag} \, C_{P,ag} \, T_{ag,in}^\circ - F_{ag} \, \rho_{ag} \, C_{P,ag} \, T_{ag}^\circ + K_T A_T \left(T^\circ - T_{ag}^\circ \right) = 0 \tag{8.52}$$

There are 11 independent variables: process inputs F_{in}, T_{in}°, C_{Ao}, $F_{ag,in}$, and $T_{ag,in}^\circ$ and process outputs F, T°, C_A, F_{ag}, T_{ag}°, and V (or residence time V/F). By taking into account that T_{in}° and $T_{ag,in}^\circ$ are fixed by the environment of the reactor, three degrees of freedom will remain available for the control system, since $N_{Variables} = 11$, $N_{Equations} = 6$, and $N_{Fixed\,variables} = 2$. Thus, according to eqs. (8.46) and (8.47), $N = 11 - 6 - 2 = 3$.

The choice of controlled variables is made according to rules and recommendations listed in Section 8.8.1. Temperature affects exponentially the rate coefficient and hence strongly modifies the conversion (operational objective related to product quality). Moreover, for exothermic reactions, extra care is required because of runaway danger (operational objective related to equipment constraints and safety).

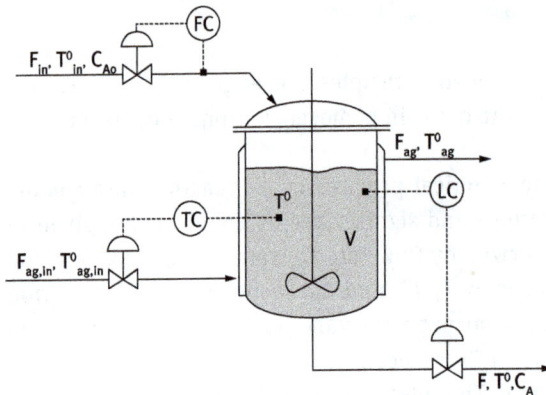

Fig. 8.23: Control system configuration for a CSTR and process described in Example 8.6 (adapted after [8]).

Thus, temperature ought to be controlled. Residence time affects both conversion (product quality) and heat transfer (operational objective related to economic issues). An option of regulating it is by means of volume, in other words, by liquid level control. Another parameter affecting the balances (8.50) and (8.51) is by means of input terms; hence, both product specification as well as overall costs are affected by the influent flow rate. It also affects the value of C_{Ao}.

The obvious manipulated variables to keep T°, V, and F_{in} at the desired values are adjusting the inlet flow rate for the jacket as well as both the reactor exit and inlet flow rates. Clearly, the measured variables should be core temperature, liquid level, and the inlet flow rate.

Tab. 8.1: Results of RGA analysis for a 3 × 3 chemical process occurring in a CSTR [8].

Manipulated variables	Controlled variables		
	F_{in}	T^o	V
F_{in}	1	0	0
$F_{ag,in}$	0	**1.3375**	−0.3375
F	0	−0.3375	**1.3375**

Table 8.1 presents the relative gain array for all three controlled and three manipulated variable pairings for the synthesis of ethyl acrylate [8]. Positive and close to 1 elements – see highlighted values in Tab. 8.1 – indicate suitable pairs (see also Chapter 5 and Section 8.8.1) [17–20]. For this particular case, the Niederlinski index is proven positive [8] and, therefore, suggests that the control system is not definitely unstable. Hence,

- Core temperature T^o is controlled by manipulating inlet flow rate $F_{ag,in}$ of the heat transfer agent; the controller uses directly measured T^o values.
- Volume V (and related residence time) is controlled by manipulating the effluent flow rate F; the controller uses an estimator block to calculate V from the measured liquid level – see eq. (8.38).
- Influent flow rate F_{in} is controlled by manipulating the feed valve; the controller uses the directly measured F_{in} values.

Example 8.7
The PFR unit.

Let us consider that the reaction takes place in a tubular plug-flow reactor and the latter has reached steady state – see Fig. 8.24. The generic model of the previous example modifies somewhat since the reaction mass is described as a plug that advances on the longitudinal axis Z of the tube (see also Section 8.2); thus the parameter values are distributed along the flow direction.

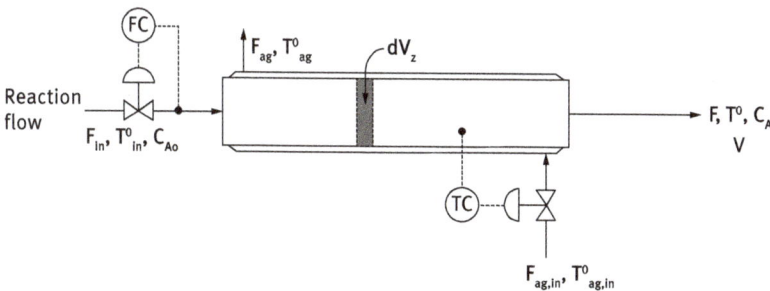

Fig. 8.24: Control system configuration for a PFR and process described in Example 8.7 (adapted after [8]).

Overall, mass balances (8.48)–(8.49) are valid for the PFR as well, but eqs. (8.50)–(8.52) are rewritten for the infinitely small volume dV, which has a $dA_{T,Z}$ heat-exchange surface. Index Z is related to position within the reactor, $dC_{A,Z}$ stands for the variation of (limiting) reactant content within the element dV, and $dT^\circ_{ag,Z}$ is the heat transfer agent temperature variation that corresponds to dV and $dA_{T,Z}$. Since each such element behaves like a CSTR, its exit temperature T°_Z also corresponds to its core and is the value at which all temperature-dependent parameters are calculated (here, r_Z and, when necessary, $C_{P,Z}$, ΔH_Z, or others). The accumulation terms are equal to zero for a steady-state model. Integration of eqs. (8.53)–(8.55) between the process-imposed limits (tube length L and V/F residence time) yields balances for the entire reactor:

$$F\,dC_{A,Z} - dV_Z r_Z = 0 \tag{8.53}$$

$$F\,\rho\,C_P(T_{in,Z}{}^\circ - T_Z{}^\circ) - \Delta H_r\,dV_Z\,r_Z - K_T\,dA_{T,Z}\left(T_Z{}^\circ - T_{ag,Z}{}^\circ\right) = 0 \tag{8.54}$$

$$F_{ag}\,\rho_{ag}\,C_{P,ag}\,dT_{ag,Z}{}^\circ + K_T\,A_{T,Z}\left(T_Z{}^\circ - T_{ag,Z}{}^\circ\right) = 0 \tag{8.55}$$

The PFR needs, like the CSTR in Example 8.6, temperature T° and residence time V/F control. The first may be achieved by the usual means of manipulating the inlet jacket flow rate $F_{ag,in}$. Since the volume V of the reaction mixture is imposed environmentally (fixed by reactor dimensions), the only means of V/F control is by manipulating the inlet flow rate F_{in} – see Fig. 8.24. The measured variables are T° and F_{in}.

---

i **Example 8.8**
The BSTR unit.
   The same first-order reaction is now considered to occur in a batch-stirred tank reactor – see Fig. 8.25. Mass and energy balances have to be written for dynamic behavior and accumulation terms are different from zero. Since the reaction mixture volume is imposed by unit dimensions, $dV/dt = 0$ and $F = 0$ (relationship (8.38) yields zero). Hence, only mass and energy balances for the reaction mass – relationships (8.56) and (8.57) – and the

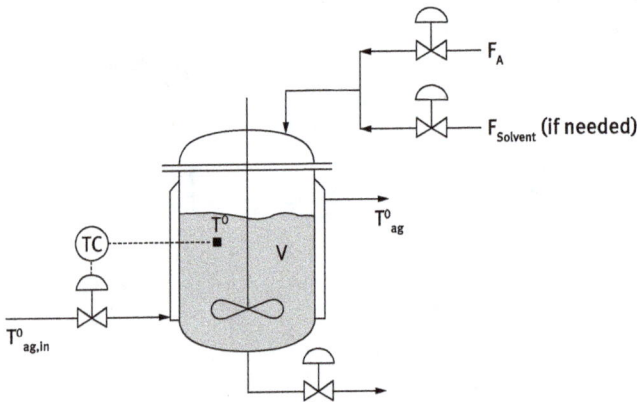

Fig. 8.25: Control system configuration for a BSTR and process described in Example 8.8 (adapted after [8]).

energy balance for the heat transfer agent – relationship (8.58) – have to be considered in control design. Integration is to be carried out for the time elapse corresponding to $t_{reaction} \cdot V_{ag}$ in eq. (8.58) stands for the fixed volume of the reactor jacket. Mass balances are expressed in kmol/s, whereas energy balances in J/s.

$$\left(\frac{d}{dt}\right)(VC_A) = -Vr = -VkC_A = -Vk_0 \exp\left(-\frac{E}{RT°}\right)C_A \tag{8.56}$$

$$\rho\, C_P \frac{d}{dt}(VT°) = -\Delta H_r Vr - K_T A_T\,(T° - T°_{ag}) \tag{8.57}$$

$$\rho_{ag}\, C_{P,ag} \frac{d}{dt}\left(V_{ag} T°_{ag}\right) = K_T A_T\,(T° - T°_{ag}) \tag{8.58}$$

The control system in Fig. 8.25 consists of a temperature control loop, based on the direct measurement of core temperature and jacket inlet flow rate manipulation. Extra features of the regulators for batch reactors are described in Section 8.3.4 (see also Fig. 8.11). During unit charging, flow rate controllers for the reactants (maybe also put on ratio adjustment – see Chapter 1) are useful but not always necessary. On the other hand, the process end-time detection is both useful and necessary. Section 8.7 goes more into this issue's details. Thus, an end-time detector, coupled with the temperature regulator, may also be imaginable.

8.8.3 Control structure design for some heterogeneous units

Example 8.9
A multibed heterogeneous catalytic column – heterogeneous PFR.
    Let us consider a gas-phase reaction occurring in a multibed heterogeneous catalytic column like the one in Fig. 8.26 (see also Fig. 4.2). The unit can be treated like a PFR. For example, the industrial methanol production from synthesis gas on copper-based catalyst occurs according to independent reaction steps (8.59) and (8.60) [34–36]:

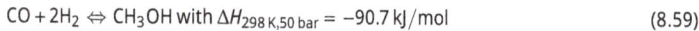

$$CO + 2H_2 \Leftrightarrow CH_3OH \text{ with } \Delta H_{298\,K,50\,bar} = -90.7\,kJ/mol \tag{8.59}$$

$$CO_2 + 3H_2 \Leftrightarrow CH_3OH + H_2O \text{ with } \Delta H_{298K,50\,bar} = -40.9\,kJ/mol \tag{8.60}$$

$$CO_2 + H_2 \Leftrightarrow CO + H_2O \text{ with } \Delta H_{298K,50\,bar} = -49.8\,kJ/mol \tag{8.61}$$

The values of the reaction enthalpies above were reported by Aasberg-Petersen et al. [35]. Other independent side reactions are [34]:

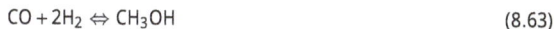

$$CO_2 + 4H_2 \Leftrightarrow CH_4 + 2H_2O \tag{8.62}$$

$$CO + 2H_2 \Leftrightarrow CH_3OH \tag{8.63}$$

A variety of rate laws have been proposed, depending on the employed catalyst, reaction conditions, reactor design, etc. [34–38]. All have, in common, the dependence of rates on both pressure and temperature; for example those employed by Machado et al. [34] in a comparative analysis of methanol production routes:

$$r_{\text{equilibrium}\,1} = \frac{K_1\, K_{CO}\left(f_{CO}\, f_{H_2}^{1.5} - \frac{f_{CH_3OH}}{f_{H_2}^{0.5}\, K_{P,\text{equilibrium}\,1}^{\circ}}\right)}{\left(1 + K_{CO}\, f_{CO} + K_{CO_2}\, f_{CO_2}\right)\left(f_{H_2}^{0.5} + \frac{K_{H_2O}}{K_{H_2}^{0.5}} f_{H_2O}\right)} \tag{8.64}$$

$$r_{\text{equilibrium}\,2} = \frac{K_2\, K_{CO_2}\left(f_{CO_2}\, f_{H_2}^{0.5} - \frac{f_{CH_3OH}\, f_{H_2O}}{f_{H_2}^{0.5}\, K_{P,\text{equilibrium}\,2}^{\circ}}\right)}{\left(1 + K_{CO}\, f_{CO} + K_{CO_2}\, f_{CO_2}\right)\left(f_{H_2}^{0.5} + \frac{K_{H_2O}}{K_{H_2}^{0.5}} f_{H_2O}\right)} \tag{8.65}$$

$$r_{\text{equilibrium}\,3} = \frac{K_3\, K_{CO_2}\left(f_{CO_2}\, f_{H_2} - \frac{f_{H_2O}\, f_{CO}}{K_{P,\text{equilibrium}\,3}^{\circ}}\right)}{\left(1 + K_{CO}\, f_{CO} + K_{CO_2}\, f_{CO_2}\right)\left(f_{H_2}^{0.5} + \frac{K_{H_2O}}{K_{H_2}^{0.5}} f_{H_2O}\right)} \tag{8.66}$$

The significance of terms in eqs. (8.64)–(8.66) is:

- $r_{\text{equilibrium},i}$ (where $i = 1$, 2, and 3) stands for the rate of individual reaction steps (8.59), (8.60), and (8.61) in this sequence (kmol/m$^3 \cdot$ s);
- dimensionless $K_{P,\text{equilibrium},i}^{\circ}$ and K_i stand for the equilibrium constants of the above reaction steps for standard and employed reaction conditions, respectively;
- dimensionless K_{CO}, K_{CO_2}, K_{CH_3OH}, K_{H_2}, and K_{H_2O} stand for the adsorption equilibrium constants of gas-phase species on the solid catalyst surface;
- dimensionless f_{CO}, f_{CO_2}, f_{CH_3OH}, f_{H_2}, and f_{H_2O} stand for the fugacity of each species involved in the main reaction steps.

Relationships (8.64)–(8.66) prove the importance of coupled pressure and temperature control. Used together with mass and energy balances like those described in Example 8.7, they complete the steady-state model of this reaction unit – see Fig. 8.26.

Fugacity in gas phase is similar to the activity in condensed phases [6, 14] (see also Section 8.6.1). Its value depends on both total gas pressure and temperature. Thus, rate laws are both pressure- and temperature-dependent because of both fugacity and the various equilibrium constant values. As a consequence, the reactor needs temperature as well as pressure control. The control structure in Fig. 8.26 not only permits temperature control at the reactor inlet via loop TC1 but also ensures optimum temperature profiles (see also Fig. 8.9) by adjusting it between catalyst beds via loops TC2 and TC3, respectively (see also Fig. 8.9 and Section 8.3.3). All loops rely on direct measurements and manipulate gas flow rates. The outlet flow can preheat the inlet reactants, thus lowering overall operating costs of the unit. This is possible by means of a three-way valve.

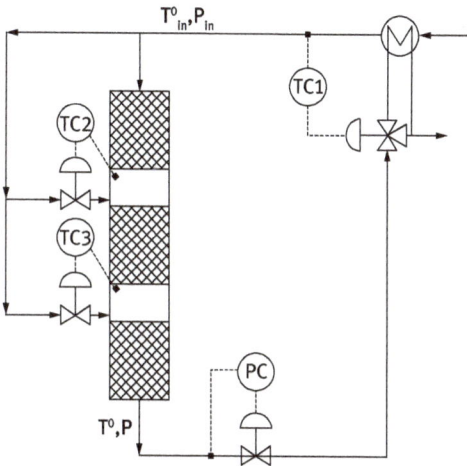

Fig. 8.26: Control system of a gas-phase reaction in a multibed heterogeneous reaction column as described in Example 8.9 (adapted after [1, 8]).

Example 8.10

A bubbling column – heterogeneous CSTR.

Let us consider a heterogeneous gas-liquid reaction occurring in a bubble column like the one in Fig. 8.27. Such units have a simple design, are suitable for continuous and semi-continuous operation, and provide high back-mixing in the liquid phase. Therefore, their model derives from that of the CSTR. One of the main advantages is that they can operate in virtually isotherm conditions because of liquid evaporation. Disadvantages are due to the limited employable temperature and the pressure ranges [11].

For the case in Fig. 8.27, the gas-phase reactant A has to dissolve first in the liquid phase B to react with it (oxidation of various organic liquids [11, 39], commercial production of acetic acid by carbonylation of methanol [40, 41], and carbon dioxide sequestration by caustic media [39, 42–45]). In other words, mass transfer by means of gas and liquid phase diffusion as well as by absorption of gas into the liquid are a prerequisite of the actual chemical reaction [4, 39]. Hence, these steps can also be rate determining.

Let us consider the process of carbon dioxide sequestration by ammonia [42–44] or other organic amines such as monoethanolamine and biethanolamine [45]. Here, diffusion and chemical reaction are much faster than absorption. Other simplifying assumptions are: the reaction occurs in the liquid bulk phase between a dissolved species A entering the unit with the gas inlet and species B contained by the liquid, according to $A_{(L)} + B_{(L)} \xrightarrow{k} P_{(L)}$; and the reaction rate law (8.67) is of second-order, the unit is perfectly stirred and operates under isotherm conditions at $T°$. If the liquid inlet temperature corresponds to the core value and the reaction heat is negligible ($\Delta H_r \approx 0$), energy balances are not necessary here:

$$r = k\, C_A\, B_A = k_0\, \exp\left(-\frac{E_a}{RT°}\right) C_A\, B_A \tag{8.67}$$

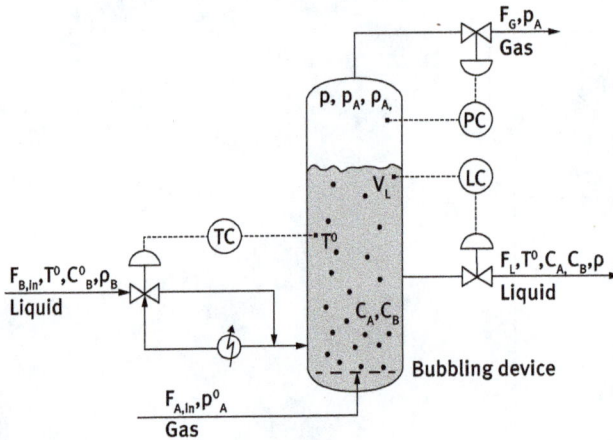

Fig. 8.27: Control system of a heterogeneous gas-liquid bubbling column as described in Example 8.10 (adapted after [8]).

The rate coefficient k in eq. (8.67) is expressed in $m^3/kmol \cdot s$ [5, 6, 14]. For the significance of the other terms, see eq. (8.6). By taking into account the listed simplifying assumptions, the generic steady-state model will consist of overall mass balances (expressed in kmol/s) for the liquid as well as gas phase in eqs. (8.71) and (8.72) respectively, mass balances (also expressed in kmol/s) for reactants A and B in eqs. (8.73) and (8.74), respectively, and relationships (8.68)–(8.70) describing the mass transfer (absorption) of species A at the gas-liquid interface. The accumulation terms in eqs. (8.71)–(8.74) are equal to zero:

$$C_{Ao} = \frac{p_A}{k_{Henry,A}}; \quad k_{Henry,A} = f(T^\circ) \tag{8.68}$$

$$p_A = \frac{n_A}{n}p; \quad \rho_A = \frac{p\,M_A}{R\,T^\circ} \tag{8.69}$$

$$N_A = k_{L,A}\,(C_{Ao} - C_A) \tag{8.70}$$

$$F_{B,in}\,\rho_B + M_A\,N_A\,A_{G-L} - F_L\,\rho = 0 \tag{8.71}$$

$$F_{A,in} - F_G - \frac{A_{G-L}N_A M_A}{\rho_A} = 0 \tag{8.72}$$

$$A_{G-L}N_A - F_L C_A - V_L k C_A C_B = 0 \tag{8.73}$$

$$F_{B,in}C_{Bo} - F_L C_B - V_L k C_A C_B = 0 \tag{8.74}$$

The significance of the terms in the above equations is (see also Fig. 8.27):
- $F_{A,in}$ and $F_{B,in}$ are the gas and liquid inlet flow rates (m^3/s);
- F_G and F_L are the gas and liquid outlet flow rates (m^3/s);

- C_{Ao}, C_{Bo}, C_A, and C_B are the liquid phase concentrations (kmol/m^3) of species A and B, respectively;
- V_L is the reaction mass volume (m^3);
- ρ_A and ρ_B are the densities of bulk gas phase A and liquid phase B, respectively (ρ stands for the density of reaction mixture, which is the exit liquid flow kg/m^3);
- p stands for the total pressure (atm), whereas p_A stands for the partial pressure of A (atm) in the unit's vapor space;
- $k_{Henry,A}$ is the temperature-dependent Henry's coefficient characterizing the mass transfer of A from gas to liquid (kmol/m^3 · atm);
- $A_{G\text{-}L}$ is the total mass transfer surface (m^2) given by bubbles;
- n is the total mole number, whereas n_A stands for the mole number of A in gas phase (kmol);
- M_A is the molecular mass of species A (kg/kmol);
- N_A is the molar flow rate of A from gas to liquid through mass transfer surface (kmol/m^2 · s);
- $k_{L,A}$ is the flow rate coefficient (m^3/m^2 · s);
- T° is the core temperature (K);
- R is the universal gas constant (m^3 · atm/kmol · K).

An operational objective such as constant product distribution in the liquid effluent requires the control system presented in Fig. 8.27. Since the temperature- and pressure-dependent mass transfer of A from gas to liquid (absorption) is assumed to be rate determining, both T° and p control are required. T° is measured directly within the core of the reaction mass and regulated via the temperature of the liquid inlet $F_{B,in}$. The pressure is also measured directly, but in the unit's vapor's space. Its value is regulated by means of the gas outlet flow valve, in other words, via value of F_G. Residence time control ensures both the above stated operational objectives and avoids flooding or drying out the column. It relies on a level measurement and a V_L estimator that accordingly manipulates the liquid effluent flow rate F_L.

It is worth mentioning that the value of $A_{G\text{-}L}$ can be raised using various bubbling devices or stirrers or by packing the unit with inert fillers [4, 11, 39].

Example 8.11

A stirred batch electrolysis cell – heterogeneous CSTR.

Electrochemical reactors are useful when high-purity products are desired that cannot be obtained otherwise with economically feasible selectivity. These units can operate continuously either like a CSTR or a PFR, semicontinuously or in batch cycles [8, 11]. Electrolysis occurs when an externally applied current flows between two solid electrodes through a usually liquid electrolyte. As a result, it mediates electron exchange between chemical species – see electrolysis law (8.75). It has the advantage of much lower temperatures than conventional thermal or catalytic procedures. The products can be solid deposits (metal refining), liquid (sodium hydroxide solution, adiponitril, etc.), or gaseous products (chlorine, hydrogen, ozone, etc.).

The process requires a minimum voltage to be applied; the thermodynamically predicted value is called *decomposition voltage*, yet the real voltage is higher (disadvantage of such units) because of electrode overpotential and ohmic losses [6, 14]. This fact has to be considered in any steady-state model – see eqs. (8.81)–(8.84) [6, 14, 46, 47]. Chemical transformations happening both on electrodes and in bulk electrolyte as well as mass transport phenomena (when occurring with significant rate [48–51]) are also parts of the model.

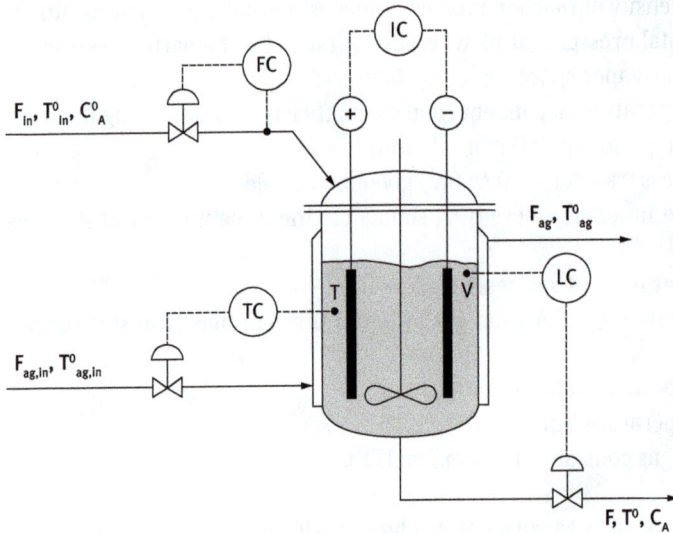

Fig. 8.28: Control system of a CSTR electrochemical unit as described in Example 8.11 (adapted after [8, 11]).

Let us consider the electrochemical unit illustrated by Fig. 8.28, in which the first-order reaction $A + z_A e^- \xrightarrow{k} P$ takes place at constant temperature $T°$. According to the first law of electrolysis [6, 14], the mass m altered at each electrode is directly proportional to the quantity of electricity transferred to it. Thus,

$$n_A = \frac{m_A}{M_A} = \frac{I\frac{V}{F}}{z_A F_{araday}} \tag{8.75}$$

The significance of terms not listed yet within this chapter is:
- n_A stands for the transformed mole number of A (mol), whereas m_A corresponds to the mass (g);
- M_A stands for molecular mass of A (g/mol);
- z_A stands for the number of exchanged electrons by A during the electrochemical process;
- I is the current intensity (A);
- terms V/F and F_{araday} stand for the *electrolysis* (residence) *time* (s) and Faraday constant (96485 C/mol), respectively.

Let us make some simplifying assumptions:
- The reactor is perfectly stirred and behaves like a CSTR [46];
- The chemical process does not affect significantly the liquid flow density ρ (kg/m^3) and its heat capacity C_P (J/kg · K); so these are considered constant;
- The reaction heat is negligible ($\Delta H_r \approx 0$), such that the reaction-generated heat will not affect energy balance (8.77).

The steady-state mass balances (8.76) and (8.77) have zero accumulation terms. Mass balances (kmol/s) (8.76) and (8.48) and (8.49) as well as energy balances (J/s) (8.77) and (8.52) are completed by eqs. (8.78)–(8.84). These describe the overall voltage, electrode potentials, various overpotentials, and other physical and chemical phenomena:

$$F_{in}C_{Ao} - FC_A - \frac{I}{z_A F_{araday}} = 0 \tag{8.76}$$

$$F_{in}\rho C_P T_{in}^\circ - F\rho C_P T^\circ - K_T A_T \left(T^\circ - T_{ag}^\circ\right) + R_{total}\frac{I^2}{2} = 0 \tag{8.77}$$

$$C_A = C_{Ao}\frac{I}{z_A F_{araday} F} \tag{8.78}$$

$$U = \varepsilon_{anode} + \varepsilon_{cathode} + \eta_{anode} + \eta_{cathode} + \eta_{ohmic} \tag{8.79}$$

$$\left(\varepsilon_{ox/red}\right)_{electrode} = \left(\varepsilon_{ox/red}^\circ\right)_{electrode} - \frac{2.303\ R\ T^\circ}{z\ F_{araday}}\lg\left(\frac{a_{ox}}{a_{red}}\right)_{electrode} \tag{8.80}$$

$$\eta_{electrode} = const_1\ T^\circ + const_2\ T^\circ \lg\left(i_{electrode}\right) \tag{8.81}$$

$$i_{electrode} = \frac{I}{A_{electrode}} \tag{8.82}$$

$$\eta_{ohmic} = I\left(R_{electrolyte} + R_{other}\right) \tag{8.83}$$

$$R_{electrolyte} = \frac{d}{\lambda_{electrolyte}\ A_{anode+cathode}} \tag{8.84}$$

Again, the significance of terms not listed yet within this chapter is:
- i stands for the current density (A/m^2);
- A_i stands for the indicated electrode surface (m^2);
- U stands for the applied potential (V);
- $\left(\varepsilon_{ox/red}^0\right)_{electrode}$ stands for the reduction potential (V) at the specified electrode (anode or cathode); according to the Nernst law (8.80) [6, 14], it is written as a function of the standard value $\varepsilon_{ox/red}^0$ (V), and activities of the oxidized and reduced forms of the species that suffer electrochemical transformation (a_{ox} and a_{red}, respectively, both expressed in kmol/m^3);

- $\eta_{electrode}$ stands for the overpotential (V) at the specified electrode (anode or cathode); according to the Butler-Volmer law (8.81) [6, 14], it is written as a function of local current density $i_{electrode}$ and temperature $T°$;
- η_{ohmic} stands for potential drop (V) caused by the resistance of all electrical charge transporting media (electrolyte, wires, contactors, etc.);
- R_i stands for the electrical resistance (Ω) of the specified media: $R_{electrolyte}$ for the electrolyte, R_{total} for electrical charge transporting media that come into contact with the electrolyte and can heat it up due to electrical current transport (Joule heating); R_{other} for the wiring, contactors, etc.;
- $\lambda_{electrolyte}$ stands for the electrolyte conductibility (S/m);
- d represents the distance between the anode and the cathode (m).

The electrochemical cell presented in Fig. 8.28 and the CSTR in Fig. 8.23 behave alike. Hence, the electrochemical reactor requires core temperature $T°$, electrolysis (residence) time V/F, and inlet flow rate F_{in} control. This is achieved as described for Example 8.6. Because current intensity I governs mass transformation, this unit needs an additional I control loop. Its value is adjusted by manipulating the applied voltage U.

There are some comments worth making for Examples 8.6–8.11. The described control schemes serve well and under stabile conditions. Yet, protection against wide variations of influent quality (C_{Ao} or p_A^o not constant, not adequate electrode material, etc.) or undesired phenomena (worsened heat transfer because of various deposits, incorrect measurement of controlled variable, detector failure or malfunction, etc.) is not guaranteed. These would require extra control loops and more complex control structures.

Control systems illustrated in Figs. 8.23–8.28, even if simple in design, can benefit from smart built-in models as well as from optimized characteristics. For example, core temperature control of the polyvinyl chloride synthesis batch reactor improves when a nonlinear model predictive controller NMPC (see Section 2.9) is used instead of a conventional PID [52]; strong disturbances and frequent changes will be handled well within smaller time windows.

References

[1] Kendall, D.C, Schlegel, D.F., Hertanu, H.I., Molnár, F., Lipták, B.G., *Chapter 8.9. Chemical Reactors: Basic Control Strategies*, p. 1664–1696 in Lipták, B.G. (Ed.), *Instrument Engineers' Handbook*, Vol. II. *Process Control and Optimization*, 4th Edition, CRC Press, Boca Raton, FL, 2006.
[2] Shinskey, F.G., *Chapter 8.10. Chemical Reactors: Control and Optimization*, p. 1697–1710 in Lipták, B.G. (Ed.), *Instrument Engineers' Handbook*, Vol. II. *Process Control and Optimization*, 4th Edition, CRC Press, Boca Raton, FL, 2006.
[3] Simándi, B. (Ed.), Cséfalvay, E., Déak, A., Farkas, T., Hanák, L., Mika, L.T., Mizsey, P., Sawinsky, J., Simándi, B., Szánya, T., Székely, E., Vágó, E., *Vegyipari műveletek II. Anyagátadó műveletek és kémiai reaktorok*, 2nd Edition, Typotex Kiadó, Budapest, 2012.
[4] Levenspiel, O., *Chemical Reaction Engineering*, 3rd Edition, John Wiley & Sons, New-York, 1999.

[5]   Marin, G., Yablonsky, G.S., *Kinetics of Chemical Reactions. Decoding complexity*, Wiley-VCH, Weinheim, 2011.

[6]   Atkins, P., de Paula, J., *Physical Chemistry*, 9th Edition, Oxford University Press, Oxford, 2010.

[7]   Agachi, P.S., Cristea, M.V., *Basic Process Engineering Control*, DeGruyter, Berlin/Boston, 2014.

[8]   Agachi, Ş, *Automatizarea proceselor chimice*, Casa Cărţii de Ştiinţă, Cluj-Napoca, 1994.

[9]   Shinskey, F.G., *Process Control Systems*, 4th Edition, McGraw-Hill, New-York, 1996.

[10] Hopkins, B., *Pressure Monitored Temperature Controlled System for a Liquid-Vapor Process*, Patent US 3708658 A, Monsato Company, 1973.

[11] Henkel, K.D., *Reactor Types and Their Industrial Applications*, in *Ullman's Encyclopedia of Industrial Chemistry*, Vol. 31, p. 293–327, Wiley-VCH, Weinheim, 2012.

[12] Imre-Lucaci, A., Agachi, P.Ş., *Optimizarea proceselor din industria chimică*, Editura Tehnică, Bucureşti, 2002.

[13] Agachi, P.Ş., Vass, E., *Optimizarea reactorului de fabricare a metanolului utilizând principiul maximului lui Pontryagin* Revista de Chimie, 6, 513–521, 1995.

[14] Motschmann, H., Hofmann, M., *Physikalische Chemie*, DeGruyter, Berlin/Boston, 2014.

[15] Csavdári, A., *Catalytic Kinetic Methods in Analytical Chemistry. Principles and Applications*, Editura MEGA, Cluj-Napoca, 2008.

[16] Hoyle, D.L., McMillan, G.K., Shinskey, F.G., *Chapter 8.32. pH Control*, p. 2044–2056 in Lipták, B.G. (Ed.), *Instrument Engineers' Handbook*, Vol. II. *Process Control and Optimization*, 4th Edition, CRC Press, Boca Raton, FL, 2006.

[17] Luyben, W.L., *Chemical Reactor Design and Control*, John Wiley & Sons, Hoboken, NJ, 2007.

[18] Seborg, D.E., Edgar, T.F., Mellichamp, D.A., *Process Dynamics and Control*, 2nd Edition, John Wiley & Sons, New York, 2004.

[19] Suliman, M.A., Ali, E.M., Alhumaizi, K.I., Ajbar, A.H.M., *Control System Design and Structure*, Lecture Notes of the "Process Control in the Chemical Industries" Workshop, King Saud University, Riyadh, Saudi Arabia, p. 39–64, 2002.

[20] Skogestad, S., *Control structure design for complete chemical plants*, Computers and Chemical Engineering, 28, 219–234, 2004.

[21] Chen, R., McAvoy, T., *Plantwide control system design. methodology and application to a vinyl acetate process*, Industrial & Engineering Chemistry Research, 42, 4753–4771, 2003.

[22] Benyahia, B., Lakerveld, R., Barton, P.I., *A plant-wide dynamic model of a continuous pharmaceutical plant*, Industrial & Engineering Chemistry Research, 51, 15393–15412, 2012.

[23] Lakerveld, R., Benyahia, B., Braatz, R., Barton, P.I., *Model-based design of plant-wide control strategy for a continuous pharmaceutical plant*, AIChE Journal, 59, 3671–3685, 2013.

[24] Bristol, E., *On a new method of interactions for multivariable process control*, IEEE Transactions on Automatic Control, 11, 133–134, 1966.

[25] Shen, Y., Cai, W.J., Li, S., *Multivariable process control: decentralized, decoupling or sparse?* Industrial & Engineering Chemistry Research, 49, 761–771, 2010.

[26] Jain, A., Babu, B.V., *A new measure of process interaction in the domain dynamics*, in *Proceedings of AIChE Annual Meeting*, San Francisco, CA, USA, Paper 202d, 2013.

[27] Tung, L.S., Edgar, T.F., *Analysis of control-output interactions in dynamic systems*, AIChE Journal, 27, 690–693, 1981.

[28] Witcher, M.F., McAvoy, T.J, *Interacting control systems: steady-state and dynamic measurement of interaction*, ISA Transactions, 16, 35–41, 1977.

[29] Kariwala, V., Forbes, J.F., Meadows, E.S., *Block relative gain: proprieties and pairing rules*, Industrial & Engineering Chemistry Research, 42, 4564–4574, 2003.

[30] Chen, Q., Luan, X., Liu, F., *RNGA Loop Pairing Criterion for Multivariable Systems Subject to a Class of Reference Inputs*, p. 4721–4726 in *Proceedings of the 19th World Congress of The International Federation of Automatic Control*, Cape Town, South Africa, 2014.

[31] Avoy, T.M., Arkun, Y., Chen, R., Robinson, D., Schnelle, P.D., *A new approach to defining a dynamic relative gain*, Control Engineering Practice, 11, 907–914, 2003.

[32] Xiong, Q., Cai, W.J., He, M.J., *A practical loop pairing criterion for multivariable processes*, Journal of Process Control, 15, 741–747, 2005.

[33] Naini, N.M., Fatechi, A., Khaki-Sedigh, A., *Input-output pairing using effective relative energy array*, Industrial & Engineering Chemistry Research, 48, 7137–7144, 2009.

[34] Machado, C.F.R., de Medeiros, J.L., Araújo, O.F.Q., *A Comparative Analysis of Methanol Production Routes: Synthesis Gas versus CO_2*, p. 2981–2990 in *Hydrogenation, Proceedings of the 2014 International Conference on Industrial Engineering and Operations Management*, Bali, Indonesia, 2014.

[35] Aasberg-Petersen, K., Stub Nielsen, C., Dybkaer, I., Perregaard, J., *Large Scale Methanol Production From Natural Gas*, Whitepaper, Haldor Topsoe, 2008.

[36] Rahman, D., *Kinetic Modeling of Methanol Synthesis from Carbon Monoxide, Carbon Dioxide and Hydrogen Over a $Cu/ZnO/Cr_2O_3$ Catalyst*, Master's Thesis, San José State University, USA, 2012.

[37] Kuzsynski, M., Browne, W.I., Fontein, H.J., Westerterp, K.R., *Reaction kinetics for the synthesis of methanol from CO and H_2 on a copper catalyst*, Chemical Engineering and Processing, 21, 179–191, 1987.

[38] Panahi, P.N., Mousavi, S.M., Niaei, A., Farzi, A., Salari, D., *Simulation of methanol synthesis from synthesis gas in fixed bed catalytic reactor using mathematical modeling and neural networks*, International Journal of Scientific & Engineering Research, 3, 1–7, 2012.

[39] Harriot, P., *Chemical Reactor Design*, Marcel Dekker. (by Taylor & Francis Group LLC), New York, 2003.

[40] Dake, S.B., Jaganathan, R., Chaudhari, R.V., *New trends in the rate behavior of rhodium-catalyzed carbonylation of methanol*, Industrial & Engineering Chemistry Research, 28, 1107–1110, 1989.

[41] Nowicki, L., Ledakowicz, S., Zarzycki, R., *Kinetics of rhodium-catalyzed methanol carbonylation*, Industrial & Engineering Chemistry Research, 31, 2472–2475, 1992.

[42] Qin, F., Wang, S., Hartono, A., Svendsen, H.F., Chen, C., *Kinetics of CO_2 absorption in aqueous ammonia solution*, International Journal of Greenhouse Gas Control, 4, 729–738, 2010.

[43] Liu, J., Wang, S., Qi, G., Zhao, B., Chen, C., *Kinetics and mass transfer of carbon dioxide absorption into aqueous ammonia*, Energy Procedia, 4, 525–532, 2011.

[44] Jeon, S.B., Seo, J.B., Lee, H.D., Kang, S.K., Oh, K.J., *Absorption kinetics of carbon dioxide into aqueous ammonia solution: addition of hydroxyl groups for suppression of vaporization*, Korean Journal of Chemical Engineering, 30, 1790–1796, 2013.

[45] Hsu, C.H., Chu H., Cho, C.M., *Absorption and reaction kinetics of amines and ammonia solutions with carbon dioxide in flue gas*, Journal of Air & Waste Management Association, 53, 246–252, 2003.

[46] Scott, K., *The continuous stirred tank electrochemical reactor. an overview of dynamic and steady-state analysis for design and modelling*, Journal of Applied Electrochemistry, 21, 945–960, 1991.

[47] Babajide, A.D., Fahidy, T.Z., *Problems in modelling and the study of electrochemical reactor dynamics*, The Canadian Journal of Chemical Engineering, 46, 253–258, 1968.

[48] Trinidad, P., Walsh, F., Gilroy, D., *Conversion Expressions for Electrochemical Reactors which Operate under Mass Transport Controlled Reaction Conditions, Part I: Batch Reactor, PFR and CSTR*, International Journal of Engineering Education, 14, 431–441, 1998.

[49] Walsh, F., Trinidad, P., Gilroy, D., *Conversion expressions for electrochemical reactors which operate under mass transport controlled reaction conditions, part II: Batch recycle, cascade and recycle loop reactors*, International Journal of Engineering Education, 21, 981–992, 2005.

[50] Thilakavathi, R., Rajasekhar, D., Balasubramanian, N., Srinivasakannan, C., Al Shoaibi, A., *CFD modeling of continuous stirred tank electrochemical reactor*, International Journal of Electrochemical Science, 7, 1386–1401, 2012.

[51] Rodriguez, G., Sierra-Espinosa, F.Z., Teloxa, J., Álvarez, A., Hernández, J.A., *Hydrodynamic design of electrochemical reactors based on computational fluid dynamics*, Desalination and Water Treatment, doi: 10.1080/19443994.2015.1114169, 2015.

[52] Nagy, Z., Agachi, Ş., *Model predictive control of a PVC batch reactor*, Computers & Chemical Engineering, 21, 571–591, 1997.

9 Control of distillation processes

Distillation is a separation technique of miscible liquids. It is based on the *boiling point* (more precisely on the *volatility*) difference of components and usually appears at the end of a technological flux. Therefore, the characteristics and behavior of upstream operations (reactions, extractions, etc.) may significantly affect its operating specifications. A well-designed and implemented automated control system (ACS) has to be robust against upstream (and any other) disturbances. Distillation columns exhibit relatively slow and delayed dynamic responses as compared to other units because of the high number of mass transfer stages that are generally involved. This fact combined with the numerous other factors affecting distillation raises special control issues and requires more attention during design of ACS configurations [1, 2].

9.1 Economic constraints of distillation

One major objective of distillation is to deliver products of prescribed purity (quality), which is a design constraint. Since distillation is associated with large heating and cooling demands, minimization of operational costs a major operating objective, that must be enforced by the control systems. Prescribed purity/quality specifications can be achieved by either using a rigorous process control (enables direct production) or by mixing products of different purities/qualities. However, the latter option is not necessarily economically profitable since the energy spent to obtain the high-purity product is lost.

   Given that most products have minimal quality requirements (think about gasoline products with different octane number categories), product quality and price are not directly correlated. Products outrunning minimal specifications do not sell better or result in improved incomes. The diagram in Fig. 9.1 demonstrates that the maximal income is obtained by the minimal required purity that still enables product marketing at the desired class. By comparing Grades 1 and 2 products, it is obvious that separation costs increase with purity monotonically, whereas the product price remains piecewise constant. Yet, if the prescribed Grade 1 quality is not achieved, the product can be marketed as Grade 2 quality, resulting in cost penalizations that can lead to overall losses (gray areas in the Fig. 9.1). Therefore, the majority of distillation columns are operated to slightly overdo the specifications so that the actual purity never falls under the prescribed value. For this reason, distillation columns may be equipped with buffer vessels that act as concentration equalizers. Expressing these facts in numbers means that if, for example, the product purity shows ±0.2% fluctuation, the concentration controller setpoint should be with 0.2% above the desired value. The benzene-toluene separation, for instance, prescribes a toluene purity of 99.5%; thus, a ±0.2% fluctuation sets the controller to 99.7%. From an economic perspective, industrial practice showed that for the setpoint of 99.7 instead of 99.5%, results in a 10% energy consumption increase. More-

https://doi.org/10.1515/9783110789737-011

over, a setpoint of 99.9% toluene consumes 36% more energy. This underscores the importance of stable, accurate concentration control system.

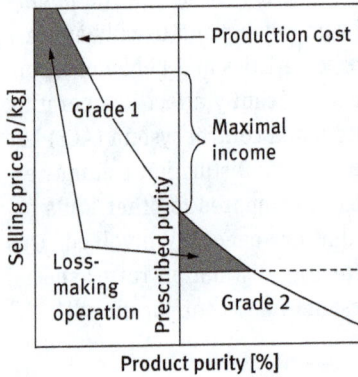

Fig. 9.1: Distillation: marketing price *vs* product purity (adapted after [1]).

Sometimes distillation yields an intermediate, not the final, product of a technological flux. Thus, the diagram in Fig. 9.1 is not valid and it makes more sense to aim for energy-saving, while approximately respecting quality requirements rather than to target rigorous specifications.

9.2 The recovery factor

Figure 9.2 illustrates schematically a *binary distillation* column. The valuable product is evacuated with the *distillate stream D*. The feed, distillate, liquid phase bottom (*waste*), and top vapor phase molar flow rates are symbolized by *F*, *D*, *B*, and *V*, respectively. The corresponding liquid phase mole fraction of the volatile compound is symbolized by *x*, whereas that of the vapor phase by *y*.

The *recovery factor r* is defined as the fraction of valuable product within the feed evacuated with the distillate stream – see eq. (9.1). Defined as such, *r* has values between 0% and 100%. The recovery decreases with increasing product purity (see explanations for Figs. 9.3 and 9.4):

$$r = \frac{D x_D}{F x_F} 100 \qquad (9.1)$$

If the feed departs the system entirely as distillate (*F* = *D*), the concentration of the valuable species within *D* and *F* streams is equal and *r* = 100%. In other words, the compound of interest has been fully recovered, but without concentrating it. If, however, $x_F = 0.5$, $x_D = 1$, and $D = F/2$, then a 100% recovery is coupled with a 50% concentration improvement.

A relationship between the desired species concentrations and the recovery factor can be established by combining eqs. (9.1) and (9.9):

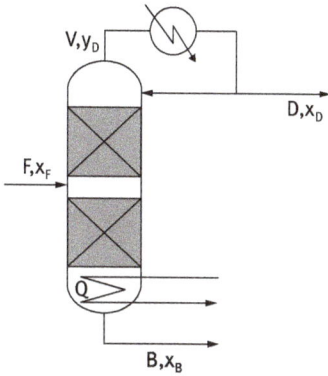

Fig. 9.2: Schematic representation of a binary distillation column.

$$r = \frac{x_D(x_F - x_B)}{x_F(x_D - x_B)}100 \tag{9.2}$$

The values of x_B and x_D can be calculated as follows [3]:

$$x_D = \frac{Sx_B}{1 + x_B(S-1)} \tag{9.3}$$

$$x_B = \frac{x_D}{S - x_D(S-1)} \tag{9.4}$$

Equations (9.3) and (9.4) contain the *separation factor S*. It represents the ratio of light to heavy components between the distillate and waste [3]. For a binary distillation it can be expressed as

$$S = \frac{\frac{x_D}{(1-x_D)}}{\frac{x_B}{(1-x_B)}} = \frac{x_D(1-x_D)}{x_B(1-x_B)} \tag{9.5}$$

The importance of separation factor in control issues will be discussed in more detail in Chapter 9.3. It connects to the *V/F* flow rate ratio [3] – see relationship eq (9.6), where n stands for the number of trays, E for the tray efficiency, and a for the average (mean) relative volatility:

$$\ln S = \frac{(nE)^{0.68} a^{1.68} \ln a}{3.5} \frac{V}{F} \tag{9.6}$$

The *D/F* flow rate ratio affects significantly the operation of a distillation column. An increase of D is followed by a decrease of its concentration x_D as well as by some enhancement of the recovery factor. Meanwhile, x_B within the bottom waste stream decreases along with *D/F* ratio. These facts are demonstrated by data presented in Fig. 9.3. It illustrates the variation of volatile species concentration in D and B streams, at various *D/F* values.

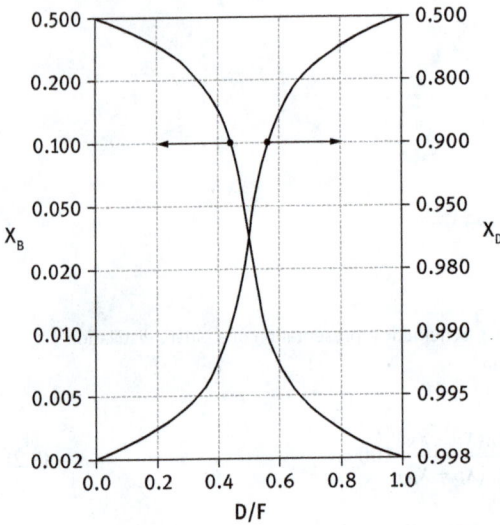

Fig. 9.3: Variation of the volatile species concentration in distillate and waste streams as a function of *D/F* ratio. The ordinates are expressed in logarithmic scales (adapted after [1]).

Another important parameter affecting distillation column performance is the V/F ratio. The vapor phase flow rate V is proportional to the heating power. For a constant F, the ratio V/F itself also reflects consumed heating power. Figure 9.4 illustrates the variation of recovery factor r against product purity at various V/F values, yet at constant $x_F = 0.5$. The curves were computed by using eqs. (9.2), (9.6), (9.38), and (9.42), respectively. A strong decrease of r can be observed as purity increases regardless of heating power.

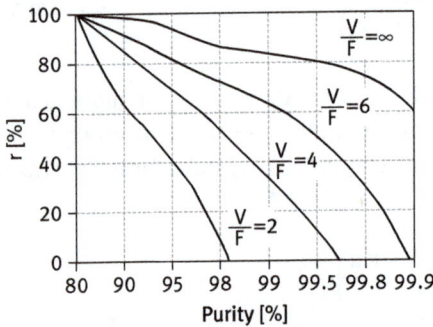

Fig. 9.4: Recovery factor *vs* product purity at various *V/F* ratios and constant $x_F = 0.5$. The abscissa is expressed in logarithmic scale (adapted after [1]).

Data in Fig. 9.4 suggest that higher purity and recovery can be achieved if V/F ratio is kept high. However, some major issues arise when applying higher vapor phase stream flow rates V:

– High vapor flow rate V causes a decrease in trail efficiency;
– Extremely high V/F may lead to flooding, an unwanted phenomena;
– Higher vapor load requires higher heating loads, undesired especially when separation of temperature-sensitive components is targeted;
– High vapor generation is usually coupled with high reflux rates. This leads to a significant increase of both heating and cooling demands, which translates into an overall cost increase.

9.3 Lowering energy demand of distillation units

Operation of distillation columns requires significant energy both for cooling and heating. The over 40,000 distillation columns across the USA in 2010 demanded approximately 40–60% of the energy consumption of chemical industry. It represented 19% of US manufacturers as well as 6% of the total US energy usage [4]. Some examples of energy-saving techniques employed in practice are:

– Improvement of tray efficiency by optimizing the vapor flow rate with the aid of suitable control configuration [4];
– Enhancing of relative volatility by working at minimal pressure. Approximately 5% energy can be saved at night time and 25% in winter by applying advanced atmospheric cooling (temperature drop) and therefore minimizing the working pressure of the column;
– Appropriate design and choice of the feeding tray; if the feeding concentration differs from that of the actual tray, the separation efficiency decreases;
– Applying internal heat integration to use the condensation heat of the distillate for heating the reboiler. Pinch analysis enables assessment of minimal heating and cooling demand as well as the extent in which it can be covered by appropriate heat integration. Figure 9.5 presents such example. Further heat integration schemes are also encountered in practice, e.g., the employment of the bottom stream heat for preheating operation of the feed [5];
– Inclusion of the distillation column in a plantwide heat integration system for advanced heat recovery [6];
– Tailored heat-integration for multicolumn systems. Figure 9.6 illustrates concurrent cascade distillation columns that separate ternary mixtures, where the heat entered into the first column is fully transferred to the second column. Here a condenser-reboiler heat pump is also employed;

Heat integration provides excellent opportunities to reduce overall energy demand. Meanwhile, it reduces process flexibility, thus rising significant controllability concerns

[7]. The choice of a reliable energy-saving technique is strongly related to cooling/heating equipment and its capital expenditures. Smart solutions may demand higher initial investments; thus capital amortization rates have to be considered as well within the early design stage of distilleries.

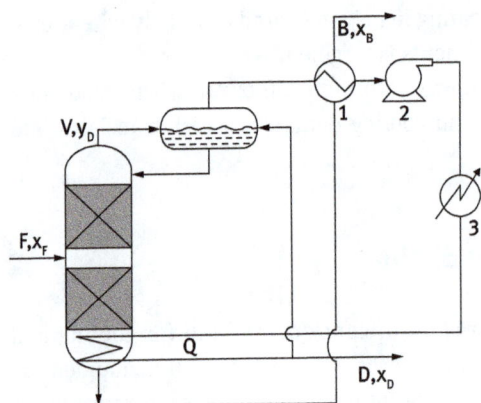

Fig. 9.5: Heat integration for a binary distillation column: an internal heat pumping system. "1" is the heat exchanger that uses heat from the bottom stream; "2" the compressor rises the temperature and the boiling point of the distillate stream; and "3" is the supplementary external heat source.

Heat integration often results in *heat exchanger networks* (HENs). The control of individual heat exchangers is well known; however control of an HEN may be challenging because of the highly nonlinear dynamic responses generated by the multiple recirculation streams. Such networks are not only of interest for thermally integrated distillation, but also consist of the core of any heat integrated plant. Mathisen et al. [8] presented rules for bypass selection in HEN control. Glemmestad and Gundersen [9] developed a freedom degree analysis to assess HEN operation optimization. Overall, the high-level HEN control strategies are rather simple:

–   Maintaining outlet temperatures bypassing the flow of process-to-process heat exchangers as well as duties of process-to-utility heat exchangers at near setpoint values;
–   Dividing flow rates using process stream splitters;
–   Keeping large material recycle streams that can store and carry energy may be beneficial for cascades of heated systems such as the one illustrated by Fig. 9.6 [10].

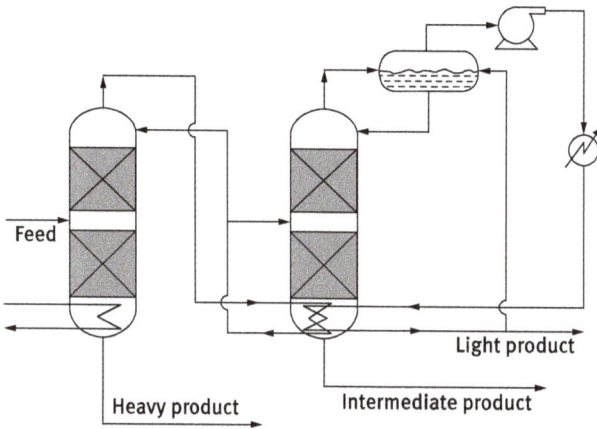

Fig. 9.6: Heat integration for concurrent cascade distillation columns.

9.4 General control of continuous distillation columns

9.4.1 Mass and energy balance imposed control issues

The design of an adequate ACS for a distillation column is based on the device's behavior analysis that is on heat and mass balances. Figure 9.2 illustrates schematically a binary distillation column and the corresponding symbols of main operating characteristics. In steady state, the global mass balance is described by eq. (9.7). According to it, input mass of the feed is completely evacuated within the output distillate and bottom streams:

$$F = D + B \qquad (9.7)$$

The mass balance for the valuable chemical species is expressed by eq. (9.8). Similarly to eq. (9.7), it states that the fed volatile compound is split between D and B streams:

$$Fx_F = Dx_D + Bx_B \qquad (9.8)$$

Replacing B from eq. (9.7) into eq. (9.8) leads to the D/F ratio as a function of concentrations:

$$\frac{D}{F} = \frac{x_F - x_B}{x_D - x_B} \qquad (9.9)$$

Nevertheless, B or F can also be expressed from eq. (9.7) and other ratios, such as B/F or B/D, may be obtained from eq. (9.8). Yet from the operator's point of view, D/F relates easier to the column efficiency.

---

i **Example 9.1**
A distillation column has a feed stream characterized by $x_F = 0.5$. The distillate is required $x_D = 0.95$ and x_B within the bottom stream should not exceed 0.05. Let us calculate the D/F ratio that ensures the minimum required purity.
   The ratio D/F can be obtained by means of eq. (9.9):

$$\frac{D}{F} = \frac{0.5 - 0.05}{0.95 - 0.05} = 0.5$$

Let us recalculate D/F if the feed concentration drops with 10%. Thus, $x_{F\,new} = 0.9 \cdot x_F$ and

$$\frac{D}{F} = \frac{0.5 \cdot 0.9 - 0.05}{0.95 - 0.05} = 0.44$$

The results show that a negative 10% disturbance demands a distillate flow rate decrease of 12% to maintain the prescribed distillate and bottom compositions.

Equation (9.9) yields the $x_D = f(x_F)$ relationship eq. (9.10) from which the distillate concentration can be computed quickly as a function of operational parameters. Derivation generates the x_D dependence on bottom concentration x_B – see (9.11):

$$x_D = x_B + \frac{F(x_F - x_B)}{D} \tag{9.10}$$

$$\frac{dx_D}{dx_B} = 1 - \frac{F}{D} \tag{9.11}$$

Equations (9.9)–(9.11) reveal that:
– the distillate concentration x_D increases linearly with the flow rate;
– a $D = 0$ value results in $x_B = x_F$, thus operating columns at $x_B > x_F$ is impossible.

The mechanism of evacuated flows D, V, and B influencing values of x_D, x_B, and y_D, respectively, is related to both heat and mass balances. For instance, the internal vapor flow rate V_i would ideally equal the external vapor flow rate V if the reflux would recirculate at the exact boiling point temperature of the tray to which it reenters the column (ideal case). In other words, neither vapor condensation nor liquid evaporation would occur. In reality, however, the reflux temperature is usually lower than that of the tray into which it is fed. Thus, a part of the internal vapor flow is condensed while the reflux stream is heated up to the boiling point. Consequently, the internal and external vapor fluxes have different flow rates and the condensed internal vapors will contribute to the internal reflux by increasing it.
   The interconnections between mass and energy balances are schematically illustrated in Fig. 9.7, where subscript i symbolizes internal flux, H stands for the liquid enthalpy, l_v is the latent evaporation heat, index n stands for the nth tray, H_n is the enthalpy at boiling point on tray n, and Q_c represents the amount of heat removed by the condenser.

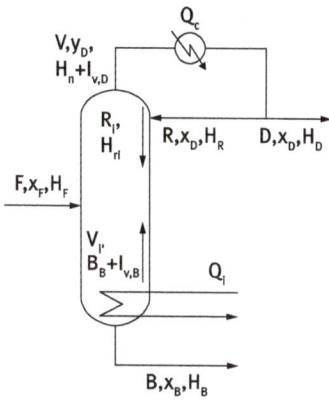

Fig. 9.7: Interconnection between energy and mass balances. Schematic representation for a binary distillation column.

The mass balance at the top of the distillation column (see Fig. 9.7) is

$$V_i + R = V + R_i \tag{9.12}$$

The left-hand side of eq. (9.12) describes input mass, whereas the right-hand side stands for the output. The corresponding energy balance is described by eq. (9.13):

$$V_i(H_n + l_{vD}) + RH_R = V(H_n + l_{vD}) + R_iH_{ri} \tag{9.13}$$

By combining information in eqs. (9.12) and (9.13), the internal reflux may be expressed as a function of external reflux in eq. (9.14). It confirms the assumption that the subcooled external reflux (when $H_n > H_R$) leads to an increase of internal reflux:

$$R_i = R\left(1 + \frac{H_n - H_R}{l_{vD}}\right) \tag{9.14}$$

The *specific enthalpy* is defined in eq. (9.15) as a ratio between the flow's heat H and boiling point enthalpy H_n difference and the vaporization heat l_v:

$$q = \frac{H - H_n}{l_v} \tag{9.15}$$

Since the liquid temperature cannot exceed the boiling point, $q \leq 0$. By replacing eq. (9.15) in eq. (9.14) a simpler form is obtained. q_R stands for the specific enthalpy of the reflux:

$$R_i = R(1 - q_R) \tag{9.16}$$

Equation (9.16) reveals that the internal reflux cannot be lower than the external reflux flow rate. The global heat balance is described by eq. (9.17), where Q_i stands for the heat entered through the reboiler and Q_c stands for the heat evacuated in the condenser. This equation shows the way energy consumption depends on flow rates F, D, and B as well as on the heat amount these carry:

$$FH_F + Q_i = Q_c + DH_D + BH_B \tag{9.17}$$

Cooling and heating demands can be calculated as a function of internal and external vapor flow rates:

$$Q_i = V_i l_{vB} \tag{9.18}$$

$$Q_c = V(H_n + l_{vD}) - VH_D \tag{9.19}$$

From eq. (9.17) to eq. (9.19), the internal vapor flow rate can be expressed as

$$V_i = \frac{V(H_n + l_{vD}) - VH_D + DH_D + BH_B - FH_F}{l_{vB}} \tag{9.20}$$

Equation (9.20) can be reformulated by introducing the external reflux, an easier measured and controlled parameter. Hence,

$$V_i = \frac{F(H_n + l_{vD} - H_F) + R(H_n + l_{vD} - H_D) + B(H_n + l_{vD} + H_B)}{l_{vB}} \tag{9.21}$$

Relationship (9.21) shows that with raising reflux, the internal vapor flow rate also elevates so as the heating energy demand.

The global heat balance (9.17) may also be expressed as a function of specific enthalpies of individual flows:

$$Fq_F + V_i \frac{l_{vB}}{l_{vD}} = V - Rq_R + Bq_B \tag{9.22}$$

To achieve constant distillate purity, the D/F ratio has to be varied independently. Otherwise, D/F values depend on both R and V. Thus, small variations of R and/or V for columns operating at high R/D generate significant variations of D/F. Moreover, V cannot be controlled independently as, according to eq. (9.23), derived from (9.22), it is a function of V_i, R, q_R, F, and q_F:

$$V = V_i \frac{l_{vB}}{l_{vD}} + Fq_F + Rq_R - Bq_B \tag{9.23}$$

If D is not directly controlled but results as the difference $D = V - R$, then eq. (9.23) becomes

$$D = V_i \frac{l_{vB}}{l_{vD}} + Fq_F - R(1 - q_R) - Bq_B \tag{9.24}$$

that yields when using the global mass balance (9.7):

$$D = \frac{V_i \frac{l_{vB}}{l_{vD}} + F(q_F - q_B) - R(1 - q_R)}{1 - q_B} \tag{9.25}$$

The extent to which variations of F and R affect the distillate flow rate D can be assessed from the differential of eq. (9.25). Relationship (9.26) is obtained under the assumption that the bottom-specific enthalpy q_B does not vary significantly since the outlet temperature is near the boiling point. Therefore, $dq_B = 0$:

$$dD = \frac{V_i \dfrac{l_{vB}}{l_{vD}} + dF(q_F - q_B) + Fdq_F - dR(1 - q_R) + Rdq_R}{1 - q_B} \qquad (9.26)$$

Division by D turns eq. (9.26) into eq. (9.27), the expression of the relative variation of distillate rate:

$$\frac{dD}{D} = \frac{\dfrac{V_i}{D}\dfrac{dV_i}{V_i}\dfrac{l_{vB}}{l_{vD}} + \dfrac{dR}{R}\dfrac{R}{D}(1 - q_R) + \dfrac{R}{D}dq_R + \dfrac{dF}{F}\dfrac{F}{D}(q_F - q_B) + \dfrac{F}{D}dq_F}{1 - q_B} \qquad (9.27)$$

A numeric analysis of the above eq. (9.27) reveals that a 1% variation of V_i for a column working at $R/D = 10$ generates an 11% variation of dD/D. If the feed-specific enthalpy q_F changes by 1%, the variation of dD/D is 2% for $F/D = 2$ and 5% for $F/D = 5$, respectively. These observations indicate that for columns working at high R/D, the R has to be controlled by a level controller of the reflux reservoir, and D should be controlled independently.

From all above it may be concluded that:

1. The V/F ratio has to be kept constant for economic (energy saving) reasons;
2. The D/F ratio has to be kept constant to ensure a desired distillate concentration x_D;
3. The bottom flow concentration x_B requires additional equations to all the above presented.

The separation factor S was already defined in eq. (9.5). It also respects condition (9.28), where a_m stands for *average relative volatility*, E for tray efficiency, and n for the number of trays, respectively. The minimum value $S = 1$ corresponds to the case for which there is no separation; in other words, the bottom and distillate concentrations are equal ($x_B = x_D$):

$$1 \leq S \leq a_m^{-nE} \qquad (9.28)$$

The relative volatility a of components is expressed as [1, 2]

$$a_{ij} = \frac{\dfrac{y_i}{x_i}}{\dfrac{y_j}{x_j}} = \frac{\dfrac{P_i^0}{P}}{\dfrac{P_j^0}{P}} = \frac{K_i}{K_i} = \frac{P_i^0}{P_j^0} \qquad (9.29)$$

where P_i^0 and P_j^0 are the partial pressures of i and j pure components and P is the total pressure in the column. $K_{i,j}$ stands for the equilibrium ratio at vaporization of i and j components (the so-called K factor): $K_i = \frac{y_i}{x_i} = \frac{P_i^0}{P}$ and $K_j = \frac{y_j}{x_j} = \frac{P_j^0}{P}$. Equation (9.29) demonstrates that for any $a = 1$, separation is impossible because the components are identically volatile.

For a binary mixture, the relative volatility (9.29) is expressed as

$$a = \frac{y(1-x)}{x(1-y)} \tag{9.30}$$

The vapor pressure P increases with temperature, obeying the Antoine eq. (9.31), in which A, B, and C are material-specific constants:

$$\ln P = A - \frac{B}{T+C} \tag{9.31}$$

The following two equations can be derived from eq. (9.29):

$$a_{ij} = \frac{P_i^0}{P_j^0} \leftrightarrow \ln a_{ij} = \ln \frac{P_i^0}{P_j^0} \tag{9.32}$$

$$\ln a_{ij} = \ln P_i^0 - \ln P_j^0 \tag{9.33}$$

Relationships (9.33) and (9.31) show that the relative volatility decreases when temperature rises.

Due to the pressure drop along the column, a varies from tray to tray. However, for basic modeling and design purposes, it can be expressed as the arithmetic average of the distillate and bottom values – see eq. (9.34). Meanwhile, for stronger variations, the geometric mean is applied – see eq. (9.35):

$$a_m = \frac{a_B + a_D}{2} \tag{9.34}$$

$$a_m = \sqrt{a_D a_B} \tag{9.35}$$

The tray efficiency E is defined in eq. (9.36) as the ratio between theoretical (n_t) and practical (n_p) tray numbers necessary to achieve a desired separation:

$$E = \frac{n_t}{n_p} \tag{9.36}$$

A typical value of tray efficiency is 0.7. If the tray efficiency decreases due to high vapor flow caused stripping, the product purity cannot be considerably improved by increasing the reflux ratio. In this situation, the attainable concentration limit of the column has been reached. Meanwhile, if the tray efficiency is small due to low vapor flow caused weeping, the energy consumption cannot be further reduced (lower V is not applicable). In this case, the lowest energy consumption has been reached.

For a binary mixture, the liquid-vapor phase equilibrium is generally described as $y = f(x)$, such as

$$y = \frac{x a_m^{nt}}{1 + x \left(a_m^{nt} - 1 \right)} \tag{9.37}$$

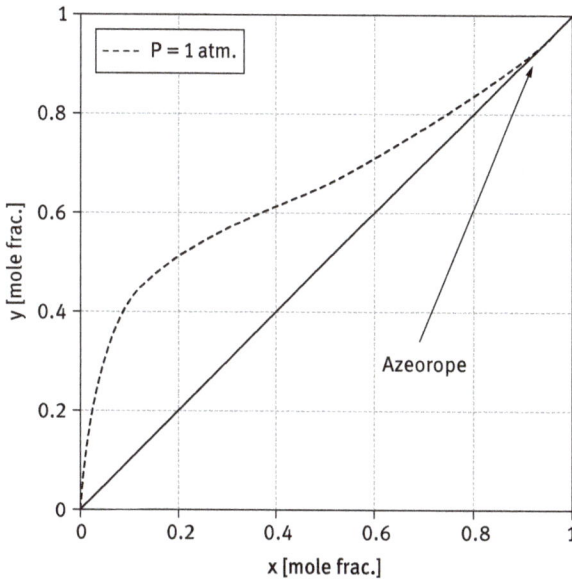

Figure 9.8 presents the vapor-liquid equilibrium curve for the ethanol-water mixture at atmospheric pressure. It represents the gas phase volatile ethanol concentration *y versus* its liquid phase concentration *x* in the mixture. The difference at any given *x* between diagonal (where $x = y$) and equilibrium curve values of *y* stands for the maximal concentration difference that can be achieved by one distillation unit having a 100% tray efficiency. The curve and diagonal exhibit closer *y* values at *x* near to 0 or to 1. This translates into more trays needed to achieve the same concentration difference. The curve crosses the diagonal in the so-called azeotropic point ($y = x$) corresponding to the azeotrope mixture. It is also known as *constant boiling mixture* because when boiling it, the vapor and liquid phases have identical compositions and therefore cannot be separated by means of simple distillation. As a result, at constant pressure distillation this $y = x$ value cannot be exceeded.

Equation (9.38) describes the vapor-liquid equilibrium (9.37) in terms of the separation factor *S*. It yields a similar relationship to that of Fenske's eq. (9.39) [11]:

$$x_D = \frac{x_B S}{1 + x_B(S - 1)} \tag{9.38}$$

$$y = \frac{x a}{1 + x(a - 1)} \tag{9.39}$$

Combining eqs. (9.9) and (9.38) gives the mole fraction of interest x_D (the more volatile species at the column top):

$$x_D = \frac{-b - \sqrt{b^2 - 4ac}}{2a} \tag{9.40}$$

The terms in eq. (9.40) are

$$a = \frac{D}{F}(S-1)$$

$$b = -\left[\left(\frac{D}{F} + x_F\right)(S-1) + 1\right] \tag{9.41}$$

$$c = Sx_F$$

Further, the value of x_B can be obtained using x_D:

$$x_B = \frac{x_D}{S - x_D(S-1)} \tag{9.42}$$

The following expression of S was shoved to be valid if the feeding stream F is introduced into the optimal tray ($q_F = 0$) [12]:

$$S = \left(\frac{a_m}{\sqrt{1 + \frac{D}{Rx_F}}}\right)^{nE} \tag{9.43}$$

Otherwise, if $q_F \neq 0$, above (9.43) becomes

$$S = \left(a_m \sqrt{1 - \frac{1 - q_F + \frac{R}{D}}{\left(1 + \frac{R}{D}\right)\left(1 - q_F + \frac{x_F R}{D}\right)}}\right)^{nE} \tag{9.44}$$

From eq. (9.44), the D/R ratio can expressed as a function of separation factor S and distillate concentration x_D – see eq. (9.45). Thus, D/R values can be calculated that keep both S and x_D at desired values:

$$\frac{D}{R} = x_F\left[\left(\frac{a_m}{S^{\frac{1}{nE}}}\right)^2 - 1\right] \tag{9.45}$$

Shinskey [11] showed that the separation factor can also be linked to the V/F – see eq. (9.6).

---

i **Example 9.2**
A binary mixture of equal 0.5 mol fractions is subjected to separation in a distillation column that has 30 theoretical trays. The average relative volatility is 1.72. If the tray efficiency is $E = 1$, let us calculate the V/F, D/F, and R/D ratios that ensure $x_D = 0.95$ and $x_B = 0.05$, respectively. Let us also calculate the recovery factor.

Equation (9.5) can be used to calculate the separation factor from given $x_D = x_B = 0.05$:

$$S = \frac{x_D(1-x_B)}{x_B(1-x_D)} = \frac{0.95 \times (1-0.05)}{0.05 \times (1-0.95)} = 361$$

Further, eq. (9.45) yields the D/R ratio, whereas D/F is computed from eq. (9.9):

$$\frac{D}{R} = x_F\left[\left(\frac{a_m}{S^{\frac{1}{nE}}}\right)^2 - 1\right] = 0.5\left[\left(\frac{1.72}{361^{\frac{1}{30}}}\right)^2 - 1\right] \approx 0.5$$

$$\frac{D}{F} = \frac{x_F - x_B}{x_D - x_B} = \frac{0.5 - 0.05}{0.95 - 0.05} = 0.5$$

Rearrangement of eq. (9.6) permits the calculation of the V/F ratio:

$$\frac{V}{F} = \frac{3.5 \ \ln S}{(nE)^{0.68} a_m^{1.68} \ln a_m} \approx 1.5$$

while the recovery factor is obtained from eq. (9.2):

$$r = \frac{x_D(x_F - x_B)}{x_F(x_D - x_B)} 100 = \frac{0.95 \times (0.5 - 0.05)}{0.5 \times (0.95 - 0.05)} 100 = 95\%$$

The result shows that 95% of the volatile component fed into the column is evacuated within the distillate stream.

Example 9.3

Let us assume that the distillation column presented in Example 9.2 is operated at constant heating power (V/F = constant), but the D/F ratio modifies to 0.48. Let us assess the changes in the concentrations and recovery factor.

The V/F ratio can be expressed based on the mass balance at the column's top. From it, the D/R ratio can be further obtained:

$$\frac{V}{F} = \frac{R_i}{F} + \frac{D}{F} \leftrightarrow \frac{V}{F} = \frac{D}{F}\left(\frac{R}{D} + 1\right)$$

$$\frac{D}{R} = \frac{\frac{D}{F}}{\frac{V}{F} - \frac{D}{F}} = \frac{0.48}{1.5 - 0.48} = 0.47$$

The result shows that the D/R value decreased as compared to that in Example 9.2. This suggests that the separation factor has to increase. From eq. (9.43), the value of S is accessible and its value proves to be indeed higher:

$$S = \left(\frac{a_m}{\sqrt{1 + \frac{D}{Rx_F}}}\right)^{nE} = \left(\frac{1.72}{\sqrt{1 + \frac{0.47}{0.5}}}\right)^{30} = 560.8$$

Equations (9.40) and (9.41) permit calculation of x_D

$$a = \frac{D}{F}(S-1) = 0.48 \times (560.8-1) = 268.7$$

$$b = -\left[\left(\frac{D}{F}+x_F\right)(S-1)+1\right] = -[(0.48+0.5)\times(560.8-1)+1] = -549.6$$

$$c = 0.5 \times 560.8 = 280.4$$

$$x_D = \frac{549.6 - \sqrt{549.6^2 - 4\times 268.7 \times 280.4}}{2 \times 268.7} = 0.974$$

Thus, the higher separation factor results in a higher distillate concentration. The value of x_B is

$$x_B = \frac{x_D}{S(1-x_D)+x_D} = \frac{0.974}{560.8\times(1-0.974)+0.974} = 0.062$$

The bottom concentration also increased. This is due to the decrease of distillate flow rate D. The recovery factor is

$$r = \frac{x_D(x_F - x_B)}{x_F(x_D - x_B)}100 = \frac{0.974\times(0.5-0.062)}{0.5\times(0.974-0.062)}100 = 93.65\%$$

Consequently, higher distillate concentration leads to lower recovery.

---

i **Example 9.4**
Let us repeat the calculations for $D/F = 0.5$, but with V/F (heating energy) increasing to 1.8.
   The D/R ratio can be calculated as follows

$$\frac{D}{R} = \frac{\frac{D}{F}}{\frac{V}{F}-\frac{D}{F}} = \frac{0.5}{1.8-0.5} = 0.385$$

The separation factor is yield by eq. (9.43) and results in the highest value of all examples so far:

$$S = \left(\frac{a_m}{\sqrt{1+\frac{D}{Rx_F}}}\right)^{nE} = \left(\frac{1.72}{\sqrt{1+\frac{0.385}{0.5}}}\right)^{30} = 2219.5$$

The distillate concentration is computed as above:

$$a = \frac{D}{F}(S-1) = 0.5 \times (2219.5-1) = 1109.3$$

$$b = -\left[\left(\frac{D}{F}+x_F\right)(S-1)+1\right] = -[(0.5+0.5)\times(2219.5-1)+1] = -2219.5$$

$$c = 0.5 \times 2219.5 = 1109.8$$

$$x_D = \frac{2219.5 - \sqrt{2219.5^2 - 4\times 1109.3 \times 1109.8}}{2 \times 1109.3} = 0.981$$

It is obvious that the highest separation factor S resulted in the highest value of x_D. Thus, increasing V/F ratios favors distillate concentration.

The value x_B of the bottom concentration is:

$$x_B = \frac{x_D}{S(1-x_D)+x_D} = \frac{0.981}{2219.5 \times (1-0.981)+0.981} = 0.0000084$$

The results show that a higher heating power almost completely stripes the volatile compound from the reboiler. The recovery factor is

$$r = \frac{x_D(x_F - x_B)}{x_F(x_D - x_B)} \; 100 = \frac{0.981 \times (0.5-0.0000084)}{0.5 \times (0.981-0.0000084)} \; 100 = 99.99\%$$

and indicates that practically the whole quantity of the feed volatile compound is evacuated with the distillate.

It can be concluded that an increasing heating power (V/F ratio) has a positive effect on both the distillate concentration x_D and recovery factor r. Nevertheless, this requires additional heating (costs) – see Sections 9.1 and 9.3. A further increase of V/F may lead to evaporation of the less volatile species, which will, in turn, decrease x_D.

9.4.2 Control solutions

When summarizing all process characteristics and economic aspects discussed so far, the following general control strategy guidelines are drawn for a distillation column:

- An ACS is required to keep V/F ratios constant by means of heating power.
- Because at constant feed composition x_F, the variations of x_D and x_B are related to those of flow rate F, a concentration control system is required at the column's top. It should manipulate independently the distillate flow rate D to keep D/F constant.
- The reflux vessel requires a level control system. This should control the D/R ratio and thus ensure the mass balance closure ($V = D + R$) within this column region.
- The column bottom requires a level control system. This adjusts the B/F ratio thus enabling bottom mass balance to close – see eq. (9.46).

Figure 9.9 presents a general control scheme of a distillation column based on above statements. Various other control solutions are employed in practice. Some of these will be briefly presented and discussed later in this chapter.

9.4.2.1 Level control

Various methods can be applied to ensure mass balance closure in both bottom and reflux vessels. The bottom mass balance requires the sum of generated vapor flow rate V_i and bottom flow rate B to be equal to the internal reflux rate entering the reboiler $R_{i,B}$:

$$R_{i,B} = B + V_i \tag{9.46}$$

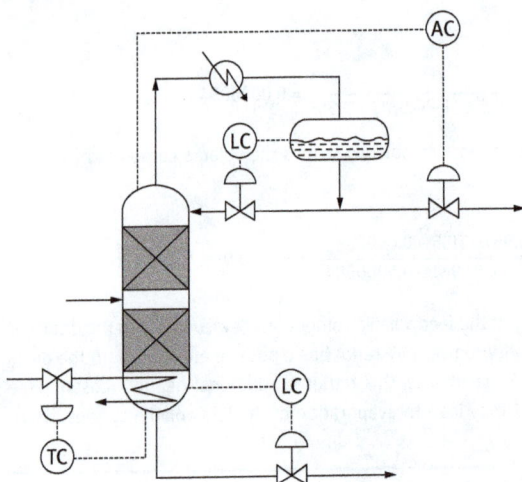

The general control scheme of a distillation column.

If B is low, the bottom liquid level control can be achieved by manipulating the heating energy and hence controlling the evaporation rate (it does not introduce significant deviation of vapor flow as compared to its prescribed value). The temperature control can be solved through the bottom flow rate B since the boiling point is a function of composition. This strategy is applied to avoid the possible inverse response of the reboiler (see also Chapter 9.5 and Fig. 9.14).

Level control in the reflux vessel can be achieved by a cascade control (see Fig. 9.10) which ensures fast response to variations of D. Because the majority of columns are operated at high(er) R/D values, this solution is handy since low disturbances in D require significant variations of R to maintain the prescribed R/D. The same dynamic effect can be achieved when manipulating flow rate D.

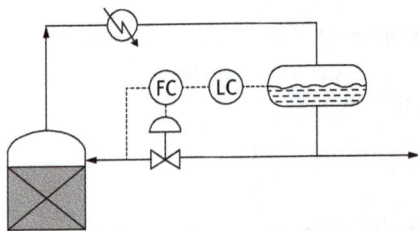

Fig. 9.10: Level control in a reflux vessel by means of a cascade control system.

9.4.2.2 Concentration control

A wide palette of instrumentation is in use for the measurement of liquid concentrations [2]. Some of the more advanced techniques involve high-pressure liquid chromatography, gas chromatography, UV/vis, and IR spectroscopy, viscosity, density-based methods, etc. However, the cost of these devices limits their wide online implementation. In contrast, low-cost manometers and thermometers are widely used in distillation columns for concentration measuring purposes. These are based on the fact that boiling point is a function of pressure and composition.

The gas phase mole fraction of a volatile component in ideal mixtures is expressed as a function of its vapor pressure P_i at a given pressure P by eq. (9.47). The Raoult eq. (9.48) links the partial pressure of the pure component P_i^* to its actual concentration x_i:

$$y_i = \frac{P_i}{P} \tag{9.47}$$

$$P_i = P_i^* x_i \tag{9.48}$$

If for a binary A--B mixture, species A is more volatile, then the combination of eqs. (9.47) and (9.48) yields for A:

$$x_A = \frac{P - P_B}{P_A - P_B} \tag{9.49}$$

The vapor pressure is known to depend exponentially on temperature. Thus, boiling point, pressure, and composition can be correlated. Figure 9.11 presents such an example for the benzene-toluene mixture, where temperature dependencies are almost linear.

Such correlations can translate a simple temperature measurement into an estimate concentration value. However, they ought to be used with caution because in some cases, the boiling point difference of pure components is rather small and falls within the measurement error domain of the thermometer. These errors can further propagate into the control system.

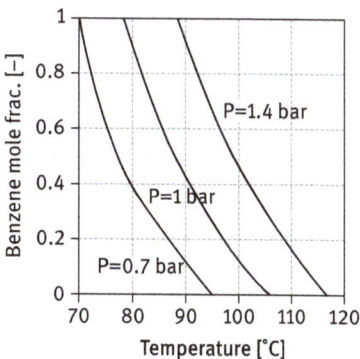

Fig. 9.11: Correlation among boiling point, pressure, and composition for the benzene-toluene mixture (adapted after [1]).

For example, a 1% molar concentration variation in a *n*-butane/*iso*-butane mixture generates just 0.25 °C boiling point variation. A similar deviation is yield by ~140 mmHg pressure increase or ~0.6% *iso*-pentane impurity. A typical 0.5% precision class thermoresistance with 40 °C maximal temperature has a 0.2 °C designed measuring error. It is obvious that this thermometer is not recommended for boiling point-based concentration determinations in this case.

Nevertheless, thermometers are widely used in distillation columns. To shorten their response time, they should always be merged into the liquid. A sensitive measurement means that any disturbance is sensed quickly and reliably. Therefore, it is recommended to install the thermometer into the tray having the maximal temperature gradient within the column's vertical temperature profile. This places it usually at the top of the column, right under the reflux tray (see Fig. 9.17(b)). However, measuring temperature and thus estimating concentration for another tray than for the one that requires control might degrade control performance. Yet, measurements directly on the controlled tray will result in more accurate concentration values but a more vulnerable to disturbances control system.

9.4.2.3 Pressure control
Because the boiling point depends on pressure, the latter has to be carefully controlled. Pressure control is carried out by taking into account both the nature and properties of the gas that generates overpressure. Two limiting cases generally occur:
- Overpressure caused only by noncondensing gases when a gas purge is added to the reflux vessel – see Fig. 9.12(a).
- Overpressure caused only by condensing gases when the coolant flow rate is manipulated – see Fig. 9.12(b). The heat transfer area of the condenser can also be varied by its partial flooding.

9.4.2.4 Recommendations for control system configuration
The conclusion of this chapter is that adequate distillation control requires *R/D*, *V/F*, and *D/F* ratios, reboiler level, and working pressure control to ensure product purity specifications and reduce process energy demand.

Numerous debates are described by the literature [3, 13] with regard to suitable control configuration for distillation columns. Yet all authors agree on the necessity of controlling bottom and distillate concentrations by manipulating the *R*, *V*, *D*, and *B*, respectively. Two of these flows are used to control concentrations and the other two for mass balance closure and recovery maximization. These four values can be paired in $C_4^2 = 6$ ways. Since *D* and *B* cannot be varied simultaneously because they are linked *via* global mass balance eq. (9.7), only five possible pairings remain to control concentrations.

Moreover, if the ratios (*R/V*, *R/D*, *R/B*, *V/D*, *V/B*, and *D/B*) are also considered as manipulated variables, a total of $C_{10}^2 = 45$ possible pairings result. Since, *D*, *B*, and *D/B* cannot be paired, that leaves 42 possibilities.

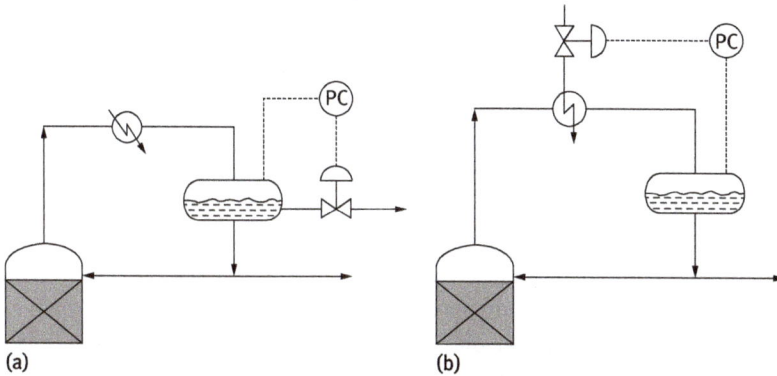

(a) (b)

Fig. 9.12: (a) Pressure control by means of a gas purge; (b) pressure control by coolant flow rate manipulation.

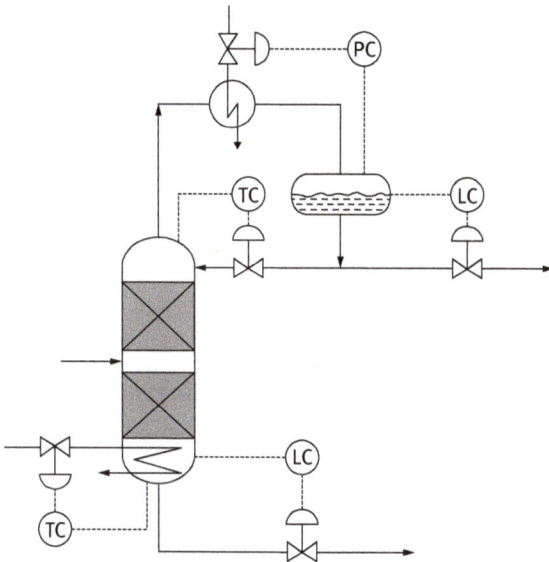

Fig. 9.13: Control structure configuration of a distillation column based on the $V - R$ pairing as manipulated variables.

Selecting adequate manipulated variables from these numerous possibilities, hence designing a well-performing control system, is strongly related to the process requirements and operating conditions. Usually, the Bristol [3] method is used (see also Chapter 5); it involves the matrix of relative gains which highlights the dynamic interactions between individual control loops. However, employment of the $V - R$ couple is most widely applied in practice (see Fig. 9.13).

9.5 Control issues of continuous distillation column dynamics

Chapter 9.4 focused on the impications of steady-state operation of distillation column behavior on the control system. Their dynamic behavior often involves significant delays that can lead to instability of the ACSs.

Any distillation column is composed of a given number of mass transfer units (trays). Thus, overall mass transfer presents capacitive dynamic behavior and the distillation column behaves like a series of capacitive elements. Yet, distinction is to be made between hydrodynamic and composition dynamics.

The trays and the reboiler contain vapor and liquid. However, the liquid mass is with some orders of magnitude higher than that of the vapor mass. Hence, the latter may be neglected when discussing column dynamics. Time delays may appear when the liquid phase is transferred from one tray to the one positioned underneath it. Such mass transfer occurs when increased internal flow rates R_i cause liquid accumulation and crossing of the tray's weir.

For a perpendicular-shaped weir, the liquid flow rate leaving the tray is given by the Francis equation below, where l is the weir width and h stands for the liquid level exceeding it [14, 15]:

$$f_i = 1.873 l h^{\frac{3}{2}} \tag{9.50}$$

The retained liquid volume within the tray is described by eq. (9.51), where f represents the evacuating flow rate over the weir:

$$V_T = Ah = A\left(\frac{f}{1.873l}\right)^{\frac{2}{3}} \tag{9.51}$$

The tray's hydraulic time constant T_h is defined as the variation of retained volume with evacuating flow rate. After doing the math, T_h is written as

$$dT_h = \frac{dV_T}{df} = \frac{0.442A}{l^{\frac{2}{3}}f^{\frac{1}{3}}} \tag{9.52}$$

When talking about hydraulic dynamics of a distillation column, it is important to mention the potential inverse response [2]. If the reboiler level is controlled by means of heating power, then increasing it may result in a quick concentration drop of the volatile compound. In some mixtures, this entails a similarly fast mixture density decrease. This inherently leads to an increased holdup volume which finally causes the *inverse response*. Naturally, due to the increased vapor generation rate, after a certain time the reboiler volume will start to decrease again (see Fig. 9.14). This time elapse is called *inversion time* t_{inv} and can be as high as some minutes or hours, in other words not negligible.

A rare consequence of high inversion times is the controller's failure in regulating the reboiler liquid level. In such cases, level control is achieved *via* bottom flow rate B. If, however, B is too low, level control must be carried out by means of reflux flow

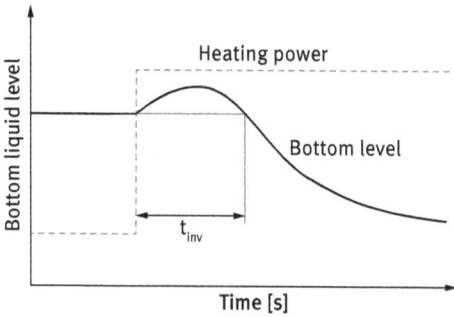

Fig. 9.14: Inverse response of bottom liquid level to increased heating power.

rate. This is a highly undesirable situation since the system's response in mass transport comes after the maximal possible time delay, which is the sum of hydraulic delays of all trays. Variation in liquid and vapor flow rates affect tray compositions.
Thus, concentration variation rate depends on flow rates as well as on liquid amounts
beholden by the tray. Therefore, concentration dynamics is slower than hydraulic dynamics.

The dynamics of a packed bed distillation column that separates methanol-water
mixtures has been assessed [12] by using Raschig rings of 25 mm diameter. The experimental results were processed by means of the following equations:

$$T_{C,ij} = \frac{H_C}{R + (R+D)\frac{P_{ij}}{P}} \tag{9.53}$$

$$T_{S,ij} = \frac{H_S}{(R+D) + (W-F-R)\frac{P_{ij}}{P}} \tag{9.54}$$

where
- T_C and T_S are the time constants of the rectifying and stripping sections, respectively;
- H_C and H_S are the liquid holdups within the rectifying and stripping sections, respectively. Their values can be calculated by means of the Schulmann, Ulrich,
  Weils, and Proulx relationships [16].
- P_{ij} is the partial pressure of i component at jth packed transfer unit temperature.

The time lapse of the response to a certain disturbance depends on the nature of disturbing parameter. A disturbance of feed flow rate F can result in up to 8 h delay in the
response of the distillate concentration. Figure 9.15 illustrates a feedforward control
loop used to attenuate such kind of disturbances by adjusting flow rate R. Meanwhile, a
distillate flow rate D disturbance causes only a 15 min delay in the value of x_D. The time
constant of a tray is of about 20 min and does not depend on the disturbance's location
(F or D).

High-purity distillation columns [10] show three time scales for their dynamic behavior:

- Fast time scale for both the condenser and reboiler heat duties. The coolant and steam flows are manipulated variables that regulate temperatures of reboiler and condenser.
- Intermediate time scale for the compositions and holdups of individual stages. Manipulated variables are the reflux rate R and the vapor flow rate V.
- Slow time scale for global input-output column behavior. The outputs are product purity and overall material balance. The manipulated inputs (for a model-based nonlinear controller) are the small bottom and distillate flows as well as the set points of level controllers (a total of four variables: two flow rates and two references for the two level regulators, the bottom and the reflux vessel).

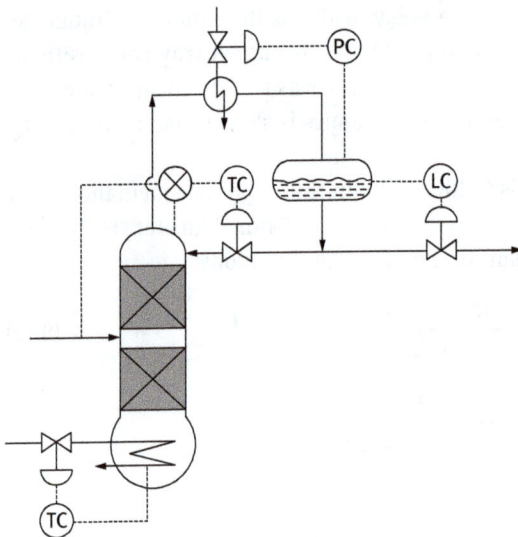

Fig. 9.15: A feedforward control loop that reduces the effect of time delays occurring in the product quality when the feed flow rate is disturbed.

Each distillation system may require individual control strategies such as feedforward, cascade, or model predictive control (MPC) [17–19]. The latter is also able to incorporate process nonlinearity aspects (see also Chapter 2). State estimators ensure good quality control under plant-model mismatch and disturbances [20].

Chapter 2 explains that MPC performance strongly relates to the process model. The transfer functions of process dynamics, eqs. (9.53) and (9.54), are not able to describe the start-up and shut-down conditions. Because first-principle models describe the process dynamics better, MPC applications often require their employment. Such a dynamic model (see also Fig. 9.17) for the pilot-scale continuous distillation column of ethanol-water mixtures illustrated by Fig. 9.18 was developed and validated by Agachi et al. [21, 22]. Model calibration [21, 23] was carried out by defining and solving a process optimization problem. The objective function is the sum-squared error (SSE) described by eq. (9.55), where θ denotes the vector of decision variables defined in eq. (9.56):

$$SSE(\theta) = \sum_{i=1}^{N} \sum_{j=1}^{P} \left(T_{ij,sim} - T_{ij,exp}\right)^2 + w \sum_{i=1}^{N} \left(x_{i,sim} - x_{i,exp}\right)^2 \overset{!}{=} min \qquad (9.55)$$

$$\theta = [E_r, E_s, h_r, h_s, Q_b, Q_c] \qquad (9.56)$$

The decision variables within eq. (9.55) are the tray efficiency of the rectifying (E_r) and stripping (E_s) sections, respectively, the liquid level of the trays within the rectifying (h_r) and stripping (h_s) sections as well as the heat losses of both bottom (Q_b) and condenser (Q_c). The latter were included because neither the bottom nor the condenser is thermally insulated. The first part of objective function (9.55) minimizes the calculated and measured temperature ($T_{ij,sim}$ and $T_{ij,exp}$) differences in each tray (P) and each measured moment of time (N). The second part is aimed to fit the time evolution of calculated and measured distillate concentrations ($x_{i,sim}$ and $x_{i,exp}$). w is a weighting factor. The bottom concentration was measured and used to verify mass balance eq. (9.8). Figure 9.16(a) presents the correlation between simulated and experimental dynamic data and demonstrates their fairly good superposition. Figure 9.16(b) depicts steady-state measured and simulated temperature profiles of distillate and feed streams [23]. Again, data fit well.

9.6 Control issues of batch distillation columns

Discontinuous distillation columns are rarely used at a large industrial scale. However, fine chemical and pharmaceutical industry often operates batch-wise; thus batch distillation may still occasionally be in use. Next to the production-scale limitations, batch distillation is unfavored by operational reasons as well. Among these are time-consuming auxiliary operations such as loading the mixture, unloading the bottom at end-point, reaching boiling temperatures, and cleaning operations. Therefore, this topic is only briefly addressed to within this chapter.

If the column is operated at constant distillate flow rate mode, its concentration varies with the bottom concentration, but the separation factor is constant in time. Thus, global mass and component balances are written as follows:

$$M_B = M_{B,0} + tD \qquad (9.57)$$

$$M_B X_B = M_{B0} X_{B0} + D \int x_D dt \qquad (9.58)$$

The bottom concentration x_B and the concentration x_D of collected distillate are hence:

$$X_B = \frac{M_{B0}X_{B0} - D \int x_D dt}{M_{B0} - tD} \tag{9.59}$$

$$\bar{x}_D = \frac{\int x_D dt}{t} \tag{9.60}$$

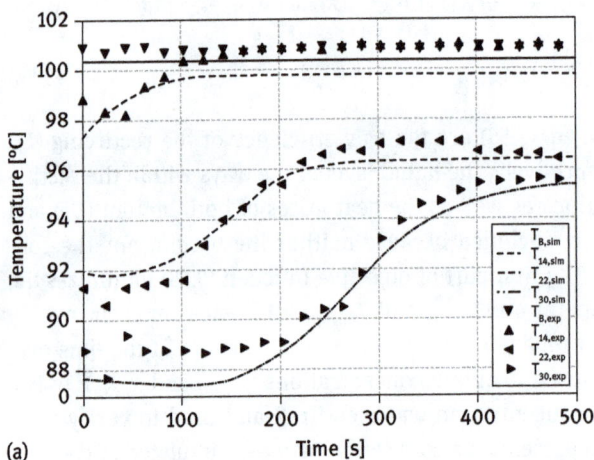

(a)

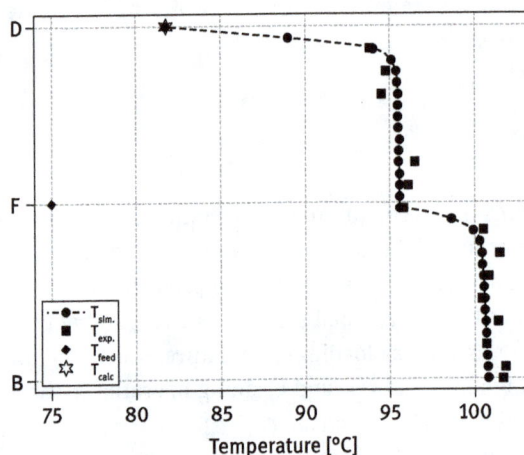

(b)

Fig. 9.16: First-principle model validation for a pilot-scale continuous distillation column of ethanol-water mixtures: (a) experimental and simulated dynamic data as a function of time after a step-change in heating power; (b) steady-state experimental and simulated temperature profiles of distillate and feed streams (reprinted from [23], with permission from Elsevier).

The process end-point implies the setpoint $\bar{x}_D$ to be reached ($\bar{x}_D = x_D$). When operating at high D and low R/D ratio, the distillation stops early. In contrast, when R/D is high and D is low the separation is good, the product quality is high but the energy demand is also high. Therefore, for a better control, the value of D should be high at the beginning of the process, then gradually declines to compensate the decreasing bottom concentration x_B. Figure 9.17 presents the typical control scheme of a batch distillation column.

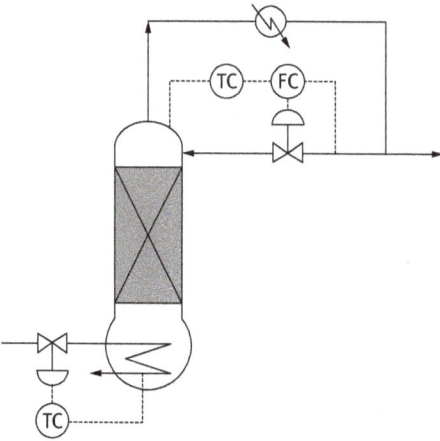

Fig. 9.17: Control scheme of a batch distillation column.

Classic operation modes of batch columns are either at constant reflux flow rate R or at constant distillate concentration x_D, respectively. When R is constant, the value of x_D will decrease in time; thus employed R has to be chosen carefully to ensure the prescribed end-point concentration. Constant x_D operation requires time-variable R; at the beginning while x_B is relatively high, R can be low. Afterward, it must be gradually increased while x_B drops.

The optimal control strategy of batch distillation columns has been a debated topic during the last decades [24, 25]. Even so, variation of R between the two limiting cases (constant reflux or concentration) [26, 27] is the most encountered control solution. Reflux optimization may be carried out by dynamic programming. An example is presented and discussed in detail in Chapter 4.2 – see also explanations for Fig. 4.5.

Model predictive control is often used to simultaneously take into account disturbances, control objectives, and process constraints. However, an earlier study [28] has proved that for more complicated separations, a significant time improvement can be achieved with two-stage distillation.

References

[1]   Agachi, Ş, *Automatizarea Proceselor Chimice*, Casa Cărţii de Ştiinţă, Cluj-Napoca, 1994.
[2]   Luyben, W.L., *Practical Distillation Control*, Springer Science & Business Media, Berlin/ Heidelberg/ Dordrecht/New York City, 2012.
[3]   Shinskey, F.G., *Distillation Control: For Productivity and Energy Conservation*. McGraw-Hill, New York, 1984.

[4]   Porru, M., Baratti, R., Alvarez, J., *Energy saving through control in an industrial multicomponent distillation column*, IFAC-PapersOnLine, 48(8), 1138–1143, 2015.

[5]   Gadalla, M., Olujic, MZ., Sun, L., De Rijke, A., Jansens, P.J., *Pinch analysis-based approach to conceptual design of internally heat-integrated distillation columns*, Chemical Engineeering Research and Design, 83, 987–993, 2005.

[6]   Linnhoff, B., Dunford, H., Smith, R., *Heat integration of distillation columns into overall processes*, Chemical Engineering Science, 38(8), 1175–1188, 1983.

[7]   Iancu, M.H., *Advanced control of the heat integrated complex plants*, PhD thesis abstract, Babeş-Bolyai University of Cluj-Napoca, Romania, 2010; http://doctorat.ubbcluj.ro/sustinerea_publica/rezumate/2010/inginerie-chimica/Iancu_Mihaela_en.pdf.

[8]   Mathisen, K.M., Skogestad, S., Wolff, E.A., *Bypass selection for control of heat exchanger networks*, Computers & Chemical Engineering, 16, S263–S272, 1992.

[9]   Glemmestad, B., Gundersen, T., *A systematic procedure for optimal operation of heat exchanger networks*, AIChE Symposium Series, 1998, 320(94).

[10]   Baldea, M., Daoutidis, P., *Dynamics and Nonlinear Control of Integrated Process Systems*, Cambridge University Press, Cambridge, 2012.

[11]   Shinskey, F.G., *Energy Conservation Through Control*, Academic Press, New York, 1978.

[12]   Douglas, J.M., Jafarey, A., Seemann, R., *Short-cut techniques for distillation column design and control. 2. Column operability and control*, Industrial & Engineering Chemistry Process Design and Development, 18(2), 203–210, 1979.

[13]   Skogestad, S., Morari, M., *Control configuration selection for distillation columns*, AIChE Journal, 33(10), 1620–1635, 1987.

[14]   Kister, H.Z., Mathias, P.M., Steinmeyer, D.E., Penney, W.R., Crocker., B.B., Fair, J.R., *Chapter 14. Equipment for Distillation, Gas Absorption, Phase Dispersion, and Phase Separation*, in Green, W.G., Perry, R.H. (Eds.), *Perry's Chemical Engineering Handbook*, 8th Edition, McGraw-Hill, New York, 2007.

[15]   Wittgens, B., Skogestad, S., *Evaluation of dynamic models of distillation columns with emphasis on the initial response*, Modeling, Identification and Control, 21(2), 83–103, 2000.

[16]   Shulman, H.L., Ullrich, C.F., Wells, N., Proulx, Z., *Performance of packed columns. III. Holdup for aqueous and nonaqueous systems*, AIChE Journal, 1(2), 259–264, 1955.

[17]   Gokhale, V., Hurowitz, S., Riggs, J.B., *A Comparison of Advanced Distillation Control Techniques for a Propylene/Propane Splitter*, Chemical Engineeering Research and Design, 34(12), 4413–4419, 1995.

[18]   Kiss, A.A., *Advanced Distillation Technologies: Design, Control and Applications*, John Wiley & Sons, Chichester, 2013.

[19]   Rewagad, R.R., Kiss, A.A., *Dynamic optimization of a dividing-wall column using model predictive control*, Chemical Engineering Science, 68(1), 132–142, 2012.

[20]   Assandri, A.D., de Prada, C., Rueda, A., Martínez, J.L., *Nonlinear parametric predictive temperature control of a distillation column*, Control Engineering Practice, 21(12), 1795–1806, 2013.

[21]   Agachi, P.Ş., Cristea, V.M., *Basic Process Engineering Control*, De Gruyter, Berlin/Boston, 2014.

[22]   Agachi, P.Ş, Cristea, V.M., Nagy, Z.K., Imre-Lucaci, A., *Model Based Control: Case Studies in Process Engineering*, John Wiley & Sons, Weinheim, 2007.

[23]   Szavuly, M., Toos, A., Barabas, R., Szilágyi, B., *From modeling to virtual laboratory development of a continuous binary distillation column for engineering education using MATLAB and LabVIEW*, Computer Applications in Engineering Education, 27(5), 1019–1029, 2019.

[24]   Macchietto, S., Mujtaba, I.M., *Design of Operation Policies for Batch Distillation*, p. 174–215 in Reklaitis, G, Sunol, A., Rippin, D.T., Hortaçsu, Ö. (Eds.), *Batch Processing Systems Engineering SE – 9*, Vol. 143, Springer Verlag, Berlin/Heidelberg, 1996.

[25]   Zavala, J.C., Coronado, C., *Optimal control problem in batch distillation using thermodynamic efficiency*, Industrial & Engineering Chemistry Reseasrch, 47(8), 2788–2793, 2008.

[26] Logsdon, J.S., Biegler, L.T., *Accurate determination of optimal reflux policies for the maximum distillate problem in batch distillation*, Industrial & Engineering Chemistry Reseasrch, 32(4), 692–700, 1993.

[27] Diwekar, U.M., Malik, R.K., Madhavan, K.O., *Optimal reflux rate policy determination for multicomponent batch distillation columns*, Computers & Chemical Engineering, 11(6), 629–637, 1987.

[28] Christensen, F.M., Jørgensen, S.B., *Optimal control of binary batch distillation with recycled waste cut*, Chemical Engineering Journal, 34(2), 57–64, 1987.

10 Control of absorption processes

Absorption is a separation process of gas mixtures based on the selective dissolution of one or more gas components in a liquid solvent system. The inverse process, the stripping of absorbed gas from the solvent, is called *desorption*. It is usually carried out in *absorbers*, chemical equipment with various configurations designed to carry out efficient gas-liquid mass transfer. The most frequent type encountered in practice is the *absorption column* (either packed or with trays – see Fig. 10.1). The polluted gas stream is fed at the column bottom in high concentration (mole fraction y_F), and after dissolving the soluble component, the purified gas stream is evacuated at the top (with a mole fraction of y_V). The liquid solvent enters at the top with a low concentration of dissolvable gas (mole fraction x_R) and flows in a countercurrent with the gas. While in contact with the solvent, the soluble gas species dissolves; thus, the evacuation becomes more concentrated (mole fraction x_B). The corresponding stream flow rates, expressed in mol/s, are liquid inlet R and outlet B and gas inlet F and outlet V, respectively.

The mass balance (mol/s) equations for an absorber under a steady state are

$$F + R = V + B \tag{10.1}$$

and

$$F y_F + R x_R = V y_V + B x_B \tag{10.2}$$

Expressing B from eq. (10.1) and substituting it to eq. (10.2), the following results:

$$F y_F + R x_R = V y_V + (F + R - V) x_B \tag{10.3}$$

The non-dissolving gases leave the column within flux V:

$$F(1 - y_F) = V(1 - y_V) \tag{10.4}$$

Expressing V from eq. (10.4), substituting it to eq. (10.3), and after rearrangement, the following is obtained:

$$y_V = \frac{y_F(1 - x_B) + \frac{R}{F}(x_R - x_B)}{(1 - x_B) + \frac{R}{F}(x_R - x_B)} \tag{10.5}$$

Equation (10.5) reveals a connection between outlet gas stream purity and operating parameters such as inlet gas and liquid flow rate as well as inlet concentrations.

Meanwhile, for dilute systems, the Kremser-Brown-Sounders [1] equation can be used to calculate the outlet gas stream concentration:

https://doi.org/10.1515/9783110789737-012

$$\frac{y_{F,i} - Y_{V,i}}{y_{F,i} - K_i X_{R,i}} = \frac{1 - A_i^{-n}}{1 - A_i^{-(n+1)}} \tag{10.6}$$

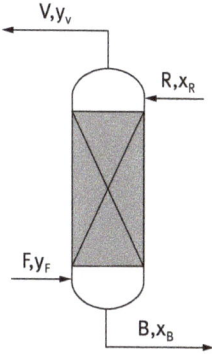

Fig. 10.1: Scheme of a continuous absorption column.

where i denotes a certain component/chemical species; $Y_{V,i} = y_{V,i}\, V/F$ is the corrected gas phase mole fraction (dimensionless since mol/mol); similarly, the liquid is characterized by $X_{B,i}$; P_i is the vapor pressure of pure component at working temperature (atm); $K_i = P_i/P$ the pressure fraction of component "i" (dimensionless); $A_i = R/(K_i V)$ is the absorption factor (dimensionless); n stands for the theoretical number of trays.

The $V - F$ difference is a result of gas absorption. Hence, if more than one species are absorbed, mass balance (10.1) becomes eq. (10.7). A similar relationship stands for the liquid stream:

$$V - F = V \sum_i (y_{F,i} - Y_{V,i}) \tag{10.7}$$

The i component mass balance (within the numerous individual mass balances) can be written instead of eq. (10.2) as

$$V(y_{F,i} - Y_{V,i}) = R(X_{B,i} - X_{R,i}) \tag{10.8}$$

After processing eqs. (10.7)–(10.8), absorber outlet stream composition characteristics in the specified species i are obtained – see eq. (10.9). These values are helpful in absorbance yield and performance assessment. They also represent controlled variables:

$$y_{V,i} = \frac{Y_{V,i}}{1 - \sum_i (y_{F,i} - Y_{V,i})}$$
$$X_{B,i} = \frac{X_{B,i}}{1 - \sum_i (X_{B,i} - X_{R,i})} \tag{10.9}$$

Since the gas is not *pure*, when assuming dilute systems, i.e., low liquid-phase concentrations, the *liquid equilibrium* concentrations (mole fractions) x_B^* and x_R^* can be expressed by Henry's law [2, 3]. T_F and T_V stand for the stream temperatures of F and V, respectively:

$$x_B^* = \frac{P_{F,i}}{H(T_F)}$$

$$x_R^* = \frac{P_{V,i}}{H(T_V)} \tag{10.10}$$

The Henry's constant's temperature dependence is described by the Van't Hoff relationship [2] as

$$\frac{d\ln H}{dT} = \frac{E_{abs}}{RT^2} \tag{10.11}$$

where E_{abs} stands for the activation energy of absorption and R is the universal gas constant.

Relationships (10.10) assume that at the top (contact point of V and R streams) and bottom (contact point of F and B streams) of the column, the phases are in equilibrium. Although thermodynamic equilibrium does not occur in normal operating regime, this assumption is widely applied in the mathematical description of absorbers. The information presented above demonstrates that the outlet gas stream concentration (value of y_V, usually it is a controlled variable) depends on inlet x_R and y_F, inlet flowrates F and R, temperature, and overall pressure.

The industrial practice of absorber operation indicates that the most significant part (~80%, according to the estimates) of total mass transfer occurs in the first and last mass transfer unit (tray). Meanwhile, the temperature profile in the column can vary due to dissolution heat release and heat transfer between the gas and liquid inlet streams. This causes a temperature gradient along the column that affects the gas-liquid mass transfer rate through both Henry's law and K_i factor values.

When inlet stream temperatures and compositions are constant, the R/F ratio controls the composition of both streams. If the feed temperature increases, solubility will decrease. Hence, the value of R/F has to increase to compensate for this negative effect. Similarly, if the amount of soluble component increases in the feed, the values of R/F should also increase (more mass has to be dissolved). These observations are quantitatively described by eqs. (10.6) and (10.9).

Consequently, the control structure of an absorption column (see Fig. 10.2) must involve the following elements:

– R/F ratio control;
–   temperature control of both inlet streams;
–   level control at the bottom (to close the mass balance);
–   pressure control at the top.

A large variety of control strategies can be applied. The choice of a reasonable controller configuration based purely on theoretical considerations requires sensitivity and parametric analyses [5]. Such model-based analysis reveals steady-state and dynamic behavior patterns and allows to identify the best manipulated and controlled variables. Control solutions more advanced than presented in Fiugre 10.2 can be employed

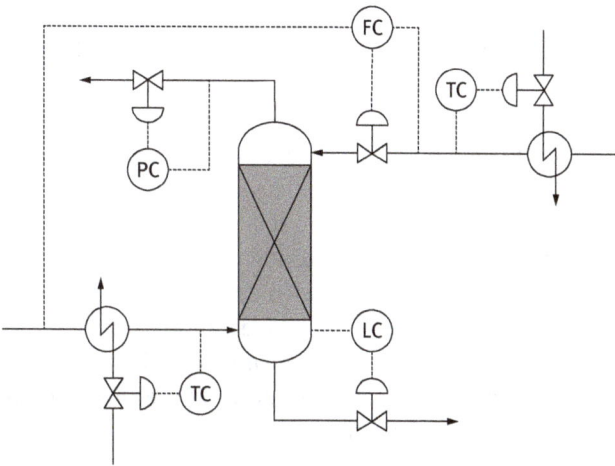

Fig. 10.2: General control configuration of a continuous absorption column. The flow rate controller FC is on ratio.

as well. Feedforward or cascade controllers can be implemented [4], for instance, to effectively reject the effects of disturbances in the temperature or composition of feed streams.

In several cases, a chemical reaction accompanies the physical dissolution (absorption). This happens, e.g., during the SO_2 absorption in water: first, SO_2 gas is dissolved physically in the water and then enters a chemical reaction with it:

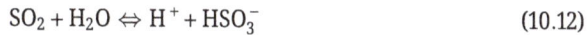

$$SO_2 + H_2O \Leftrightarrow H^+ + HSO_3^-$$ (10.12)

In such situations, the *chemical equilibrium is* described by *equilibrium constant K_{eq}*. For diluted solutions, the water concentration is considered unchanged and the equilibrium constant becomes

$$K_{eq} = \frac{[H^+][HSO_3^-]}{[SO_2]}$$ (10.13)

where K_{eq} depends on temperature and pressure; these functions are described by thermodynamic and kinetic equations [2]. Equation (10.13) can be used to calculate the amount of SO_2 consumed by chemical reaction. The reaction leads to an apparent increase in the amount of dissolved SO_2; the total quantity of SO_2 transferred from the gas into liquid will increase as compared to the thermodynamically expected value (Henry's law). Nevertheless, the absorber is operated based on similar considerations (see Figs. 10.1 and 10.2).

As presented in Fig. 10.1, the absorbed gases leave the absorber in the stream *B*. These gases have to be stripped from the liquid in a subsequent column to allow recirculating the solvent into the absorber. Stripping occurs in a desorption column. The desorber operates under higher temperatures and lower pressures than the absorber (conditions that favor stripping).

Continuous absorption and desorption columns are generally operated in couples. Such a system is presented in Fig. 10.3 and has multiple advantages. Among these, the following aspect can be mentioned:
- The liquid phase leaving the desorber is recycled to the absorber. This reduces necessary fresh solvent amounts. Some specific absorption processes that use expensive and/or toxic solvents are deliberately conducted under such a strategy.
- The simultaneous operation opens new doors for heat integration. This allows reusing the thermal energies of the internal streams, which inherently reduces the external energy demand of the process. Deeper heat integration results in lower energy demand but makes the process more difficult to control *via* the additional internal couplings. Meanwhile, model predictive controllers (see Chapter 2) can improve the control performance of such rigid systems [8].

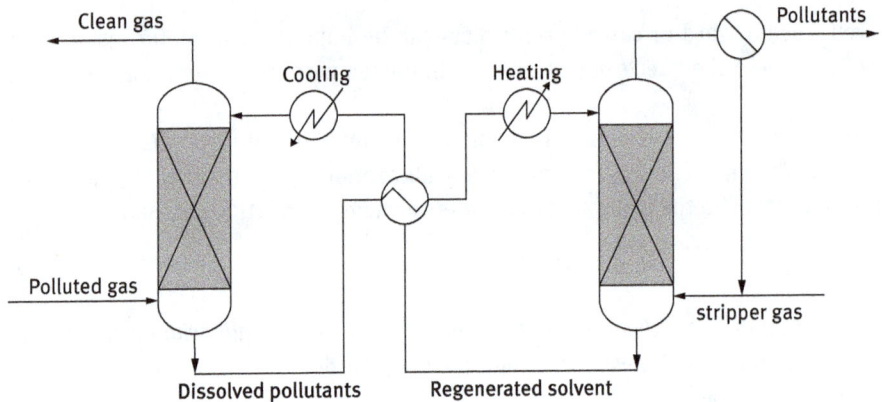

Fig. 10.3: Absorption-desorption system with partial heat integration.

During the last decade, CO_2 absorption from post-combustion gases was granted special attention [6]. The absorber works under pressure and employs an organic liquid (often aqueous mono-ethanol amine solution [7]) for capturing CO_2. This strategy is primarily applied to reduce emissions of fossil power plants. CO_2 absorption processes are carried out in couped absorption-desorption systems (similar to that of Fig. 10.3). The pure CO_2 released in the desorber is either used as a marketed product or starting material or it is stored in underground or underwater reservoirs. The absorption process relies on the following elementary steps:

1. Physical dissolution of CO_2:

$$CO_{2,gas} \Leftrightarrow CO_{2,Sol} \tag{10.14}$$

2. A chemical reaction between solute CO_2 and mono-ethanol amine and/or water, respectively:

$$CO_{2,Sol} + RNH_2 \Leftrightarrow RNH^+ + HCO_3^- \tag{10.15}$$

$$CO_{2,Sol} + H_2O \Leftrightarrow H^+ + HCO_3^- \tag{10.16}$$

A several control strategies were proposed for CO_2 absorption systems in constant load regime, for example, lean solvent flow or loading variation strategy. However, some fossile fuel power plants, especially the natural gas fired plants, are being operated in grid-balancing mode, meaning that their power output is varied to match the actual power demand under the varying power output of weather-dependent power plants. Hence, the flexibility of CO_2 absorption plant has to be ensured. One way to enhance the flexibility through control is the application of MPCs. Gaspar et al. [9] executed a control-oriented dynamic study of such systems and achieved enhanced flexibility by the partial decoupling of the absorber and desorber column *via* solvent buffer tank approach (Fig. 10.4). This underscores that improved controllability often seeks additional equipment.

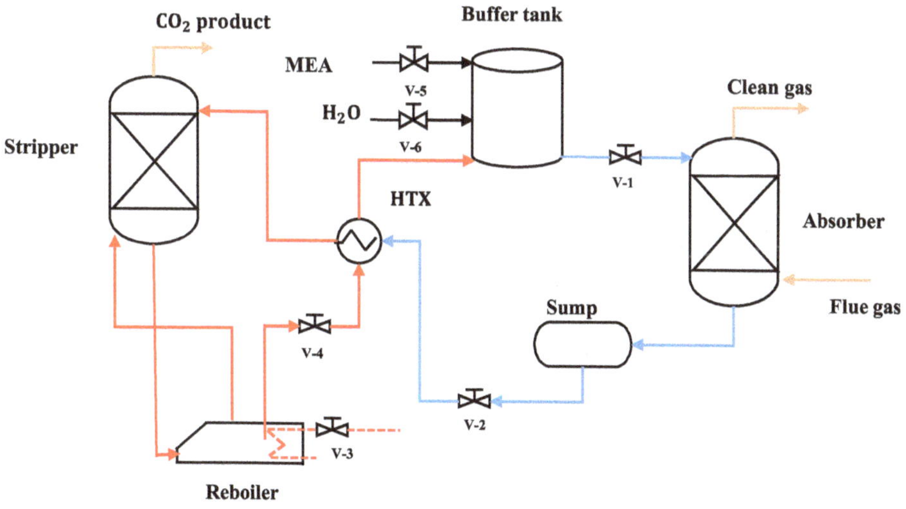

Fig. 10.4: CO_2 absorption process flowsheet with control structure (adapted from [9], Fig. 1).

The absorption/desorption/compression system can be thermally integrated with the power plant by carrying out pinch analysis. This yields significantly higher overall power-plant efficiency compared to the isolated thermal operation of the absorption/desorption/ compression system. However, this also seeks more advanced control solutions [10].

i **Example 10.1**

An absorption column with four theoretical units works at 5 bar pressure and 38 °C temperature. The feed and reflux stream compositions as well as K_i values (see eq. (10.6)) are given in Tab. 10.1. Let us calculate the variation in the composition of outlet streams if the pressure increases with 5% for $R/V = 0.5$.

Tab. 10.1: Feed and reflux stream compositions of an absorption column as well as the K_i values.

	y_F	x_R	K_i	
			$P = 5$ bar	$P = 5.25$ bar
Propane (1)	0.070	0.0002	2.15	2.13
Isobutane (2)	0.900	0.0004	1.00	0.95
n-Butane (3)	0.030	0.0005	0.74	0.70

The right-hand side of eq. (10.6) can be computed for all compounds from the above-listed data:

$$A_1 = \frac{0.5}{2.25} = 0.222 \qquad A_2 = \frac{0.5}{1} = 0.5 \qquad A_3 = \frac{0.5}{0.74} = 0.675$$

Further

$$\frac{1-A_1^4}{1-A_1^5} = 0.221 \qquad \frac{1-A_2^4}{1-A_2^5} = 0.483 \qquad \frac{1-A_3^4}{1-A_3^5} = 0.662$$

The corrected gas composition is according to eq. (10.6):

$$Y_{V,1} = 0.07 - 0.221 \times (0.07 - 0.221 \times 0.0002) = 0.054$$
$$Y_{V,2} = 0.9 - 0.483 \times (0.9 - 0.483 \times 0.0004) = 0.465$$
$$Y_{V,3} = 0.03 - 0.622 \times (0.03 - 0.622 \times 0.0005) = 0.011$$

Further, the sum on the right-hand side of eq. (10.7) becomes

$$\sum_i (y_{F,i} - Y_{V,i}) = (0.07 - 0.054) + (0.9 - 0.465) + (0.03 - 0.011) = 0.47$$

This result coupled with the use of eq. (10.9) easily yields the outlet gas stream composition:

$$y_{V,1} = \frac{0.054}{1 - 0.47} = 0.102 \qquad y_{V,2} = \frac{0.465}{1 - 0.47} = 0.877 \qquad y_{V,3} = \frac{0.011}{1 - 0.47} = 0.020$$

A similar algorithm leads to the liquid phase composition. From eq. (10.8), the corrected liquid values can be calculated as follows:

$$X_{B,1} = \tfrac{1}{0.5} (0.07 - 0.054) + 0.0002 = 0.0322$$
$$X_{B,2} = \tfrac{1}{0.5} (0.9 - 0.465) + 0.0004 = 0.8704$$
$$X_{B,3} = \tfrac{1}{0.5} (0.03 - 0.01) + 0.0005 = 0.0385$$

Further, the sum appearing in eq. (10.9) is

$$\sum_i (X_{B,i} - X_{R,i}) = (0.0322 - 0.0002) + (0.8704 - 0.0004) + (0.0385 - 0.0005) = 0.94.$$

Outlet liquid mole fractions are computed using eq. (10.9):

$$x_{B,1} = \frac{0.0322}{1+0.94} = 0.016 \qquad x_{B,2} = \frac{0.8704}{1+0.94} = 0.4486 \qquad x_{B,3} = \frac{0.0385}{1+0.94} = 0.019$$

The pressure variation involves the modification of K_i values (see Tab. 10.1). After similar calculations, the modified mole fraction values are

$$y_{V,1} = 0.102 \qquad y_{V,2} = 0.875 \qquad y_{V,3} = 0.019$$
$$x_{B,1} = 0.0172 \qquad x_{B,2} = 0.0459 \qquad x_{B,3} = 0.02$$

Tab. 10.2: Calculated outlet mole fraction values in liquid and gas streams at 5 and 5.25 bar and the net percentage change after the pressure variation.

Parameter	Composition					
	$y_{V,1}$	$y_{V,2}$	$y_{V,3}$	$x_{B,1}$	$x_{B,2}$	$x_{B,3}$
$P = 5$ bar	0.102	0.877	0.020	0.0160	0.4486	0.019
$P = 5.25$ bar	0.102	0.875	0.019	0.0172	0.4590	0.020
Net change (%)	0	−0.228	−5.000	−7.5000	+2.3180	+5.000

Table 10.1 summarizes the calculated values in Example 10.1. The results demonstrate that increased pressure shifts the gas-liquid equilibrium of species i toward the liquid. Thus, gas solubility increases with pressure [2].

References

[1] Agachi, Ş., *Automatizarea Proceselor Chimice*, Casa Cărţii de Ştiinţă, Cluj-Napoca, 1994.
[2] Tosun, I., *The Thermodynamics of Phase and Reaction Equilibria*, Elsevier, Amsterdam, 2013.
[3] Kister, H.Z., Mathias, P.M., Steinmeyer, D.E., Penney, W.R., Crocker, B.B., Fair, J.R., *Chapter 14. Equipment for Distillation, Gas Absorption, Phase Dispersion, and Phase Separation* in Green, W.G., Perry, R.H. (Eds.), *Perry's Chemical Engineering Handbook*, 8th Edition, McGraw-Hill, New York, 2007.
[4] Govindarajan, A., Jayaraman, S.K., Sethuraman, V., Raul, P.R., Rhinehart, R.R., *Cascaded process model based control: packed absorption column application*, ISA Transactions, 53, 391–401, 2014.
[5] Ungureanu, S., *Stabilitatea Sistemelor Dinamice*, Editura Tehnică, Bucureşti, 1988.
[6] Salvinder K.M.S., Zabiri H., Taqvi S.A., Ramasamy M., Isa F., Rozali N.E.M., Suleman H., Maulud A., Shariff A.M., *An overview on control strategies for CO_2 capture using absorption/stripping system*, Chemical Engineering Research and Design, 147, 319-337, 2019.
[7] Lv B., Guo B., Zhou Z, Jing G., *Mechanisms of CO_2 capture into monoethanolamine solution with different CO_2 loading during the absorption/desorption processes*, Environmental Science & Technology, 49, 10728-10735, 2015.

[8] Hossein Sahraei, M., Ricardez-Sandoval, L.A., *Controllability and optimal scheduling of a CO_2 capture plant using model predictive control*, International Journal of Greenhouse Gas Control, 30, 58–71, 2014.

[9] Gaspar J., Ricardez-Sandoval L, Jørgensen J.B., Fosbøl P.L., *Controllability and flexibility analysis of CO_2 post-combustion capture using piperazine and MEA*, International Journal of Greenhouse Gas Control, 51, 276-289, 2016.

[10] Vega F., Baena-Moreno F.M., Gallego Fernández L.M., Portillo E., Navarrete B., Zhang Z., *Current status of CO_2 chemical absorption research applied to CCS: Towards full deployment at industrial scale*, Applied Energy, 260, 114313, 2020.

11 Control of extraction processes

Extraction is a complete or partial separation process of a chemical called *solute*, from a liquid or solid mixture called *matrix*, by means of another species called *solvent*. It is useful when liquid individual components cannot be separated by means of distillation because of their close boiling points, low volatility, thermal instability, or formation of an azeotrope. In case of solid mixtures, it is a handy tool for the recovery of useful components.

If the feed matrix **F** contains the solute species **A** to be separated by employing solvent **S**, then the following requirements have to be fulfilled for the extraction to be possible [1, 2]:

- Liquid solvent **S** is immiscible (or partially miscible) with **F** if this is a liquid, or **S** does not dissolve **F** in case the latter is a solid;
- Solute **A** dissolves in solvent **S**;
- At equilibrium, the concentration of solute **A** in solvent **S** should be (considerable) higher than in initial matrix **F**.

Any extraction demands the following sequence of steps:
- Contact between matrix and solvent;
- Transfer of solute from the matrix into the solvent until (ideally) the equilibrium concentrations are reached due to partition of solute between matrix and solvent;
- Separation of resulting extract from the raffinate or residue, respectively.

The raffinate is further processed to recover both solute (when desired) and solvent. The transfer of solute from the loaded solvent into another phase is called *stripping* or *back-extraction* and serves the purpose of obtaining it. Meanwhile, recovery of the solvent is compulsory because of economic as well as environmental reasons.

If both the matrix and the solvent are liquids, the procedure is called *liquid-liquid (L-L) extraction, solvent extraction* or *refining*. The solute loaded solvent is called *extract* and the impoverished matrix *raffinate*, respectively. If the matrix is a solid, the technique is called *solid-liquid (S-L) extraction* or *percolation* [1, 2]. The resulting liquid and solid phases are called *extract* and *residue*, respectively. In the case of L-L extraction, the raffinate may not contain any solvent, whereas for S-L extraction, the solid residue will always be wetted by it. The extract may also require post-extraction treatment because of solid residue particles.

Liquid-liquid extraction is suitable for continuous operation [1, 2] even at throughputs of 100,000 m^3/h [3] and is therefore applied at large scale in the oil industry [3, 4]. Meanwhile, solid-liquid extraction is applied mainly batchwise and at much smaller scale in the food and medical industry. Therefore, this chapter focuses mainly on L-L extraction.

https://doi.org/10.1515/9783110789737-013

The most widely employed techniques involve either *crosscurrent* (see Fig. 11.1(a)) or *countercurrent* (see Fig. 11.1(b)) L-L extraction. Both involve a cascade of *stages* or *extractors*. Within the first option, fresh solvent is added to the raffinate of each stage; the extract is collected from each stage. The second option implies the solvent and the raffinate to flow in countercurrent through the entire sequence of stages [1].

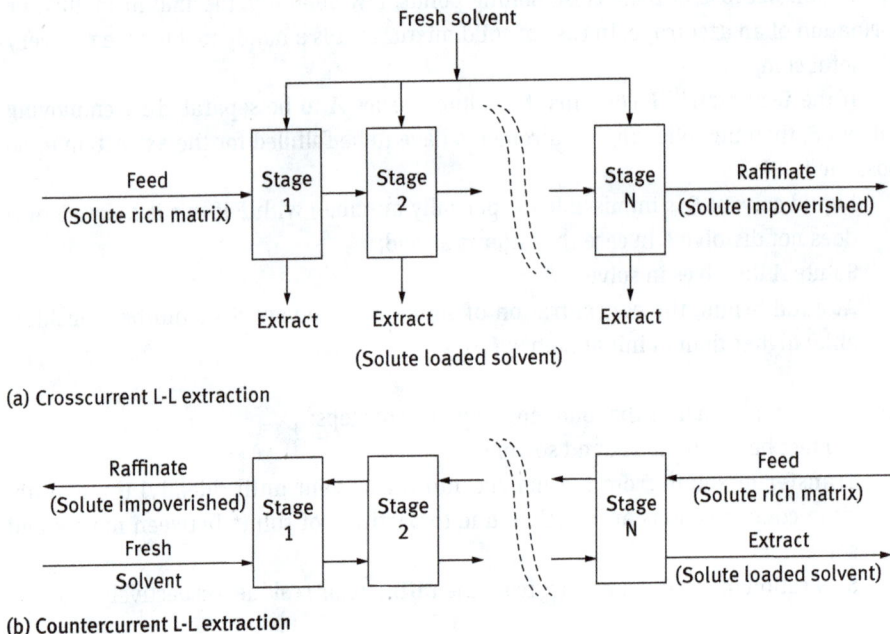

Fresh solvent

| Feed (Solute rich matrix) | Stage 1 | Stage 2 | Stage N | Raffinate (Solute impoverished) |

Extract   Extract   Extract

(Solute loaded solvent)

(a) Crosscurrent L-L extraction

| Raffinate (Solute impoverished) | Stage 1 | Stage 2 | Stage N | Feed (Solute rich matrix) |
| Fresh Solvent | | | | Extract (Solute loaded solvent) |

(b) Countercurrent L-L extraction

Fig. 11.1: (a) Crosscurrent and (b) countercurrent liquid-liquid extraction (adapted after [1]).

Mass transfer between the feed (or raffinate for a multistage extraction) and the solvent (or extract for the case of multistep countercurrent extraction) includes droplet formation/dispersing, transport, and settling of the phases. The feed can be dispersed into the solvent (Fig. 11.2(a)) or the solvent dispersed in it (Fig. 11.2(b)) [3]. Figure 11.2 reveals these differences and depicts the feed matrix as a light-phase that accumulates at the top of the heavy one. Dispersion, and settling, of phases is based on the difference between the densities of feed and solvent. Density difference is also responsible for the spontaneous countercurrent flows. Thus, light-phase droplets will move upward and heavy ones in the opposite direction along the height of an *extraction tower*. The types of dispersion remain as depicted in Fig. 11.2 for the case of a heavy-phase feed entered at the top of the tower. A widely employed industrial scale L-L extractor is the agitated extraction column. It requires, as a limiting condition, at least a 0.05 kg/m^3 density difference between the feed and the solvent [3, 4]. Figure 11.2 also illustrates the differences imposed for the heavy liquid (extract of interest) level control (see also explanations for Fig. 11.4).

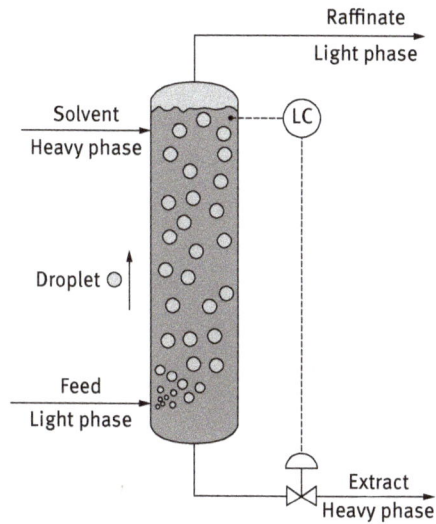

(a) Light phase dispersion

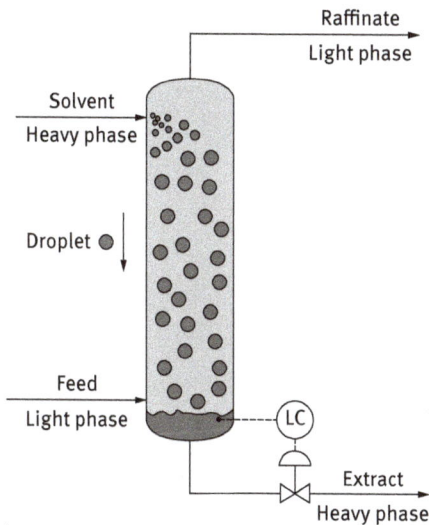

(b) Heavy phase dispersion

Fig. 11.2: Liquid level control for extraction towers with (a) light phase and (b) heavy phase dispersion. Light gray stands for the light/low density and dark gray stands for the heavy/high density liquid (adapted from [3]).

Figure 11.3 presents the scheme of mass balance for a countercurrent L-L packed extraction tower, with light-phase dispersion and in continuous steady-state operation mode. It describes a *binary mixture* that shares *a single* solute. The species of interest is *A* to be extracted from feed matrix *F* into solvent *S*. The flow rates (m³/h or kg/h) are symbolized with *F*, *S*, *E*, and *R* for the feed, solvent, extract, and raffinate, respectively. The dimensionless mass or molar fraction of solute *A* in each of these streams is symbolized with x_{AF}, x_{AS}, x_{AE}, and x_{AR}, respectively. Ideally, $x_{AR} = 0$ (100% extraction efficiency) and $x_{AS} = 0$ (pure solvent, 100% stripping efficiency).

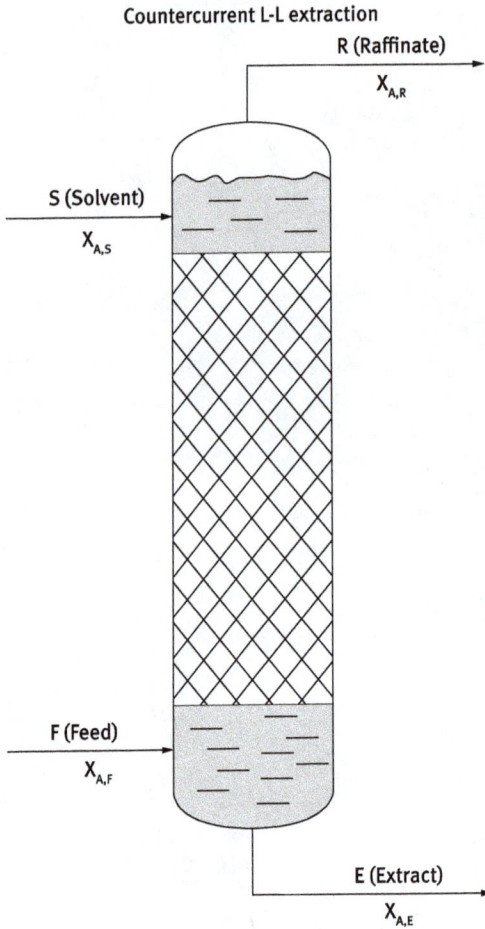

R (Raffinate)

$X_{A,R}$

S (Solvent)

$X_{A,S}$

F (Feed)

$X_{A,F}$

E (Extract)

$X_{A,E}$

Fig. 11.3: Mass-balance of a countercurrent L-L packed extraction tower, with light phase dispersion and in continuous steady-state operation mode.

The total mass balance and those of solute A and solvent S are expressed by eqs. (11.1) and (11.2)–(11.3), respectively:

$$F + S = E + R \tag{11.1}$$

$$Fx_{AF} + Sx_{AS} = Ex_{AE} + Rx_{AR} \tag{11.2}$$

$$S(1 - x_{AS}) = E(1 - x_{AE}) \tag{11.3}$$

Operational objectives of continuous L-L extraction control include, among others, extract quality specifications. This translates into maintenance of a desired x_{AE} value for variables F, x_{AF}, and x_{AS}. The mass balances above lead to eq. (11.4) and further to the expression (11.5) of x_{AE}, the mass/molar fraction of solute A in extract stream E:

$$x_{AE}[S(1-x_{AR}) + F(x_{AF} - x_{AR})] = Sx_{AS}(1-x_{AR}) + F(x_{AF} - x_{AR}) \tag{11.4}$$

$$x_{AE} = \frac{\frac{S}{F}x_{AS}(1-x_{AR}) + (x_{AF} - x_{AR})}{\frac{S}{F}(1-x_{AR}) + (x_{AF} - x_{AR})} \tag{11.5}$$

The flow rate ratio S/F affects x_{AE}; hence, input stream rates have to be put on ratio control (see Fig. 11.4). Thus, S will automatically adjust to disturbances of F. Flow rate control also ensures productivity specifications such as desired hourly throughputs. In addition, it avoids drying out of the extractor or its flooding by either the light or heavy phase. Liquid level control (see Figs. 11.2 and 11.4) helps to ensure the latter operational constraint but also provides constant (steady-state) volumes of light and heavy phases, respectively. These contribute to the desired x_{AE} and x_{AR} values.

The driving force of mass transfer in L-L extraction is the solute concentration gradient between the steady-state bulk phase (x_{AE} and x_{AR}) and the L-L interface (x^i_{AE} and x^i_{AR}). The solute mass stream N (kg/h) between the two liquid phases is expressed by relationship (11.6), where k_E and k_R stand for global mass transfer coefficients (kg/h) in the extract and solvent, respectively. They depend on the nature and composition of streams E and R as well as temperature. The end-point for batch extraction implies $N = 0$; in other words, $x_{AE} = x^i_{AE}$ and $x_{AR} = x^i_{AR}$ (bulk solute concentrations in E and R correspond to the equilibrium values):

$$N = k_E\left(x^i_{AE} - x_{AE}\right) = k_R\left(x_{AR} - x^i_{AR}\right) \tag{11.6}$$

Separation by L-L extraction (i.e. values of $K_A = \frac{x^i_{AE}}{x^i_{AR}}$ and $\beta = \frac{K_A}{K_B}$) is governed by the thermodynamic partition *equilibrium* of the solute between the two liquid phases in contact [1–2, 5–6]. It is usually expressed by the Nernst distribution law below – see (11.7), where K_A is the dimensionless equilibrium constant called *partition coefficient*:

$$K_A = \frac{x^i_{AE}}{x^i_{AR}} \tag{11.7}$$

Equation (11.7) describes a linear correlation between the solute contents in the resulting extract and raffinate streams, respectively. It is valid only for diluted solutions. Defined as such, good performance extraction implies $K_A > 1$ since $x_{AE} > x_{AR}$ ought to be true. It is determined experimentally for each solute and feed/solvent pair at various temperatures and vapor pressures. Therefore, partition relationships are also important in the choice of correct S/F ratios [1, 2].

The temperature dependence of K_A can be expressed by the Van't Hoff equation [5, 6] (11.8), where ΔH is the equilibrium phase transfer enthalpy of the solute (J/mol), $T°$ is the temperature (kelvin), and R is the universal gas constant:

$$\frac{d\ln(K_A)}{dT^\circ} = \frac{\Delta H}{R(T^\circ)^2} \quad \text{or} \quad \ln(K_A) = -\frac{\Delta H}{RT^\circ} \tag{11.8}$$

Depending on ΔH value, K_A is more or less strongly affected by temperature (see Tab. 11.1). Therefore, the necessity of temperature control has to be checked before investing in expensive equipment (see Examples 11.1 and 11.2), whereas feed/solvent flow rate ratio and heavy-phase level control are always recommended (see Fig. 11.4).

Figure 11.4 presents a complete control system for a countercurrent L-L packed extraction tower operated continuously under steady-state conditions [7, 8]. The loops control the S/F flow rate ratio, temperatures of both F and S streams, and extract volume. Only the latter is based on a liquid level reading; the others measure directly the controlled variable. All manipulated variables are flow rates. Input streams benefit of temperature adjustment before entering the extractor. Improved performance control can be achieved using various model-based techniques [9–12].

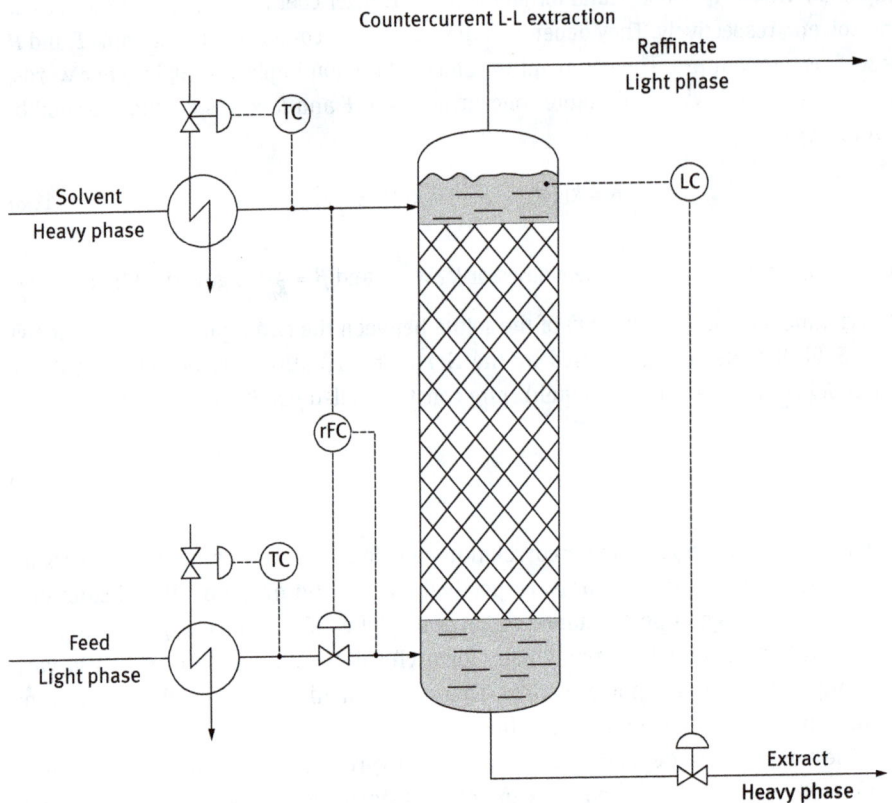

Fig. 11.4: Control scheme for the extraction tower in Fig. 11.3 (adapted after [7]).

Figures 11.3 and 11.4 as well as relationships (11.1)–(11.5) refer to binary mixtures of immiscible or partially/totally miscible liquids that share a single solute. Sometimes, such a mixture can share *two or more solutes*. If the feed matrix contains, for example, species *A* and *B*, both soluble and extractable by the solvent, another indicator of process performance is also used in practice. It is called *selectivity β* or *relative separation* and refers to the capacity of the solvent to extract preferentially a certain solute from the feed [2, 5]. *β* is defined as the ratio between the partition coefficients K_A and K_B of *A* and *B*, respectively – see eq. (11.7); thus, it is dimensionless:

$$\beta = \frac{K_A}{K_B} \tag{11.9}$$

If *β* > 1, the solvent has a higher affinity toward *A*, whereas if *β* < 1, it prefers *B*. If *β* = 1, the employed solvent is nonselective, and separation of *A* or *B* from matrix *F* cannot be carried out by means of extraction. According to relationship (11.8), both K_A and K_B are temperature-dependent; hence, adjustment of *β* to a desired value can only benefit from temperature control (see Fig. 11.4).

Multicomponent liquid mixtures are often encountered in practice [1]. In these cases, mass balances have to consider all components. Partition equilibrium will only seldom obey linear relationships such as that in eq. (11.7). Therefore, complex partition graphs are in use. Some examples are the Gibbs, Janecke, or tetrahedral diagrams for ternary or quaternary systems. These are drawn at constant temperature and vapor pressure [1, 2, 5].

Example 11.1　　　　　　　　　　　　　　　　　　　　　　　　　　　　　　　　　　　　　 <kbd>i</kbd>

Let us consider an extraction tower such as presented in Figs. 11.3 and 11.4. The goal is to extract at 300 K ethylamine from an aqueous solution, available at $F = 100$ m³/h flow rate, by using methylbenzene (density of $\rho_s = 870$ kg/m³). The dispersed phase is the solvent (see Fig. 11.2(a)). The feeding streams are characterized by $x_{AF} = 0.15$ and $x_{AS} = 0.01$, respectively [7]. The distribution of solute between aqueous and organic phases is described by the Nernst equation (11.7) by $K_A = 1.360$ at 300 K. The setpoint is at $x_{AE} = 0.05$.

Let us estimate the change in the necessary solvent flow rate as well as in the resulted hourly extracted ethylamine quantity if the temperature increases by 25 K.

Equation (11.5) will lead to the value of S/F ratio. By replacing in it x_{AR} according to eq. (11.7), the following will be obtained:

$$\frac{S}{F} = \frac{(x_{AF} - x_{AR})(1 - x_{AE})}{(x_{AE} - x_{AS})(1 - x_{AR})} = \frac{\left(x_{AF} - \dfrac{x_{AE}}{K_A}\right)(1 - x_{AE})}{(x_{AE} - x_{AS})\left(1 - \dfrac{x_{AE}}{K_A}\right)} \tag{11.10}$$

Hence

$$\frac{S}{F} = \frac{\left(0.15 - \dfrac{0.05}{1.36}\right)(1 - 0.05)}{(0.05 - 0.01)\left(1 - \dfrac{0.05}{1.36}\right)} = 2.79$$

and $S = 2.79\, F = 2.79 \times 100 = 279$ m³/h at 300 K.

The extract flow rate can be computed by means of mass balance (11.3):

$$E = \frac{S(1 - x_{AS})}{(1 - x_{AE})} = \frac{279(1 - 0.01)}{(1 - 0.05)} = 290.75 \text{ m}^3/\text{h} \tag{11.11}$$

The hourly extracted ethylamine quantity is calculated under the simplifying assumption that the solvent density is not affected significantly by low solute contents:

$$Ex_{AE}\rho_S = 290.75 \text{ m}^3/\text{h} \cdot 0.05 \cdot 870 \text{ kg/m}^3 = 12{,}647.63 \text{ kg/h} \tag{11.12}$$

Hence, hourly productivity is of ≈ 12.65 ton/h.

At 325 K (25 K temperature increase), the partition coefficient will be computed by using eq. (11.8). The necessary ΔH value is obtained from K_A at 300 K:

$$\Delta H = -RT° \ln (K_A) \tag{11.12}$$

Thus

$$\Delta H = -8.314 \text{ J/mol K} \cdot 300K \cdot \ln (1.36) = -766.93 \text{ J/mol} \approx -767 \text{ J/mol}$$

Further

$$\ln (K_A) = -\frac{-767 \text{ J/mol}}{8.314 \text{ J/mol K} \cdot 325K} = 0.284 \text{ and } K_A = e^{-0.284} = 1.328$$

Values at 325 K lead to $\frac{S}{F} = 2.77$, $S = 277$ m³/h, and $E = 288.66$ m³/h, and a productivity of ≈ 12.56 ton/h removed solute.

With increased temperature, K_A decreases slightly (with 2.35%) and less solute will be removed by the solvent. Accordingly, to keep the setpoint x_{AE}, the flow rate S has to be slightly adjusted (with 0.7%). The system responds with the same 0.7% less productivity. If these fairly small deviations fall within the range of admitted specifications, then temperature control is not necessary. However, if the solute is expensive or raises environmental/health concerns, the setpoint x_{AE} might be tightly controlled and temperature regulation is necessary. However, the example illustrates the importance of S/F flow rate ratio control.

Example 11.2

Let us consider the extractor described in the previous example but for a 10-fold higher as well as lower value of phase transfer enthalpy: $\Delta H_1 = -7670.0$ J/mol and $\Delta H_2 = -76.7$ J/mol. The input streams are set on constant ratio. The data are $S/F = 3$, $x_{AF} = 0.15$, and $x_{AS} = 0.01$, respectively. Depending on the season, the temperature of the input streams can vary between 5 and 60 °C. Let us assess whether temperature control is a necessity with respect to setpoint x_{AE}.

Equation (11.8) is employed for the calculus of K_A at the extremities of the temperature disturbance interval. By replacing eq. (11.7) in eq. (11.5), the value of x_{AE} can be obtained for each set of data. The results are cumulated in Tab. 11.1.

Tab. 11.1: Effect of temperature on the solute mass fraction of the resulting extract.

Temperature		$\Delta H_1 = -7.670$ J/mol		$\Delta H_2 = -76.7$ J/mol	
°C	Kelvin	K_A	X_{AE}	K_A	X_{AE}
5	278	27.619	0.0566	1.034	0.0453
60	333	15.965	0.0562	1.028	0.0452

Higher K_A, as defined in eq. (11.7), favors extraction of solute and increases x_{AE}. Temperature changes from 5 to 60 °C bring about less than 1% variation in x_{AE} values. Thus, the system is virtually temperature-independent and its regulation is unnecessary.

References

[1]   Robbins, L.A., Cusak, R.W., *Section 15: Liquid-Liquid Operations and Equipment*, p. 15-1–15-47 in Perry, R.H., Green, W.G., Maloney, J.O., (Eds.), *Perry's Chemical Engineering Handbook*, 7th Edition, McGraw-Hill, New York, 1997.

[2]   Tudose, R.Z., Ibănescu, I., Vasiliu, M., Stancu, A., Cristian, Gh., Lungu, M., *Procese, operaţii, utilaje în industria chimică*, Editura Didactică şi Pedagogică, Bucureşti, 1977.

[3]   Schultz + Partner Verharenstechnik Gmbh, *Liquid-liquid extraction*, prospectus, http://www.schulzpartner.com/en/downloads/

[4]   Sulzer Chemtech, *Liquid-liquid extraction technology*, prospectus, http://website-box.net/site/www.sulzer.com

[5]   Atkins, P., de Paula, J., *Physical Chemistry*, 9th Edition, Oxford University Press, Oxford, 2010.

[6]   Motschmann, H., Hofmann, M., *Physikalische Chemie*, DeGruyter, Berlin, 2014.

[7]   Agachi, Ş, *Automatizarea proceselor chimice*, Casa Cărţii de Ştiinţă, Cluj-Napoca, 1994.

[8]   Weinstein, O., Semait, R., Lewin, D.R., *Modeling, simulation and control of liquid-liquid extraction columns*, Chemical Engineering Science, 53, 325–329, 1998.

[9]   Mjalli, F.S., Abdel-Jabbar, N.M., Fletcher, J.P., *Modeling, simulation and control of a Scheibel liquid-liquid contactor. Part 1. Dynamic analysis and system identification*, Chemical Engineering and Processing, 44, 543–555, 2005.

[10] Mjalli, F.S., Abdel-Jabbar, N.M., Fletcher, J.P., *Modeling, simulation and control of Scheibel a liquid-liquid contactor. Part 2. Model-based control synthesis and design*, Chemical Engineering and Processing, 44, 531–542, 2005.

[11] Mjalli, F.S., *Neuronal network model-based predicted control of liquid-liquid extraction contactors*, Chemical Engineering and Processing, 60, 239–253, 2005.

[12] Djurovic, J., *An inverse control of the extraction column*, International Journal of Mathematical Models and Methods in Applied Sciences, 5, 67–76, 2011.

12 Control of evaporation processes

Evaporation is a separation process of a solid substance dissolved in a liquid solvent. The procedure is to heat the initial solution till the solvent is evaporated to the desired degree. This process is a very old one, and it is being used to obtain salt or fresh water from sea water. Usually, it is employed to concentrate the solution where the solid product is more valuable than the solvent (see NaOH dissolved in H_2O from the electrolysis of the brine) [1].

The evaporation takes place in *evaporators* with *simple (single)* or *multiple evaporation stages (effects)* (Fig. 12.1) [2]. The evaporators usually contain a fascicule of tubes through which the solution to be concentrated is circulated through natural or forced convection. These are surrounded by a steam chest. In the multiple effect evaporators, the temperature, pressure, and heat transfer coefficients vary with each consequent effect, decreasing with each effect.

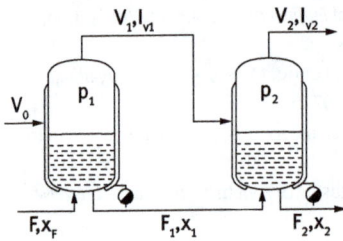

Fig. 12.1: Evaporator plant with two effects. F, $F_{i(1,2)}$ are the mass flows to or from the effects; $V_{i(0,1,2)}$ are the steam and vapor mass flows; $p_{i(1,2)}$ are the vapor pressures inside the evaporators; $x_{i(F,1,2)}$ are the solid mass fractions in the feed and evaporator's solution; $l_{vi(0,1,2)}$ is the latent heat of vaporization in each evaporator either steam or solvent (the vapor flow is not always water, the solution may contain another solvent).

To save energy, vapors that result from the evaporation of the solution in each effect are used as heating agent for the following ones.

To establish the control solution, one has to describe the process through a steady-state mathematical model that links the output variable x_2 that is the final solid product mass fraction, to the input variables, by using the following hypotheses:
- the evaporators are always at the boiling temperature of the solution by manipulation of pressures p_1 and p_2;
- the feed concentration is kept at the steady-state value of the concentration in the first effect;
- the vapors do not entrain solid particles; this is not entirely true if we are looking at the deposit of solid on the vapor outlet piping;
- the steam used for heating always condenses completely at the condensing point by transferring only the latent heat of vaporization;

https://doi.org/10.1515/9783110789737-014

- the heat losses to the environment are considered null; in reality, a good insulation has a heat loss of 5% from the heat content of the equipment;
- the solution level in the evaporators is constant.

With these simplifying assumptions, the mathematical model is formed of mass (eqs. (12.1) and (12.2)) and heat balances (eq. (12.3)) expressed for both evaporators:

$$F = V_1 + F_1 \text{ and } F_1 = V_2 + F_2 \tag{12.1}$$

$$Fx_F = F_1 x_1 \text{ and } F_1 x_1 = F_2 x_2 \tag{12.2}$$

$$V_0 l_{v0} = V_1 l_{v1} \text{ and } V_1 l_{v1} = V_2 l_{v2} \tag{12.3}$$

These mass/heat balances are valid only in the conditions imposed by the constraints (simplifying hypotheses) above. Equation (12.1) can be written as such if the level is constant (hypothesis 6); eq. (12.2) respects hypotheses 2 and 3; eq. (12.3) respects conditions 1, 3, 4, and 5; the entire heat transferred by the heating agent $(V_0 l_{v0})$ is used to vaporize the solution in the first effect, and the heat transferred by the heating agent to the second effect $(V_1 l_{v1})$ is used to vaporize the solution in it.

By replacing in the second part of eq. (12.2) the expression of $F_2 = F_1 - V_2$ (eq. (12.1)), one obtains

$$x_2 = x_1 \frac{F_1}{F_2} = x_1 \frac{F_1}{F_1 - V_2}, \tag{12.4}$$

and by further replacing $V_2 = V_1 \frac{l_{v1}}{l_{v2}}$ (eq. (12.3)), one obtains

$$x_2 = x_1 \frac{F_1}{F_1 - V_1 \frac{l_{v1}}{l_{v2}}}, \tag{12.5}$$

Consequently, by using the same logic

$$x_2 = x_F \frac{F_1}{F - V_0 l_{v0} \left(\frac{1}{l_{v1}} + \frac{1}{l_{v2}} \right)} \tag{12.6}$$

Generalizing for an *n-effect* evaporator leads to

$$x_n = x_F \frac{1}{1 - \frac{V_0}{F} l_{v0} \sum_{i=1}^{n} \frac{1}{l_{vi}}}. \tag{12.7}$$

By analyzing the expression of the final mass fraction, one observes that it depends on x_F, which is considered to be an uncontrolled variable (disturbance) as well as on the ratio $\frac{V_0}{F}$. However, since the mass balance in steady state is written in the form of eq. (12.1), the level control in both evaporators has to be ensured by two-level control loops. The other issue is that of keeping the boiling temperature constant and appropriate; this is carried out by using a pressure control loop. An additional control loop for the feed temperature control can be added to the entire automation scheme. Optionally, a ratio control can be applied to the $\frac{V_0}{F}$ flow rate ratio (Fig. 12.2(a) and (b)) if the RGA, or any sensitivity analysis, mentions the ratio as being an important control variable, as expressed in eqs. (12.7) and (12.15). If the disturbance x_F is really important and frequent, the ratio can be calculated by an inferential scheme with

$$\frac{V_0}{F} = \frac{x_{2\,set} - x_F}{x_{2\,set} l_{v0} \sum_{i=1}^{n} \frac{1}{l_{vi}}} \tag{12.8}$$

Relationship (12.8) is obtained from eq. (12.6) or by means of a feedforward controller. Especially the feedforward controllers are requested for small holdup evaporators that are difficult to control with independent control loops. At these evaporator plants, variations in steam and feed, as in feed concentration, cause high loss of valuable product.

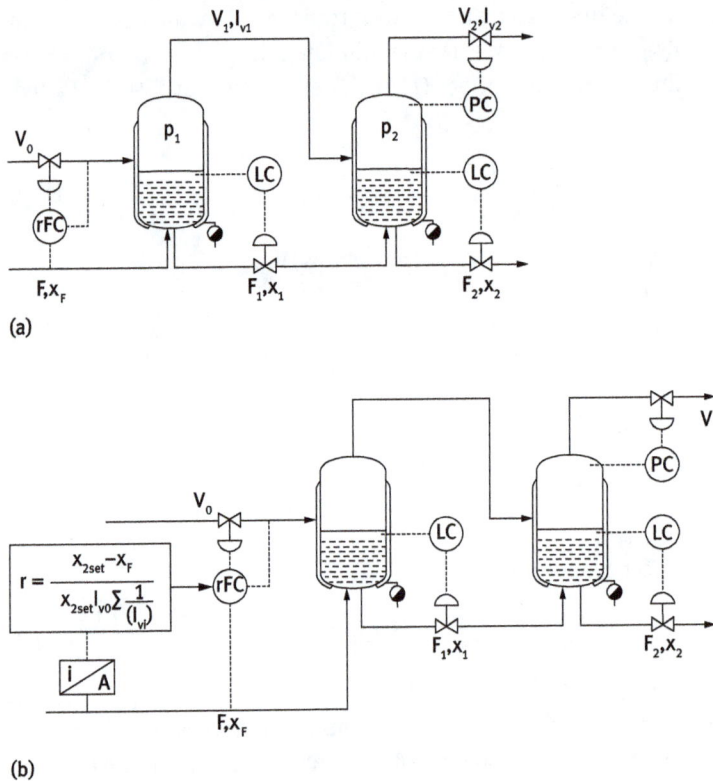

(a)

(b)

Fig. 12.2: Possible control schemes for the evaporation process: (a) classic evaporation control scheme; (b) advanced control scheme.

12.1 Sensitivity analysis relative to pressure, steam/feed flow rates, and feed concentration variations

In order to adopt a control solution for an evaporating system, a sensitivity analysis has to be done [3].

Example 12.1

Consider a multitubular evaporator for a concentrated NaOH solution (Fig. 12.3). It has the following characteristics (technological parameters are given at nominal operation point):

Number of pipes of the boiler: $n = 1860$
Pipelines length: $l = 8.5$ m
Inner diameter: $d_i = 0.033$ m
Outer diameter: $d_o = 0.038$ m
Solution level in the evaporator: $h = 7.1$ m
Inner tubes wall thermal conductivity: $\lambda_w = 50$ W/m · K
Input temperature: $T_F^o = 140$ °C
Input NaOH mass fraction: $x_F = 0.1$ kg/kg
Output NaOH mass fraction: $x_1 = 0.1568$ kg/kg
Operation pressure: $p = 5.5$ bar
Solution feed flow rate: $F = 37$ kg/s
Steam flow rate: $V_0 = 10$ kg/s
Steam pressure: $p_{st} = 12$ bar
Solution thermal conductivity: $\lambda_S = 0.48$ W/m · K
Steam thermal conductivity: $\lambda_{st} = 0.65$ W/m · K
Solution viscosity: $\eta_S = 3 \cdot 10^{-3}$ Pa · s
Steam viscosity: $\eta_{st} = 1.4 \cdot 10^{-5}$ Pa · s

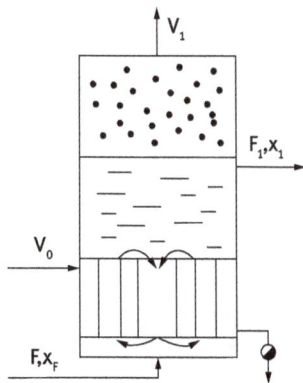

Fig. 12.3: Multitubular evaporator with single effect.

The change in output concentration with the variation of the pressure in the evaporator, with ±0.5 bar relative to the operational pressure, has to be calculated. Same evaluations are to be done for a steam pressure change from 12 to 8 bar and for a steam flow increase from 10 to 12 kg/s. Comments on the control system in these conditions are required.

The sodium hydroxide solution circulates inside the tubes while the steam flows outside them. The equations describing the operation of the evaporator are:

$$F c_{pF,sol} T_F^\circ - F_1 c_{p,sol} T_{b,sol}^\circ - V_1 i_{v,sol} + K_T A_T \left(T_{st}^\circ - T_{b,sol}^\circ \right) = 0$$

$$F - F_1 - V_1 = 0$$

$$F x_F - F_1 x_1 = 0$$

$$c_{p,NaOH} = \left(\frac{10^3}{40} \right) \left(7.34 + 125 e^{-T_{b,sol}^\circ} + \frac{13.88 \cdot 10^5}{T_{b,sol}^\circ{}^2} \right)$$

$$c_{p,sol} = c_{p,NaOH} x_{NaOH} + (1 - x_{NaOH}) \cdot 4180$$

$$i_{v,sol} = c_{p,sol} T_{b,sol}^\circ + l_v$$

$$T_{b,sol}^\circ = T_{b,H_2O}^\circ + \Delta + T_{b,H_2O}^\circ + 0.0162 \frac{T_b^{\circ 2}}{l_v} \Delta^0$$

where $c_{p,sol}$ is the heat capacity of the NaOH solution, $T_{b,sol}^\circ$ is the boiling temperature of the solution at a certain pressure p, $i_{v,sol}$ is the enthalpy of the solution, K_T and A_T are the heat transfer coefficient and heat transfer area, respectively, T_{st}° is the temperature of the steam at 12 bar, l_v is the water latent heat of vaporization, $x_{i(F,1,NaOH)}$ is the mass fraction of the components in different points, Δ is the ebullioscopic temperature increase at pressure p, Δ^0 is the ebullioscopic temperature increase at atmospheric pressure, and T_{b,H_2O}° is the boiling temperature of water.

Parameters calculated by means of a computer program are given in Tab. 12.1.

Tab. 12.1: Operational parameters of the evaporator in Example 12.1 when operating under steady-state conditions.

No.	x_F (%)	p(bar)	$F \left(\frac{kg}{s} \right)$	$F_{st} \left(\frac{kg}{s} \right)$	p_{st} (bar)	T_F°(°C)	x_1(%)	$V_1 \left(\frac{kg}{s} \right)$	$F_1 \left(\frac{kg}{s} \right)$
1	10	5.5	37	10	12	140	15.68	13.40	23.60
2	8	5.5	37	10	12	140	12.71	13.70	23.30
3	10	6.0	37	10	12	140	15.32	12.85	24.15
4	10	5.0	37	10	12	140	16.06	13.95	23.05
5	10	5.5	40	10	12	140	14.91	13.17	26.83
6	10	5.5	37	12	12	140	18.04	16.49	20.51
7	10	5.5	37	10	8	140	10.84	2.87	34.13
8	10	5.5	37	10	12	140	15.69	13.42	23.68

The nominal operation regime parameter values are presented in the first row (row 1) of the table.

It is clear that a change of 20% (row 2) of the input concentration is affecting with approximately 19% the output concentration. At the same time, if a change of the feed flow rate occurs (+8% – row 5), the concentration decreases with 5%. This demonstrates the importance of a ratio control system with inferential recalculation of the ratio F_{st}/F when such variations occur.

At the same time, when a change in pressure occurs (±9% – rows 3 and 4), the decrease/increase in concentration at the output is of ±2.4%. To evaporate at this pressure the same quantity of solvent in order to obtain the desired nominal concentration, the additional consumption of steam (row 3) is of +0.55 kg/s. This translates into a supplementary heat consumption of 1 gcal/h (at l_v = 525 kcal/kg). At an amount of 7200 h of operation/year and at a cost of US $70/gcal, this means an additional expenditure of ~US $500,000/year. The importance of a tight pressure control is hence demonstrated.

The steam pressure variation influences substantially the quality of the product (row 8): 25% of steam pressure decrease produces a reduction of mass concentration from 15.68% to 10.84% (ca. 30%), which means the steam station has to have a tight output pressure control.

One has to look at the important aspect of technical specifications. Let us suppose that the plant has to deliver NaOH solution at 15.68%. If an accidental increase of 20% of the steam flow rate occurs (row 7), the consequence is the unnecessary concentration of the solution to 18.04% with a supplementary heat consumption of 1837 kcal/s. If the producer is obliged to dilute the solution at the required mass fraction, the supplementary energy is lost together with its costs. The sensitivity to disturbances is an important issue. The efficiency of the control scheme resides not only in the decrease of the energy consumption but also in the stability of the quality of the final product in the conditions of variable feed rate and heat and/or composition input.

By considering the general case of n effects, and by generalizing eq. (12.2), the following is obtained

$$Fx_F = F_n x_n,$$ (12.9)

The overall material balance includes all vapor flows removed from each effect, $\sum_{i=1}^{n} V_i$; thus

$$F = F_n + \sum_{i=1}^{n} V_i.$$ (12.10)

The total vapor flow to be extracted in order to reach the desired concentration is obtained when combining eqs. (12.9) and (12.10):

$$\sum_{i=1}^{n} V_i = F\left(1 - \frac{x_F}{x_n}\right),$$ (12.11)

so that

$$x_n = x_F \left(1 - \frac{\sum\limits_{i=1}^{n} V_i}{F} \right) \tag{12.12}$$

The differential of above expression as a function of $\frac{\sum_{i=1}^{n} V_i}{F}$ indicates the sensitivity of the final concentration relative to the steam/feed ratio $\left(\frac{V_0}{F} \right)$

$$\frac{dx_n}{d \left(\frac{\sum\limits_{i=1}^{n} V_i}{F} \right)} = \frac{x_F}{\left(1 - \frac{\sum\limits_{i=1}^{n} V_i}{F} \right)} = \frac{x_n^2}{x_F} \tag{12.13}$$

If one wants to see which is the sensitivity of the concentration at a given percentage of steam/feed ratio, one may calculate as follows:

$$\frac{dx_n}{d \left(\frac{\sum\limits_{i=1}^{n} V_i}{F} \right) \left(\frac{\sum\limits_{i=1}^{n} V_i}{F} \right)} = \frac{x_n(x_n - x_F)}{x_F} \tag{12.14}$$

At the same time, the sensitivity of the process can be determined not only by the steam/feed ratio but also by the input/output (final) concentration ratio $\frac{x_n}{x_F}$ as well. This is the main reason for the variation of product quality. The differential as a function of concentration is

$$\frac{dx_n}{dx_F} = \frac{1}{1 - \frac{\sum\limits_{i=1}^{n} V_i}{F}} = \frac{x_n}{x_F} \tag{12.15}$$

It shows that the feed concentration change has a greater influence than that of the steam supply change. In such cases, the inferential control scheme illustrated by Fig. 12.2(b) can be used.

The extended approach of evaporation process control is described in [4].

Evaporators are used not only in concentrating "orange juice" or "caustic soda", but also quite recently in air conditioning (AC) as well in automotive industries and in households. The reason is simple: in Europe, 75–80% of the new cars are fitted with AC systems, as compared to 12% in 1990, while in the USA "surveys by the Department of Energy in USA suggest that the yearly energy expenditure for residential air-conditioning in 1997 was 0.42 Quadrillion BTU (~0.1 Quadrillion kcal)" and still continues to rise [5]. Chapter 5 of this volume presents new methods of multivariable adaptive control. More recent work of the center for AC at the University of Illinois is described in both [5] and [6]. In [7], a fuzzy control algorithm is used in the same field of interest.

In Chapter 1 of this work, we presented the combinations of inferential – ratio, inferential – cascade, inferential – feedforward control which gave significantly better results applied to different processes including the evaporation process [8]. These new combinations avoid the delays induced in the chain of evaporators, using feedforward and Kalman filter estimates to prevent the lags in the propagation of the control action.

References

[1]   Agachi, P. Ş., Constantinescu, D. M., Macedon, D., Oniciu, L., Topan, V., Neacsu, I., *Contributions at the diminishing energetic consumption of the brine electrolysis process*, in *Paper in the volume of the 11th Meeting the Scientific Research Center Rm.Vilcea 10–12 October 1995*, p. 1–12.

[2]   McCabe, L. W., Smith, J. C., Harriot, P., *Unit operations of Chemical Engineering*, p. 465, McGraw-Hill Book, 1993.

[3]   Shinskey F. G., *Energy conservation through control*, p. 195, Academic Press, (1978).

[4]   Shinskey F. G., *Energy conservation through control*, p. 183, Academic Press, (1978).

[5]   Shah, R., Alleyne, A. G., Bullard, C. W., Rasmussen, B. P., Hrnjak, P. S., *Dynamic Modeling And Control of Single and Multi-Evaporator Subcritical Vapor Compression Systems*, p. 1. Air Conditioning & Refrigeration Center, Mechanical & Industrial Engineering Dept. University of Illinois, 2003.

[6]   Rasmussen, B. P., Alleyne, A. G., *Dynamic Modeling and Advanced Control of Air Conditioning and Refrigeration Systems*, p. 207, ACRC TR-244, 2006.

[7]   Wua, C., Xingxib, Z., Shiminga, D., *Development of control method and dynamic model for multi-evaporator air conditioners (MEAC)*, Energy Conversion and Management, 46(3), 451–465, 2005.

[8]   Karimi, M., Jahanmiri, A., Azarmi, M., Inferential cascade control of multi-effect falling-film evaporator, Food Control, 18, 1036–1042, 2007.

13 Control of drying processes

Drying is the separation of *moisture* from a solid substance through exposure to heated air or nonsaturated gas in (most cases). Another option is to use the heat produced in electric dryers (e.g. washing machines and drying chambers) by respecting Joule's first law. There are other nonconventional drying procedures [1] such as freezing, microwave exposure, and dielectric drying. These are not approached here. There are multiple options for drying, depending on the specific characteristics of the material to be dried [1]. Hereby, we discuss processes for which the drying agent is air or gas, but also mention some recent developments.

In terms of energy consumption, drying is a highly energy-intensive operation and represents 10% to 25% of the national industrial energy production in developed countries [1]. Therefore, it is natural to find control solutions to save money. The reports mention savings between 1.5% and 30% of total costs, depending on how close to the optimum the process is operated [7].

Notions of *psychrometry* are given in [2]. The drying is either continuous (e.g. tile-drying tunnels) or in batches (e.g. electrical insulator or fruit drying chambers). The heat necessary for drying is transferred to the drying agent before it meets the humid solid to be dried. In this situation, the drying is adiabatic.

The relative humidity depends on the pressure at which the drying process takes place [2]. Consequently, the measurements of relative humidity through measuring the dew point can be applied only at constant pressure. Once the pressure decreases (e.g. an atmospheric front), the relative humidity increases for a given dew point. The humidity content of solid materials is in equilibrium with the relative humidity of the air. Inside drying chambers, the environment is far from being in equilibrium with the solids to be dried, hence creating the drive for drying. From the thermal transfer point of view, the solvent/water evaporation rate is proportional with the temperature difference between the solid and its drying agent. The temperature of the drying agent is equal to the dry-bulb temperature (T_{db}°) while that of the solid is closer to the wet-bulb temperature (T_{wb}°) because the wet bulb is a wetted solid on which the liquid evaporates adiabatically. Thus, (T_{wb}°) is significant from the point of view of the drying rate.

It is easy to measure (T_{wb}°) of the ambient, but it is a real problem to measure it at continuously high temperatures as in the drying chambers/tunnels because they favor the fast evaporation of water in the wet-bulb. Therefore, (T_{wb}°) can be estimated either when knowing (T_{db}°) and the absolute humidity by means of an iterative process [3] or by empirically found expressions based on measurement data [4]:

$$T_{wb,c}^{\circ} = T^{\circ}\tan^{-1}\left[0.151977\,(\varphi\% + 8.313659)^{\frac{1}{2}}\right] + \tan^{-1}(T^{\circ} + \varphi\%) - \tan^{-1}(\varphi\% - 1.676331) +$$

$$0.00391838(\varphi\%)^{\frac{3}{2}}\tan^{-1}(0.023101\varphi\%) - 4.686035$$

$$\tag{13.1}$$

https://doi.org/10.1515/9783110789737-015

$T° - T°_{db}$ is the ambient temperature. The relative humidity can be found from psycho-metric charts as in Fig. 7.61 from [2]. For this calculus, a first guess of $T°_{wb,0}$ is required to estimate the difference $T° - T°_{wb,0}$ and thus the first approximation of $\varphi_0\%$. If $T°_{wb,c} - T°_{wb,0}$, the calculus ends; if not, the value of $T°_{wb,0}$ is replaced with the calculated value and the wet-bulb temperature is recalculated once again until the difference between $T°_{wb,c}$ at step j and $T°_{wb,c}$ at step $j - 1$ is less than 0.1 °C [3].

The drying rate R depends on the air speed along the solid, the dimension of its particles, its water content, the dryer characteristics, and the temperature difference $T° - T°_{wb}$. When a solid particle absorbs solvent, its surface is wet. In a dry ambient, the liquid starts to evaporate. If $T° - T°_{wb}$ is kept constant, the evaporation rate of the liquid is constant until dry spots on the particle appear. This is the constant rate-drying period that lasts until the critical moisture content w_c is reached. It represents the humidity of a system's atmosphere above which a crystal of a water-soluble salt will always become damp (absorb moisture from the atmosphere) and below which it will always stay dry (release moisture to the atmosphere; Fig. 13.1).

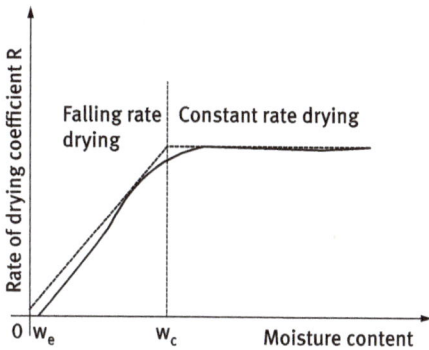

Fig. 13.1: Evolution of the drying rate R as a function of the moisture of the solid.

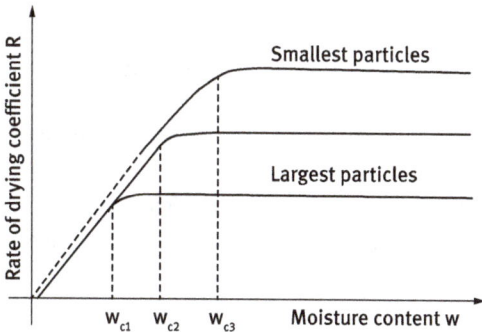

Fig. 13.2: The drying rate as function of the particle size of dried material.

Since dry spots appear on the surface, the drying rate decreases because the liquid, in order to be evaporated, has to diffuse from the interior to the solids surface. The rate decreases to 0 when the absolute humidity of the solid reaches the equilibrium state with the environment, w_c. The critical moisture content depends on the dimension of the particles [5, 6]. Thus, for particles with a diameter of $d_1 = 90–160\mu$, $w_c = 5\%$; for particles with $d_2 = 60–90\mu$, $w_c = 10\%$; and for particles with $d_3 < 60\mu$, $w_c = 21\%$, respectively (Fig. 13.2).

13.1 Batch drying control

13.1.1 Conventional batch drying control

Many products are dried batch-wise: fruits and vegetables, construction materials (sand or bricks), electrical insulators, pharmaceutical products, etc. An example of a batch dryer is given in Fig. 13.3.

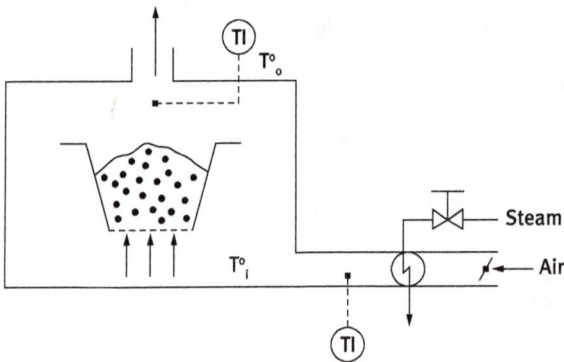

Fig. 13.3: Scheme of a batch dryer.

During a batch drying, both solid and outlet air temperatures increase in time (Fig. 13.4).

If the solid's temperature is smaller than T_{wb}°, part of the received heat is used to raise the temperature to the value of T_{wb}°. This phenomenon causes a slight increase of the outlet temperature T_o°, and after that, during the constant rate drying period, the outlet temperature stays at the value imposed by the surface evaporation, T_{oc}°. During the last period of variable drying rate, the temperature of the heating agent increases asymptotically to the dryer's inlet temperature T_i°.

The heat balance during the constant rate drying period is

$$F_g C_{pg}\left(T_i^\circ - T_o^\circ\right) = V l_v \tag{13.2}$$

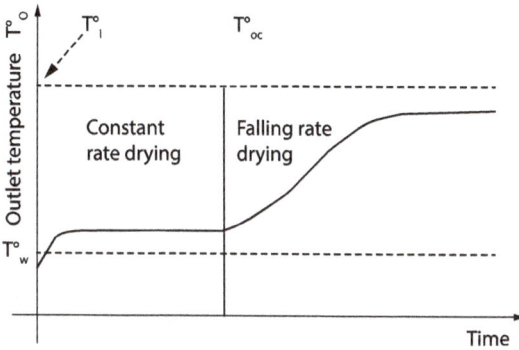

Fig. 13.4: Drying agent temperature profile during batch drying.

where F_g is the drying agent mass flow (air), c_{pg} is the drying agent specific heat, V is the vapor mass flow, and l_v is the latent heat of vaporization.

The vaporization rate dV from a particle may be written as

$$dV = k_m R \cdot dA \left(T^\circ - T_{wb}^\circ \right) \tag{13.3}$$

and is proportional with the drying rate R, heat transfer area dA, mass transfer coefficient k_m, and with the temperature difference between T° and T_{wb}°.

When the air passes over the particle to be dried, the temperature decreases due to the evaporation:

$$F_g c_{pg} dT^\circ = - dV l_v \tag{13.4}$$

From both eqs. (13.3) and (13.4) it results that

$$dA = \frac{F_g c_{pg}}{k_m R l_v} \cdot \frac{dT^\circ}{T^\circ - T_{wb}^\circ}, \tag{13.5}$$

After integration the relationship above becomes

$$A = \frac{F_g c_{pg}}{k_m R l_v} \ln \left(\frac{T_i^\circ - T_{wb}^\circ}{T_0^\circ - T_{wb}^\circ} \right) \tag{13.6}$$

In the period of variable drying rate (Fig. 13.4), $R = kx \le kx_c$ and from (13.6) the following can be obtained:

$$x = \frac{F_g c_{pg}}{k_m k A l_v} \ln \left(\frac{T_i^\circ - T_{wb}^\circ}{T_0^\circ - T_{wb}^\circ} \right) \tag{13.7}$$

Equation (13.7) contains impossible to evaluate constants, k_m and k. Therefore, in the constant drying rate zone, $x = x_c$ and

$$\frac{k_m kA l_v}{F_g C_{pg}} x_c = Kx_c = \ln\left(\frac{T_i^\circ - T_{wb}^\circ}{T_{oc}^\circ - T_{wb}^\circ}\right) \tag{13.8}$$

Thus, eq. (13.7) can be solved for any output temperature at which a certain humidity x^* can be attained:

$$x^* = \frac{1}{K} = \ln\left(\frac{T_i^\circ - T_{wb}^\circ}{T_o^\circ - T_{wb}^\circ}\right) \text{ or} \tag{13.9}$$

$$T_{oc}^{\circ*} = T_{wb}^\circ + e^{-Kx^*}\left(T_i^\circ - T_{wb}^\circ\right) \tag{13.10}$$

Eliminating T_{wb}° from eq. (13.8) it results that

$$T_o^{\circ*} = T_{oc}^\circ\left(\frac{1 - e^{-Kx^*}}{1 - e^{-Kx_c}}\right) + T_i^\circ\left(1 - \frac{1 - e^{-Kx^*}}{1 - e^{-Kx_c}}\right) \tag{13.11}$$

or

$$T_o^{\circ*} = K^* T_{oc}^\circ + (1 - K^*)T_i^\circ \tag{13.12}$$

This way, the temperature control system should implement, eq. (13.11), or its simplified version, eq. (13.12), to stop the drying process at the desired value $T_o^{\circ*}$. One condition to be satisfied is that the inlet temperature, T_i°, has to be kept constant (meaning an inlet temperature control).

Example 13.1

Let us consider the batch drying of a material with $x_c = 0.1$ to the desired moisture content of $x^* = 0.02$. The inlet temperature is $T_i^\circ = T_{i,db}^\circ = 93\,°C$ and $T_{wb}^\circ = 34\,°C$. The outlet temperature during the constant rate drying period is $T_{oc}^\circ = 40\,°C$. What is the value of the temperature, T_{o1}°, at which drying should be stopped? What happens if the material-specific area is twofold smaller? What happens if, in this second case, the temperature at which the drying is stopped is T_{o1}°? [7]

From eq. (13.7)

$$K = \frac{1}{0.1}\ln\left(\frac{93 - 34}{40 - 34}\right) = 22.81$$

From eq. (13.11)

$$K^* = \frac{1 - e^{-22.81 \cdot 0.02}}{1 - e^{-22.81 \cdot 0.1}} = \frac{0.367}{0.898} = 0.408$$

From eq. (13.12)

$T_{o1}^{\circ*} = 0.408 \cdot 40 + (1 - 0.408) \cdot 93 = 72\,°C$. This is the temperature at which we stop drying.

If the specific area of the solid material diminishes twofold, K (eq. (13.7)) decreases to its half. Thus (eq. (13.10)) becomes

$T_{oc,2}^{\circ*} = 34 + e^{-\frac{22.81}{2}} \cdot 0.1(93 - 34) = 52.8\,°C$. K^* stays at the same value and then

$T_{oc,2}^{\circ*} = 0.408 \cdot 52.8 + (1 - 0.408) \cdot 93 = 76.6\,°C$

This is the temperature at which the drying has to be stopped in the second case.
The attained moisture content [eq. (13.8)] is

$$x_2^* = \frac{2}{22.81} \cdot \ln\left(\frac{93-34}{76.6-34}\right) = 0.028$$

If the "stopping" temperature is kept at 72 °C in the second case, then

$$x^* = \frac{2}{22.81} \cdot \ln\left(\frac{93-34}{72-34}\right) = 0.038,$$

and a higher than desired moisture content is retained by the solid.

Shinskey suggests that the control system should be inferential and should calculate $T_0^{\circ*}$ in order to interrupt the drying process when $T_0^\circ \geq T_0^{\circ*}$ (Fig. 13.5). It has to be mentioned that in addition to the "stopping" scheme, the input temperature has to be kept constant by means of a temperature control system [to respect eq. (13.12)].

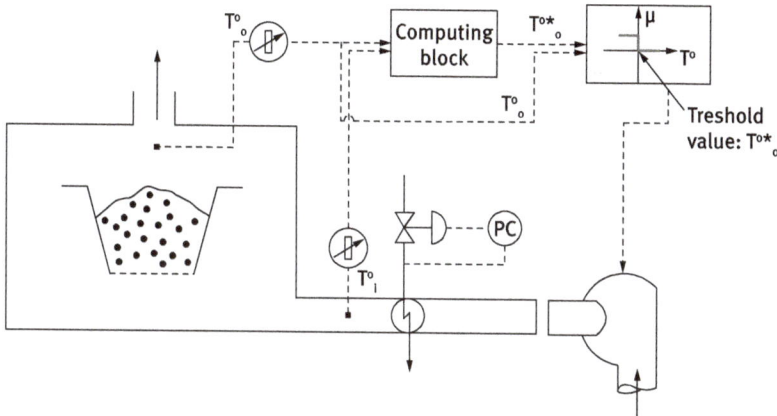

Fig. 13.5: Control scheme for stopping the drying process at a certain desired value of the moisture in the solid to be dried.

Several suggestions of control are given in [8]. Dufour identified control solutions that are currently in industrial use. Our research in advanced control drying [9, 10] proposed solutions that targeted cost reduction while preserving the standard quality of the dried material.

13.1.2 Advanced batch drying control

---

i **Example 13.2**

The high-voltage electric insulator production implies a two-stage batch drying process. During the first step, the moisture content of the target product is reduced from 18 to 20% to 0.4% in special gas-heated chambers. The second step is carried out in high-temperature ovens to achieve an even lower moisture content. In [9], two advanced algorithms (fuzzy and)MPC) are proposed to control the drying process of the insulators. The original notations are kept within this example [9].

In Fig. 13.6, both the structure and the operation of the drying chamber are described.

Nomenclature:

x Moisture content of the air coming from the insulator

x_f Moisture content of the burned gases

x_o Moisture content of the outflow gases

x_{ext} Moisture content of the air absorbed in the burner

w Moisture content of the solid

w_c Critical moisture content of the solid

$\dot{m}_a$ Mass outflow rate of burned gas

$\dot{m}_{ai}$ Mass flow rate of fresh air

$\dot{m}_{st}$ Mass flow of evaporated steam

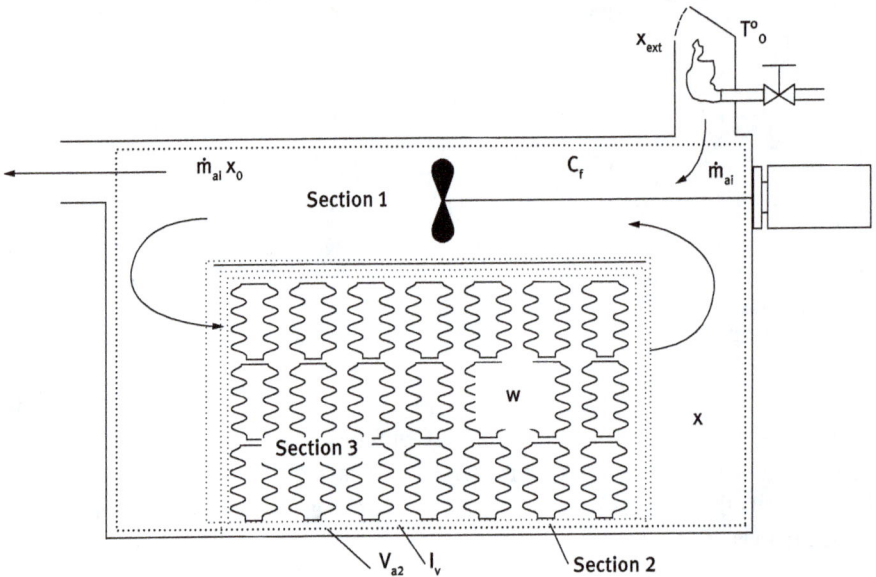

Fig. 13.6: Scheme of the drying chamber. Section 1 represents the air volume within the drying chamber; Section 2 represents the direct surroundings of the drying product; Section 3 represents the drying product itself.

m_s Mass of the dried solid
V_{ach} Free volume of the drying chamber
V_{a2} Air volume in Section 2
A_s Evaporation area of the solid
A_{ch} Heat transfer area of the drying chamber
T_i° Input temperature of the burned gases
T_0° Output temperature of the mixture of gases and steam from the drying chamber
T_{ext}° Outer temperature
K_A Heat transfer coefficient through the walls of the drying chamber
ρ Density of the gas mixture in the immediate vicinity of the solid subjected to drying
ρ_a Density of the mixture of burned gas and vapor in the drying chamber
l_v Latent heat of vaporization
V_F Flow rate of the methane gas burned in the burner
H_F Heat of combustion

Mass and energy balance equations are used to describe the dynamic behavior of the system. The main studied outputs of the model are the moisture content of the drying product w, the outlet air temperature T_0°, and the air humidity x_o. The input variables are the natural gas flow rate V_F and the mass flow rate of fresh air $\dot{m}_{ai}$. The mass balance of steam within Section 1 is described by

$$\dot{m}_{ai}x_f + \dot{m}_a x - (\dot{m}_a + \dot{m}_{ai})x_o = V_{ach}\rho_a \frac{dx_o}{dt}$$

with V_{ach} being the volume of air in Section 1. In Section 2, the steam flows around the drying product are modeled by

$$\dot{m}_a(x_o - x) - m_s\frac{dw}{dt} = \frac{d}{dt}(V_{a2}\rho x)$$

with V_{a2} being the infinitesimal small volume of air in Section 2 and $m_s\frac{dw}{dt}$ the steam flow coming from the solid. Because of this, the last term of the equation above can be neglected, which results in the differential equation:

$$\frac{dw}{dt} = (x_o - x)\frac{\dot{m}_a}{m_s}$$

In Section 3, the behavior of the drying good itself is described with a normalized diagram by means of the following equation [10, 11]:

$$\frac{dw}{dt} = -\frac{\dot{m}_a}{m_s}A_s$$

The drying velocity during the three periods of the entire drying process of a hygroscopic material is characterized by the diagrams given in Fig. 13.7 [11].

Diagram (a), only available by experiments and valid for certain conditions, can be normalized to (b) according to

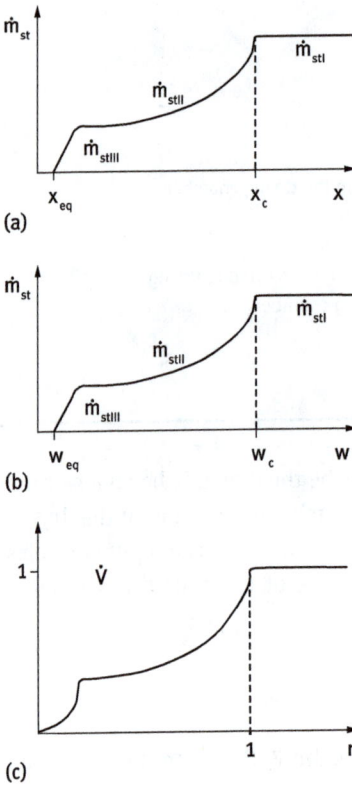

(a)

(b)

(c)

Fig. 13.7: Drying rate depending on the moisture content of the drying product: (a) absolute and (b) normalized.

$$\dot{v}(\eta) = \frac{\dot{m}_{st}}{\dot{m}_{stI}} \text{ and } \eta = \frac{w - w_{equ}}{w_c - w_{equ}}$$

It is assumed that w_c is constant and does not depend on the drying conditions and that w_{equ} only depends on the relative air humidity, without being affected by other factors. It is also assumed that all diagrams of the drying velocity for different drying conditions are geometrically similar.

The equilibrium humidity w_{equ} dependence on the relative air humidity ϕ was described for clay by means of a correlation equation. The saturation humidity of the air, x_{sat}, is dependent on the temperature T_o°. For low partial pressures of steam, a simple equation for $\dot{m}_{stI}$ (Fig. 13.7) was considered:

$$\dot{m}_{stI} = k(x_{sat} - x)$$

where the mass transfer coefficient k is determined experimentally.

Two energy balance equations, one for the drying chamber:

$$\dot{m}_{ai}\left[c_{pa}\left(T_i^\circ + T_o^\circ\right) + x_f\left(l_v + c_{pst}T_i^\circ\right) - x_0\left(l_v + c_{pst}T_o^\circ\right)\right] + m_s\frac{dw}{dt}\left(l_v + c_{pst}T_o^\circ\right)$$
$$- K_A A_{ch}\left(T_o^\circ + T_{ext}^\circ\right) = V_{ach}\rho_a\left[\left(c_{pa} + x_0 c_{pst}\right)\frac{dT_o^\circ}{dt} + c_{pst}\frac{dx_0}{dt}T_o^\circ\right]$$

and the other for the burner:

$$\dot{m}_{ai}\left[c_{pa}T_{ext}^\circ + x_{ext}\left(l_v + c_{pst}T_{ext}^\circ\right)\right] + \left(c_{pF}T_{ext}^\circ + H_F\right)\frac{M_{FpF}}{R\left(T_{ext}^\circ + 273\right)}\dot{V}_F$$
$$= \dot{m}_{ai}\left[c_{pa}T_i^\circ + x_f\left(l_v + c_{pst}T_i^\circ\right)\right]$$

are used to describe the outlet temperature change.

The fuzzy controller proposed for the batch drying chamber receives a measured value from the system, it fuzzifies it (assigns it a membership value), applies the system's rules, computes an overall result of all the rules, and then defuzzifies the result, converting it into a number which is an appropriate command for the system it controls (see also Chapter 3). The mission of the model predictive controller (see also Chapter 2) is accomplished by anticipating (predicting) the outputs of the system with the aid of the mathematical model. The controller output is generated based on the anticipated behavior of the system. All control methods investigated in this paper obey the current control practice, i.e., driving the evolution of the moisture content of the drying product in the desired way by means of controlling the air temperature inside the chamber. Usually, the desired decreasing profile of the drying product moisture content is obtained by imposing an increasing ramp-constant profile on the air temperature. The setup of the simulated system is shown in Fig. 13.8.

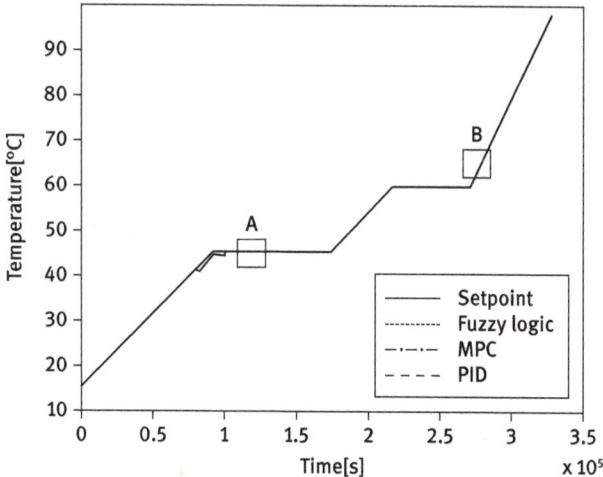

Fig. 13.8: Comparative behavior of FL, MPC, and PID control in the presence of the heating power disturbance. A and B are below the magnified windows corresponding to constant temperature and to ramp temperature change, respectively.

The scaled dynamic sensitivity analysis of the output variables with respect to the studied inputs pointed out the natural gas flow rate as the most important manipulated variable (about 10 times more important than the mass flow rate of fresh air). The control system was designed accordingly. The control scheme is presented in Fig. 13.9.

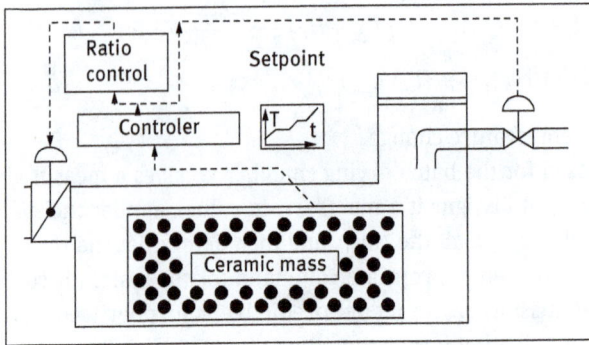

Fig. 13.9: Structure of the control system of the drying chamber described in Example 13.2.

The authors studied the comparative behavior of three control algorithms: PID, fuzzy logic (FL), and MPC; the results are presented in Figs. 13.10–13.12. The performance tests were carried out for three significant disturbances typically occurring in industrial practice: a 10 °C inlet air temperature T_{ext}° drop (from 16 to 6 °C), a 10% heating power capacity H_F drop of the natural gas, and a 10% rise in the moisture content of the inlet air. The disturbances were applied as steps at time $t = 116,000$ s.

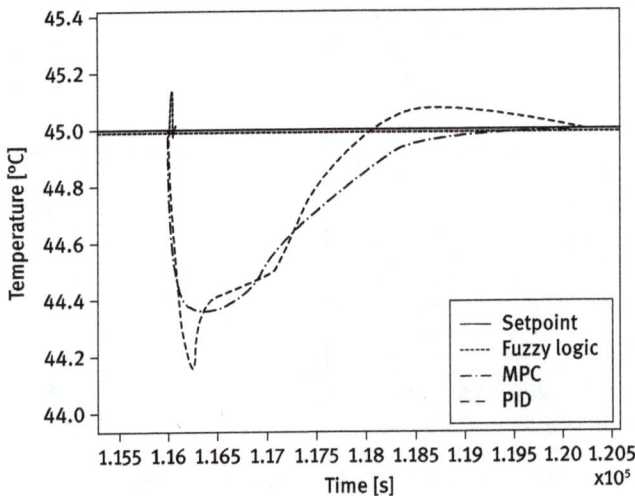

Fig. 13.10: Comparative behavior of FL, MPC, and PID control in the presence of a heating power disturbance (detailed).

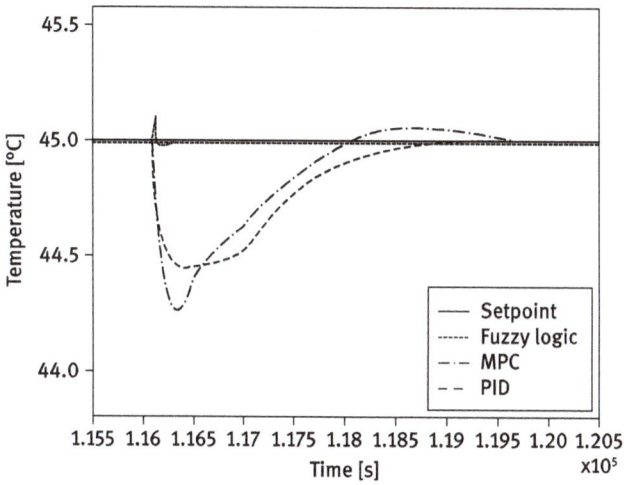

Fig. 13.11: Comparative behavior of FL, MPC, and PID control in the presence of an air inlet temperature drop disturbance (detailed).

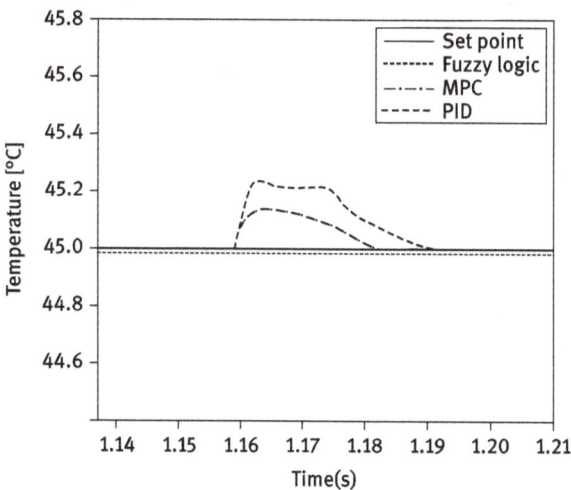

Fig. 13.12: Comparative behavior of FL, MPC, and PID control in the presence of an air inlet humidity increase disturbance (detailed).

With respect to setpoint tracking performance, the results reveal a good behavior in the case of PID and MPC, yet FL control proves superior abilities. The figure below illustrates that FLC is very accurate; it follows with precision both the constant and the ramp sections of the temperature setpoint scheduling function (Fig. 13.13).

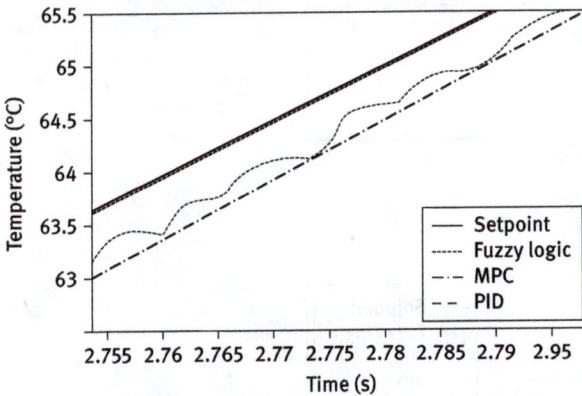

Fig. 13.13: Detailed presentation of the ramp setpoint following performance of FLC, MPC, and PID control.

All control methods exhibit a low-offset behavior for the constant parts of the set-point function. For the ramp sections, as in Fig. 13.13 (detail B on Fig. 13.8), MPC and PID control proved to be less accurate than FL by showing a larger offset. The conclusion of the authors, when referring to the FL controller performances, is that the accuracy of FL control is largely due to the asymmetrical membership function definition. It takes into account the need for an asymmetric amplitude of the manipulated variable change (i.e. a controller response of higher amplitude to a negative error compared to a lower amplitude response for a positive error) in the ramp section of the setpoint function. With respect to disturbance rejection performance, FL control showed a considerably shorter (more than 10 times) response time and smaller (more than five times) overshoot than other control strategies. It is worth mentioning that these new techniques are extremely valuable when food is processed so that nutritive proprieties of goods are preserved [12, 13].

13.2 Continuous adiabatic drying

Continuous drying can be carried out in fluidized beds or in longitudinal dryers. Figure 13.14 presents the control scheme of a fluidized bed dryer [14].

The air flow is maintained in a way that ensures particle's fluidization. At the same time, the air flow rate should be kept at a value that does not entrain solid particles to the exit. The differential pressure is a measure of fluidization and it is used to control the air flow rate. The humid product enters the dryer, and it is homogenized inside it and leaves the equipment at the opposite side in a dry form. If the product to be dried has at exit a higher moisture content than the critical value, then the dryer is not self-regulated; consequently, it functions in the constant drying rate regime, and according to eq. (13.8), the output temperature depends on the critical moisture content. In this situation, the control system presented in Fig. 13.14 cannot control the product humidity.

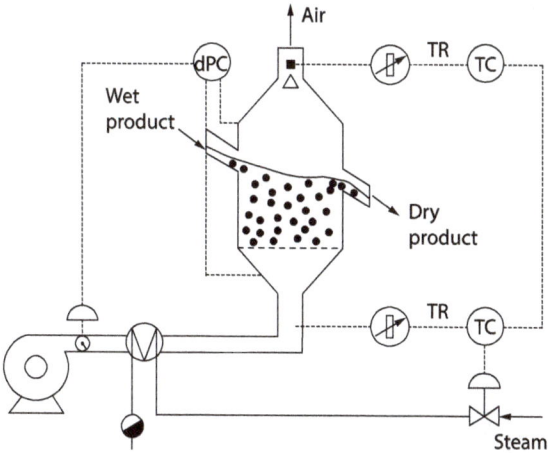

Fig. 13.14: Control scheme of a continuous fluidized bed dryer.

Normally, the product leaves the dryer with $w < w_c$ in such a way that the output humidity corresponds to the value expressed in eq. (13.7). It is necessary to ensure the residence time imposed by diagram 13.15.

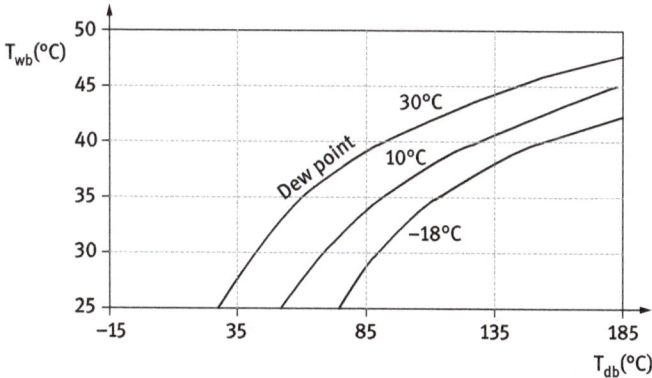

Fig. 13.15: Wet bulb temperature (T_{wb}) as a function of the dry bulb temperature (T_{db}) and the dew point.

The increase of the material inflow rate, or of its moisture content, will induce an increase of the air humidity x. The temperature controller manipulates the butterfly valve to increase T_i° with the consequence of increasing T_O° to the desired value. The deficiencies of this control system are explained through eq. (13.7): at a higher evapo-ration load (rate), both T_i° and T_{wb}° are increased. If T_O° is kept constant, x must in-crease to satisfy eq. (13.7). To keep x at x^*, T_O° has to be controlled at the value of $T_O^{\circ*}$ expressed by eq. (13.12), but for which value of T_{wb}° has to be either measured or esti-mated [see eq. (13.10) and approximation formula (13.1)]. For water, the diagrams in Fig. 13.15 can be used; the curves cover the temperatures of the dew point in the tem-perate zone.

The moisture of the fluidized bed can be controlled either through the heat inflow or through the solid material flow rates; usually, the solid material is more difficult to be manipulated and thus the input temperature is controlling the moisture content.

The curves in Fig. 13.16 show that the output temperature can be approximated with a straight line:

$$T_o^{\circ*} = aT_i^\circ + b \tag{13.13}$$

with the value of b adjusted to the values of the dew point [15]:

$$b = b_0 + f\left(T_{dew}^\circ\right)$$

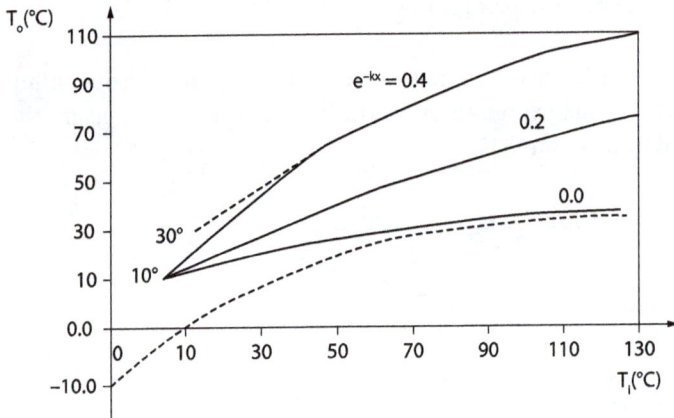

Fig. 13.16: The solutions of eq. (13.11) for $T_{wb}^\circ = 10°C$ and different values of e^{-Kx*}.

To implement the control scheme, a calibration to determine the values of e^{-Kx} in the operating conditions of the dryer is needed:

$$e^{-Kx} = \frac{T_o^\circ - T_{wb}^\circ}{T_i^\circ - T_{wb}^\circ} \tag{13.14}$$

The automatic control system capable to keep constant the desired moisture content is illustrated in Fig. 13.17.

As one may see, the control of the moisture content is based on indirect measure-ments and temperatures, and moreover, some of them have to be estimated. Recently,

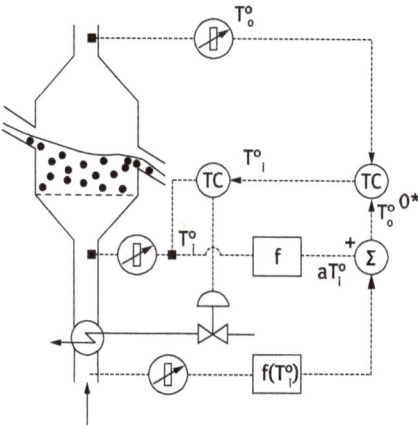

Fig. 13.17: Automatic control system of the moisture content for a solid material.

researchers from UMIP [16] have developed a tool to allow the online measurement and consequently control of moisture in a fluidized bed dryer. The system uses electrical capacitance tomography to take images of gas-solid distribution and to provide online moisture measurements. This way, the setpoint can be fixed directly to the desired value of w^*.

More recent developments refer to freeze-drying for sensitive biological substances. Researchers from Italy, USA, and Japan [17, 18] reported the monitoring and control of freeze-drying for pharmaceuticals at very low temperatures (around −40 °C).

Drying, a quite intensive energy consumption process, is used on a large scale not only in ceramic or cement industry but also in food industries [19, 20], timber/wood industry [21], polymer industry [22], etc. The control of the process, involving new techniques as tomography-assisted control with Kalman filter estimation and linear quadratic Gaussian control, MPC, and optimization using genetic algorithms, give superior results in stabilizing the controlled variable (moisture generally) with impact on the quality of the dried material and important saving of energy.

References

[1]   Mujumdar, A. (Eds), *Handbook of Industrial Drying*, p. 4, 4th Edition, CRC Press, Taylor & Francis Group,, 2015.
[2]   Agachi, P.S., Cristea, V.M., *Basic Process Engineering Control*, p. 217, Walter de Gruyter Gmbh, Berlin/ Boston, 2014.
[3]   Martinez, A.T., *On the evaluation of the wet bulb temperature as a function of dry bulb temperature and relative humidity*, Atmosfera, 7, 179, 1993.
[4]   Stull, R., *Wet-bulb temperature from relative humidity and air temperature*, American Meteorological Society, 2267, 2011.
[5]   Fonyó, Z., Fábry, G., *Vegyipari művelettani alapismeretek (Basic Knowledge in Operations In Industrial Chemistry)*, p. 905, Nemzeti Tankönyvkiadó, Budapest, 2004.

[6] Perry, R., Green, D., Maloney, J. (Eds.), *Perry's Chemical Engineer's Handbook*, McGraw-Hill, Ch. 12–34, Table 12.7, 1999.

[7] Shinskey, F.G., *Energy Conservation Through Control*, Academic Press, New York, 212, 1978.

[8] Dufour, P., *Control engineering in drying technology: review and trends*. Special Issue of Drying Technology on Progress in Drying Technologies (5), 24(7),889–904, 2006.

[9] Bâldea, M., Cristea, V.M., Agachi, P.S., *A fuzzy logic approach to the control of the drying process*, Hungarian Journal of Industrial Chemistry, 30, 167–179, 2002.

[10] Van Meel, D.A., *Adiabatic convection batch drying with recirculation of air*, Chemical Engineering Science, 9, 36–44, 1958.

[11] Krischer, O., Kast, W., *Die wissenschaftlichen Grundlagen der Trocknungstechnik*, Springer-Verlag, Berlin, 1992.

[12] Cristea, V.M., Irimiţă, A., Ostace, G., Agachi, P.Ş., *Control of forced convection drying in food slabs*, in *Proceedings of the 22nd European Symposium of Computer Aided Process Engineering, 17–20 June*, London, 932–936, 2012.

[13] Yuzgeca, U., Beceriklib, Y., Turkerc, M., *Nonlinear predictive control of a drying process using genetic algorithms*, ISA Transactions, 45(4),589–602, 2006.

[14] Shinskey, F.G., *Energy Conservation through Control*, Academic Press, New York, 226, 1978.

[15] Senbon, T., Hanabuchi, F., *Instrumentation Systems: Fundamentals and Applications*, p. 598, Berlin, Chapter 9, 1991.

[16] Yang, W., Wang, H., *Methods and apparatus relating to fluidized beds*, University of Manchester, UMIP, US Patent 8,461,852 B2, 2013.

[17] Barresi, A., Pisano, R., Rasetto, Valeria, Fissore, D., Marchisio, D., *Model-based monitoring and control of industrial freeze-drying processes: Effect of batch nonuniformity*, Drying Technology: An International Journal, 28(5), 577–590, 2010.

[18] Patel, S., Doen, T., Pikal, M., *Determination of end point of primary drying in freeze-drying process control*, AAPS PharmSciTech, 11(1), 73–84, 2010.

[19] De Temmerman, J., Dufour, P., Nicolai, B., Ramon, H., *MPC as control strategy for pasta drying processes*, Computers and Chemical Engineering, 33, 50–57, 2009.

[20] Wongphaka, W. Abdunasser, Y., Elkamel, A., Douglas, P., Lohi, A., *Control vector optimization and genetic algorithms for mixed-integer dynamic optimization in the synthesis of rice drying processes*, Journal of the Franklin Institute, 348, 7, 1318–1338, 2011.

[21] Ge, L., Cheng, G-Sh., *Control modeling of ash wood drying using process neural networks*, Optik, 125, 6770–6774, 2014.

[22] Hosseini, M., et.al., *Tomography-assisted control for the microwave drying process of polymer foams*, Journal of Process Control, 114, 16–28, 2022.

14 Control of crystallization and filtration processes

The separation and purification of materials *via* crystallization results in a slurry where the particle phase contains the compound of interest in desired chemical/polymorphic purity and particle properties, and the liquid phase consists of the solvent system as well as dissolved contaminants and uncrystallized solute. The crystallization is predominantly followed by a solid-liquid separation step – often the filtration. The filtration time is inversely proportional to the particle sizes; therefore, the crystallization process directly impacts the subsequent filtration. Hence, from a technological perspective, it is desired to produce crystals with a larger mean size and low length to width ratio, which of course must remain within the particle property domain of the product (particle size and shape distributions, polymorphic form, etc.). Therefore, the good operation of filtration cannot be achieved without a good operation of crystallization. Hence, the crystallization and filtration processes are discussed together in this chapter.

14.1 The process of crystallization

Crystallization is a separation, purification, and particle formation technique. It is widely used in food and chemical industries, with particularly important role in the manufacturing of fine chemicals and pharmaceuticals. The driving force of crystallization is the thermodynamic instability caused by *supersaturation*. A solution is said to be supersaturated if the actual solute concentration is higher than the solubility under given thermodynamic conditions. Depending on the way supersaturation is generated, different types of crystallization are distinguished [1]:

- *Crystallization by cooling* is based on the fact that solubility generally increases with temperature and thus decreases during cooling. If a warm saturated solution is cooled, it becomes supersaturated – see the temperature drop from 90 to 60 °C in Fig. 14.1(a).
- *Antisolvent crystallization* is based on the different solubility of the same species in various solvents or solvent mixtures. If another significantly lower solubility solvent is added continuously into a saturated solution prepared in a high solubility solvent, the mixture becomes supersaturated (see Fig. 14.1(b)) and the excess solute "crystallizes". The low solubility solvent is called antisolvent, and the high solubility solvent is said to be the "solvent".
- *Reaction crystallization (precipitation)* occurs during a fast (usually ionic) chemical reaction that yields a hardly soluble product. Its solubility is generally some orders of magnitude lower than the expected product concentration; thus, at the reaction endpoint, the mixture is highly supersaturated and the excess precipitates.

https://doi.org/10.1515/9783110789737-016

– *Evaporation:* In this case, the solvent is continuously removed *via* evaporation, which triggers supersaturation and so crystallization. The crystallization occurs at the boiling point of the solution, and the control input is the heating power: more heat input translates to faster evaporation, hence, to higher supersaturation in the remaining solution.

The supersaturation is the main driving force of crystallization, and it is quantitatively characterized either by the *supersaturation ratio (S)* or the *relative supersaturation (σ)* – see eq. (14.1). Defined as such, both (S) and (σ) are dimensionless C and C_s stand for the actual and the saturation solute concentrations, respectively:

$$\sigma = S - 1; \quad S = \frac{C}{C_s} \tag{14.1}$$

Regardless of the supersaturation generation method, the crystallization process consists of two main mechanisms: *nucleation* and *crystal growth* (but further mechanisms, such as fragmentation or agglomeration may also appear in some systems). These coexist in the slurry and occur either successively and/or competitively.

Nucleation is the birth of new crystals. From an industrial point of view, there are two types: *primary* and *secondary* nucleation [2]. Primary nucleation occurs when a new crystal is formed directly from the bulk solution from individually solvated molecules. Secondary nucleation is a collective name for the phenomena in which existing crystals generate new nuclei under the action of shear stresses or crystal-impeller/wall and crystal-crystal collisions. The primary nucleation occurs at the beginning of crystallization until the formation of the first crystals and it is further gradually replaced by the secondary nucleation. In crystallizers of industrial importance, the secondary nucleation is governing [3].

14.2 Traditional operation and control of batch and continuous crystallization processes

Crystals obtained in a crystallization process may be an end-product, or it may undergo further processing steps such as mixing with excipients and tableting. Nevertheless, from an economic perspective, it is desired to recover the largest possible part of the material, i.e., to maximize the yield. The yield is governed by the solubility domain in which the process is operated (Fig. 14.1). In a cooling crystallization, solubility depends on temperature, which dependency may be described by the Van't Hoff equation, or an empirical polynomial equation, as it is shown below:

$$C_s[\text{kg/kg}] = \sum_{i=0}^{N} a_i T^i, \quad N = 2 \text{ or } 3 \tag{14.2}$$

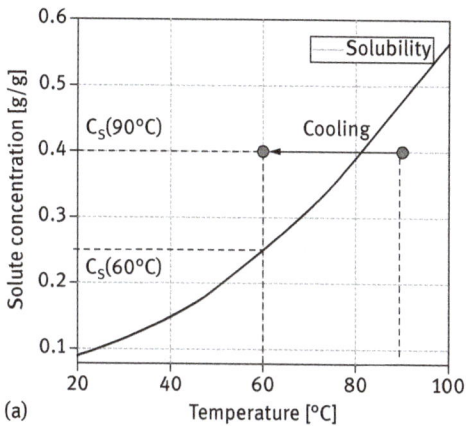

(a)

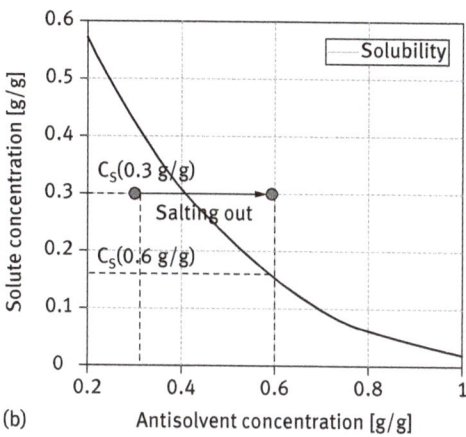

(b)

Fig. 14.1: Typical temperature (a) and solvent composition (b) dependencies of solubility.

The yield is determined by the operating temperature range. Assuming that the process is started and terminated in equilibrium, the yield of a batch cooling crystallization is calculated as

$$\eta = \frac{C_s(T_i) - C_s(T_f)}{C_s(T_i)} \tag{14.3}$$

The final temperature (T_f) is set based on various techno-economic reasons, but it is imperative to synchronize it with the filtration temperature to prevent filtration issues (related to heating or cooling the slurry by the filter media). The highest initial temperature (T_f) maximizes the yield, but limitations exist in the choice of initial temperature as well: in some cases, the solubility may be very high, which would result in excessive solid concentration toward the end of the batch, leading to, e.g., stirring difficulties, including damage of agitator or the motor. The too-high initial temperature may also lead to excessive solvent evaporation or, for thermolabile compounds, may trigger thermal decomposition. The cooling time is set based on preliminary experimentation: too fast cooling leads to high supersaturations which deteriorate the product quality, and too slow cooling hampers the productivity.

Beyond the yield, productivity $(p, \text{kg}/(\text{m}^3\text{h}))$ is a key economic indicator, which is the amount of crystal produced in a unit of time. The total batch time (t_b) is the sum of cooling, equilibration, and auxiliary times (i.e. the filling, discharging, and cleaning time of the crystallizer). Then, the productivity is expressed as

$$p = \frac{C_s(T_i) - C_s(T_f)}{t_b} \tag{14.4}$$

Primary nucleation is known to be highly uncertain, acting as an inherent disturbance in self-nucleating batch crystallizers. To mitigate this disturbance, industrial crystallizers are predominantly operated in seeded mode. Assuming that the crystal growth is the dominant mechanism, in a batch crystallization process the size of the product crystals (L_p) can be estimated in the knowledge of the size of seed crystals (L_s), seed crystal weight (m_s), and product crystal weight (m_p):

$$L_p = L_s \sqrt[3]{\frac{m_p}{m_s}} \tag{14.5}$$

Equation (14.5) relies on the mass conservation law and assumes that the seed and product size distribution is monodisperse and the crystallized material is distributed evenly amongst the seed crystals. The weight of product crystals is calculated based on the solubility and the amount of the slurry, including also the weight of seeds.

The design eqs. (14.1)–(14.5) holds for seeded crystallization with negligible nucleation. To ensure this in a reproducible way, the process should be kept within the metastable zone, meaning that the initial solvent composition and seed quality and quantity must be identical from batch to batch, and the temperature should be controlled precisely. The latter is a process control problem.

Figure 14.2 presents the two widely used control strategies: natural and linear cooling. In the former, the crystallizer is cooled with a cooling agent at a constant flow rate: at the beginning of the process the temperature declines quickly with slowing cooling rates toward the end of the batch. The time at which the desired final temperature is reached can be adjusted *via* the coolant flowrate. In contrast, linear cooling employs a feedback temperature control loop that follows a preset temperature profile. Often PI(D)-based control is employed to adjust the flow rate of coolant based on the actual temperature of the crystallizer. Since the temperature is directly controlled, this strategy results in better reproducibility, hence, more consistent product quality compared with the natural cooling.

The crystallization is accompanied by a latent heat release that has to be absorbed by the coolant. Usually, the crystallization dynamics and the resulting heat release rate are slower than the dynamics of heat transfer between the crystallizer and the jacket.

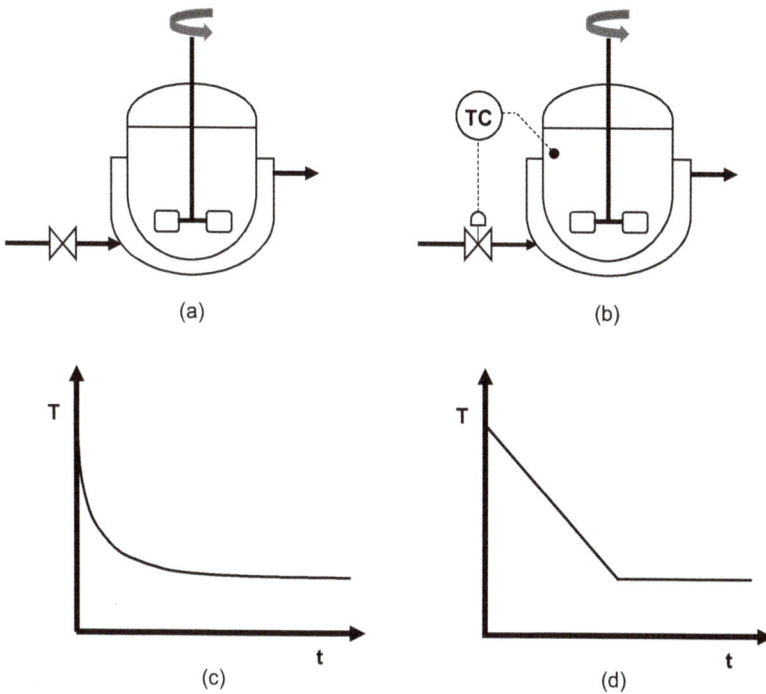

Fig. 14.2: Operation schemes for natural (a) and controlled (b) operation of cooling batch crystallization processes and the implemented temperature profiles (c and d).

This means that the load on the temperature controller remains low. Therefore, tuning the temperature controller of a batch cooling crystallizer is a standard controller tuning task. It is noteworthy that intense nucleation and the accompaining intense crystal growth may result in temporary but noticeable deviation between the actual temperature and the setpoint, which may be used to identify the beginning of crystallization.

The analog of the well-known continuous stirred tank reactor (CSTR) concept is the continuous mixed suspension mixed product removal (CMSMPR) crystallizer. The liquid phase is assumed to be perfectly mixed, and the suspension removal is unclassified (the CSD and solution composition of the outlet stream is identical to that of the crystallizer slurry). The scheme of a cooling CMSMPR crystallizer is presented in Fig. 14.3. The unit is often cooled *via* a jacket and is continuously fed with an inlet stream (F_i, expressed in m^3/ s) of high temperature (T_i, expressed in either °C) and solute concentration (C_i, expressed in kg/m^3). The value of C_i is close to that of the solubility, but still somewhat lower to avoid fouling in the feeding stream (undesired deposition in pipes). CSD in the inlet stream is denoted with ($n_i(L)$). The crystallizer temperature (T) is considerably lower than that of the feeding stream; thus supersaturation is generated as illustrated by Fig. 14.1(a).

As the solubility decreases with temperature (see Fig. 14.1(a)) higher overall yields may be achieved when operating at lower temperatures. In a continuous tank crystal-

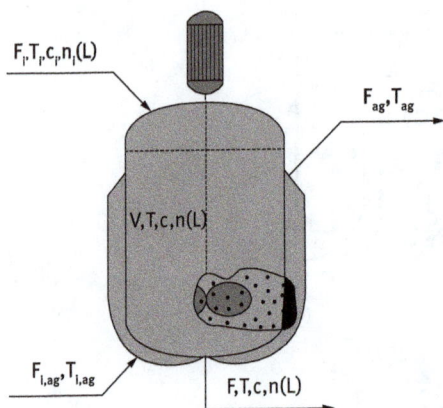

Fig. 14.3: Scheme of a CMSMPR crystallizer.

lizer, the steady-state yield can be calculated as a function of feeding and evacuation solute concentrations (C_i and C, respectively) as

$$\eta = \frac{C_i - C}{C_i} \tag{14.6}$$

Tank crystallizers are often operated at low supersaturations to promote the crystal growth over the nucleation and improve the yield, meaning that $C \approx C_s(T)$. As the crystal growth is a relatively slow process, this approximation is valid for relatively long mean residence times (i.e. high V/F). This enables a simple approximation of the overall yield without the necessity of concentration measurements and connects the overall yield to the performance of the temperature controller directly: oscillations in temperature (X °C oscillation amplitude) lead to oscillations in solubility, but through the nonlinear function of solubility [see eq. (14.1)]. According to Fig. 14.1, solubility is a concave-up curve, hence, the following equation holds:

$$C_s(T+X) - C_s(T) > C_s(T) - C_s(T-X) \tag{14.7}$$

Equation (14.7) means that an oscillatory temperature control always results in global yield loss as the yield gain realized during the temperature undershoot period [right-hand side of eq. (14.7)] cannot be counterbalanced by the yield loss of the temperature overshoot [left-hand side of eq. (14.7)].

The productivity of the crystallizer is calculated in terms of the mean residence time as well as feeding and evacuation concentration:

$$p = \frac{F(C_i - C)}{V} \tag{14.8}$$

This underlines that flow control can be employed to manipulate the productivity of the crystallizer. Since crystallization may be accompanied by a small change in the slurry density, the evacuation pumps cannot be controlled in an open-loop manner, but active level control is necessary. Such a general control structure of CMSMPR systems is presented in Fig. 14.4.

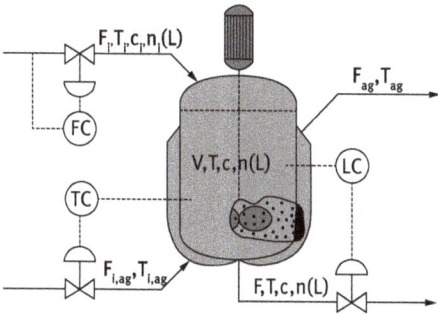

Fig. 14.4: General control structure of a CMSMPR crystallizer.

In some configurations, the level control is solved passively, for example, using an overflow as a slurry withdrawal system.

Beyond the yield and productivity, controlling the particle properties became important in numerous industries, from food to pharmaceutical products, which must be addressed [4]. This involves advanced control solutions.

14.3 Particle size control in batch crystallization

Several macroscopic properties of particulate products depend on crystal size distribution: specific surface, porosity, dissolution rate, flowability, bulk density, etc. The CSD affects the filtration performance as well. Therefore, achievement of target CSD in the crystallization eliminates the need for downstream crystal size adjustment operations, hence, reduces production cost and lowers environmental impact. It also ensures higher-quality crystals since these secondary size-modifying operations do not result in flat and clear crystal facets. Consequently, CSD control is an important aspect of the modern chemical industry.

For effective particle size control it is important to understand the fundamental processes governing particle sizes during crystallization. The nucleation, i.e., the formation of new, usually small particles, and the growth of existing crystals are present in virtually every industrial crystallization process. While the main driving force, the supersaturation, would increase both the nucleation and growth rates, the relative magnitude is different: usually, the higher supersaturation favors the nucleation rate more than the growth rate. This enables to find operating condition windows under which the desired product particles are obtained. Beyond the nucleation and crystal growth, other secondary processes such as breakage and agglomeration may appear. Furthermore, if the slurry becomes undersaturated, either intentionally as a part of the operating procedure or accidentally, the existing crystals will undergo partial or total dissolution.

The crystal size control strategies can be divided into two major groups: model-free and model-based control techniques. The former utilizes online particle size measurements to manipulate the operating conditions such as to match the target particle size, whereas the latter relies on process models to determine the optimal control moves in real time.

14.3.1 Model-free crystal size distribution control

The quick spread of *process analytical technologies* (PAT) enables online crystallization process monitoring [5]. The real-time measurements cover the tracking of liquid and solid-phase properties using one or more of the following PAT instruments:

1. *Laser backscattering based in situ relative size measurement:* The principle of laser backscattering permits the construction of a specific transducer of particulate systems. It emits a laser beam from a sensor inserted into the slurry. The beam rotates with a specified angular velocity during which it intersects (meets) the crystals. The duration of this intersection, widely called *chord length* (CL) is then detected by the sensor (see Fig. 14.5). Up to a few thousand CLs are measured per second. The number of intersections, called counts, is proportional to the crystal number per volume unit ($\#/m^3$). The CL distribution (CLD) can be constructed from the length of individual CL values and is proportional to the CSD. CLD is a fingerprint of the particulate system. The focused beam reflectance measurement (FBRM) technology utilizing laser backscattering was developed in the early 90s and it gained significant attention.

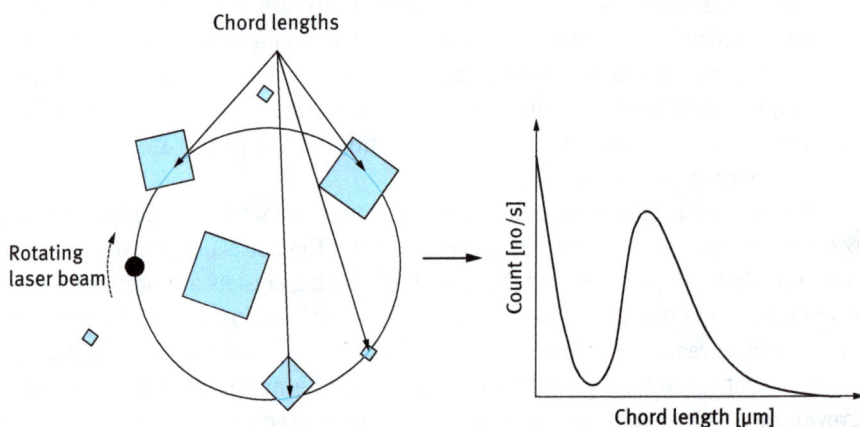

Fig. 14.5: Measuring principle of FBRM and a typical chord length distribution (CLD) (reprinted with permission from [15]). Copyright 2019 American Chemical Society.

2. *In situ microscopy and image analysis:* The spread of imaging devices infiltrated into the process monitoring technologies. Modern in-process process microscopes are able to detect particles from ~1 µm, some of them having detection limits in the submicron range under certain circumstances (optical properties of the particles and the solvent system). The recording frequency of these tools is in the range of tens of images per second. The real-time analysis of these images not only allows to obtain relative crystal number and size as laser backscattering but shape information can be gathered, and other phenomena such as agglomeration can be identified. A typical in situ image is presented in Fig. 14.6.

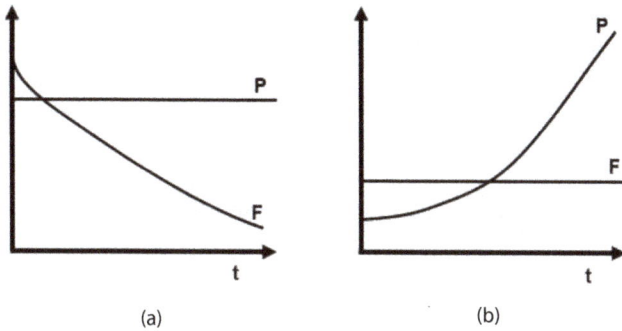

(a)   (b)

Fig. 14.6: Typical in situ microscope image, captured during the crystallization of an inorganic material, Loughborough University, 2014.08.12.

3. Depending on the nature of the solid and liquid phases, the solute concentration is usually assessed *via* online UV/VIS/NIR or FTIR.
4. The in situ discrimination of polymorphic forms can be carried out with in-line RAMAN spectrometry or, if the shape of polymorphs significantly differs, by in situ imaging and image analysis.

14.3.1.1 Direct nucleation control (DNC)
The DNC relies on the mass conservation law. Assuming that a certain mass of crystal is produced, variation of the crystal number results in the adjustment of the crystal size: if fewer crystals are produced, they will inevitably be bigger and *vice versa*. The controlled variable is the relative particle number density, which is tracked in real time.

The working principle of DNC is simple [6]. Crystallization is started with a high-temperature saturated solution, and then it is cooled at a specified rate. When supersaturation is reached (see Fig. 14.1(a)), nucleation and growth of crystals begin. Meanwhile, the CSD is monitored *via* an online particle sizing tool. When the upper relative number density limit is exceeded (i.e. nucleation produced an excessive number of particles), a heating stage begins. Because of the heating, the solution becomes undersaturated and the dissolution of crystals commences. This reduces the crystal number, and thus the measured relative number density. According to the Oswald ripening phenomenon, the smaller crystals dissolve faster, and thus the relative number density decreases faster as compared to the global (mass) dissolution. When the relative number density lower limit is reached (i.e. the excess particles were removed *via* internal dissolution), the second cooling loop is started. It triggers again supersaturation. Within this second cooling stage, in contrast to the first, some crystals already exist in the slurry and start to grow immediately. As it has already been discussed, the lower supersaturation favors the growth; thus, as long as the growth of these crystals consumes supersaturation, the nucleation is not favored. However, too slow cooling prolongs the batch time; hence, higher cooling rates are recommended for improved productivity. Yet, this may lead to enhanced nucleation, which in return leads again to a relative particle number density increase. Consequently, heating and cooling are alternatively repeated until the final target temperature and a relatively stable relative number density (translated into CSD) are reached. Figure 14.7 presents the typical time diagram of a DNC control.

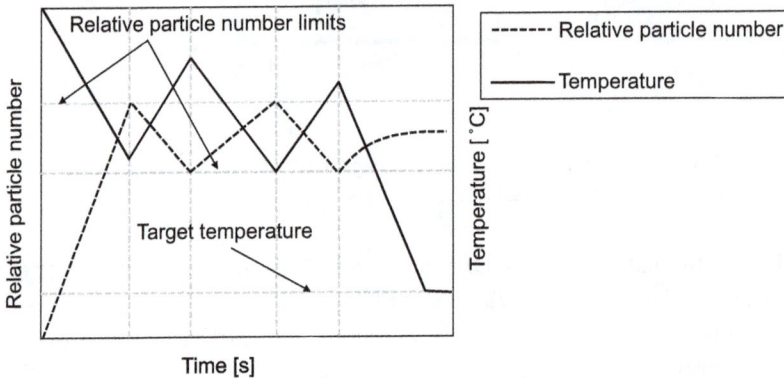

Fig. 14.7: Presentation of the DNC concept: the controller maintains the FBRM count between the predefined limits by heating and cooling cycles.

The main advantage of DNC is that if the appropriate settings are applied (cooling and heating rates), a stable control can be achieved in most cases without any prior system information (solubility, nucleation, and growth kinetics). It is robust to disturbances generally caused by impurities or by deposit detachments from the crystallizer wall or stirrer. However, in some situations, high batch times may result and CSD is indirectly affected *via* relative crystal number. In other words, the average crystal size and number can be satisfactory, but the dispersion of crystal sizes may be high, and the crystal shape can hardly be controlled.

14.3.1.2 The supersaturation control (SSC)

Figure 14.8 illustrates in other terms the practical observation that low supersaturation favors crystal growth: a second characteristic line can be defined and tracked in the concentration-temperature phase diagram. It is called the *metastable limit*. Under it, nucleation is negligible, but crystal growth occurs. The zone between the metastable limit and the solubility line is called the *metastable zone*.

The main idea behind the supersaturation control (SSC) is the addition of *seed crystals*, in a well-specified quantity and size distribution, to a solution characterized by the metastable zone. The supersaturation is further controlled by manipulating the temperature or the antisolvent amount to keep the process in this zone [7]. The improved growth rate is ensured by keeping supersaturation near the metastable limit. Thus, similarly to the DNC, crystal size distribution is controlled indirectly.

SSC conducts the process in the vicinity of the maximum allowed supersaturation and thus eliminates the formation of small crystals and shortens batch time as compared to the DNC. On the contrary, it requires preliminary knowledge of solubility and metastable limit as well as requires in situ concentration measurement. The accurate and proper determination of these may be experimentally time-consuming, and the calibration is associated with numerous practical issues, which have to be readjusted from time to time.

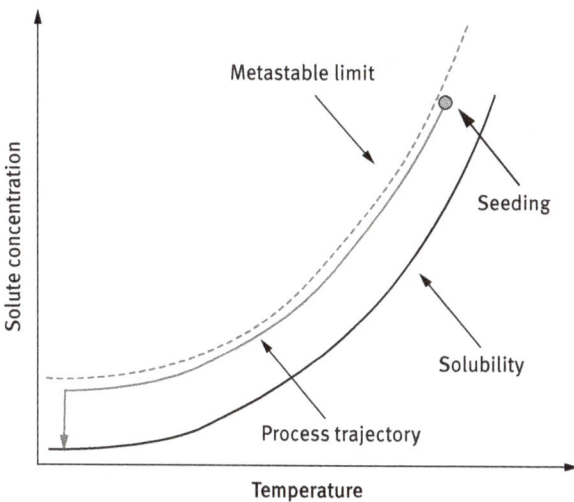

Fig. 14.8: The SSC concept: the process is conducted near the metastable limit, which favors crystal growth without generating significant nucleation.

14.3.2 Model-based crystal size distribution control

The model-based crystal size distribution control strategies use the mathematical model of crystallization process to find the optimal temperature profile or antisolvent addition rate for the achievement of target crystal quality (yield, purity, CSD, shape, etc.) with respect to given constraints. A crystallization model incorporates the nucleation and growth rates into a population balance equation (PBE) that allows calculating the temporal evolution of CSD [8].

Numerous model equations have been proposed to describe the nucleation rates. Among them, two popular equations for the primary and secondary nucleation rates are

$$B_p = k_p \exp\left(-\frac{k_e}{\ln^2(S)}\right) \exp\left(-\frac{E_{a,p}}{RT}\right) \tag{14.9}$$

$$B_s = k_s \, \sigma^b V_C^j \exp\left(-\frac{E_{a,s}}{RT}\right) \tag{14.10}$$

The significance of terms in eqs. (14.9)–(14.10) is
- B_p, primary nucleation rate (#/m$^3 \cdot$ s);
- B_s, secondary nucleation rate (#/m$^3 \cdot$ s);
- k_p, primary nucleation rate constant (#/m$^3 \cdot$ s);
- k_s, secondary nucleation rate constant (#/m$^3 \cdot$ s);
- k_e, a constant characterizing B_p dependence on supersaturation (#/m$^3 \cdot$ s);
- b, a constant characterizing B_s dependence on supersaturation (dimensionless);
- j, a constant characterizing B_s dependence on volume fraction of crystals within the suspension (dimensionless);
- V_c, fraction of total slurry volume occupied by formed crystals (m^3/m^3);
- $E_{a,p}$ and $E_{a,s}$, activation energy of primary and secondary nucleation, respectively (J/mol $\cdot$ K).

Equations above show that B_p and B_s depend on supersaturation ratio ($\sigma = S - 1$), temperature, and some material/system-dependent constants. The latter may be sensitive to mixing conditions and sometimes to very small amounts (ppm order of magnitude) of impurities.

The nuclei and existing crystals are growing as long as the solution is supersaturated. The deposition of solvated molecules (or atoms/ions) on the crystal surface consists of three mass transfer steps:
- Convective transfer from the bulk solution toward the mass transfer boundary layer;
- Conductive transfer (diffusion) through the boundary layer toward the crystal surface;
- Integration into the crystal lattice (deposition).

Deposition occurs according to two major types of surface integration mechanisms:
- *Monolayer* crystal growth: new layers are built only after the previous ones are finished;
- *Polylayer* crystal growth: more than one crystal layer is being built at the same time. This mechanism occurs at higher supersaturation ratios.

The precise mathematical description of crystal growth is complicated because it strongly depends on liquid phase mass transfer as well as on the nature of surface deposition. However, the following empirical formula can successfully be applied for the macroscopic description of numerous systems in certain supersaturation domains:

$$G = k_g \sigma^g \exp\left(-\frac{E_{a,g}}{RT}\right) \tag{14.11}$$

where the significance of terms is:
- G, linear crystal growth rate (m/s)
- k_g, crystal growth rate constant (m/s)
- g, a constant characterizing G dependence on supersaturation (dimensionless);
- $E_{a,g}$, activation energy of crystal growth (J/mol · K).

Equation (14.11) shows that similarly to nucleation, crystal growth depends on super-saturation, temperature, and system/material-specific constants. Generally speaking, the supersaturation dependency of crystal growth is lower than that of nucleation; hence, operating in lower supersaturation regimes favors crystal growth over nucleation. For diffusion-controlled growth, the exponent g is around 1, but for a surface integration-limited growth, the g is near 2.

The global crystallization rate is determined by both the nucleation and growth rates. If the nucleation rate is small and that of the growth is high, fewer crystals of bigger dimensions are formed. The reverse situation yields a high crystal number having smaller sizes. The goal of the crystallization process control is to adjust both nucleation and growth to achieve the desired purity and CSD of the product.

The crystal size distribution is characterized by the ($n(L)$) size density (expressed in # m/m^3) which gives the number of crystals in the $[L, L+dL]$ size interval (expressed in m) per unit volume of slurry. Assuming a constant slurry volume V (expressed in m^3), perfect mixing, and negligible agglomeration and breakage, the PBE governing the size distribution of crystals is described by [2]:

$$\frac{\partial n(L)}{\partial t} + \frac{\partial Gn(L)}{\partial L} = (B_p + B_s)\,\delta(L - L_n) \tag{14.12}$$

where the initial condition is set by

$$n_{in}(L|t=0) = n_0(L) \tag{14.13}$$

In eq. (14.12), L_n stands for nuclei size. The subscript *in* denotes the initial values. The first term on the left side of eq. (14.12) stands for the time evolution of CSD and the second term takes into consideration the effects of crystal growth. The right side of eq. (14.12) describes the effects of both nucleation rates. The Dirac-delta function $\delta(L - L_n)$ of the last term in eq. (14.12) implies that nucleation affects only the number of crystals having sizes of L_n, but has no effect on the number of bigger crystals. In other words, the value of the function $\delta(L - L_n) = 1$ if $L = L_n$ and $\delta(L - L_n) = 0$ if $L \neq L_n$ (has the role of a boundary condition).

The solute mass balance in a batch crystallizer is written as follows:

$$\frac{dC}{dt} = -3k_V\rho_c \int_0^\infty L^2 Gn(L)dL \tag{14.14}$$

The initial condition is given by

$$C(0) = C_{in} \tag{14.15}$$

As mentioned above, solute concentrations are symbolized by C. ρ_c in eq. (14.16) stands for the crystal density (expressed in kg/m³) and k_V is a dimensionless volume shape factor used to compute the volume of crystals. The volume of a single crystal can be expressed in the function of its size L as

$$V_c = k_V L^3 \tag{14.16}$$

Defined as such, $k_V = 1$ for the cube and $k_V = \pi/6$ for a sphere. The shape factor can be deduced for other geometries easily.

Estimation of kinetic parameters of an industrial crystallization process can be carried out based on experimental data with various techniques. A popular technique is the model-based estimation, in which case the unknown kinetic parameters are adjusted such as the deviations between the measurements and simulations are minimal [9]. Depending on the crystallization process, measurements may include solute concentration variations over time, particle size distributions, and others.

Once the model is available, crystallization control can be divided into open-loop and closed-loop model-based control. Various goals can be formulated, for example, for the batch crystallizer (also see Chapter 2 on MPC): batch time minimization, crystal size distribution optimization, cost reduction, etc. A typical goal function realizes smooth temperature profile $(f(\Delta T))$ and realizes the product specification with minimal deviation $(f(\Delta Q))$:

$$\min_T[w\, f(\Delta T) + (1 - w)f(\Delta Q)] \tag{14.17}$$

where w is a weight factor that allows to adjust the relative importance of operating conditions and the product properties. Various practical constraints can be implied for physical realizability, for example, the cooling rate must stay within the technical limits of the cooling system $(c_{r,c}, °C/s)$:

$$\frac{\Delta T}{\Delta t} < c_{r,c} \tag{14.18}$$

Multiobjective goals an multiple constraints can also be formulated. Due to the nonlinear feature of crystallization, often nonlinear MPC (N-MPC) is required. Chapter 2 is dedicated to model predictive control; thus only a general description of these concepts' application to batch crystallization control will be subsequently presented.

14.3.2.1 Open-loop model-based control

In the case of any open-loop control, the process output of interest is taken into account in the off-line operating policy determination, but in the implementation stage it is not controlled directly. The control signal (temperature profile or the antisolvent addition rate) is determined by solving a process optimization problem that employs the model. By optimizing the control signal, the constraints are explicitly taken into account and the CSD can be directly manipulated. The advantage of open-loop model-based (optimal) control over the PAT-based technique is the possibility of constraint satisfaction, direct CSD manipulation, and cost reduction (no need for expensive instrumentation). The major disadvantage of this approach is that the process may be extremely sensitive to disturbances; without feedback, the process can easily depart from the optimal path after relatively small disturbances (the control signal is predefined and optimization is carried out before the process starts).

14.3.2.2 The closed-loop online MPC

In the case of closed-loop MPC, the process model is used in *real time* to optimize the control signal with respect to global performance requirements. This permits the incorporation of harmful effects of disturbances as well as model uncertainties into the control strategy. This generally results in better control performance. Yet, the process simulation has to be fast enough to enable real-time optimization.

During the duration of a batch cycle, data are available from the initiation of the process until the certain moment when the measurement is carried out. These can be used in the MPC algorithm to improve its performance. The state model incorporates CSD values, yet these may not be measured directly the PAT tool presented in Section 14.3.1. Thus, state estimators are needed to assess the unmeasurable system states. For a batch crystallizer that operates with a fixed batch time, the prediction horizon is decreasing, but the quantity of experimental data is increasing during the duration of a batch cycle. Therefore, the MPC is referred to as *shrinking horizon* MPC and the state estimator as to *growing horizon estimator* (GHE). Figure 14.9 presents this principle of a closed-loop MPC for a cooling batch crystallization with state estimation [10].

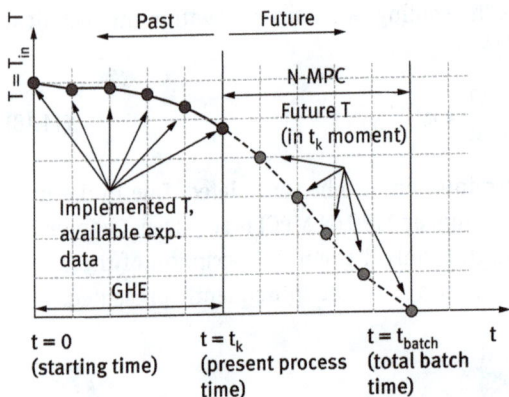

Fig. 14.9: The scheme of batch crystallization MPC: a shrinking horizon MPC with GHE (reprinted with permission from [15]). Copyright 2019 American Chemical Society.

Since the kinetic parameters are sensitive to various factors that are difficult to control, for a successful model-based control, the kinetic constants need to be readjusted in real time to reduce the plant-model mismatch. Beyond estimating the unmeasured system states, the GHE solves a kinetic parameter readjustment problem in real time as eq. (14.19) presents:

$$\min_{p} f(\Delta Y) \qquad (14.19)$$

where **P** stands for the vector of kinetic parameters that are readjusted such as the deviation between the simulations and measurements (ΔY) over the available horizon (see Fig. 14.9) is minimal. Note that (ΔY) may include various measurements, e.g., solute concentration, relative particle number, and relative particle shape. f is a performance metrics function, e.g., mean squared deviation. The block diagram of such a controller is presented in Fig. 14.10.

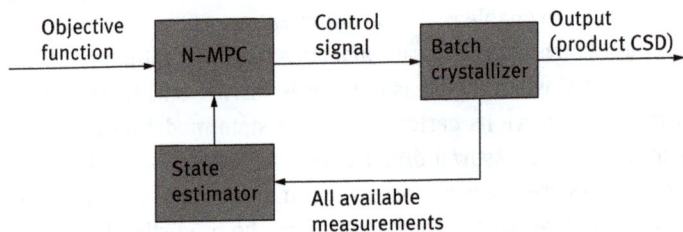

Fig. 14.10: The block diagram of a shrinking horizon MPC with receding horizon RHE applied to the control of batch crystallizers.

It is noteworthy that sophisticated model-free and model-based techniques are both able to control not only crystal size distribution but also polymorph transformations and cocrystallization. These strategies are based on the difference between the solubility of various polymorphic forms of the same species or various solutes, respectively.

14.4 The process of batch filtration

Filtration is a physical separation process that removes solid particles from a slurry by using a filter medium through which ideally only the liquid can pass. Filtration equipment can be divided into threenmajor groups: (i) cake filters (including the Nutsche filters, filter presses, and liquid bag filters), (ii) rotary drum filters, and (iii) horizontal rotary/belt-type filters. Filter selection depends on the throughput, particle properties (size, shape), and operational mode (continuous or batch). This section is aimed to present the principles and control solutions of discontinuous filtration processes.

   In a batch filtration, the separated particles form a cake on the top of the filter. The liquid and the small particles that pass the filter medium are said to be the filtrate. After completing the filtration, the filter cake is washed with a small amount of pure solvent to remove any liquid residues from the surface of crystals in the cake, as residual liquid may contain dissolved impurities. The main driving force of filtration is the pressure difference between the two sides of the filter medium, which helps the liquid pass the filter. Increasing pressure difference translates to an increasing overall filtration rate. During the filtration, there is a certain level of slurry at the top of the filter medium that alone generates hydrostatic pressure. In practice, the filtration is intensified by enhancing the driving force such as (Fig. 14.11) [11]

- applying a vacuum on the filtrate side that will such the liquid through the medium or
- applying pressure on the cake side to push the liquid through the filter.

The attainable pressure gradient is limited when vacuum filtration is applied as the pressure gradient cannot exceed the atmospheric pressure. Beyond this technical limitation, under the action of the high vacuum evaporation of solvent residues may occur, leaving the dissolved impurities in the surface of solids. High-pressure filtration may generate greater driving force and reduce solvent evaporation simultaneously.

   The filtration process is described using the Darcy's law, which expresses that the filtrate flow rate or flux $(F[m^3/s])$ is directly proportional to the applied pressure gradient $(\Delta P[Pa])$ and filters medium area $(A[m^2])$ and inversely proportional to the viscosity of the filtrate $(\mu[Pa\ s])$, the medium resistance $(R_m[m^{-1}])$ and the cake resistance $(R_c[m^{-1}])$ [12]:

$$F(t) = \frac{dV_f}{dt} = \frac{\Delta P \cdot A}{\mu \cdot (R_c + R_m)} \tag{14.20}$$

where $(V_f[m^3])$ is the filtrate volume. According to Darcy's law the filtration rate is directly proportional to the pressure gradient and the filtration area and inversely

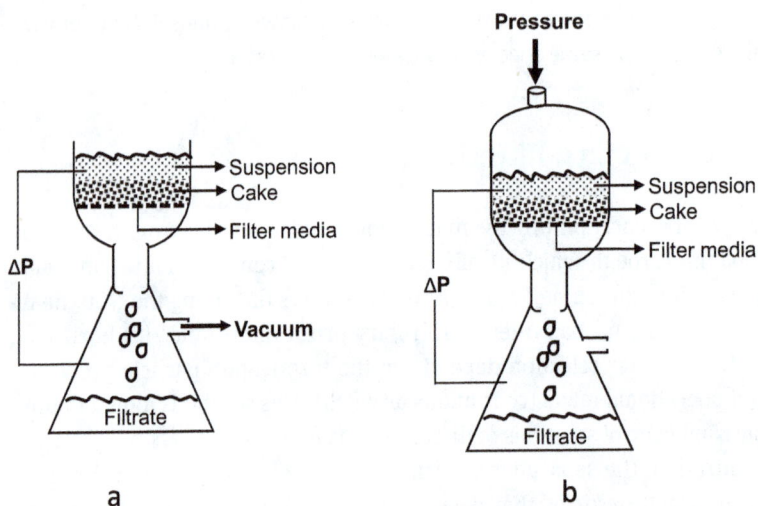

Fig. 14.11: Schematic representation of discontinuous vacuum (a) and high-pressure (b) filtration.

proportional to the viscosity and overall resistance. This underscores the importance of temperature in an integrated crystallization-filtration system lowering the terminal temperature of the crystallization process improves the yield. However, at low temperatures the viscosity of solvents increases, which results in longer filtration times. The denominator of eq. (14.20) contains the sum of resistances. The filter medium resistance (R_c) is roughly constant as it depends on the opening and number of holes (although some holes may get blocked by particles during the filtration, resulting in some increase in (R_m). In contrast, the cake builds up during the process, translating to continuously increasing cake resistance (R_c), as described by eq. (14.21):

$$R_c = \frac{a \cdot s \cdot V_f}{A} \qquad (14.21)$$

where s is the mass of crystals deposited per unit filtrate volume [kg/m^3] and a is the specific cake resistance [m/kg]. Specific cake resistance depends on particle size and shape: small particles have significantly higher cake resistance than the large particles, as it is easier for the liquid to flow between larger bodies. This makes another connection point between the filtration rate and the outcome of the crystallization process. Combining eqs. (14.20) and (14.21) yields the following expression:

$$F(t) = \frac{\Delta P \cdot A^2}{\mu \cdot \left(A R_c + a \cdot s \cdot V_f \right)} \qquad (14.22)$$

Assuming that the filter medium resistance is negligible compared to the cake resistance, i.e., $a \cdot s \cdot V_f \gg A R_c$, $A R_c$ vanishes and the filtration rate becomes a square function of the cross-section area of the filter medium. In the beginning, however, there is no cake, therefore, $\alpha \cdot s \cdot V_f = 0$, resulting in linear cross-section area dependency.

14.5 Controlling a discontinuous filtration process

Batch filtration may be operated in constant pressure or constant flowrate mode.

In constant pressure filtration mode, the pressure (or vacuum) is adjusted to keep the driving force constant. In a feedback control mode this may happen through the application of a feedback pressure control loop (see Fig. 14.12(a)). As the driving force is kept constant and due to the cake buildup throughout the process, the filtrate flowrate decreases during the process as illustrated in Fig. 14.12(c). The advantage of this strategy is the technical simplicity and cost-effectiveness, as the pump may be hooked to the filter without the pressure feedback loop. In the case of constant flowrate filtration the pressure is manipulated in order to match a desired filtrate flowrate (Fig. 14.12(b)). Applied pressure has to be continuously increased with cake thickness (Fig. 14.12(d)), which is realized through PI(D) controllers. Constant rate filtration is implemented in high precision applications, and it is also a more reliable method for data acquisition, especially when determining the filter medium resistance [13].

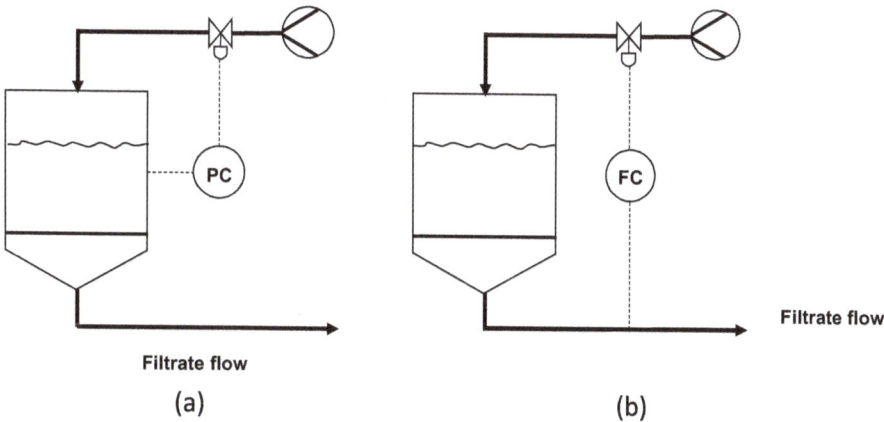

(a)   (b)

Fig. 14.12: Typical filtration control: constant pressure (a) and constant filtrate flow rate (b) modes and the corresponding pressure and filtrate flow rate profiles (c and d).

Utilizing eq. (14.22) to calculate a filtration pressure profile to realize a desired filtrate flowrate profile and then to implement it in an offline manner is discouraged due to the sensitivity of cake resistance on particle size, shape, and other particle properties. As this equation also illustrate, the upper rate of filtration that can be generated with pressure difference (ΔP) is defined by the filter medium properties (R_c), filter size (A), and the cake resistance (R_c). The former two are related to the filter selection, whereas the latter largely depends on the particle properties. Therefore, in an integrated crystallization-filtration process for the good operability of the filtration, extra effort is made in the crystallization process control for the realization of suitable cake resistance and hence to broaden the flowrate control space of the filtration [14]. An important closing remark is necessary here. If the slurry contains impurities (as, for example, the otherwise pure crystals were obtained from a reaction mixture), a carefully designed washing protocol must be applied during the filtration to prevent trapping liquid residues, and, with that, impurities into the crystal cake. In those cases, the impure liquid must be removed as much as possible from the cake during pre-filtration to get rid of the contaminated solution. However, excessive pre-filtration times, especially in vacuum filtration, must be avoided as this can lead to solvent evaporation and, hence, deposition of solid impurities. After removing the impure liquid from the cake but before excessive solvent evaporation, the cake must be washed with an appropriate amount of well-chosen washing liquid to flush the residual solvent and dissolve the eventually solidified impurities. Such a strategy can significantly improve the purity of the final product, and a failure to implement adequate washing can hamper the purity despite otherwise well-operated crystallization and filtration processes being put in place. The exception is when the solvent is trapped within strongly bonded agglomerate particles. When crystal agglomeration generated impurity degradation is significant, the attainable purification effect of cake washing may be limited, and anti-agglomeration strategies must be implemented during the crystallization.

☐ **Example 14.1**

A chemical species having temperature-dependent solubility is crystallized in a batch unit. The solubility is described with the following equation. It is valid within 20 and 80 °C:

$$C_s \,[\text{kg/kg}] = 0.223 - 9.6 \times 10^{-4} T + 1.17 \times 10^{-4} T^2 \qquad (14.23)$$

The unit is charged with a 100 kg saturated solution at 60 °C. Let us calculate the expected mass of crystals if the filtration and drying losses are of 5% each and the final batch temperature is 20 °C.

According to eq. (14.12), the solubility concentrations at the initial and final operating temperatures are

$$\text{for } T = 20 \,°\text{C}; \ C_s = 0.223 - 9.6 \times 10^{-4} \times 20 + 1.17 \times 10^{-4} \times 20^2 = 0.2506 \text{ kg/kg}$$

$$\text{for } T = 60 \,°\text{C}; \ C_s = 0.223 - 9.6 \times 10^{-4} \times 60 + 1.17 \times 10^{-4} \times 60^2 = 0.5866 \text{ kg/kg}$$

The difference between the two values is

$$\Delta C_s = 0.366 \text{ kg/kg}$$

It represents the quantity of crystallized matter, when 100% crystallization yield is assumed, for 1 kg solution that is cooled from 60 to 20 °C. Then the quantity of crystallized matter for the entire batch unit is

$$m_c = 100 \times 0.366 = 33.6 \text{ kg}$$

After taking into account the losses of filtration and drying (5% each), the final product quantity is obtained:

$$m_{\text{prod}} = 33.6 \times 0.95^2 = 30.324 \text{ kg}$$

Example 14.2

A CMSMPR (see Fig. 14.3) crystallizes a species for which the solubility is described by eq. (14.1(a)). After a total loss of 15% in the downstream operations, the plant yields a 250 kg/h product. The feed flow rate is 1000 kg/h, has a temperature of 80 °C, and a solute concentration 5% lower than the corresponding solubility. The unit operates at 30 °C. Let us calculate the yield of crystallization. What operational changes can be proposed to improve its value?

The solubility concentrations at the input and operating temperatures are for

$$\text{for } T = 30\,^\circ C; \ C_s = 0.223 - 9.6 \times 10^{-4} \times 30 + 1.17 \times 10^{-4} \times 30^2 = 0.2995 \text{ kg/kg}$$

$$\text{for } T = 80\,^\circ C; \ C_s = 0.223 - 9.6 \times 10^{-4} \times 80 + 1.17 \times 10^{-4} \times 80^2 = 0.895 \text{ kg/kg}$$

The inlet solute concentration at 80 °C can be calculated based on the inlet solubility:

$$C_{in} = 0.95 \ C_s = 0.95 \times 0.895 = 0.801 \text{ kg/kg}$$

The inlet mass is

$$m_{in} = C_{in}F = 0.801 \left[\frac{kg}{kg}\right] \times 1000 \left[\frac{kg}{h}\right] = 801 \left[\frac{kg}{h}\right]$$

The hourly crystal mass production can be calculated as follows:

$$m_{out} = \frac{250}{0.85} = 294.12 \left[\frac{kg}{h}\right]$$

If a 100% crystallization yield is assumed, the productivity at 30 °C would be

$$m_{out,100} = (C_{in} - C_s) \ F = (0.801 - 0.2995) \times 1000 = 501.5 \left[\frac{kg}{h}\right]$$

By considering the real production of 294.12 kg/h, the value of crystallization yield can be computed:

$$\eta = \frac{m_{out}}{m_{out,100}} \cdot 100 = \frac{294.12}{501.5} = 58.65\%$$

Yield improvement can be achieved by
- reducing the operating temperature to increase supersaturation;
- increasing the average residence time, either by reducing the feed flow rate or by increasing the crystallizer volume.

Example 14.3

A slurry is filtered with a laboratory filter with a filtering surface area of 0.05 m² to determine the specific cake and medium resistance using a vacuum giving a pressure difference of 0.7 bar. The volume of filtrated collected in the first 5 min was 250 mL, and after a further 5 min, an additional 150 mL was collected. The filtrate viscosity is 0.001 Pa · s, the slurry contains 5 vol% of solids with a density of 3000 kg/m³. Calculate the specific cake resistance and the medium resistance.

Equation (14.20) is rearranged and modified to discrete time as

$$\frac{dV_f}{dt} = \frac{\Delta P \cdot A}{\mu \cdot (R_c + R_m)} \rightarrow \frac{\Delta t}{\Delta V_f} = \frac{\mu \cdot R_c}{\Delta P \cdot A} + \frac{\mu \cdot R_m}{\Delta P \cdot A} \tag{14.24}$$

Equations (14.24) and (14.21) can be combined to bring in the specific cake resistance:

$$\frac{\Delta t}{\Delta V_f} = \frac{\mu \cdot a \cdot s \cdot \Delta V_f}{\Delta P \cdot A^2} + \frac{\mu \cdot R_m}{\Delta P \cdot A} \tag{14.25}$$

Equation (14.25) can be interpreted as a linear equation ($y = ax + b$), from where the medium and specific cake resistance can be expressed:

$$y = \frac{\Delta t}{\Delta V_f}$$

$$a = \frac{\mu \cdot a \cdot s}{\Delta P \cdot A^2} \rightarrow a = \frac{a \cdot \Delta P \cdot A^2}{\mu \cdot s}$$

$$x = \Delta V_f$$

$$b = \frac{\mu \cdot R_m}{\Delta P \cdot A} \rightarrow R_m = \frac{b \cdot \Delta P \cdot A}{\mu}$$

y_1 and y_2 can be calculated as

$$y_1 = \frac{\Delta t_1}{\Delta V_{f,1}} = \frac{5 \cdot 60}{250 \cdot 10^{-6}} = 1.2 \cdot 10^6 \text{ s/m}^3$$

$$y_2 = \frac{\Delta t_2}{\Delta V_{f,2}} = \frac{10 \cdot 60}{400 \cdot 10^{-6}} = 1.5 \cdot 10^6 \text{ s/m}^3$$

The slope and intercept of the line in the $x - y$ coordinate system (a and b) can be also calculated as

$$a = \frac{y_2 - y_1}{\Delta V_{f,2} - \Delta V_{f,1}} = \frac{y_2 - y_1}{\Delta V_{f,2} - \Delta V_{f,1}} = \frac{1.5 \cdot 10^6 - 1.2 \cdot 10^6}{400 \cdot 10^{-6} - 250 \cdot 10^{-6}} = 2 \cdot 10^9 \text{ s/m}^6$$

$$b = y_2 - a \cdot \Delta V_{f,2} = 1.5 \cdot 10^6 - 2 \cdot 10^9 \cdot 400 \cdot 10^{-6} = 7 \cdot 10^5 \text{ s/m}^3$$

The medium and specific cake resistance can be calculated from the slope and the intercept:

$$a = \frac{a \cdot \Delta P \cdot A^2}{\mu \cdot s} = \frac{2 \cdot 10^9 \cdot 7 \cdot 10^4 \cdot 0.05^2}{0.001 \cdot (0.05 \cdot 3000)} = 2.33 \cdot 10^{12} \text{ m/kg}$$

$$R_m = \frac{b \cdot \Delta P \cdot A}{\mu} = \frac{7 \cdot 10^5 \cdot 7 \cdot 10^4 \cdot 0.05}{0.001} = 2.45 \cdot 10^{12} /\text{m}$$

References

[1]   Mersmann, A. *Crystallization Technology Handbook*, Marcel Dekker Inc, New York, Basel, 2001.
[2]   Randolph, A., Larson, M. *Theory of Particulate Processes*, Academic Press, Salt Lake City, 1973.
[3]   Myerson, A.S. *Handbook of Industrial Crystallization*. Elsevier, 2002.
[4]   Nagy, Z.K., Braatz, R.D. *Advances and new directions in crystallization control*. Annual Review of Chemical and Biomolecular Engineering, 3(1), 55–75, 2012.
[5]   Simon, L.L. et al. *Assessment of recent process analytical technology (PAT) trends: A multiauthor review*. Organic Process Research & Development, 19(1), 3–62, 2015.

[6]   Abu Bakar, M.R., Nagy, Z.K., Saleemi, A.N., Rielly, C.D. *The impact of direct nucleation control on crystal size distribution in pharmaceutical crystallization processes.* Crystal Growth & Design, 9(3), 1378–1384, 2009.

[7]   Zhou, G.X. et al. *Direct design of pharmaceutical antisolvent crystallization through concentration control.* Crystal Growth & Design, 6(4), 892–898, 2006.

[8]   Hulburt, H.M., Katz, S. *Some problems in particle technology. A statistical mechanical formulation.* Chemical Engineering Science, 19, 555–574, 1964.

[9]   Rawlings, J.B., Miller, S.M., Witkowski, W.R. *Model Identification and Control of Solution Crystallization Processes – a Review.* Industrial & Engineering Chemistry Research, 32, 1275–1296, 1993.

[10]   Szilágyi, B., Borsos, Á., Pal, K., Nagy, Z.K. *Experimental implementation of a Quality-by-Control (QbC) framework using a mechanistic PBM-based nonlinear model predictive control involving chord length distribution measurement for the batch cooling crystallization of L-ascorbic acid.* Chemical Engineering Science, 195, 335–346, 2019.

[11]   Rushton, A., Ward, A.S., Holdich, R.G. *Filtration Fundamentals*, in *Solid-Liquid Filtration and Separation Technology*, Wiley, Weinheim, 1996.

[12]   Whitaker, S. *Flow in porous media I: A theoretical derivation of Darcy's law.* Transport in Porous Media, 1, 3–25, 1986.

[13]   Mahdi, F.M., Holdich, R.G. *Laboratory cake filtration testing using constant rate.* Chemical Engineering Research and Design, 91, 1145–1154, 2013.

[14]   Nagy, B., Szilágyi, B., Domokos, A., Tacsi, K., Pataki, H., Marosi, G., Nagy, Z.K., Nagy, Z.K. *Modeling of pharmaceutical filtration and continuous integrated crystallization-filtration processes.* Chemical Engineering Journal, 413, 127566, 2021.

[15]   Szilágyi, B., Agachi, P.S., Nagy, Z.K. *Chord length distribution based modeling and adaptive model predictive control of batch crystallization processes using high fidelity full population balance models.* Industrial & Engineering Chemistry Research, 57, 3320–3332, 2018.

15 Case studies

15.1 Cement manufacturing control

Cement is one of the most important building materials in the world. Global cement production is expected to increase from 3.27 billion tons in 2010 to 4.83 billion metric tons in 2030 as stated in the U.S. Geological Survey (2018). The cement industry is one of the industries with the highest thermal energy consumption in the world, representing about 5% of global anthropogenic CO_2 emissions [1].

There are two main cement manufacturing processes: the dry and the wet process. In the dry cement manufacturing (DCM) process, the raw materials are crushed, conditioned, proportionally weighed, mixed, then fed into the kiln; the clinker obtained is treated and fined again. In the wet cement manufacturing (WCM) process, the raw materials are crushed, conditioned, and then directly mixed in the presence of water to make a fine thin paste known as slurry. The slurry is fed to the kiln. The WCM is a huge energy-consuming one and therefore, nowadays, it is rarely used. Our discussions will be related to the DCM.

Going to some details, DCM process requires mining and quarrying limestone, clay, marl, or shale. Actually, there are many receipts used by several manufacturers. The conditioning is done in the first stages after mining (crushing, blending, grinding). In some processes, this "raw meal" goes to pre-calcination and then, to calcination producing the clinker. The clinker is evacuated at about 1400–1450 °C, cooled, grinded again in cement mills, and processed again adding gypsum, fly ash, and other additives (Fig. 15.1) [2]. A detailed description of the cement production process can be found in [3].

The main subprocesses extremely interesting for monitoring and control are the blending and calcination process.

Through the whole process, the general transformation, from oxides to composites, is described in Fig. 15.2 [3].

15.1.1 Crushing process control

The limestone brought from the quarry is crushed in crushers, which are very high energy consumers! These crushers should be usually big and powerful enough to crush the big boulders; a possible solution is not to oversize the crushers, but to crush the boulders in several stages, needing more crushers. Another more intelligent one is to use a single equipment that is equipped with a speed control function of the dimensions and hardness of the stone (Fig. 15.3).

The output of the crusher load control system is the electric current absorbed by the crusher motor (CM). When the mechanical load of the crusher increases, the cur-

https://doi.org/10.1515/9783110789737-017

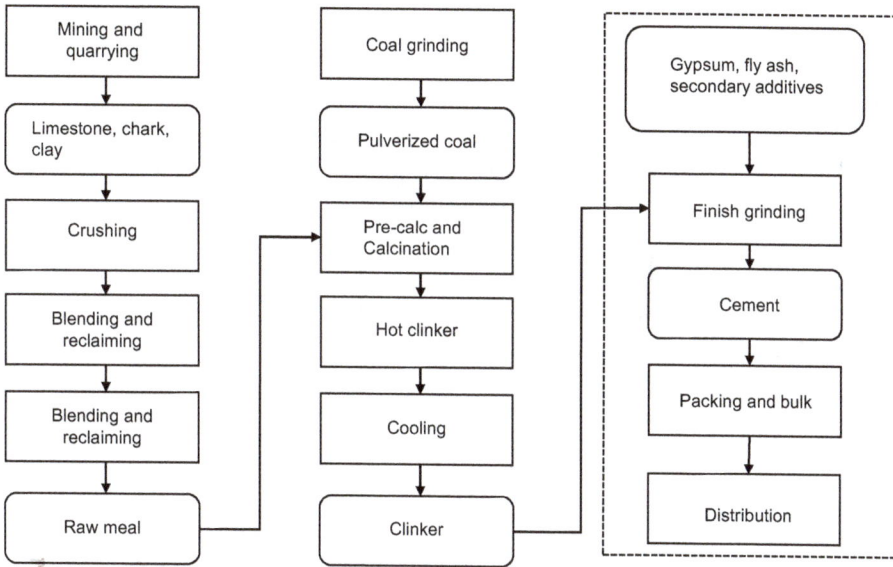

Fig. 15.1: Stages in cement manufacturing.

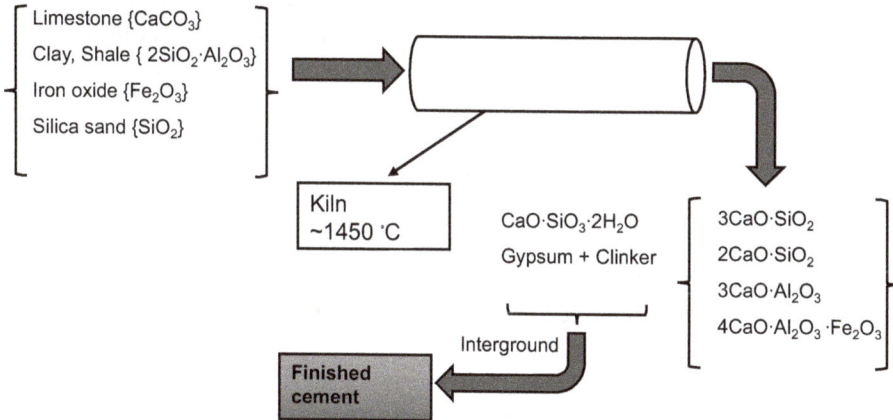

Fig. 15.2: General composition of cements.

rent demand increases, and the current load transducer (T) transmits the signal that "the supply speed has to be reduced" to the controller. The variable speed drive (VSD) will slow down the conveyor belt speed (circuit a). Usually, the controller is a PI controller with PB $\approx 30\%$ and $T_i = 1 - 2\,\text{min}$ [4]. This control loop can usually face load increases up to 25%; for the protection of the motor, a safety circuit (b) is added. In the controller structure of the loop a, the derivative action is not usually needed since no abrupt load disturbances are expected.

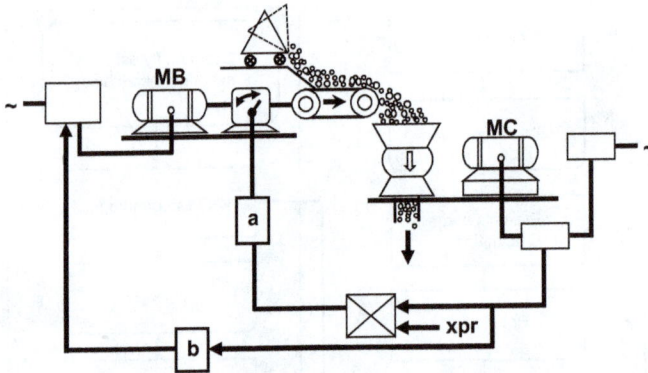

Fig. 15.3: Crusher control system.

15.1.2 Blending process control

The main components of the raw materials presented above are oxides: CaO (60–67%) in limestone mainly, SiO_2 (17–25%) in sand or shale/schist, Fe_2O_3 (0.5–6.0%) in pyrite, and Al_2O_3 (3–8%) in bauxite. There are other materials less important but each having a certain role (magnesium oxide (MgO), sulfur trioxide (SO_3), and alkalis (K_2O, Na_2O)) in conferring certain properties to the final product. It is very difficult to find only one raw material having all these oxides in the desired proportion. Thus, a mixture of several raw materials is used, and the proportioning of the oxides is done based on their content in each raw material [5, 6].

In Fig. 15.4, the raw material blending stage of the cement manufacturing is presented. The steady-state and dynamic characteristics of the process are as follows:
a. Big dead time of the blending stage
b. Big time constants
c. Important oscillations of the values of the quality indices and components of the raw material
d. Same oxides are in the composition of each raw material in different proportion
e. The changes in the process are caused by disturbances (oxides' content, humidity, caloric content, etc.)

The solution to these problems resides in applying computer control of the process which assures:
a. Quick reaction to disturbances (due to the anticipation of changes through a mathematical model of the process)
b. Constant quality of the blending
c. Decrease of the blending volumes (smaller investment and faster dynamic reaction)

d. Determining and signaling of the operational limits of the plant, transducers, of the parameters with slow feedback reaction, filtering the output signals, and calculation of averages of the measured parameters.

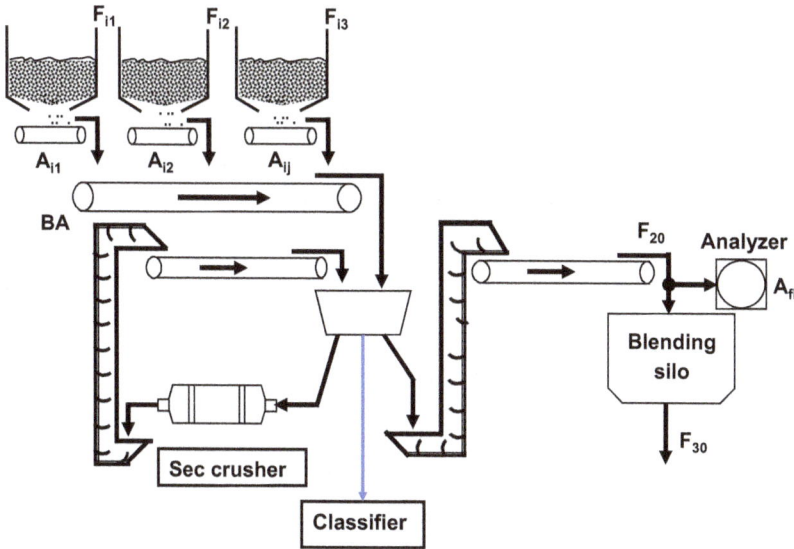

Fig. 15.4: Blending of raw materials.

N raw materials (depending on the receipt, e.g., limestone, clay, and marl) are available in N feeder tanks, in which S is the silica oxide content, A is the alumina oxide content, F is the iron trioxide content, and C is the calcium oxide content. S_i. A_i, F_i, C_i, where $i = \in 1, N$ are weight fractions of the oxides in each raw material [kg/kg]; S_m, A_m, F_m, C_m are the contents of the desired oxides in the mill before calcination [kg/kg]; and M_1, M_2, ..., M_N are the quantities delivered from each raw material tank to the mill [kg]:

$$
\begin{bmatrix} S_m \\ A_m \\ F_m \\ C_m \end{bmatrix} = \begin{bmatrix} S_1 \ S_2 \ldots \ldots S_N \\ A_1 \ A_2 \ldots \ldots A_N \\ F_1 \ F_2 \ldots \ldots F_N \\ C_1 \ C_2 \ldots \ldots C_N \end{bmatrix} \begin{bmatrix} M_1 \\ M_2 \\ M_3 \\ . \\ . \\ . \\ M_N \end{bmatrix}
\tag{15.1}
$$

For example, the composition of raw materials in one Lafarge cement factory is reported in [5].

Tab. 15.1: Example of composition of raw materials and final content of oxides in the blend.

Raw materials	SiO$_2$	Al$_2$O$_3$	Fe$_2$O$_3$	CaO	Weight
Quarry material	3.84	0.54	1.04	47.50	90.0
Yellow schist	75.30	7.07	3.49	4.40	7.60
Coal schist	35.57	18.44	4.28	2.86	1.30
Spain schist	6.70	0.75	32.45	0.70	1.10
Product composition	22.36	3.16	1.05	63.14	100.0

The composition in oxides is measured with an XRD taking samples from all raw materials and the meal. We are obliged to suppose that the compositions of the N raw materials are constant between samples, from t_{i-1} to t_i.

A possible system of control of the blending process is presented in Fig. 15.5 [7]. There are i raw materials in feed tanks $i = 1 \div N$ and j oxides in each tank, $j = 1 \div K$. Totally, there are N tanks and K oxides. In the example from Tab. 15.1, $N = 4$ and $K = 4$.

The mass balance on each component is given by the set of eqs. (15.2):

$$F_1 A_{11} + F_2 A_{12} + \cdots + F_N A_{1N} = F_{20} A_{1f}$$
$$F_1 A_{21} + F_1 A_{22} + \cdots + F_N A_{2N} = F_{20} A_{2f} \tag{15.2}$$
$$\ldots\ldots\ldots\ldots\ldots\ldots\ldots\ldots$$
$$F_1 A_{K1} + F_2 A_{K2} + \cdots + F_N A_{KN} = F_{20} A_{Kf}$$

where F_i are the flows from each feed tank, $i = 1 \div N$;

F_{20} is the summing flow from all feed tanks flows, going to the blending silo and then to the kiln;

A_{ij} are the concentrations of the jth oxide in the ith tank, $j = 1 \div K$; and

A_{if} are the final concentrations of the oxides before the blending silo.

The system of eqs. (15.2) can be written in matrix form as in (15.3):

$$\begin{bmatrix} A_{11} & A_{12} & \ldots & A_{1N} \\ A_{21} & A_{22} & \ldots & A_{2N} \\ & \ldots & & \\ A_{K1} & A_{K2} & \ldots & A_{KN} \end{bmatrix} \begin{bmatrix} F_1 \\ F_2 \\ \ldots \\ F_N \end{bmatrix} = F_{20} \begin{bmatrix} A_{1f} \\ A_{2f} \\ . \\ A_{kf} \end{bmatrix} \tag{15.3}$$

Thus, the changes in flowrates of the conveyor belts for each raw material corresponding to the deviation from the prescribed values can be calculated as follows:

$$
\begin{bmatrix} \Delta F_1 \\ \Delta F_2 \\ \ldots \\ \Delta F_N \end{bmatrix} = F_{20} \begin{bmatrix} A_{11} \ A_{12} \ \ldots \ A_{1N} \\ A_{21} \ A_{22} \ \ldots \ A_{2N} \\ \ldots\ldots\ldots\ldots\ldots \\ A_{K1} \ A_{K2} \ \ldots\ldots \ A_{KN} \end{bmatrix}^{-1} \begin{bmatrix} A_{1pr} - A_{1f} \\ A_{2pr} - A_{2f} \\ . \qquad . \\ A_{kpr} - A_{kf} \end{bmatrix} \tag{15.4}
$$

Consequently, a control scheme of the blending process measures the concentrations of the components in the raw material and in the raw meal, and the flow-rates of the raw materials and of the collector conveyor to the blending silo, calculates the composition using eq. (15.2), compares it with the desired one A_{ipr}, and corrects the influx F_i. Practically, it is a little more sophisticated cascade control system (Fig. 15.5).

The models:
Model (1) is the mathematical model of the collector conveyor belt.
Model (2) is the calculation of the concentrations of the oxides $C_j = F_{20}A_{jf}$.
Model (3) is the calculation of the changes of the flows of raw materials $[\Delta F_i] = F_{20}[A_{ji}]^{-1}[\Delta C_j]$ based on the error between the prescribed composition and the real measured one and $\Delta C_j = A_{jpr} - A_{jf}$.

Introducing the deviation of the concentration $A_{jpr} - A_{jf}$, through model IT (3), the change of the reactive signal of the outer controllers C_{1j} is calculated. These controllers fix the setpoints of the inner ones C_{2j} in the inner loop of the cascade (Fig. 15.6).
   PLANT$(\tau_i, \ T_i)$ is the controlled process with several conveyor belts.

Transducers:
T_j and T_A are analyzers measuring the composition of all oxides in the raw materials before blending and in the blend.
T_F measures the flows on the conveyor belts of each raw material tanks and of the collector conveyor belt.

Controllers:
C_{1j} are outer loop controllers controlling composition and giving a flow set point signal to the conveyorbelts' inner controllers.
C_{2j} are inner controllers controlling the raw materials' conveyor belts.

Final control elements:
FCE$_i$ is the conveyor belt with VSD for each raw material tank.

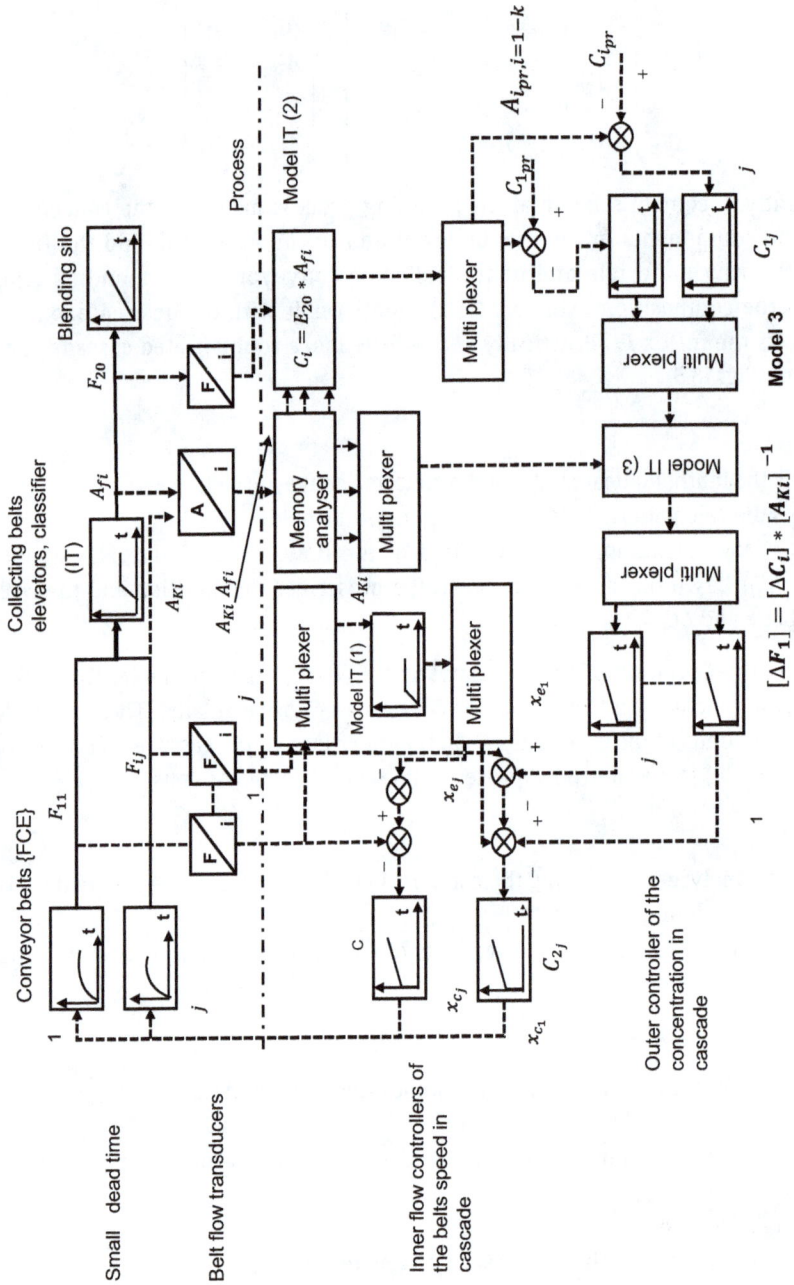

Fig. 15.5: The control system of the blending stage of the cement manufacturing [7].

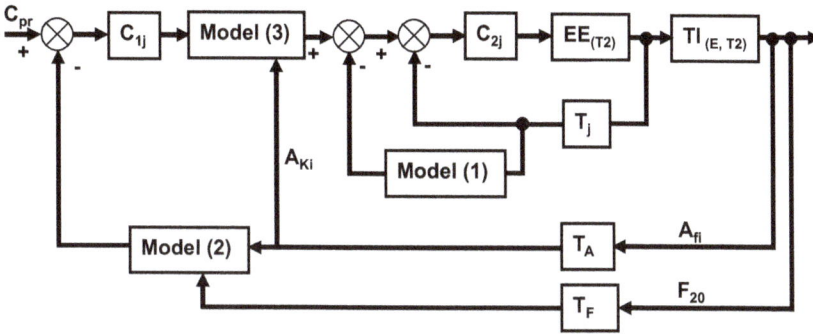

Fig. 15.6: Equivalent simplified cascade scheme using mathematical models IT(l) with $l = 1, 3$.

Mathematical models:

Model 1 – dynamic model of raw materials' conveyor belts. They may be of the form $\frac{Ke^{-\tau s}}{1 + Ts}$.

Model 2 – steady-state model of the collector conveyor belt $C_j = F_{20}A_{jf}$.

Model 3 – steady-state model calculating the changes needed for the setpoints of the inner controllers of the raw materials flows $[\Delta F_i] = F_{20}\left[A_{ji}\right]^{-1}\left[\Delta C_j\right]$.

To speed up the response of the inner loop, a predictive dynamic model, Model (1), of the conveyor belts (including the presence of the dead time and the time constant of each belt) is used. Its role is to prove the changes proposed by the C_{1j} controllers for reaching the appropriate concentration values, using F_i. In case Model (1) confirms the return of the concentrations at the prescribed values, the computer does not give any correction signal for the prescribed values to the C_{2i} controllers. If the values calculated are not the correct ones, the computer gives the necessary corrections for the setpoints of the inner flow loops.

The control system presented is a multivariable/cascade/inferential/adaptive control system.

In some applications, the setpoints are not only simple values of parameters, but also calculated variables (15.5), or even sometimes, some quality indices [8]. Such indices, known in the cement industry as cement moduli, are defined in terms of the proportion of limestone (CaO), silica SiO_2, iron Fe_2O_3 and bauxite Al_2O_3 as follows:

Lime saturation factor: LSF and MI have to be written in the same way as MS and MF with CaO at the numerator of the fraction

$$LSF = CaO2.8\ SiO_2 + 1.18Al_2O_3 + 0.65Fe_2O_3$$

Hydraulic modulus:

$$MI = CaOSiO_2 + Al_2O_3 + Fe_2O_3 \tag{15.5}$$

Silica modulus:

$$MS = \frac{SiO_2}{Al_2O_3 \; + \; Fe_2O_3}$$

Alumina modulus:

$$MF \; = \; \frac{Al_2O_3}{Fe_2O_3}$$

The control system presented in [9] is focused on one module, the LSF. The paper reports an adaptive control system to cope with this situation. But the difference is only at the setpoint of the outer controller: in the previously presented situation, the prescribed concentrations $A_{ipr}, i = 1 - K$ of the components are the setpoints; in the last presented application, the setpoint is LSF_{pr}.

Model (3) is replaced with the calculation of the prescribed composition (the concentrations of the four oxides in the blend, S, F, A, C) to assure the prescribed value of the LSF, manipulating the K flows of raw material.

15.1.3 Clinker manufacturing process control

This paragraph refers to the processes taking place mainly in the rotary cement kiln.

A quite recent paper [10] exposes the principles of modern control of temperature of the rotary cement kiln. The "modern" control principles do not differ very much from the traditional ones but implement newer control technologies as adaptive or Model Predictive Control (MPC). In all applications, it is very important to understand and describe, using a mathematical model, the process of clinker manufacturing. In this paragraph we introduce the burning process first, the original control technology, very largely used even today, and later, the newer control approaches, which are very valuable energetically and qualitatively.

After grinding and blending, the feed material is conveyed through a drying chamber to the kiln (Fig. 15.7).

The raw meal enters first in a drying chamber (7) and then, on a metallic conveyor in the preheater (8); these two chambers are heated by the exhaust gas passing through the material in its path to the rotary kiln. The dust laden flue gas is passed through a dust collector with filter bags (5) and, in the end, through an electrostatic filter (6). In this way, most of the dust is retained in the bags and filter and the pollution is minimized.

If the thermal inflow of the fuel, the primary and secondary air (air blown in the kiln for burning and from the grate cooler) are mainly constant, the rotation speed of the rotary kiln and the transporting speed of the grates at the input and output of the kiln are constant, the heat balance (Fig. 15.8) remains constant, and the quality of the clinker is assured. Therefore, the control systems of the burning process are mainly focused on cancelling all possible disturbances.

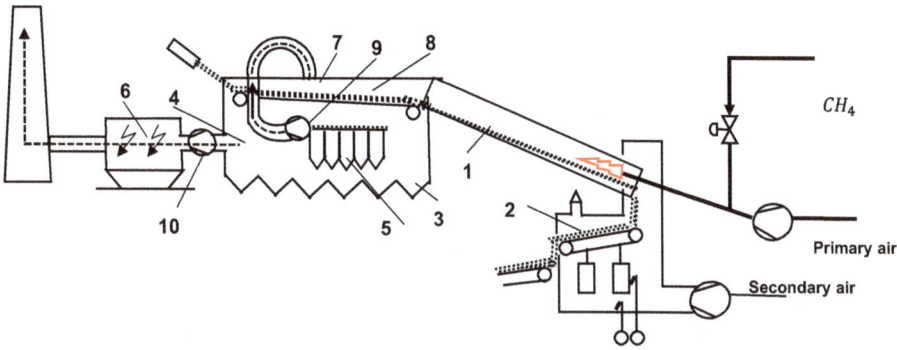

Fig. 15.7: The rotary kiln with drying and preheating chambers, cyclones, electrostatic filters, and cooling chamber: 1, rotary kiln; 2, clinker cooling chamber; 3, 4, suction (draught) chambers; 5, de-duster; 6, electrostatic filter; 7, drying chamber; 8, preheating chamber; 9, 10, blower; 11, chimney.

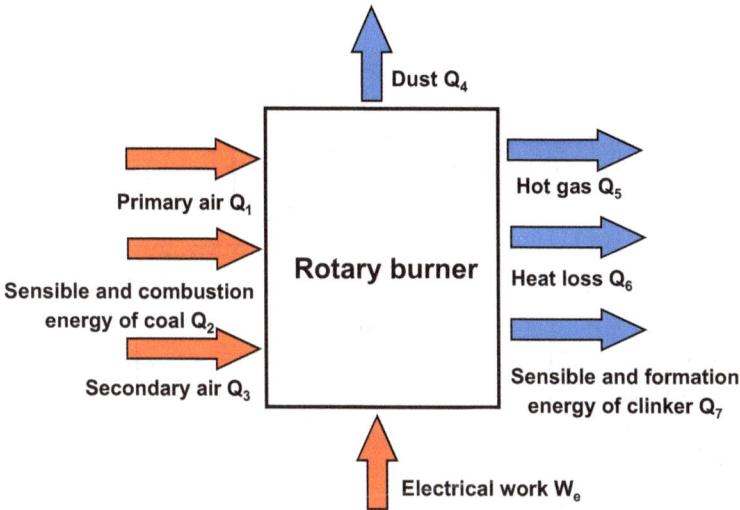

Dust Q_4

Primary air Q_1

Sensible and combustion energy of coal Q_2

Secondary air Q_3

Rotary burner

Hot gas Q_5

Heat loss Q_6

Sensible and formation energy of clinker Q_7

Electrical work W_e

Fig. 15.8: The energy balance of the rotary kiln [11]. Farine is the raw material for the cement manufacturing.

What happens in the rotary kiln [11, 12]?

There are three temperature zones in the rotary kiln: heating zone, burning zone (BZ), and cooling zone. All behavior in these zones is influenced by the secondary processes taking part in the pre-burning and post-burning process. It is vital to control the process in the BZ, but we must understand that the control of the clinker manufacturing is not limited to the burning zone temperature (BZT). But what happens in and before the kiln [12]?

Free water of the raw meal evaporates between 70 and 110 °C Between 400 and 600 °C, the minerals are decomposed in their constituent oxides, mainly SiO_2 and

Al_2O_3. Dolomite $(CaMg(CO_3)_2)$ decomposes to $CaCO_3$, MgO and CO_2. Between 650 and 900 °C, $CaCO_3$ reacts with SiO_2 to form belite (Ca_2SiO_4). Between 900 and 1050 °C, $CaCO_3$ decomposes to CaO and CO_2. At temperatures between 1300 and 1450 °C, 20–30% of the material melts, and belite reacts with CaO, to form alite $(Ca_2O \cdot SiO_4)$. Alite is the characteristic constituent of the cement. At 1400–1450 °C, the reaction is complete, and the partial melting produces nodules of typically 1–10 mm diameter. This is the clinker. According to Ferdoush and Gonzalez [12] and Ostergaard et al. [13], the finished clinker is 65% alite (tricalcium silicate) and 13% belite (dicalcium silicate). There are other chemical compounds in its composition including 1.2% free CaO. At the outlet, the clinker is cooled in a cooler. The clinker leaves the cooler as up-to-fist-sized sinter and is then mixed with additives such as gypsum and ground again to produce cement. The clinker phase Alite determines the main properties of cement, as it hardens rapidly in combination with water.

In the following sequence, the traditional and still largely used control system is presented:

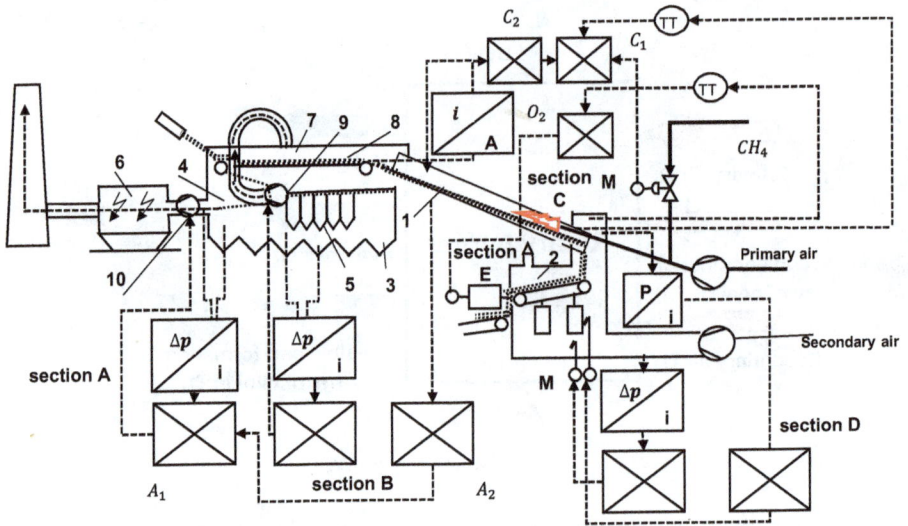

Fig. 15.9: Clinker manufacturing control system.

A. Draught between chamber 4 and flue gas channel control
Function of exhaust gas flow (see Fig. 15.8 of the heat balance), the burning diagram (Fig. 15.10) is respected or not. A smaller-than-needed draught makes the preheating of the material insufficient; on the contrary, if the draught is bigger, the preheating of the raw meal is steeper, producing unintended side effects. At the same time, the dust retention in the bag filters (5) is less efficient. The control of the draught is achieved by measuring the differential pressure between the chamber 4 and the exhaust gas pipe and manipulating the exhaust gas blower 10. At a constant draught, the burning

becomes incomplete if the oxygen content varies around its optimal value (Fig. 15.11) with unpleasant consequences on the fuel consumption and the quality of the clinker plus higher pollutant emissions. Therefore, the setpoint of the draught is controlled in a cascade with the outer oxygen analyzer (Figs. 15.9 (section A) and 15.12).

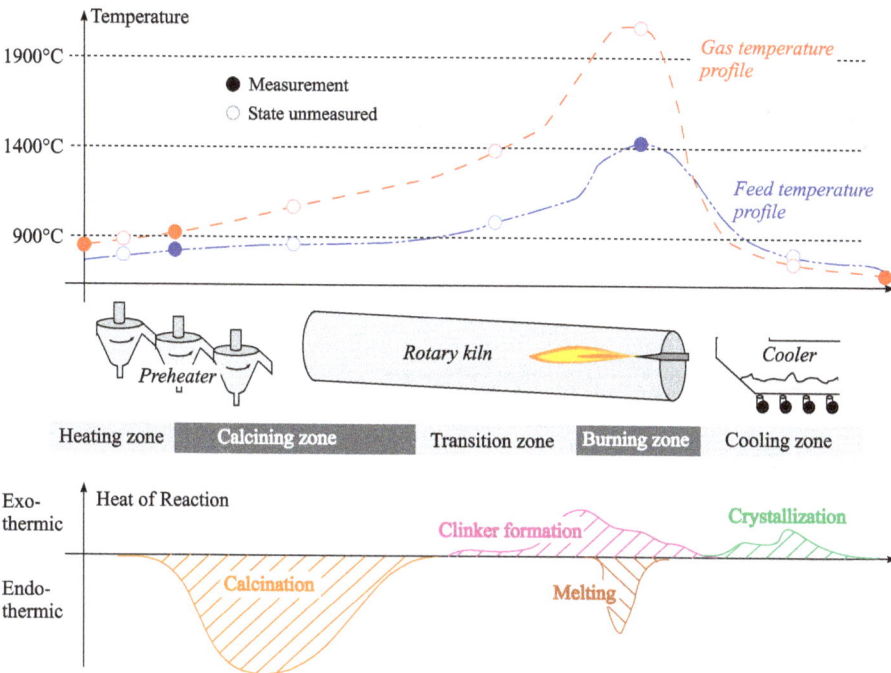

Fig. 15.10: Temperature diagram and processes in a cement kiln [14]. (Reprinted with the permission of Elsevier)

The ultimate control variable is the oxygen content of the flue gas.

B. Differential pressure control between chambers 3 and 4 (Fig. 15.9)
An additional measure to stabilize the burning profile is controlling the differential pressure between chambers 3 and 4 and manipulating the blower 9 speed. But, assuring a constant differential pressure between the two chambers, the undesired transversal currents are avoided; these transversal currents cause a very unpleasant effect of transporting a high dust quantity from chamber 3 in chamber 5 (disturbing the dust retention). On the other hand, the passage of the flue gas with lower temperature from chamber 4 to chamber 3 and thus through the drying chamber 8 disturbs the drying process causing deviations of the humidity of the raw material from the desired value (<1% for the dry process) and the input kiln temperature (about 900 °C); see Fig. 15.9 (section B).

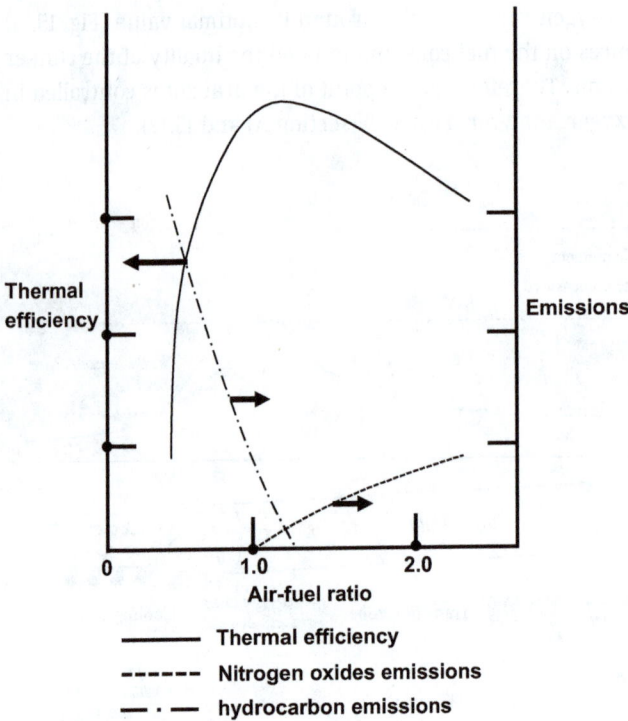

Fig. 15.11: Burner efficiency depends on the air/oxygen to fuel ratio [15].

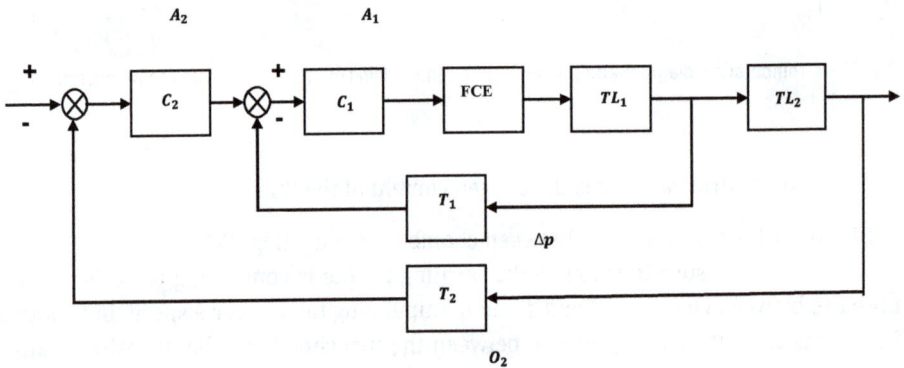

Fig. 15.12: Cascade system for the cement kiln draught control.

C. Control of the temperature diagram along the rotary kiln

The determining temperature for the temperature profile is that from the burning/ sintering zone, near the burner. The usual control is done through measuring the temperature in this zone and acting on the fuel valve or in a ratio of fuel/primary air control system (Fig. 15.9, section C). But, because the temperature change is detected with delay (slow process with big time constant), sometimes, when possible, the measured variable is the oxygen content of the flue gas. The oxygen transducer is common for the control systems A and C. To accelerate the behavior of this very slow process (big quantities of clinker involved, huge refractory brick mass of the kiln walls accumulating heat), the adopted system is usually a cascade one (Fig. 15.13).

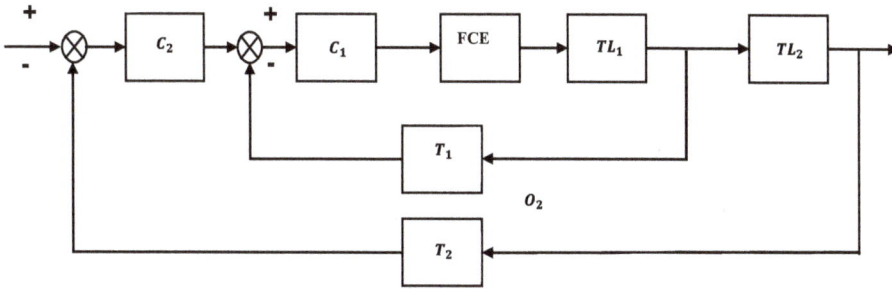

Fig. 15.13: Cascade control of the temperature in the burning zone.

But this is a peculiar cascade control scheme, since the fast-changing variable (O_2 content) is the external one, imposing the setpoint for the internal (temperature) loop. But, as we stated before, some controls do not manipulate the fuel flow, but the ratio of fuel/air.

D. Pressure control in the kiln vault.
The pressure in the vault is important because it determines the length and position of the flame in the BZ. Its control is done measuring the pressure in the vault and manipulating the secondary air flow passing through the clinker cooling chamber using butterfly valves (Fig. 15.9, section D).

E. Temperature control in the kiln vault.
This temperature not only influences the BZT but also indicates the heat quantity lost with the clinker; if the vault temperature is smaller than the prescribed one, this means that the clinker is evacuated too hot outside, in the surrounding environment. This can be controlled reducing the speed of the grate conveyor belt (Fig. 15.9, section E) passing through the clinker cooler (2). In general, the clinker cooler fulfils the following functions [11]: evacuates the clinker from the kiln, allowing its continuous operation; cools the clinker to the handling temperature; preheats the secondary and tertiary air; and recovers energy from the clinker. The temperature of evacuation in the clinker cooler negatively affects the specific energy consumption.

15.1.4 Grinding and milling process control

After the clinker is obtained, there are other processes that influence its quality. Grinding and milling of the clinker [16, 17]:

The ball mill is the most common process for cement grinding (Fig. 15.14). The clinker, possibly crushed in advance in a roll crusher to achieve a preliminary size reduction, is mixed with up to 5% gypsum and/or natural anhydrite. Fine grinding using ball mills is in general extremely energy inefficient. Many plants use a roll crusher to achieve a preliminary size reduction of the clinker and gypsum. Just 4% of the energy available is efficiently used for grinding [16]. Usually, PID control or linear predictive control is used [17].

The reason for such a low efficiency is either the under load (causing energetically inefficient use) or overload (causing an insufficient comminution degree). The optimum load is indicated by the inner noise of the mill. The noise intensity is around 100–110 Phons. Using a microphone (M), the load control system can identify variations of 0.2 Phons relative to the usual noise level. If the noise intensity varies from the prescribed value, a VSD of the feeding disc is addressed by the PID controller to accommodate the feed with the optimal load of the ball mill.

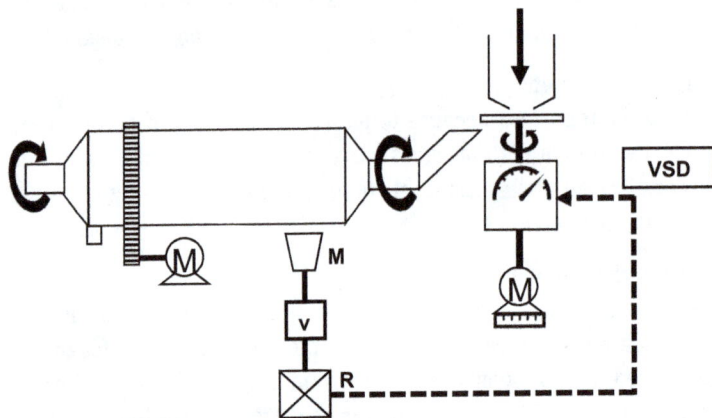

Fig. 15.14: The grinding process control in ball mills.

These are the traditional most used control systems for the manufacturing of cement.

Although we noticed how complicated the whole process of cement manufacturing is, some of the authors researched especially new developments of the cement temperature control philosophy, which use the techniques of MPC, adaptive, fuzzy algorithms [2, 6, 10, 14,18–20]. Reference [20] is especially interesting since it approaches a robust, resilient control in the case of sensor failure.

From several relevant papers on MPC, we chose to describe the approach from [14] because the results were also tested in practice.

For MPC, one must have a dynamic mathematical model.

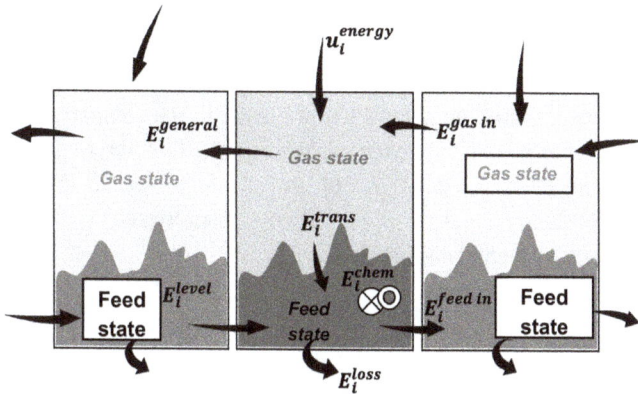

Fig. 15.15: Virtual sections of the rotary kiln. The feed and gas temperature are denoted by feed and gas state.

The rotary kiln can be virtually described by three sections (Fig. 15.15).

The behavior of the compartments is described by an energy and mass balance. It has energy sources and energy sinks related to feed and gas transport, combustion of fuels (where appropriate), losses to the ambient, energy sources, or sinks related to the chemical reactions (Fig. 15.8).

The thermodynamic processes considered here are nonlinear and are described by the set of eqs. (15.6)–(15.9).

It is easier to section the kiln in several zones/compartments (those in Fig. 15.10): heating zone (HZ), calcining zone (CaZ), transition zone (TZ), burning zone (BZ), and cooling zone (CoZ). Each compartment is divided into solid and gas phase with corresponding mass and heat balance equations. Homogenous mass distribution is assumed inside the compartments. The average mass density $\rho[t/m]$ is supposed to be the same in one compartment. For each compartment, equations are written. Solid mass transport dynamics:

$$L_i \frac{d\rho}{dt} = F_{mi} - \rho v \tag{15.6}$$

where $\rho_s[t/m]$ is the average solid mass density, $L_i[m]$ is the compartment length, $F_{mi}[t/h]$ is the solid feed input, $v[m/h]$ is the velocity of the material in the compartment and is proportional to the rotation speed of the kiln, and $\rho_s v[t/h]$ is the feed output rate from the compartment.

The dynamic behavior depends on the length of the compartment, the input and output feed rate; all terms are expressed in t/h.

Solid flow temperature dynamics:

$$L_i c_{ps} \rho_s \frac{dT_s^\circ}{dt} = L_i c_{ps} F_{mi}(T_{si}^\circ - T_{so}^\circ) + L_i K_{gs}(T_g^\circ - T_s^\circ) + v\rho r \Delta H + K_T A_T (T_s^\circ - T_{ext}^\circ) \qquad (15.7)$$

where $c_{ps}[MJ/t \cdot {}^\circ C]$ is the specific heat capacity of the solid feed; T_{si}°, T_{so}°, and T_s° [°C] are the solid input, output temperature, and temperature of the solid in the compartment, respectively; T_g°[°C] is the gas temperature in the compartment; T_{ext}°[°C] is the temperature of the exterior of the compartment; $K_{gs}[MJ/m \cdot h \cdot {}^\circ C]$ and $K_T[MJ/m^2 \cdot h \cdot {}^\circ C]$ are the gas-solid and kiln-outer environment heat transfer coefficients, respectively; $A_T[m^2]$ is the heat transfer area of the compartment considered.

Gas temperature dynamics:

$$L_i c_{pg} \rho_g \frac{dT_g^\circ}{dt} = L_i c_{pg} F_g (T_{gi}^\circ - T_g^\circ) + L_i K_{gs}(T_g^\circ - T_s^\circ) + v\rho r \Delta H + K_T A_T (T_g^\circ - T_{ext}^\circ) + \Delta H_{fuel}$$

$$(15.8)$$

where $c_{pg}[MJ/t \cdot {}^\circ C]$ is the specific heat capacity of the gas or air; $\rho_g[t/m]$ is the average gas density; T_{gi}°, T_g°[°C] is the gas input and output temperature in the compartment; $F_g[t/h]$ is the gas flow (related to the exhaust blower speed); and $\Delta H_{fuel}[MJ/h]$ is the heat produced from burning the fuel (related to the fuel type and flow).

Heat transfer in rotary kilns is usually done through radiation described by Stefan-Boltzmann law and depending mainly on the difference $\left(T_g^\circ\right)^4 - \left(T_s^\circ\right)^4$. Conduction and convection can be modeled using the electrothermal approximation assimilating the heat transport with the electricity transport through resistors.

Oxygen dynamics:
This simple oxygen model, calculating the oxygen content in the exhaust gas is added to ensure the correct combustion:

$$\frac{dC_{O_2}}{dt} = a^{oxy} C_{O_2} + k^{oxy} F_g - c^{oxy} F_{fuel} \qquad (15.9)$$

where C_{O_2}[%] is the oxygen concentration in the exhaust burning gas; a^{oxy}, k^{oxy}, c^{oxy} are parameters describing the oxygen transducer dynamics (a^{oxy}), the relationship between the oxygen source and the gas/air draft through the kiln system (k^{oxy}), and the oxygen sink related to combustion (c^{oxy}).

Thus, the whole model is linearized and with the "technology" presented in Chapter 2 of the present book, mainly in Section 2.4, the control algorithm calculates the input move vector for the predicted horizon (eq. 2.48). In reference [14], the above-mentioned procedure is detailed in control problem formulation: model linearization, moving horizon state estimation, capturing model uncertainty, reducing the estimation problem, and optimal control problem.

The results of the MPC applied at a Holcim cement plant in Switzerland (2009) are presented in Fig. 15.16 [14, Fig. 5].

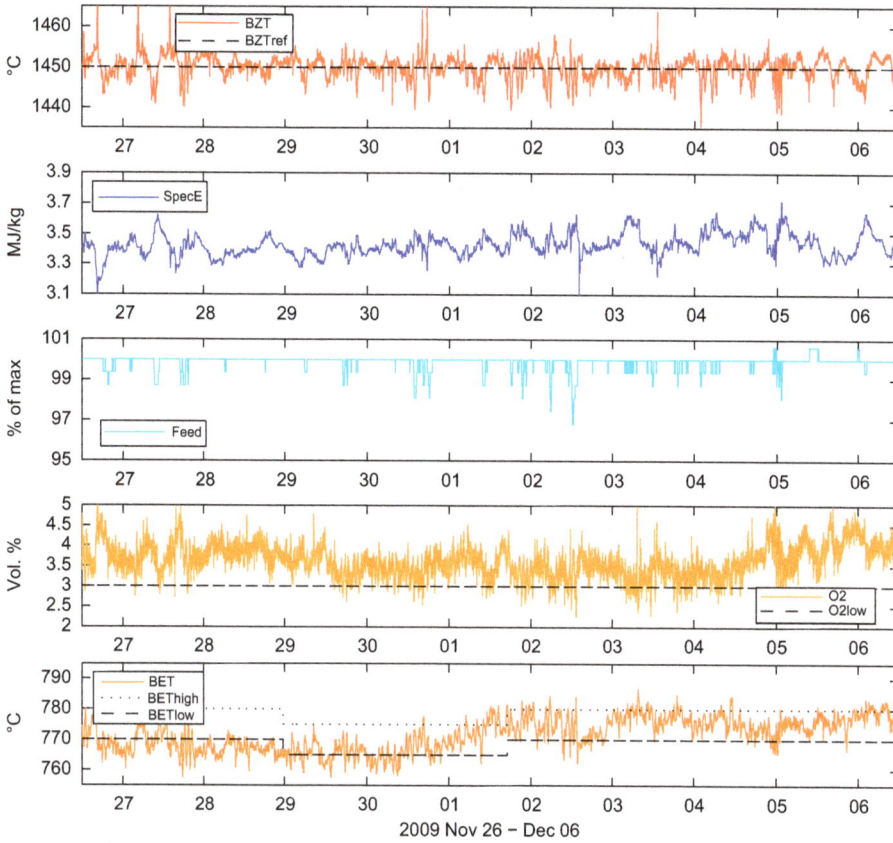

Fig. 15.16: The results presented by Stadler et al. in [14] regarding the efficiency of the MPC. Top subplot: BZT temperature and BZT reference – black; second subplot: feed rate in percentage of maximum feed rate; third subplot: specific energy input; fourth subplot: oxygen level at the exit – solid line and lower. Oxygen soft constraint value – dashed line; lower subplot: back-end temperature (BET) – solid line, and lower and higher constrained values – dashed and dotted, respectively. (Reprinted with the permission of Elsevier [14])

A comparison was also made at the same time between manual and MPC and it is shown in Fig. 15.17.

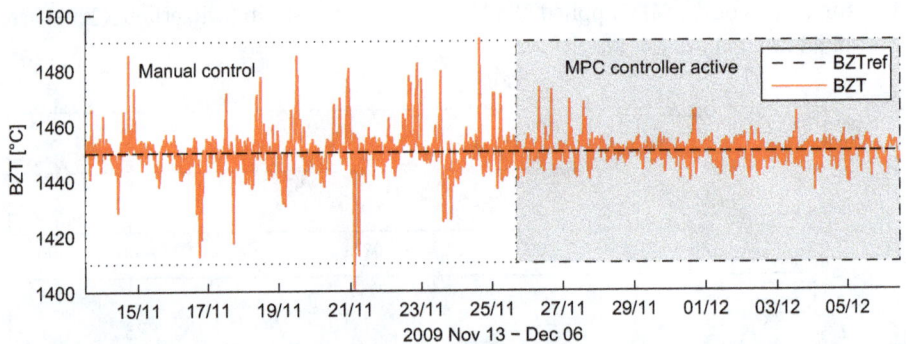

Fig. 15.17: BZT – solid line, and BZT reference – dashed line in the two cases of manual and MPC. The results of MPC are obviously superior reducing the temperature variability in a range between practically −10 and +15 °C. (Reprinted with the permission of Elsevier [14])

Bibliography

[1] Atmaka, A., and Yumrutas, R., *Analysis of the parameters affecting energy consumption of a rotary kiln in cement industry*. Applied Thermal Engineering, 66, 435–444, 2014.

[2] Sellito A.M., Balugani, E., Gamberini, R., Rimini, B., *A Fuzzy Logic Control application to the Cement Industry*, IFAC PapersOnLine, 51(11), 1542–1547, 2018.

[3] Chatterjee, A.K., *Cement Production Technology*, CRC Press, 2018

[4] Agachi, P.S, *Automatizarea proceselor din Industria Materialelor d e Construcții (The Automation of the Processes in the Building Materials Industry)*, Ed. University Babes-Bolyai, p. 272, 1976.

[5] Lin, P., Yun, Y.S., Barbier, J.P., Babey, Ph., Prevot, P., *Intelligent tuning and adaptive control for cement raw meal blending process*, in *IFAC Conference of Intelligent Tuning and Adaptive Control*, Singapore, 1991

[6] Lin, P., Miquel, M., Prevot, P., Barbier, J.P., *Two-level adaptive control system for a cement raw material mixing process*, in *IFAC Conference in Mining, Mineral and Metal Processing Conference*, Buenos Aires, Argentina, 317–320, 1989.

[7] Agachi, P.S., *Automatizarea proceselor din Industria Materialelor d e Construcții (The Automation of the Processes in the Building Materials Industry)*, Ed. University Babes-Bolyai, p. 251, 1976.

[8] Bittanti, S., Frachini, A., Lovera, M., Manigrasso, R., *Adaptive control of raw material mix in cement plants*, pp. 1269–1274 in *Conference IFAC Control of Industrial Systems*. Belfort, France, 1997.

[9] Wurzinger, A., Leibinger, H., Jakubek, S., Kozek, M.,*Data driven modeling and model predictive control design for a rotary cement kiln*, IFAC PapersOnLine, 52(16), 759–764, 2019.

[10] Anjana C., Dinakar D., *Adaptive fuzzy logic controller for rotary kiln control*, International Research Journal of Engineering and Technology (IRJET), 03(10), Oct 2016

[11] Atmaka, A., Yurumtas, R., *The effects of grate clinker cooler on specific energy consumption and emissions of a rotary kiln in cement industry*, International Journal of Exergy, 18(3), 371–386, 2015.

[12] Locher, F.W., *Zement: Grundlagen der Herstellung und Verwendung*. Verlag Bau+ Technik, 2015.

[13] Gazzani, M., et al, *CEMCAP framework for comparative techno-economic analysis of CO2 capture from cement plants*, May 2017, EU project, Horizon, 2020

[14] Stadler, K.S., Poland, J., Gallestey, E., *Model predictive control of a rotary cement kiln*, Control Engineering Practice, 19, 1–9, 2011.

[15] Edgar, T.F., Himmelblau, D., M., Lasdon, L.S, *Optimization of chemical processes*, 2nd Edition, McGraw
Hill, p. 11, 2001.

16   Prasath, G., Recke, B., Chidambaram, M., Jørgensen, J.B, *Soft Constrained based MPC for Robust
Control of a Cement Grinding Circuit*, pp. 475–480 in *10th IFAC International Symposium on Dynamics
and Control of Process Systems, The International Federation of Automatic Control*. Mumbai, India,
December 18–20, 2013.

[17] Chen, X.S., Li, Q., Fei, S.M. *Constrained model predictive control in ball mill grinding process*, Powder
Technology, 186, 31–39, 2008).

[18] Feliu-Battle, V., Rivas-Perez, R., *Design of a robust fractional order controller for burning zone
temperature control in an industrial cement rotary kiln*, IFAC papers OnLine, 53(2), 3657–3662, 2020.

[19] Mujumdar, K.S., Ranade, V.V., *Simulation of rotary cement kilns using a one-dimensional model*,
Chemical Engineering Research and Design, 84(3), 165–177, 2006.

[20] Veersamy, G., Amirtharajan, R., et al., *Integration of genetic algorithm tuned adaptive fading memory
kalman filter with model predictive controller for active fault-tolerant control of cement kiln, under sensor
faults with inaccurate noise covariance*, Mathematics and Computer Simulation, 191, 256–277, 2022.

15.2 Control of fluid catalytic cracking process

Fluid catalytic cracking (FCC) is the last step in the evolution of cat cracking processes – also introduced in 1942 – just like TCC or thermafor cat cracking, during the
Second World War in an effort to make high-octane number gasoline to produce high
power related to high compression ratios in the combustion engines [1]. There are
several versions of FCC units (FCCU), with different configurations, researched and
implemented by several petrochemical companies like Kellogg Brown and Root,
Exxon Mobil Research and Engineering, Shell Global Solutions International, Institut
Français du Petrole (IFP), and Honeywell Universal Oil Products (UOP) [1].

FCCU is composed of the cracking reactor (with the feed preheater), the regenerator
with the air blowers). The reactor and the regenerator are connected by the regenerated
and spent catalyst pipelines. Usually, this setup is accompanied by a fractionator column
and a wet gas compressor for the tail gas collected at the top of the column (Fig. 15.18).
The fractionator separates several products from top to bottom: gas, gasoline, light cycle
oil, heavy cycle oil, and, finally, the heaviest fractions, decant oil (slurry).

Different configurations of the FCCUs are presented in [1].

The feed, that is gas oil preheated to about 150 °C, is mixed with the slurry from the
main fractionator and introduced into the reactor with steam. The riser part of the reactor where the hot fresh catalyst (fine ground zeolite particles) meets the fresh feed temperature that is around 520 °C. The cracking reactions on zeolites flowing with the
reactants takes place in seconds, having fast kinetics. The products are sent to a fractionator after going through a series of cyclones, obviously, to separate the fluid products
from the particles of the catalyst. The products are separated by the main fractionator in
gas, gasoline, light cycle oil, heavy cycle oil, and, finally, the heaviest fractions, slurry.

At the same time, in the catalytic cracking reactor, the burnt hydrocarbons are
transformed in coke deposit on the catalyst, deactivating it. The so-called spent cata-

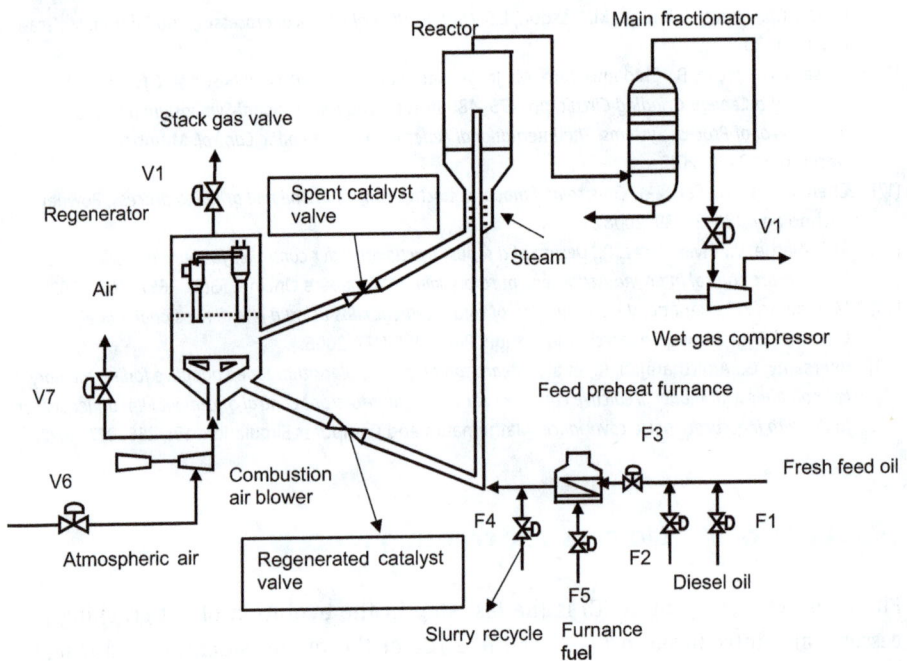

Fig. 15.18: FCCU, UOP design [2].

lyst (zeolite covered with coke), is sent through the spent catalyst pipeline to regenerator, where the coke, containing carbon and hydrogen, from the catalyst surface is burnt in air and the regenerated catalyst is sent back to the reactor, closing the catalyst cycle.

The temperatures in the regenerator could reach 700–750 °C. The combustion products or flue gases from the regenerator could be sent to a CO boiler because the gas may contain significant amount of carbon monoxide, which could be burnt to CO_2 to provide additional heat used further in the process.

So the catalysts that are now regenerated are sent to the reactor to close the catalyst cycle and to meet the fresh feed for the cracking reaction.

The reactions taking place in both units discussed are [3]:

Reactor:
In the catalytic cracking reactor, two types of catalytic reactions take place:
–   Primary cracking of the gas oil molecules
–   Secondary rearrangement of re-cracking of cracked products

The products are summarized in Tab. 15.2.

To define the controlled variables, we took the simplest description of the process from Luyben and Lamb [4, 5]. Although the model is quite simplified it predicts the

Tab. 15.2: Products of catalytic cracking.

Hydrocarbon type	Catalytic cracking
n-paraffins	C_3 – C_6 is the major product; few olefines above C_4
Olefins	Large amount of aromatics formed with aliphatics at 500 °C
Naphtenes	Crack at the same rate as paraffins
Alkyl-aromatics	Crack next to ring

changes of the dependent variables and indicates the order of magnitude estimate of these variables.

The authors assume that the reaction represented by eq. 15.10 occurs in the reactor:

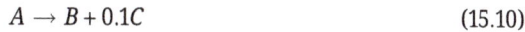

$$A \rightarrow B + 0.1C \tag{15.10}$$

where A, B, and C represent gas oil, gasoline, and coke involved in the reaction, respectively. The coke combustion taking place in the regenerator is described by (15.11):

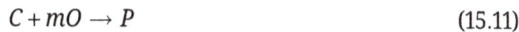

$$C + mO \rightarrow P \tag{15.11}$$

where O and P represent oxygen and combustion products and m is a stoichiometric coefficient depending on the degree of completion of carbon combustion and the percentage of hydrogen in coke.

The simplifying assumptions of the model are:
– The reactor and the regenerator are perfectly mixed.
– The heat capacities are independent of composition and temperature.
– The catalyst holdup remains constant in both reactor and regenerator.
– The reaction rates are first order, and the rate of coke combustion is independent of the amount of coke.
– The multitude of reactions presented later in this chapter (Fig. 15.23) is considered formed of one reaction.

The mass and heat balance equations of both reactor and regenerator are the sets (15.12) and (15.13).

Reactor:
Mass and component balance:

$$\frac{dN_r}{dt} = F_{gfr} - F_{gr}$$

$$\frac{dN_r y_2}{dt} = F_{gfr} y_{2f} - F_{gr} y_2 + N_r (1 - y_2) A_1 e^{-\frac{E_1}{RT_r}}$$

$$M_r \frac{dX_r}{dt} = F_c X_{rg} - F_c X_r + 0.1 N_r (1 - y_2) A_1 e^{-\frac{E_1}{RT_r^\circ}} \tag{15.12}$$

Heat balance:

$$c_{ps} M_r \frac{dT_r^\circ}{dt} = c_{pgfr} F_{gfr} T_{ir}^\circ - c_{ps} F_{sc} T_r^\circ - c_{pgr} F_{gr} T_r^\circ + c_{ps} F_{regc} T_{reg}^\circ - (\Delta H_1) N_r (1 - y_2) A_1 e^{-\frac{E_1}{RT_r^\circ}}$$

Regenerator:

Mass and component balance:

$$\frac{dN_{reg}}{dt} = F_{afreg} - F_{greg}$$

$$\frac{dN_{reg} y_O}{dt} = F_{afreg} y_{Of} - F_{greg} y_O - m N_{reg} y_O A_c e^{-\frac{E_1}{RT_{reg}^\circ}}$$

$$M_{reg} \frac{dX_{rg}}{dt} = F_{sc} X_r - F_{regc} X_{reg} + N_{reg} y_O A_c e^{-\frac{E_c}{RT_{reg}^\circ}} \tag{15.13}$$

Heat balance:

$$c_{ps} M_{reg} \frac{dT_{reg}^\circ}{dt} = c_{pareg} F_{afreg} T_{afreg}^\circ + c_{ps} F_{sc} T_r^\circ - c_{pgreg} F_{greg} T_{reg}^\circ - c_{ps} F_{regc} T_{reg}^\circ$$

$$- (\Delta H_2) N_{reg} y_O A_c e^{-\frac{E_c}{RT_{reg}^\circ}}$$

where N_r, N_{reg} represent gas holdup in the reactor and the regenerator, respectively (mol); $F_{gfr}, F_{gr}, F_{afreg}, F_{greg}$ represent molar gas flowrates of fresh feed in the reactor, gas flowrate from the reactor, air flowrate to the regenerator, gases flowrate from the regenerator (mol/s); F_{sc}, F_{regc} represent spent and regenerated catalyst flowrate (kg/s); A_i, A_c represent preexponential factor in Arrhenius equation; $i = 1$ for cracking to gasoline; index c for coke burning; E_i, E_c represent the energy of activation; $i = 1$ for cracking to gasoline; index c for coke combustion (J/mol); $T_r^\circ, T_{ir}^\circ, T_{reg}^\circ, T_{afreg}^\circ$ represent temperature in the reactor, temperature of the gas oil in the reactor, temperature of the regenerator and temperature of the air feed to regenerator (K); M_r, M_{reg} represent catalyst holdup in the reactor and the regenerator (kg); X_r, X_{reg} represent coke deposited on catalyst in the reactor and the regenerator (mol/kg cat); $c_{ps}, c_{pgfr}, c_{pgr}, c_{pareg}, c_{pgreg}$ represent the heat capacity of solids, of oil feed in the reactor, of gas in the reactor, of air to the regenerator, of gases in the regenerator (J/kg K); ΔH_i is the heat of coke combustion; $i = 1$ for coke combustion; $i = 2$ for cracking to gasoline (J/kg); y_{2f}, y_2, y_{Of}, y_O represent mole fraction of components in feed and in the reactor, mole fraction of oxygen in air feed to regenerator and in the regenerator; and m is the stoichiometric coefficient.

The constraints of the process are generally similar in all plants and are presented in Tab. 15.3 [5].

Tab. 15.3: The constraints of the process with average values depending on the type of the process.

Variables	Significance	Constraint type	Constraints	Limit
$T_r\circ$	Riser output temperature	technological	$T_r\circ < T_r^{omax}$	500 °C
$T_{reg}\circ$	Regenerator temperature	technological	$T_{reg}\circ < T_{reg}^{omax}$	800 °C
Δp_{r-reg}	Pressure drop Reactor-Regenerator	safety	$\Delta p_{r-reg} < \Delta p_{r-reg}^{max}$	3 bar
M_r	Catalyst level in the Reactor	Technological	$h_r^{min} < h_r < h_r^{max}$	2–6 m

It is obvious from the set of eqs. (15.12) and (15.13) which output parameters are to be kept constant: holdups in the reactor and regenerator, temperatures in the reactor and regenerator, differential pressure between the reactor and regenerator, which determines the catalyst circulation; to stabilize the inputs, additional controls of feed stock to the preheater, temperatures of the feedstock at the entrance of the riser, and steam flow to the stripper of the reactor.

The traditional control scheme is presented in Fig. 5.19.

The spent catalyst holdup in the reactor is maintained by a level control loop with the manipulating variable, the flow of spent catalyst. The holdup of the regenerator is kept constant by the standpipe (Fig. 15.20) functioning as an overflow.

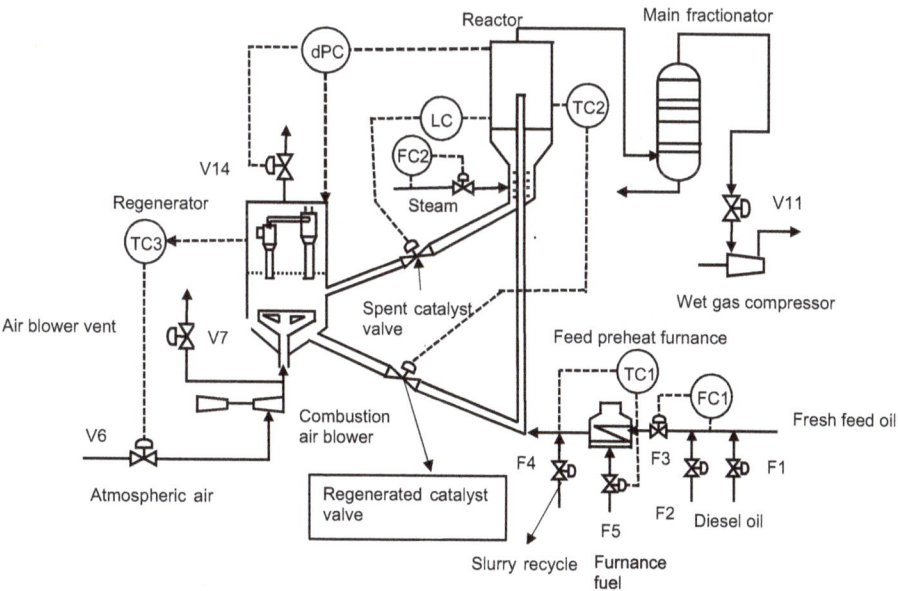

Fig. 15.19: Classic controls of the FCCU.

This control scheme has some disadvantages because of the strong interactions between the reactor and regenerator. Moreover, the main fractionator operation influences the good or bad behavior of the reactor.

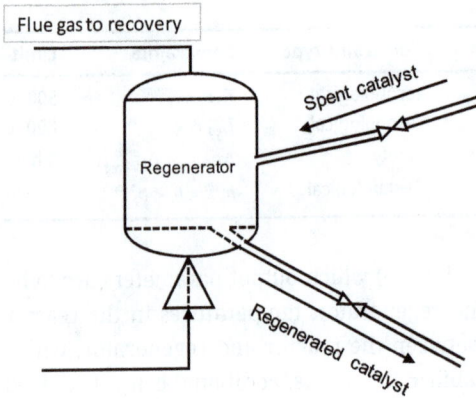

Fig. 15.20: Regenerator with the standpipe controlling naturally the holdup.

Other control schemes are similar as being proposed in [6, 10, 11], but the parameter control suffers from all the same disadvantages.

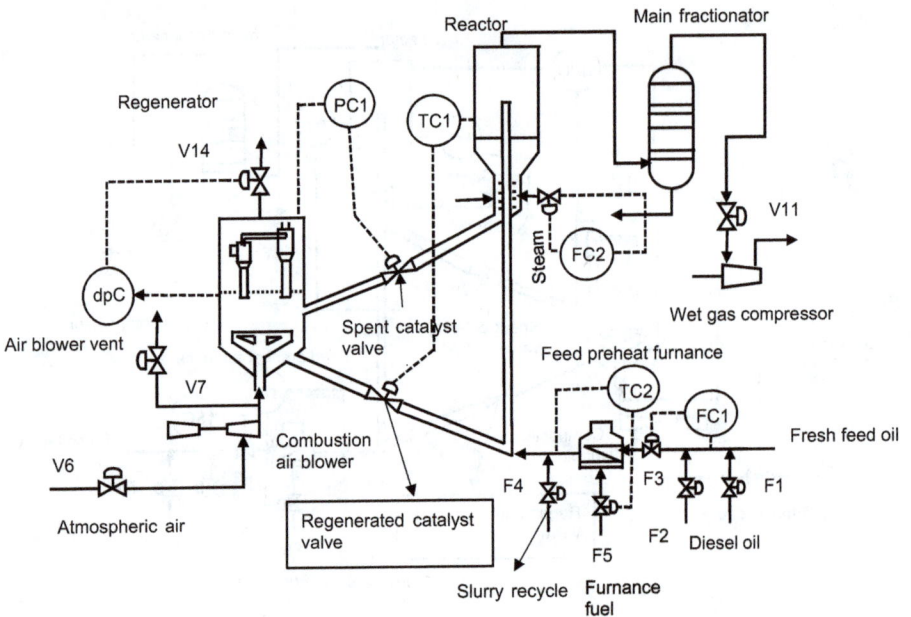

Fig. 15.21: An alternative control scheme of the tandem reactor-regenerator [6].

It can be observed, as a common feature, the control of the reactor temperature is done using the very important influx of heat with the regenerated catalyst flow. The stabilization of the inputs (temperature and flow of the fresh feed) is done with independent control systems, the pressure of gases in the regenerator, similar to the classic controls in Fig. 15.20.

In 1991, Chevron R&D mentioned the instability of the parameters (temperature of the reactor, flows of catalysts were running wildly at ±10% around the setpoint), at their FCCU at El Segundo Refinery in Los Angeles. The instability brought important negative quality and safety consequences. The company asked California Institute of Technology to find a solution to stabilize the process [7]. The main complaint was the repeated violation of the metallurgical limit of the temperature in the reactor, which, in the end, led to an important safety issue!

After modelling the process and simulating it in MATLAB-Simulink programming environment, we could observe a possible stabilization of the process, using the operation with the flooded standpipe [8] (Fig. 15.20) and preventing, in this way, the strong fluctuations of the spent catalyst flow, due to the constant level of the catalyst in the standpipe. In addition, a new MPC scheme was designed to improve the stability and the research was continued and reported in [2, 9].

This time, a more complex approach for the modeling and control purposes was chosen: a five-lump model (Fig. 15.22), the crude oil being transformed in gas, gasoline, diesel, and coke [4], each transformation being characterized by certain kinetics with k_{1-6} – the global reaction constants.

The reactions considered are presented in Fig. 15.22 for the reactor and in eq. (15.15) for the regenerator.

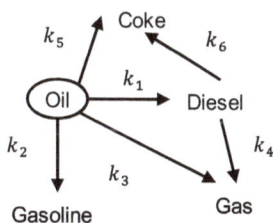

Fig. 15.22: The five-lump model for the catalytic cracking.

Regenerator:

$$4H + O_2 \rightarrow 2H_2O \ (1)$$

Carbon in the coke reacts with oxygen brought by the air blowers to produce CO and CO_2:

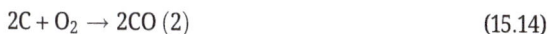

$$2C + O_2 \rightarrow 2CO \ (2) \tag{15.14}$$

$$C + O_2 \rightarrow CO_2 \ (3)$$

Carbon monoxide reacts to produce carbon dioxide according to:

$$2CO + O_2 \rightarrow 2CO_2 \quad (4)$$

The general structure of the reactor and regenerator mass, heat, and pressure balances model used in the elaboration of the MPC controller is presented below:

A. Reactor:
–   The three-lump model (Fig. 15.23) was considered for the global description of the phenomena in the reactor.
–   The reactor is divided into two main parts, riser and stripper, under the following assumptions:
    –   In the riser, there is an ideal plug flow with a very short (seconds) residence time and is considered as a diluted solid phase transport line in which hot catalyst from the regenerator mixes with crude preheated oil from the preheater and slurry (recycle oil) from the main fractionator.
    –   The mass balance describes the gasoline and coke + gases production using Weekman's triangular kinetic model [9].
    –   The hot catalyst provides sensible heat, latent heat of vaporization, and heat of reaction for the endothermic cracking reactions (Tab. 15.4).
    –   The yield of coke on the catalyst in the riser is assumed to be influenced by the velocity of the particles, the residual carbon on the regenerated catalyst, and catalyst residence time in the riser.
    –   No gasoline overcracking takes place; diesel only cracks to gas, not to gasoline.

The reactor model considers all reactions taking place in the riser, which was approached like a plug flow reactor. All five pseudo components from Fig. 15.22 are described with appropriate mass balance equations. The coke balance is added together with the riser energy balance. The deactivation factor for coke formation is also considered.

B. Regenerator:
–   The regenerator has the dense phase fluidized bed at the bottom and the diluted phase (entrained catalyst zone) in the upper part of the equipment; at the top. Above this, there is the so-called disengaging zone, with very little catalyst, where only one reaction, 15.14 (4), is considered to take place.
–   The catalyst phase is presumed to be perfectly mixed; the $T_{reg}°$, temperature of the regenerator is considered to be the same in the whole equipment.
–   Coke deposited on the catalyst particles in the reactor consists of carbon and hydrogen; all hydrogen is presumed to be burnt in the dense phase of the regenerator.
–   The amount of catalyst decreases towards the top of the regenerator in the disengaging zone.
–   Heat is generated by (2) and (3) steps of reaction (15.14).
–   In the region above the disengaging zone, only (4) of reaction (15.15) takes place.

The regenerator model consists of mass balance equations for O_2, CO, CO_2, and coke and the heat balance for the solid and gaseous phases.

Besides these two important pieces of equipment, the model of the FCCU contains: the feed system model, the airblower of the regenerator, the wet gas compressor attached to the main fractionator and also the circulation of the spent and regenerated catalyst in the catalyst pipelines. The main fractionator unit was a continuous distillation column with 38 stages and the model was a 114th- order model with three pseudo components: gasoline, diesel, and slurry; pressure balance in the column was also modeled. The global model developed is a high-order differential algebraic equations system consisting of 933 differential equations and more than 100 algebraic equations and allows capturing the main dynamic characteristics of the process and becoming the internal part of the Nonlinear Model Predictive Control (NMPC) controller (Section 2.9 of this book).

In [9], a comparison of the behavior of the FCCU subjected to classic PID control, MPC, and NMPC controls is done for a 5 × 5 control structure. The five controlled variables were: the reactor temperature, the main fractionator pressure, the reactor catalyst inventory (holdup), and gasoline composition at the top and the bottom of the main fractionator. The five manipulated variables were: slide valves positions for spent and regenerated catalysts (svsc and svrgc), stack valve position V_{14}, condenser liquid flowrate LT, and liquid flow rate of the reboiler, VB. The MPC scheme is presented in Fig. 15.23.

A typical disturbance, the coking rate K_c step change of 1% was chosen for depicting the behavior of the system (Fig. 15.24) controlled by an NMPC.

A comparison between the three modalities of control (PID, LMPC, and NMPC) in the case of a change of pressure difference between the reactor and the main fractionator is presented in Fig. 15.25.

The proposals seem to be benefitting the smooth running of the FCCU with favorable consequences in quality and energy consumption of products.

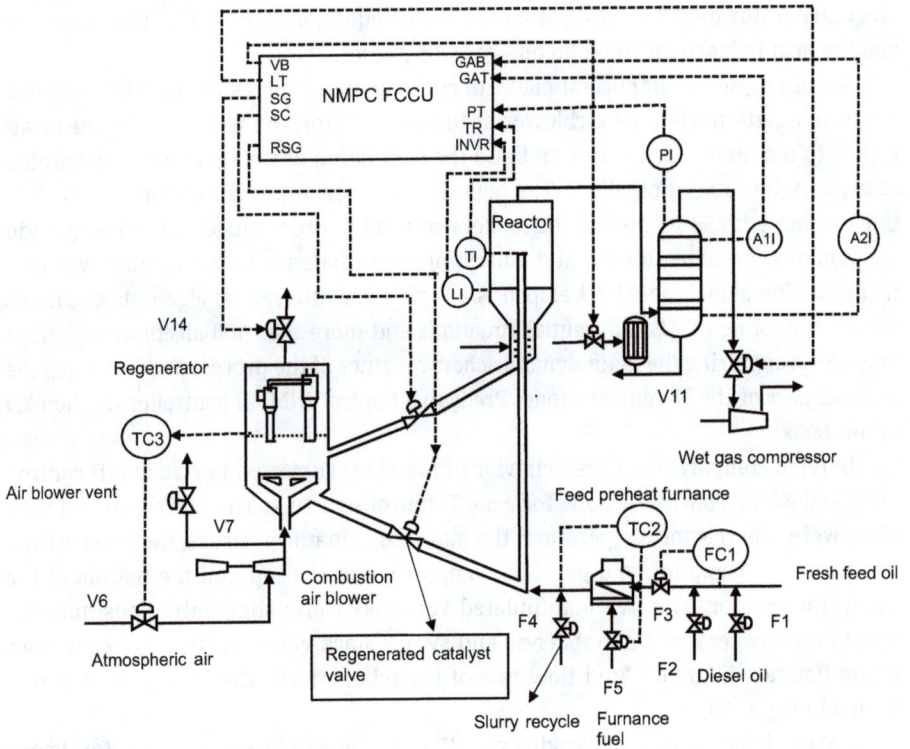

Fig. 15.23: NMPC for FCCU proposed in [9]: VB, liquid flowrate of reboiler; LT, condenser's liquid flowrate; SG, stack gas flowrate; SC, spent catalyst flowrate; RGC, regenerated catalyst flowrate; GAB, gasoline concentration at the bottom of main fractionator; GAT, gasoline concentration at the top of main fractionator; PT, pressure at the top of main fractionator; TR, reactor temperature; INVR, catalyst inventory in the reactor.

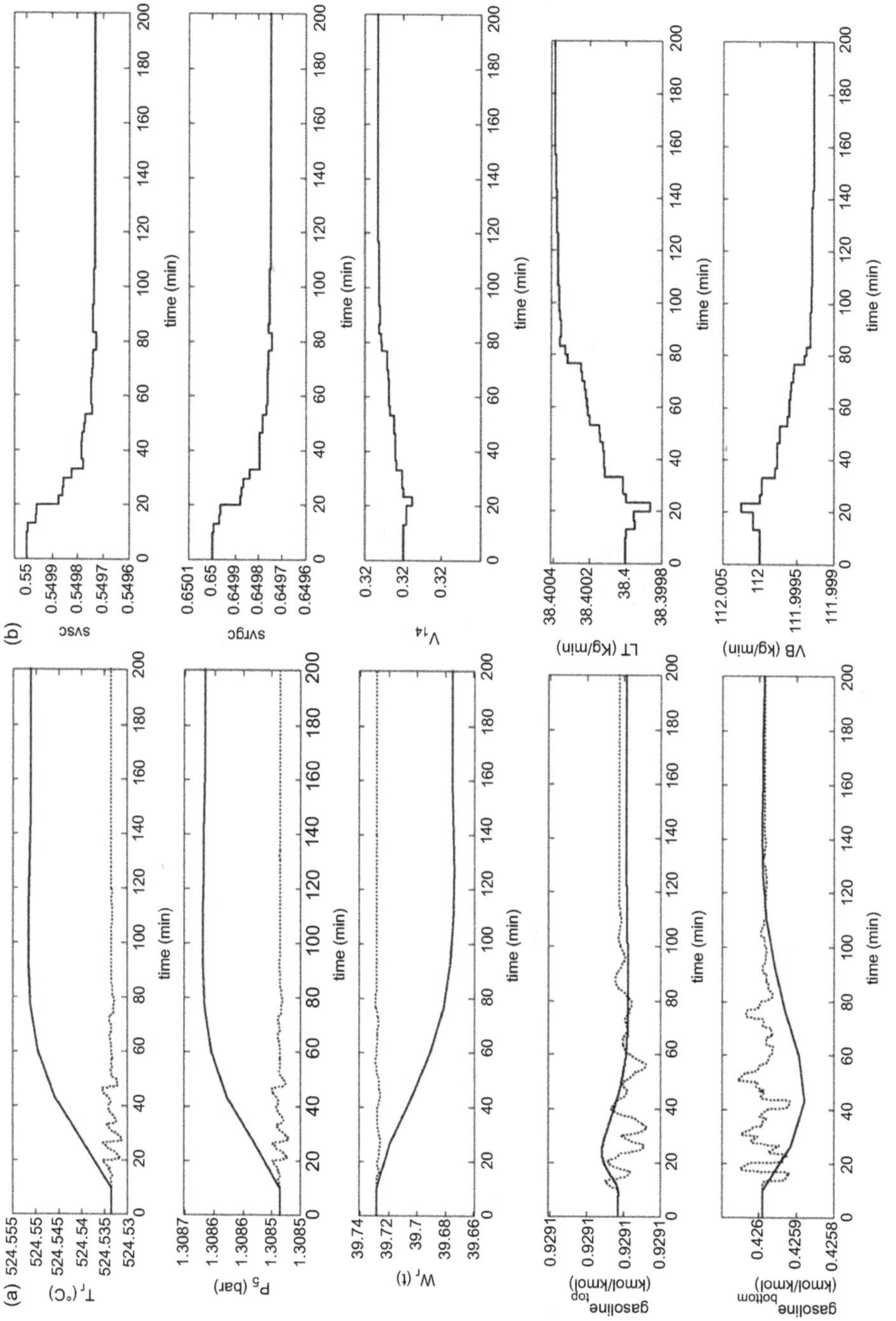

Fig. 15.24: Simulation of FCCU dynamic behaviour in the presence of the coking rate disturbance (1% step increase t = 10min); NMPC results (dotted line) and open loop modelling results (solid line); (a) controlled variables; (b) manipulated variables. (Reprinted with the permission of Elsevier [9])

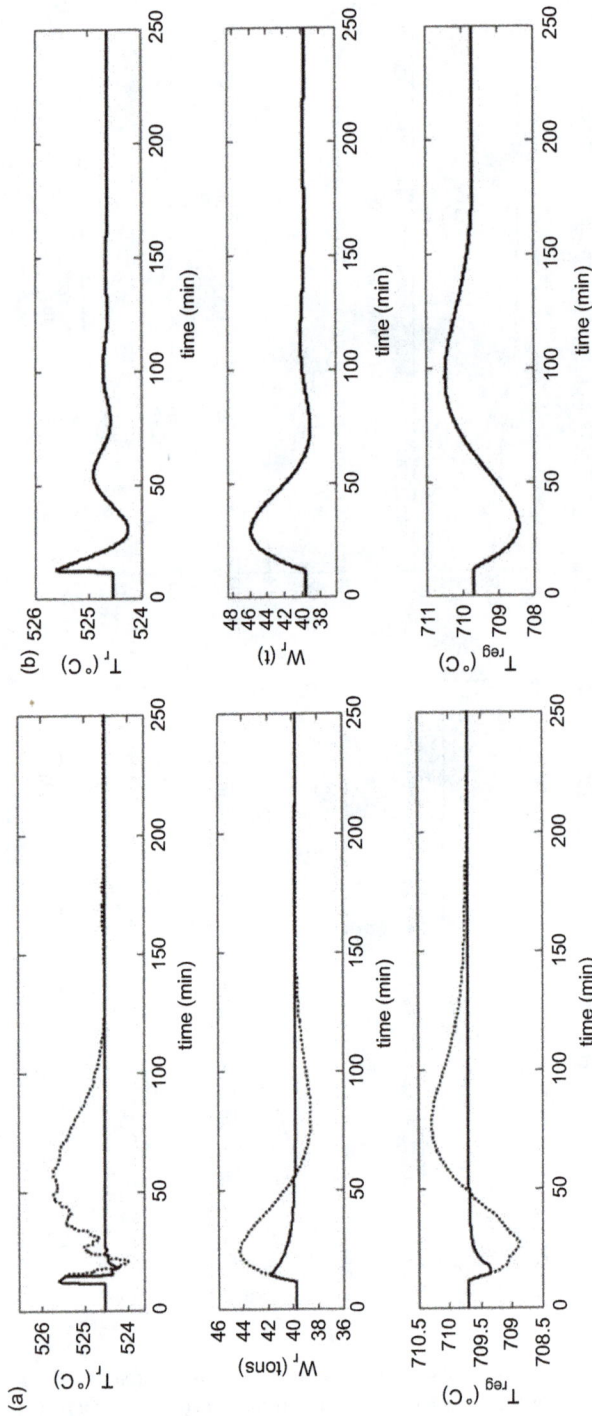

Fig. 15.25: Control approaches for the FCCU plant. Comparison between different FCCU control approaches in the presence of the pressure drop disturbance (27% step decrease at $t = 12$ min), the controlled variables; (a) LMPC (dotted line); NMPC (solid line) and (b) PID control. (Reprinted with the permission of Elsevier [9])

Bibliography

[1]   Dutton, J.A, *Fluid Catalytic Cracking (FCC)*, FSC 432 Module Petroleum processing, Lesson 7, Penn State College of earth and Mineral Sciences, https://www.e-education.psu.edu/fsc432/content/fluid-catalytic-cracking-fcc

[2]   Cristea, V.M, Agachi, P.S., Marinoiu, V., *Simulation and model predictive control of a uop fluid catalytic cracking unit*, Chemical Engineering and Processing, 42, 67–91, 2003.

[3]   Sadeghbeigi, R., *Fluid Catalytic Cracking Handbook*, 2nd Edition, Butterworth-Heinemann, p. 129, 2000.

[4]   Luyben, W., Lamb, D.E., *Feed-forward control of a fluidized reactor-regenerator system*, Chemical Engineering Progress Symposium Series, 59, 165–171, 1963.

[5]   Elnashaie, S.S.E.H., Elshishini, S.S., *Comparison between the different mathematical models for the simulation of Fluid Catalytic Cracking (FCC) Units*, Mathematical Modeling and Computing, 18(6), 91–110, 1993.

[6]   Ramachandran, R., Rangaiah, G.P., Lakshminarayanan, S., *Data analysis, modeling and control performance enhancement of an industrial Fluid Catalytic Cracking Unit*, Chemical Engineering Science, 62, 1958–1973, 2007.

[7]   Morari, M., Huq, I., Agachi, P.S., Bomberger, J., Donno, B., Zheng, A., *Modelling and control studies on model IV FCCUs*, Joint research project between Chevron Research and Technology Company and California Institute of Technology, 1993

[8]   Huq, I., Morari, M., Sorensen, R.C., *Improving dynamic performance of Model IV FCCUs*, p. 215–220 in *IFAC Symposium of Integration of Process Design and Control*, Baltimore, USA, 1994.

[9]   Roman, R., Nagy Z.K., Cristea, M.V., Agachi, P.S., *Dynamic modeling and nonlinear model predictive control of a fluid catalytic cracking unit*, Computers and Chemical Engineering, 33, 605–617, 2009.

[10]   Popa, C., *Application of plantwide control strategy to catalytic cracking process*, Procedia Engineering, 69, 1469–1474, 2014.

[11]   Duraid F.A., *Modeling and simulation of a catalytic cracking unit*, Journal of Chemical Engineering and Process Technology, 12(7), 1–6, 2016.

15.3 Process control and optimization of pyrolysis and gasification plants

15.3.1 Pyrolysis and gasification process

Pyrolysis and gasification are thermochemical conversion processes extracting different products as oil/petrol, char, and syngas from carbonaceous products like coal or solid waste (e.g., plastic, sawdust, and wastewater treatment sludge). Since waste is at the horrendous level of 75% of the planet's resources [1], pyrolysis and gasification are important processes for the circular economy. But Botswana has the second largest reserve of coal in the world [2] and therefore it deserves to exploit the coal properly to obtain added-value products (syngas to fuels, olefins, ammonia, fertilizers, etc.), not only to simply burn it in power production manufacturing or exporting it. In Botswana International University of Science and Technology (BIUST), the group of Resources Beneficiation approaches both existing resources, coal and waste products, aiming at developing a Botswana Chemical Industry.

Pyrolysis takes place in the absence of oxygen, the coal/solid waste structure breaking at mobile aliphatic and ether bridges. The quality and quantity of the products can be changed, changing the process parameters like temperature, heating rate, and residence time, depending on the characteristics of the raw materials used and the target products.

The gasification process takes place at higher temperature and pressure and also adding steam or oxygen.

Although the processes have been known for thousands of years [2], the latest developments led to many different pyrolysis/gasification reactor designs having the main characteristics.

For pyrolysis:
a.   Temperatures (low: below 400 °C, medium – up to 600 °C, high – above 600 °C)
b.   Heating rates (slow, medium, fast, and flash)
c.   Pressures (vacuum to several bars)
d.   Different catalytic processes favoring a product or another one

For gasification [3]:
a.   Temperatures are higher (750–900 °C), even above 1000 °C
b.   Heating rates (°C/min)
c.   Oxygen pressure (bars)
d.   Steam flow and pressure (depending on gasifier)

The advantages of lower temperatures in pyrolysis versus gasification are that they allow the reutilization of wasted energy for running the process, without additional source of energy. It is, energetically, a self-sustainable process.

15.3.2 Control and optimization of the coal/waste pyrolysis/gasification plants

There is literature to be found on coal/waste pyrolysis/gasification modeling and control [5–9, 13, 20]. The content of the chapter is based mainly on the research reported in [2–4, 10–12].

The pyrolysis plant we discuss built by Pyrocarbon Energy Ltd. is presented in Fig. 15.26. Its destination is to pyrolyze the poultry litter for the poultry farms, making them energetically self-sustainable. But the authors used the same plant for other purposes, pyrolyzing and gasifying coal to produce electricity for remote consumers not connected to the high voltage power grid. Its simplified scheme (Fig. 15.27) presents the main equipment.

Fig. 15.26: Proof-of-concept chicken litter pyrolysis plant (by courtesy of Pyrocarbon Energy Ltd.).

The block diagram of the simplified pyrolysis plant in BIUST, with the inputs, outputs, and transfer paths is presented in Fig. 15.28.

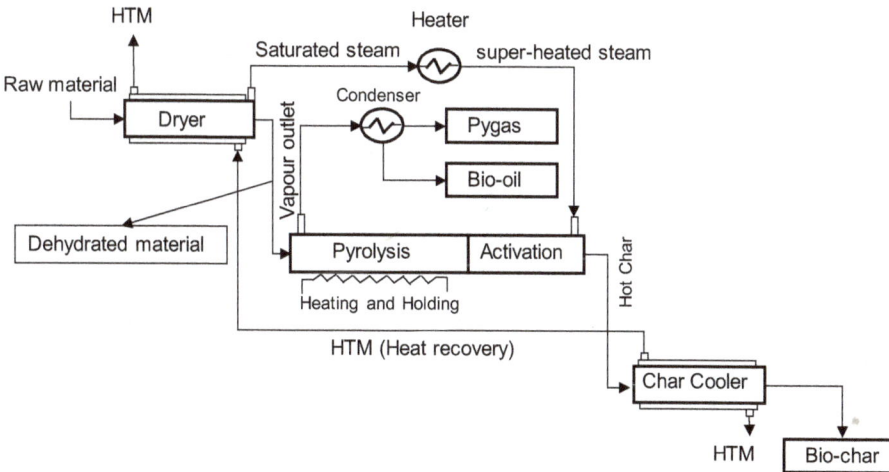

Fig. 15.27: Process flow diagram of a 1 t/day-poultry litter pyrolysis plant [10].

The model is constructed based on the scheme given in Fig. 15.27 and on the block diagram given in Fig. 15.28. In the plant scheme (Fig. 15.27), there are three auger systems: the first is a dryer of the chicken litter aimed to reach a certain water content, the second is the pyrolysis reactor aimed to pyrolyze the carbonaceous content to produce oil and activated carbon, and the third, also an auger transporter, is in fact a tubular heat ex-

Pyrolysis Reactor

Manipulated variables

Disturbances

F_m, Feed flow of raw material   [kg/h]

Q Heat flow delivered   [kW]

F_{sg}, Feed flow of sweeping gas   [kg/h]

$T^o_{od} = T^o_{in}$, Output temperature from the dryer = Input temperature of the reactor   [C]

VM_i, Input volatile material content   [%]

C_i, Input carbon content   [%]

VM_o, Output volatile material content   [%]

C_o, Output carbon content   [%]

F_{sm} Output flow of V solid material   [kg/h]

F_{vm} flow of gaseous material   [kg/h]

T^o_{ch}, Temperature of char   [C]

T^o_{vm}, Temperature of gaseous material   [C]

Dryer

Manipulated variables

Disturbances

F_m, Feed flow of raw material   [kg/h]

F_{st}, Steam flow   [kg/h]

T^o_i, input feed temperature   [C]

W, Moisture of raw material   [%]

$T^o_{st,i}$, Input temperature of the steam   [C]

w_d, Output moisture from the dryer   [%]

T^o_{od}, Output temperature from the dryer   [C]

F_m, Flow of raw material   [kg/h]

Char cooler

Manipulated variables

Disturbances

F_{cw}, Cooling water feed flow   [kg/h]

T^o_{ch}, Temperature of char   [C]

F_{SM}, Char flow   [kg/h]

T^o_{cw}, Cooling water temperature   [C]

T^o_{SM}, Temperature of solid material   [C]

F_{SM}, Flow of solid material   [kg/h]

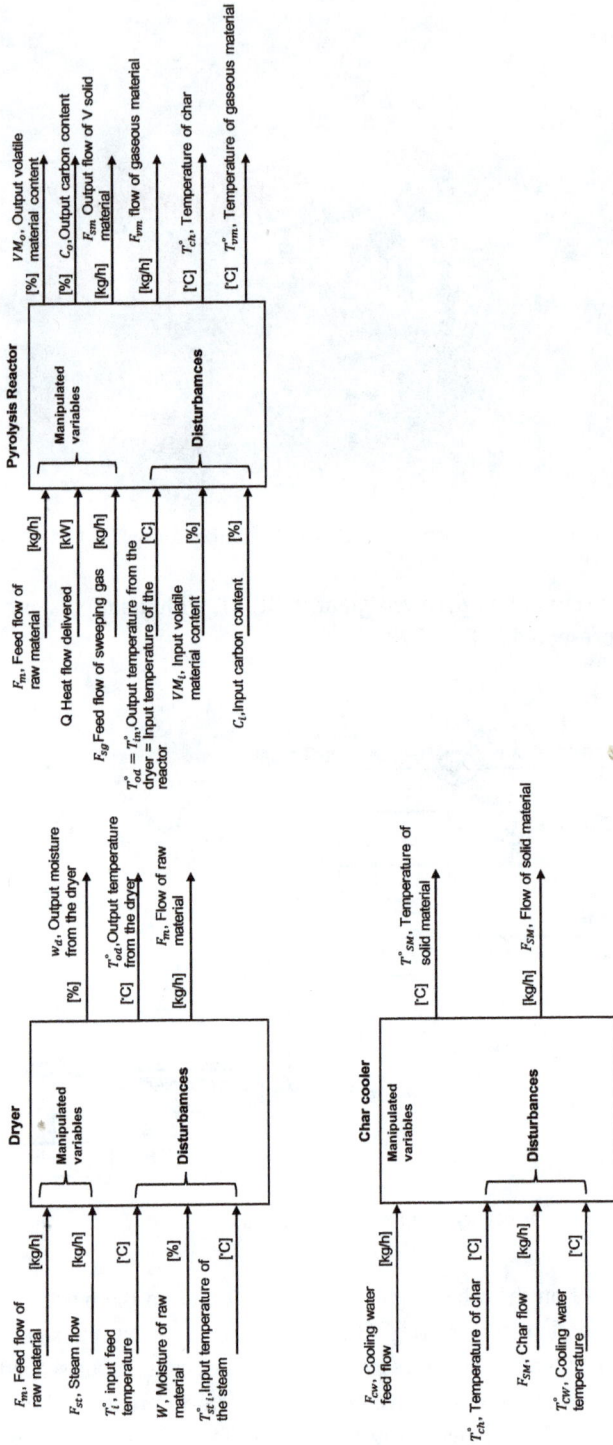

Fig. 15.28: Block diagram of the pyrolysis plant: (a) dryer, (b) reactor, and © char cooler.

changer, cooling the char. The pyrolysis plant is a heat-integrated system, using the heat recovered from the hot char to be cooled. This brings more instability to the parameters and the parameters must be tuned carefully.

The plant is equipped with units to monitor pressure and temperature profile in the auger reactor, gas outlet, char outlet, the condenser and at the gasometer. Solid's feed and discharge are controlled with pneumatic valves and feed rates and residence times of the dryer, reactor, and char cooler are controlled by variable speed drives via PI controllers. Heating is applied by means of induction heating, which is controlled using an industrial PID controller. The unit also has provision to monitor material flow rates, and all these were used to design and optimize a process control scheme. The way the control system was conceived is explained in the following paragraphs of modeling and control.

As usual, a control scheme has as basis, a steady-state mathematical model showing the influences of the different manipulating variables/process parameters on the performances of the process. Referring to [10], such a steady-state simplified model is constructed in this section.

15.3.3 The dryer

The first auger is a tubular heat exchanger with only one inner tube in which the auger is transporting the solid material, the chicken litter (a mix of straw and chicken manure). According to Hiessl et al. [10], the moisture of the raw material influences definitely the quality of the biochar obtained, not including the energy consumed in the process.

From Chapter 13 of this book, the setpoint temperature at the exit of the dryer can be calculated as a function of the desired water content (eq. 13.11):

$$T_0^{\circ *} = T_{oc}^{\circ} \left(\frac{1 - e^{Kw*}}{1 - e^{Kw_c}} \right) + T_i^{\circ} \left(\frac{1 - e^{Kw*}}{1 - e^{Kw_c}} \right) \tag{15.15}$$

where $T_0^{\circ *}$ is the desired output temperature at the end of the dryer for the desired moisture content w^*; T_{oc}° is the constant value expressing the temperature during the constant rate drying period; T_i° is the input temperature in the material (usually the ambient temperature); K is a constant (depending on the temperatures of the dry and wet bulbs measuring the humidity of the water evaporated, the temperature during the constant rate drying – these are characteristics of the solid material, specific area as well); w_c is the moisture in the constant drying rate zone (Fig. 13.1).

According to Badr and Sugiyama [11], the heat transfer in the tubular heat exchanger, a distributed parameters' system, is described by the mass and heat balance eqs. (15.16) and (15.17) (equivalent of eqs. (3.13) and (3.9) from [11]):

$$\frac{\partial}{\partial z}\left(v\rho c_p T^{\circ} \right) + \frac{4K_T}{D} \left(T_{steam}^{\circ} - T^{\circ} \right) = \frac{\partial}{\partial z} \left(k_T \frac{\partial T^{\circ}}{\partial z} \right) \tag{15.16}$$

where v is the velocity of the material in the dryer; ρ is the average density of the material; c_p is the average specific heat of the material; $T°$ is the temperature of the material in the dz element; K_T is the heat transfer coefficient from the induction heating coil to the material through the dryer's wall; D is the diameter of the dryers' tube; k_T is the heat diffusion coefficient in the solid material:

$$D_w \frac{\partial^2 w}{\partial z^2} - \frac{\partial}{\partial z}(vw) = 0 \qquad (15.17)$$

where v is the velocity of the material in the dryer; w is the water content of the solid material; and D_w is the mass diffusion coefficient of water in the material.

The control scheme of the moisture at the outlet of the dryer (entrance of the pyrolysis reactor) is derived from the model (eqs (15.14)–(15.17)) and is shown in Fig. 15.29. The calculations in the "computer block" are done offline in the real plant. To keep the output moisture content at the exit of the dryer constant, one has also to control the velocity of the conveyor.

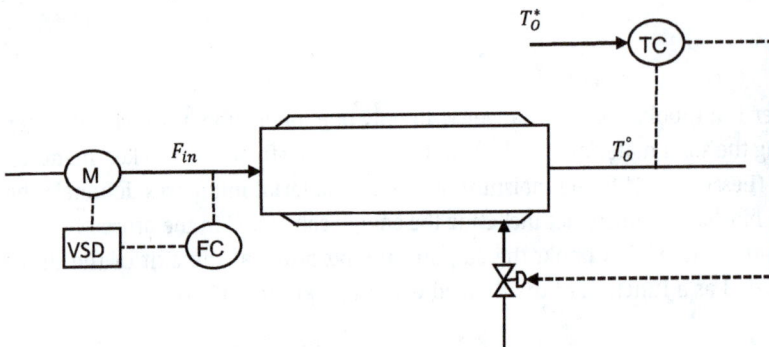

Fig. 15.29: The control scheme of the moisture content of the chicken litter at the exit of the dryer. Velocity of the conveyor included with VSD. (a) Control scheme of the moisture and (b) control scheme of the speed of the auger/conveyor.

Since the dryer is the first equipment in the chain, the feed rates in the reactor and char cooler must be coordinated with the drying rate and feed supply in the dryer. Therefore, for a simple operation, three independent flow control systems are proposed.

15.3.4 The reactor

The reactor operation can be described using a Plug Flow Reactor (PFR) reaction, transport, and heating model. The reaction is endothermic. The products are the petrol (oil) and char (activated carbon). The kinetics and thermodynamics of the process are described in [13, 17–19] and the similar equations adapted from 3.9 and 3.13 from [14] for the pyrolysis reactor are 15.19 and 15.20.

In the model, we consider two pseudocomponents, the bio-oil and the char.

The auger pyrolysis reactor very closely resembles the classical PFR type [14], represented in Fig. 15.30. The figure is particularized for the chicken litter pyrolysis.

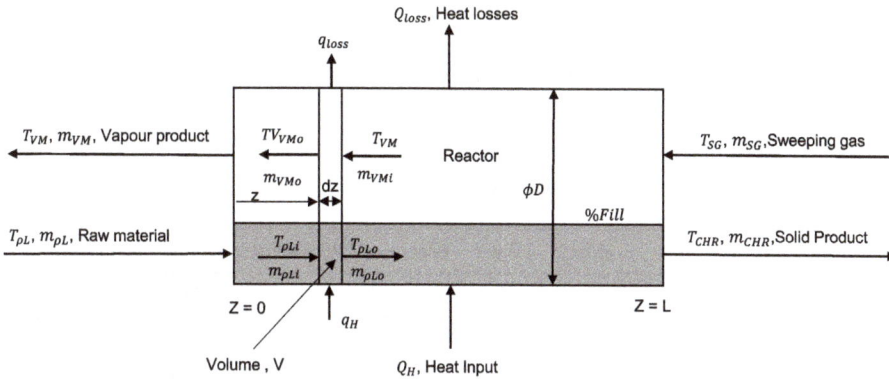

Fig. 15.30: Pyrolysis reactor modelled as a plug flow reactor.

PFR models can be solved by approximating each infinitely small volume, dV, as a continuous stirred tank reactor (CSTR) [14] and integrating the whole system of equations.

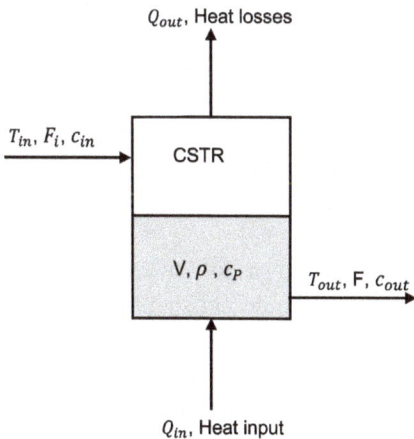

Fig. 15.31: Infinitesimal volume of PFR approximated as a CSTR.

On the other hand, experimental data is available for the product yields of batch pyrolysis processes. If each infinitesimal element (with thickness dz) of the PFR in Figs. 15.30 and 15.31 represent a small batch reactor, which progresses along the PFR from position $z = 0$ to $z = L$, then, with certain assumptions, experimental data can be used to model the steady-state production of the reactor. Applying the experimental results from the batch reactor at the corresponding temperatures along the PFR will enable the calculation of the resulting cumulative production of the PFR. Following this approach, the reaction can be simplified significantly to the following endothermic reaction where poultry litter (PL) is converted to char and volatile matter (VM):

$$PL \rightarrow CHAR + VM$$

The average composition of the poultry litter before being processed is given in Tab. 15.4 [19].

Tab. 15.4: Botswana poultry litter characteristics.

Average	Broiler 1	Broiler 2	Layers	Breeders	Pinewood shavings	Sunflower husk
Moisture	9.36	13.85	4.85	11.18	7.31	7.52
Volatile matter	60.29	55.94	47.33	47.77	69.01	65.34
Ash	10.14	11.14	29.86	22.07	0.33	2.15
Fixed carbon	20.21	19.07	17.97	18.98	23.34	24.99

Around these average values, all input parameters vary, depending on the poultry litter type processed.

The infinitesimal volume of the PFR in Fig. 15.28, which is approximated as a CSTR can be represented as shown in Fig. 15.29. The material entering the CSTR consists of a mixture of solid material and volatile matter (VM). The mass percentage of VM can be equated to the "concentration", C_{VM}, of product formed in classical CSTR theory, starting with zero at $z = 0$ and increasing to its final value at $z = L$.

The following assumptions can be made:

- Steady-state conditions are considered (time derivative is equal to 0).
- The volume and density in the CSTR remain constant (i.e. the fill % does not vary at any given position, z, in the reactor).
- The sweeping gas is inert and only acts as a carried gas for the VM.
- The carrier (sweeping) gas flow rate and composition do not significantly alter the results. In practice, as volatiles are produced, they join the sweeping gas stream and therefore the sweeping gas flow rate and composition vary along the length of the PFR. The batch reactor does not take these effects into account. Heat losses are 0 (very good reactor insulation).
- The mass flow rate is considered constant.

For a CSTR the steady-state total mass balance is given in eq. (15.18) and for VM eq. (15.19) can be derived in terms of mass flow in [kg/s] and concentration in mass [%]:

$$F_{min} - F_{mout} = 0 \tag{15.18}$$

that is, the total mass flow is constant through the reactor, F_m, and

$$F_m C_{VMin} + r \cdot dV = F_m \cdot C_{VMout} \tag{15.19}$$

where F_m is the mass flow rate of the material through the reactor [kg/s]; dV is the volume of the CSTR [m³]; ρ is the average density of the material [kg/m³]; C_{VMin} and C_{VMout} are the concentrations of the VM at the inlet and outlet, mass [%]; and r is the rate of reaction [kg/(m³·s)] and is defined as

$$r = -k(T^\circ) \cdot (C_{VM})^m \tag{15.20}$$

$$k(T^\circ) = k_0 \cdot e^{-\frac{E_a}{R \cdot T^\circ}} \tag{15.21}$$

where k_0 is the pre-exponential factor [s⁻¹]; m is the order of reaction; E_a is the activation energy [kJ/kmol]; R is the gas constant, 8314 [kJ/kmol · K]; and T° is the temperature [K].

The reaction kinetics is not as simple as might seem because of the complexity of the reaction system; the composition depends on the nature of the manure (it is a mixture of organic matter and can differ, function of the chicken feed being very heterogeneous), of the litter (it can be straw or saw dust, etc.). In addition, poultry manure contains mainly organic matter, approximately 85% [17].

In [17–19], it is very difficult to determine the order of reaction m, the rate constant $k(T^\circ)$, the energy of activation E_a, and the preexponential factor k_0. Practically, the kinetics is a guess depending on too many unknown and variable factors.

Therefore, in the component balance equation, the kinetic term depends on the type of chicken litter processed and thus, the controllers' parameters' are adjusted.

For the sake of poultry litter conversion, it is evident that the temperature and the mass flow are the important parameters to be controlled.

The component balance for the char is similar, the input concentration being that of the fixed carbon (Tab. 15.1).

The heat balance (15.22), adapted from eq. (3.13) from [3], is

$$\frac{\partial}{\partial z}(v\rho c_p T^\circ) + r\Delta H_r + \frac{4K_T}{D}(T^\circ - T^\circ_{coil}) = \frac{\partial}{\partial z}\left(k_T \frac{\partial T^\circ}{\partial z}\right) \tag{15.22}$$

Indicating that temperature can be controlled using the quantity of electricity delivered to the heating coil. The heating system is based on induction; the power generated and the temperature of the coil have been calculated in [13]

$$P = \frac{\pi^2 \cdot B_p^2 \cdot t^2 \cdot f^2}{6 \cdot \rho \cdot D}$$

where B_p is the peak magnetic field in [T]; t is the thickness of the reactor shell in [m]; f is the frequency in [Hz]; ρ is the resistivity of the reactor shell material in [Ω·m]; D is the density of the coil material in [kg/m³].

Both heating rate and residence time, important parameters for the chicken litter conversion, depend on the mass flow and power delivered. Thus, the control system of the reactor is presented in Fig. 15.32.

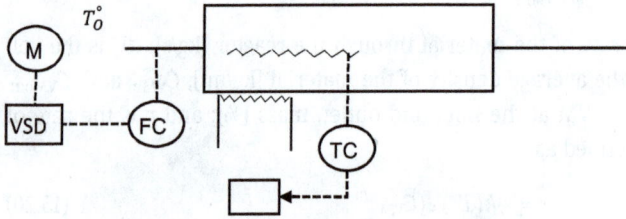

Fig. 15.32: The control system of the reactor.

15.3.5 The cooler

The cooler is a unitubular heat exchanger with a cooling jacket having cooled raw water recirculated as a cooling agent. (Fig. 15.27).

Writing the equations for the heat exchanger:

$$F_m c_{pchar} T^\circ_{ichar} - F_m c_{pchar} T^\circ_{ochar} - K_T A_T (T^\circ_{ochar} - T^\circ_w) = 0 \tag{15.23}$$

$$F_w c_{pw} T^\circ_{iw} - F_s (c_p T^\circ_w + l_v) + K_T A_T (T^\circ_{ochar} - T^\circ_w) = 0 \tag{15.24}$$

where c_{pchar} and c_{pw} are the specific heat of the char and of the cooling water ... T°_{ichar} and T°_{ochar} are the input and output char temperatures respectively .. ., K_T is the heat transfer coefficient from the cooled internal mass to the cooling agent .., A_T is the heat transfer area of the heat exchanger . . ., T°_{iw} and T°_w are the input and output temperatures of the colling water . . .

l_v is the water latent heat of vaporization.

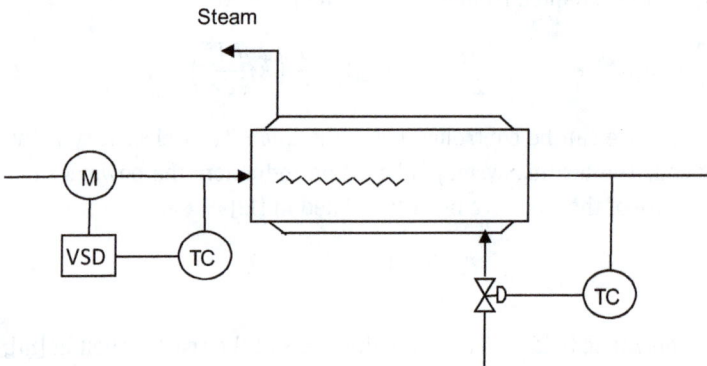

Fig. 15.33: The control system of the char cooler.

One can observe that the output temperature of the char is a function of the mass flow of the solid material and the cooling capacity of the jacket.

One important conclusion from the synthesis of the control system of the pyrolysis plant is that the mass flows in the dryer, reactor, and char cooler must be synchronized. Another conclusion is that the tuning of the reactor temperature controller takes into consideration the type of litter pyrolyzed.

The pyrolysis oil yield is obtained through the condensation of the VM obtained from the reactor and mist formation is one of the primary concerns during the condensation process. Once mist has formed it becomes a colloid in the vapor stream and is difficult to separate. An important part of pyrolysis process control, not presented here, is therefore to establish parameters that will prevent mist formation.

The proof-of-concept plant manufactured by Pyrocarbon Energy Ltd. (Fig. 15.26) is containerized and contains all control systems described (Fig. 15.34).

Fig. 15.34: The proof-of-concept plant for the pyrolysis of the chicken litter. The control panel and automation equipment (by courtesy of Pyrocarbon Energy Ltd.).

15.3.6 Optimization of the coal gasification reactor

The sub chapter is based of the BIUST research reported in [3]. Depending on different types of gasification processes and adjacent equipment, there are several possibilities of optimizing the process [20, 21]. We preferred to present the approach used in our own experiments in BIUST; as a matter of fact, similar procedures were used in [6]. The gasification reactor is practically the same type of reactor (Fig. 15.27) adding the steam injection. The plant used is presented in Figs. 15.35–15.37. The results obtained can be scaled up, using the known procedures, to industrial reactors.

The optimization of the gasification reactor was done using a factorial design for the experiments.

Different operating parameters were varied to understand how the process variations affected the response of interest. While steam flow rate, oxygen flow rate, temperature, and pressure were fixed as in Table 15.5: residence time, granulation, and temperature were varied to see how they affect the quality and quantity of the gas produced. Experiments were performed using the 2^3-factorial method.

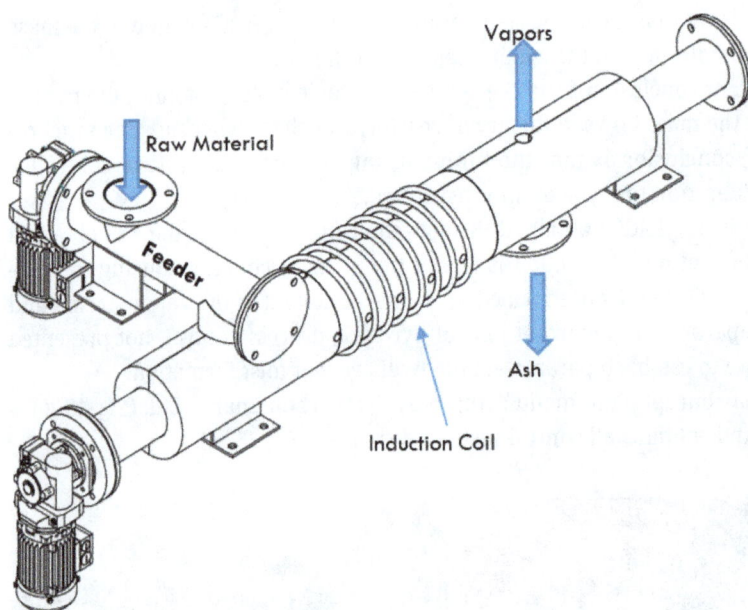

Fig. 15.35: A 5 kg/h bench-scale auger reactor [3, 8].

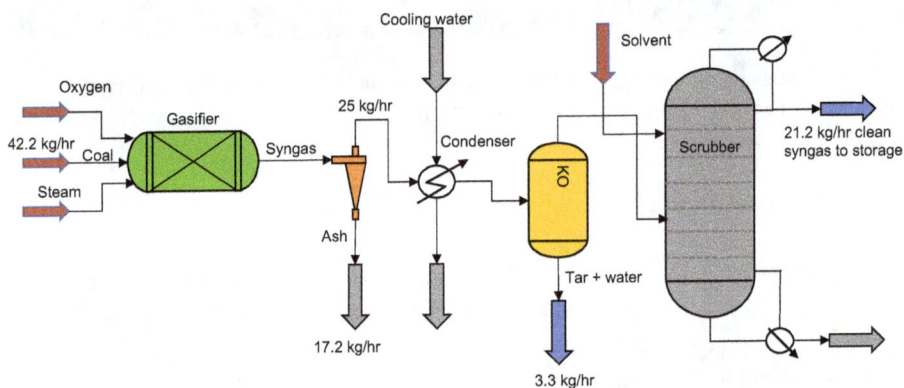

Fig. 15.36: The gasification plant in BIUST pyrolysis/gasification laboratory. Blueprint.

The design of the experiment was focused on factors that may have large effects on the output of the process (quality and quantity of the product gas). The factorial design was used to study and select the best operation parameters for induction-heated fixed bed reactor. Some plots were generated to evaluate the significance of the effects. The factorial design method also helped in determining the factors that are important in improving production rate of synthetic gas from coal.

Fig. 15.37: BIUST bench-scale fixed bed reactor used in coal gasification (by courtesy of BIUST).

Tab. 15.5: Gasification process input parameters.

Parameter	Range
Particle size range	1–10 mm and 10–40 mm
Residence time	1 and 2 hours
Temperature	750 and 900 °C
Oxygen pressure	4 bars
Oxygen flow	3.42×10^{-4} kg/s
Steam flow	13.7 g/min
Steam pressure	4 bars

The experiment consisted of two levels and three factors, which are residence time, granulation, and temperature (Tab. 15.6).

Tab. 15.6: Levels and factors used for the factorial design experiments in this study.

Parameter	(–) = low limit	(+) = high limit
Res. time (h)	1	2
Granulation (mm)	1–10	10–40
Temperature (°C)	750	900

In this work, a (2^3) experimental design matrix was formed, according to the capability of the fixed bed gasification plant used. The experimental conditions for the initial screening of the affecting parameters are given in Tab. 15.7. The repeatability of the gasification tests was evaluated by performing three repeat tests of the upper and the lower limits of the factorial design tests.

The first level was residence time, which was varied between 1 h and 3 h, followed by granulation with size ranges 1–10 mm and 10–40 mm, and lastly temperature variation between 750 and 900 °C. There were eight experiments repeated three times for accuracy of results.

Tab. 15.7: Factorial combinations for volume of gas produced from coal through gasification in a fixed bed reactor.

A	B	C				Interactions	
Res. time (h)	Granulation (mm)	Temp (°C)	AB	BC	AC	ABC	Quantity of gas (L)
−1	−1	−1	1	1	1	−1	50
1	−1	−1	−1	1	−1	1	82
−1	1	−1	−1	−1	1	1	42

Analysis of regression model:

Excel software was used to generate the runs, Tab. 15.7.

The regression obtained in the mentioned research [3] was:

$$y = 63.625 + 6.458A - 9.042B + 1.375C - 0.0417AB - 6.792BC - 5.625AC + 0.375 \quad (15.25)$$

showing the highest importance of the granulation and in order of the residence time; temperature has a smaller coefficient of 1.375, which means temperature has a smaller significant and positive impact on the quantity of gas produced compared with the residence time. Increasing the temperature by 1 °C will increase the yield by 1.375 L, while increasing granulation by 1 will lower the yield by 9.042 L according to the regression model.

From a total of eight sets of experiments each having eight runs, the highest amount of product gas (82 L/kg coal) was observed to be obtained at temperature of 900 °C, particle size range of 1–10 mm, and at residence time of 1 h.

In conclusion, this section introduces possible methods of control and optimization of the pyrolysis/gasification process used extensively in processing carbonaceous materials to produce fuel mainly, but not only. The process is very much used in the circular economy, where waste is transformed in useful products.

References

[1] Charpentier, J.C., *"Among the trends for a modern chemical engineering": Cape an efficient tool for process intensification and product design and engineering*, Plenary Lecture at ESCAPE, 17, 27–30 May 2007, Bucharest, Romania.

[2] Makoba, M., et al., *A review on botswana coal potential from pyrolysis and gasification perspective*, Periodica Polytechnica Chemical Engineering, 65(1), 80–96, 2021.

[3] Makoba, M., *Synthetic gas production from botswana coal through gasification in a fixed bed reactor*, PhD thesis, BIUST, Botswana, 56, 2023.

[4] Makoba, M., Agachi, P.S., Muzenda, E., Mamvura, T., *Evaluation of botswana coal for syngas-production*, Bulletin of Romanian Chemical Engineering Society, 9(2), 2022, 67–81.

[5] Zhang, B., et.al, *Automatic control system of a biomass pyrolysis gas carbon compound furnace based on PLC*, pp. 435–442 in 3rd in World Conference on Mechanical Engineering and Intelligent Manufacturing, IEEE, 4–6 Dec. 2020.

[6] Gobel, B., Henriksen, U., Jensen, T.K, Qvale, B, Houbak, N., *The development of a computer model for a fixed bed gasifier and its use for optimization and control*, Bioresource Technology, 98, 2043–2052, 2007.

[7] Zhang, J., Hou, J., Zhang, Z., *Real-time identification of out-of-control and instability in process parameter for gasification process: integrated application of the control chart and Kalman filter*, Energy, 238, 121845, 2022.

[8] Peng, Q., Dai, Z., Xu, J., *On-line reset control of a commercial-scale opposed multi-burner coal-water slurry gasification system using dynamic reduced-order model*, Computers and Chemical Engineering, 143, 107074, 2020.

[9] Cui, Z., Tian, W., Zhang, H., Guo, Q., *Multi-scale modeling and control of chemical looping gasification coupled coal pyrolysis system for cleaner production of synthesis gas*, Journal of Cleaner Production, 299, 126903, 2021.

[10] Botha, D.E., *Poultry litter pyrolysis to address agricultural and energy challenges in Botswana*, PhD thesis, BIUST, Botswana, 2023, chapter. 2.

[11] Botha, D.E., Agachi, P.S., *Poultry litter pyrolysis to address agricultural and energy challenges in Botswana*, Bulletin of Romanian Chemical Engineering Society, 9(1), 103–121, 2022.

[12] Botha, D.E., Agachi, P.S., *Design and construction of a proof-of-concept Poultry litter pyrolysis plant*, Studia Ubb Chemia, LXVIII(1), 105–118, 2023.

[13] Doti, B., Nyaanga, D.M., Nyakach, S., Nyaanga, J., Ingasia, O., *Effect of selected pyrolysis parameters on the production and quality of biochar and pyroligneous acid from biomass*, Journal of Engineering in Agriculture and the Environment, 8(2), 15, 2022.

[14] Agachi, P S., Cristea, V.M, Makhura, E.P, *Basic Process Engineering Control*, 2nd Edition, De Gruyter GmbH, Berlin, Boston, 71, 2020.

[15] *Ibidem, 70.*

[16] Botha, D.E., *Poultry litter pyrolysis to address agricultural and energy challenges in Botswana*, PhD thesis, BIUST, 66, 2023.

[17] Kim, S., Agblevor, F., *Pyrolysis characteristics and kinetics of chicken litter*, Waste Management, 27, 135–140, 2007.

[18] Kim, S., Kim, J., Park, Y.H, Park, Y.K., *Pyrolysis kinetics and decomposition characteristics of pine trees*, Bioresource Technology, 101, 9797–9802, 2010.

[19] Shomana, T., Ntuli, F., Agachi, P.S., *Kinetics models of different types of poultry litter*, article under preparation.

[20] Okolie, J., Epelle, E., Nanda, S., Castello, D., Dalai, A., Kozinski, J., *Modeling and process optimization of hydrothermal gasification for hydrogen production: A comprehensive review*, The Journal of Supercritical Fluids, 173, 105199, 2021.

[21] Imre-Lucaci, A., Agachi, P.S, *Optimizarea Proceselor în Industria Chimică (Process Optimization in the Chemical Industry)*, Ed. Tehnică, București, 2002

15.4 Automatic control of mineral processing plants

In many countries rich in minerals, the mineral processing sector is an extremely important national endowment. In these countries, for different reasons, this is the only "chemical/process industry" that exists. Unfortunately, concentrate production is usually the last processing stage of the supply chain as no further value addition processes are carried out.

Minerals, metallic or nonmetallic (e.g., iron, copper, gold, mica, graphite, and diamonds) are usually contained in very small proportions in the ore. The ore is a mixture of valuable mineral (e.g., chalcopyrite) and valueless material called *gangue*. The valuable material is in form of *native* ore where the metal is present in its elementary form or in *sulfide or oxidized* ores, where the metal is in a chemical compound and must be separated afterwards through chemical/physical/chemophysical treatment.

The raw ore needs to be treated to concentrate the valuable mineral, which can be in quantities as low as grams/ton of ore, in the form of a metal, an alloy, or a compound.

A generic mineral processing material flow is presented in Fig. 15.38 [1]. The content of this chapter is inspired in an important proportion by Juran [1] and Blanke et al. [3].

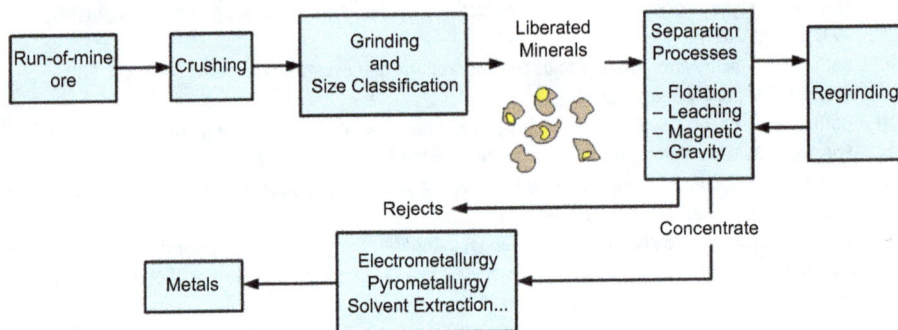

Fig. 15.38: Major stages in mineral processing (reprinted with the permission of Elsevier from [1]).

One of the major performance indicators of these processes is recovery monitoring. Recovery is measured in percentage, and it represents the fraction of the valuable/target mineral in plant feed that has been successfully separated to the concentrate stream. In many instances, for example Botswana's BCL nickel concentrator and smelter, plant control to obtain target recovery is done manually when samples are taken around the plant every hour and sent to the laboratory for metal content analysis. This practice incurs control delay time of up to 2 h for a plant that processes about 500 tons/h of solid material.

Most of the plants that installed automatic control reported metal recoveries increments between 0.5% and 3%. More important are the reductions in consumption of reagents, reported to be between 10% and 20% [15].

In this chapter, the authors mainly discuss the control in the stages up to the electrometallurgy/solvent extraction stage with accent on grinding and separation (leaching, flotation, etc.).

Crushing and grinding are common in other industries as well, and they have been treated in Chapter 15.1 (cement manufacturing control). There are differences discussed in this chapter.

The separation stage of the valuable component from the *gangue* is based on differences in physical/chemical properties like surface hydrophobicity, specific gravity, magnetic and electrical properties, etc.

Thus, the processes in the mineral processing are [1]:
- Minerals liberation processes (crushing, grinding, and size classification)
- Minerals separation processes (flotation, magnetic or gravimetric separation, sorting, leaching, etc.)
- Concentrate pretreatment for subsequent mineral extraction (solid-liquid separation, drying, agglomeration, sintering, etc.)

15.4.1 Process variables

To configure a control system for any process, it is useful to define the variables involved in that process and their status: input, output, state, and from these, which ones are manipulating and disturbance variables. Figs. 15.39 and 15.40 show the variables of the grinding and flotation processes, which are the ones the authors are focusing on: u is used for the manipulated variables (control variables), y for the output (controlled) variables, d for the disturbances, and x for state variables. The x variables are process states that are dependent on inputs and are specified at design stage.

In Fig. 15.39, the main components of a grinding circuit are shown, together with the process variables [2, 3].

As can be observed from Fig. 15.39, the major disturbances of milling are:
1. Ore hardness (function of changing ore mineralogy)
2. Feed particle size distribution (function of crushing and screening efficiency)

The output of this circuit goes to the flotation section.

The primary role of the grinding section is ore liberation so that concentrating processing can have adequate access to separate valuable mineral from the *gangue.* Grinding is the final stage of *size reduction* or *comminution,* which starts with crushing, and it is the biggest energy consumer in mineral processing, accounting for up to 50% of concentrator section energy consumption.

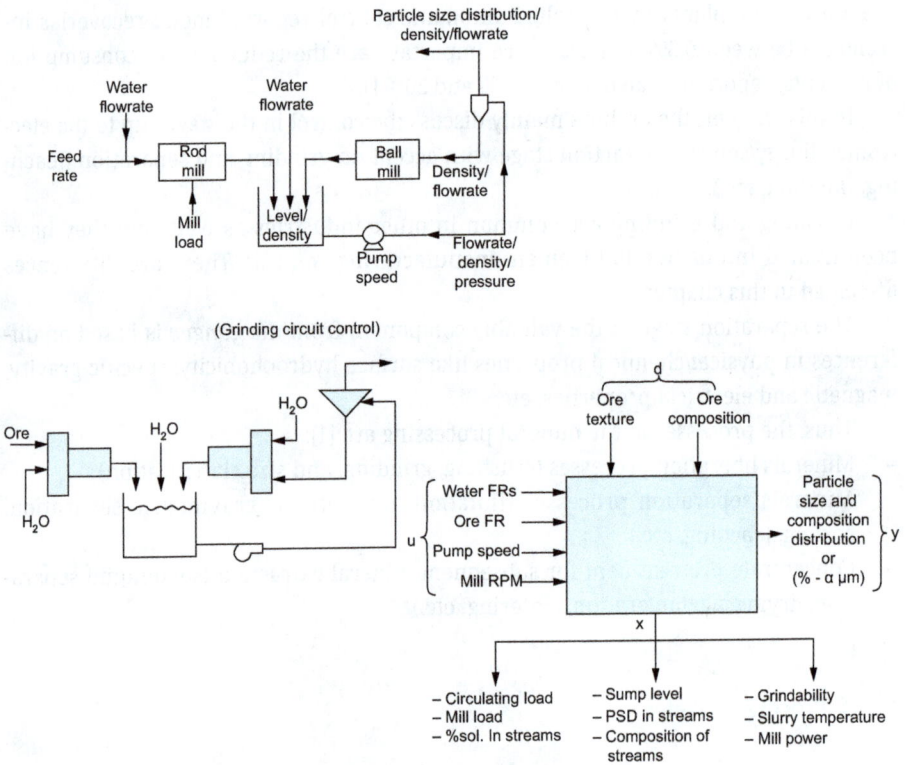

Fig. 15.39: Process variables for a grinding circuit (reprinted with the permission of Elsevier [1]).

Considering the grinding process, the objectives are to maintain a certain particle size distribution to be constant at a certain design output; to maintain constant the feed rate; and to coordinate the grinding stage production with the flotation stage.

The disturbances presented in Fig. 15.39 are the changes in the ore characteristics, particularly ore hardness and feed particle size distribution and variation of the input flow. The grinding control strategy is to minimize the influence of the disturbances. Usually, this is obtained through:

1. Ratio of solid/water control in the feed slurry; this is important because it controls the slurry density (a proxy for the product size distribution) and the load of the mill or mill retention time (solids dispersion and transportation) and ultimately mill power draw, which is a function of slurry viscosity.
2. Sump level control determines the pumping flow from mill sump to cyclone.
3. Product size distribution control determines the flotation operation.
4. Load of the grinding circuit control; important to know for retention time and for material balance of the plant.

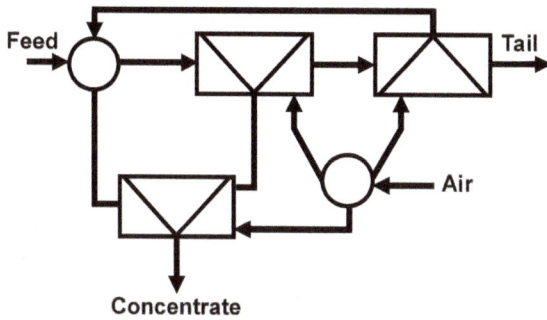

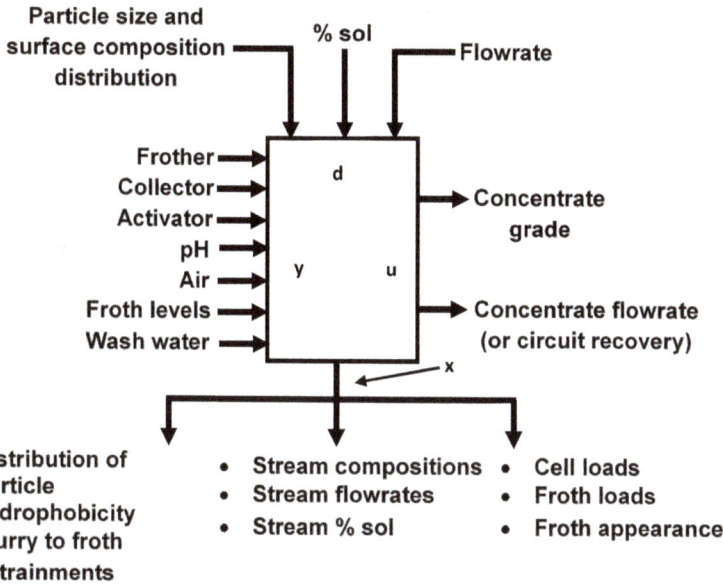

Fig. 15.40: Process variables for a flotation circuit [4].

The control structure of the grinding section is presented in Fig. 15.41 [5, 6]. The control focus is on keeping the particle size constant by regulating the crushed ore feed to the rod mill, and on stabilizing the cyclone feed density by regulating the water addition to the cyclone pump sump:

- The rod mill feed is measured by means of an electric belt weight transducer and is controlled by varying the speed of the conveyor (VSD).
- Water addition to the mill is controlled according to the density setpoint and changing its feed rate; in this way, a constant slurry density is maintained.
- The rod mill discharge is fed to a sump where it joins the circulating load discharge from closed-loop grinding circuit. The sump level is monitored by means of a level (pressure) transducer or a float level transducer and controlled by a variable speed pump.

- When pumping slurry to a hydrocyclone, the slurry flow rate and density in the feedline as well as the cyclone pressure are monitored critically. These three parameters affect cyclone efficiency. The cyclone underflow is fed to a mill and is measured as mill recirculating load. The circulating load is typically maintained around 150–300%, with the overflow proceeding to the flotation plant. The particle size distribution and density being the most important output parameters of the grinding circuit, are measured directly by an online particle size analyzer and a density meter. The particle size is typically controlled at 60–100 μm while density is maintained around 30% solids (a calculated figure using eq. (15.26)). The coarse product from the cone classifier was fed to a pebble mill, the pebble feed being controlled according to the power consumed.

A control strategy of this first stage of the process of separating minerals is proposed by Nunez et al. [5] implying a feedforward approach (Fig. 15.40):

$$X = \frac{100s(D - 1000)}{D(s - 1000)} \tag{15.26}$$

where D is slurry density, s is solid's specific gravity, and x is solid % weight.

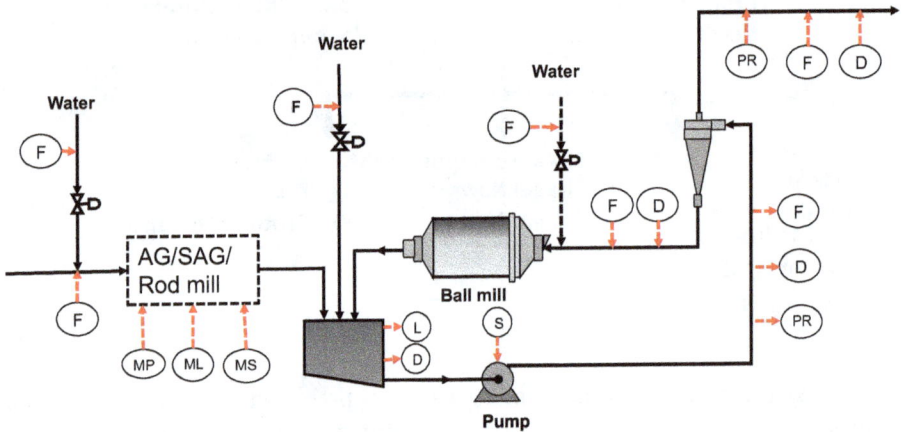

Fig. 15.41: Grinding circuit control variables. Notations: F - Flowrate; D - Density; L - Level; ML - Mill Load; MP - Mill Power; MS - Mill Speed; PR - Pressure; PS - Particle Size; S - Pump Speed.

The control scheme presented in Fig. 15.42 aims to maintain the solid/water ratio constant in the condition of the main disturbance, which is the solid feed. The disturbance transducer (WT – weight transducer) placed on the solid conveyor belt addresses the signal to a feedforward controller (FFC) before the effect of the delays (dead time) is felt at the entrance of the rod mill. The feedforward ratio controller changes the setpoint of the water flow controller.

Due to the delays produced by conveyor belts, the performance of the feedback control and even of the feedforward control is limited in [7], the authors applied this strategy named "Smith predictive controller" to semiautonomous grinding (SAG) mills plant. The predictor is in fact an adaptive structure, presented in Fig. 15.43, like those presented in [8].

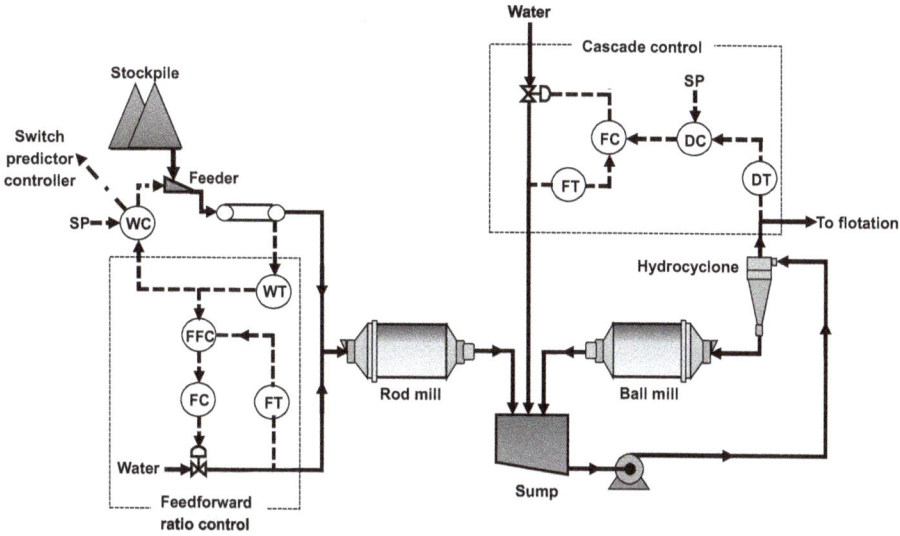

Fig. 15.42: Feedforward/ratio control configuration for the grinding circuit. Notations: WT - Weight Transducer; FT - Flow Transducer; DT - Density Transducer; WC- Weight Controller; FC - Flow Controller; DC - Density Controller.

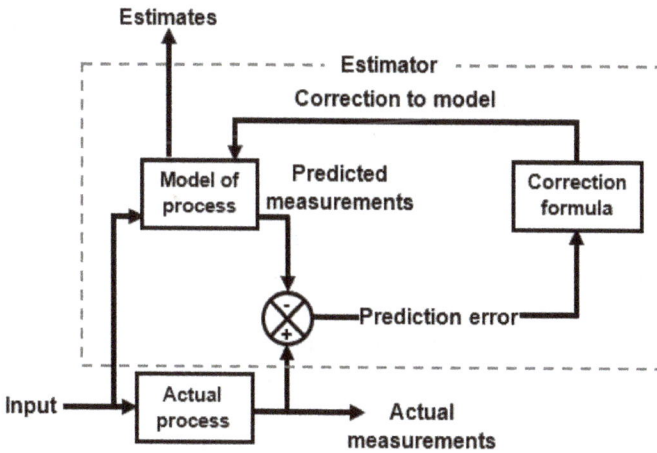

Fig. 15.43: Smith predictive controller operating principle. Used in control strategy of grinding processes [6].

A particular interest should be given to the hydrocyclone. It has the role of separating the fines (at the top overflow) from the coarse (at the underflow) that is sent back to regrinding.

A single parameter, defining the classifying efficiency of a hydrocyclone, is the d_{50} value. The control strategy is based on maintaining the d_{50} value at a certain point. The d_{50} represents the separation point of a particular particle size, 50% in the underflow and 50% in the overflow. The change of the d_{50} value is due to change in feed characteristics; additionally, mechanical corrosion affects cyclone lining, with jeopardizing cyclone shell, which is more expensive to replace. The variations in the particle sizes of outlet streams affect flotation processes having as result poor overall performance and economic losses (related to recovery and product grade) [9]. The d_{50} is not a directly monitorable parameter, but it can be a calculated function of construction and operation parameters. In [9], eq. (15.27) is proposed:

$$d_{50} = 23.36 \frac{e^{(-0.0125D_u + 0.1031)\phi}}{e^{0.2721D_u}} \cdot \left[(-0.0229D_u - 0.0211) \frac{F}{F_{min}} \cdot 0.0739D_u + 0.9138 \right] \cdot$$
$$\left[-0.426 \frac{H}{H_D} + 1.42 \right] \cdot \left[0.2 \frac{T_n^\circ}{T^\circ} + 0.81 \right] \tag{15.27}$$

where D_u is the outlet diameter of the pipe; ϕ is the density of the slurry; F is the volumetric flow of the liquid; F_{min} is the minimum volumetric flow through the hydrocyclone; H is the height of the vortex finder from top of the spigot; H_D is the height; T° is the temperature of the slurry in operation; and T_n° is the normal temperature of 25 °C.

From eq. (15.27) one can calculate the setpoint for the input flow F to the hydrocyclone (Fig. 15.44), changing the speed of the pump motor, using a variable speed drive (VSD) [FC].

The flotation stage (Fig. 15.45) is a more complex separation process [1, 3, 11, 12]. Flotation principle is based on the surface properties of the minerals in relation with water. Some minerals are wetted by water – the *hydrophilic* ones (e.g., oxides, hydroxides, and silicates); others repel the water and they are the *hydrophobic* ones (e.g., graphite, sulfur, and talc). The system is complex, involving solid, gas, and liquid phases in contact.

The operation of the froth flotation cell (Fig. 15.45) is the following:

The slurry coming from the hydrocyclones is treated with flotation reagents in order to increase or to create hydrophobicity. To do that, *surfactants (collectors)* are added to the pulp. Depending on the kinetics of adsorption, a certain residence time is needed for the adsorption of the surfactants on the minerals in the cell. The air pumped in the cell assures not only the floatability but also a good mixing in order to keep solids suspended in the pulp; this is achieved in most flotation cells by using an agitator. The hydrophobic minerals attach to the air bubbles and float to the upper part of the cell and are collected in the overflow. They form the concentrate which, further on, is processed through pyro and/or hydrometallurgical processes for further

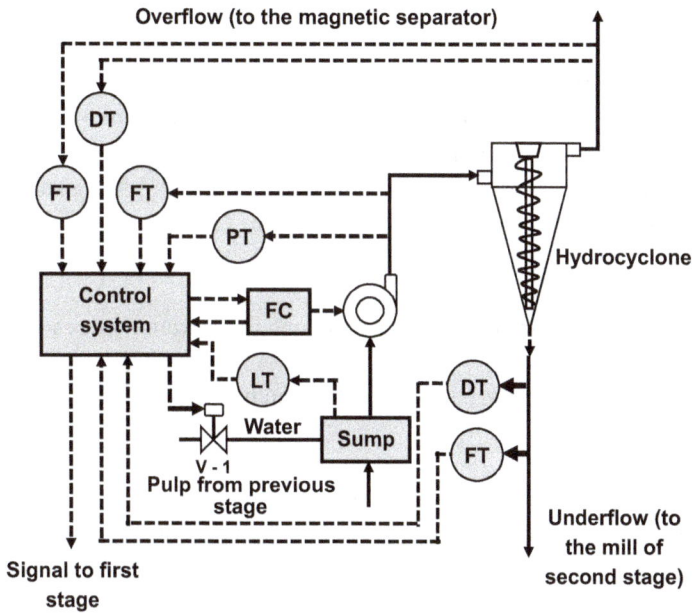

Fig. 15.44: Scheme of the control of hydrocyclone separating process [10]. Notations are the same as before.

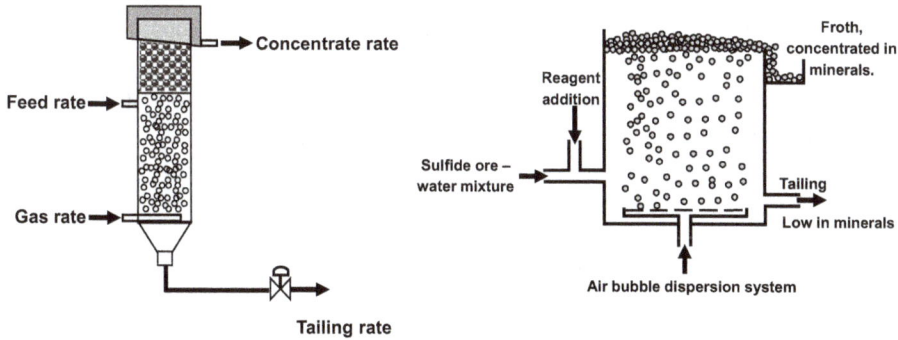

Fig. 15.45: Froth flotation cell/column [8].

metal concentration. Some target minerals report to the froth layer by entrainment in between bubbles as well as by water passing through the froth.

Together with collectors, *frothers* are added. They have the main functions of helping the formation of small bubbles (increasing thus the area of contact bubbles-minerals and thus the efficiency of the separation), of reducing the bubble rise velocity (increasing the residence time in the flotation cell and giving time to collectors to be adsorbed on the mineral particles), and to help the formation of the froth, which hinders the burst of the bubbles at the exit from the cell.

There are other additives too:

- *Regulators* are used to modify the action of the collectors, increasing, or decreasing, at request, the water-repellant effect on the mineral surface.
- *Depressants* are used to increase the selectivity of the flotation system, intentionally preventing some minerals to float.
- **pH has an important role in controlling the depression.** Generally, the flotation takes place in alkaline condition, the *pH window* (e.g., pH 10–12) depending on each mineral.

This cocktail of reagents must be carefully balanced and quantities from each reagent closely controlled.

Figure 15.40 presents the process variables for a flotation plant.

Major disturbances of flotation are:

1. Ore mineralogy
2. Particle size distribution (a function of milling and classification)
3. Solids concentration (also measured as density – a function of classification efficiency)

Models proposed for supporting the control philosophy of the process [11–14].

Early, in 1982, Bascur [14] proposed semiempirical model equations to determine the rate constant of the attachment and detachment subprocesses in the froth phase. These model equations were developed as a function of operating conditions such as the volumetric air flowrate (F_{air}), particle size (d_p), bubble size in the froth (d_b, froth), froth depth (h_f), as well as linear (K_{ij}^{FAT} and K_{ij}^{FDT}) and exponential (n_c) fitting parameters, as follows:

$$K_{ij}^{FAT} = K_i^{FAT} F_{air} \left(\frac{d_p}{d_{b,froth}} \right) \left(\frac{h_f}{d_{Bf}} \right) \tag{15.28}$$

$$K_{ij}^{FDT} = K_i^{FDT} \rho_i d_p^{n_c} v_\infty$$

The term ρ_i denotes the specific gravity of mineralogical class i and v_∞ is the bulk fluid velocity due to drainage. From the same study, the concentrate flowrate, F_c is also estimated by using the empirical equation presented as follows:

$$F_c = a_c l_{lip} \left(h_f - h_T \right)^{1.5} \left(1 - \varepsilon_g \right) \tag{15.29}$$

where a_c is a fitting parameter, l_{lip} is the overflowing lip length of the cell, h_f is the froth depth, h_T is the total flotation cell height.

However, this simple model does not explain the complex interrelated processes occurring in the flotation cell.

Phenomenological models for flotation control for the froth phase have been presented largely in [12]. For example, Zaragoza and Herbst (1989) [15] elaborated a kinetic model for the solid mass (M_f) in the froth phase, defined in terms of operating conditions including the concentrate flowrate (F_{conc}), tailings flowrate ($F_{tailings}$), the entrainment water flowrate (F_E), and the air flowrate (F_{air}):

$$\frac{dM_f}{dt} = -\left[F_R K_R + F_{conc}\left(1+a_f\right)\right]\frac{M_f}{\left(1+a_f\right)V_{LF}} + \left(F_{tailings} + F_{air}\frac{1-\varepsilon_g}{\varepsilon_g}a_p\right)\frac{M_p}{\left(1+a_p\right)V_{LP}}$$

(15.30)

where the terms a_p and a_f are the equilibrium constants between the attachment and detachment in the pulp and froth phases, respectively; ε_g is the gas hold-up; V_{LP} is the volume of the liquid in the pulp; V_{LF} is the volume of the liquid in the froth; F_R is the water flowrate draining back; K_R is a froth stability constant.

The dependence of the quantity and quality of the concentrate (grade and flowrate) is synthetically expressed in [1, 13] (Fig. 15.46).

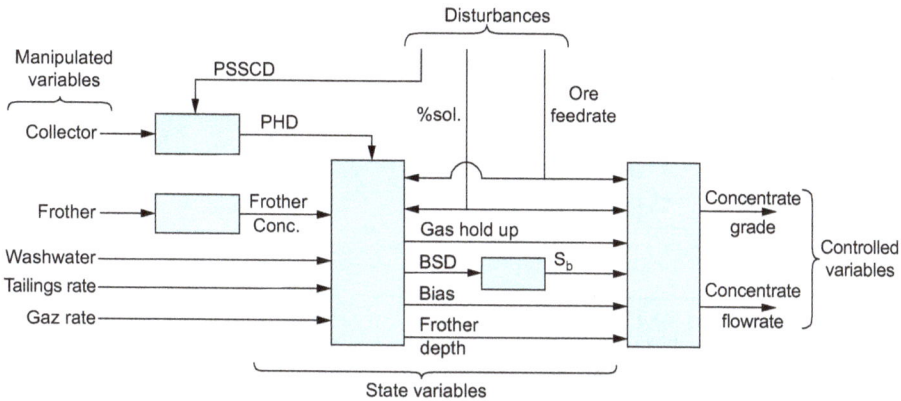

Fig. 15.46: A synthesis scheme of variables involved in the flotation section of mineral processing (reprinted with the permission of Elsevier [1]).

The aim of the control strategy of the flotation stage is:
–   to improve the metallurgical efficiency, by maintaining a specific optimum design point on grade–recovery curve (Fig. 15.47), and
–   to ensure desired mass pull or concentration ratio (C/F) is achieved.

Consequently, explained in this chapter in the Modelling section (eqs 15.28–15.30), the froth flotation control is illustrated in [16].

The fundamental controlled variables in Flotation are: pulp level, air rate, pH, and reagents addition.

An example of classic control for a feedforward strategy of a sulfide rougher bank is that from Fig. (15.48) [3, 15]

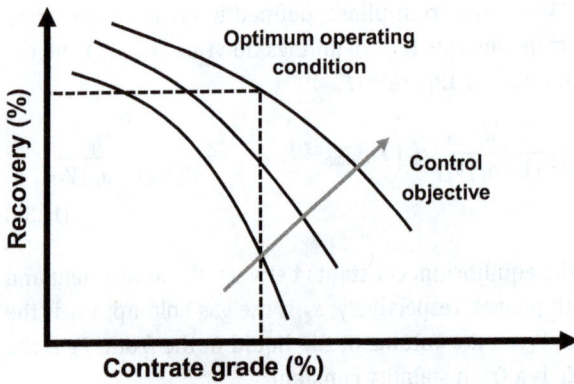

Fig. 15.47: Flotation control objectives.

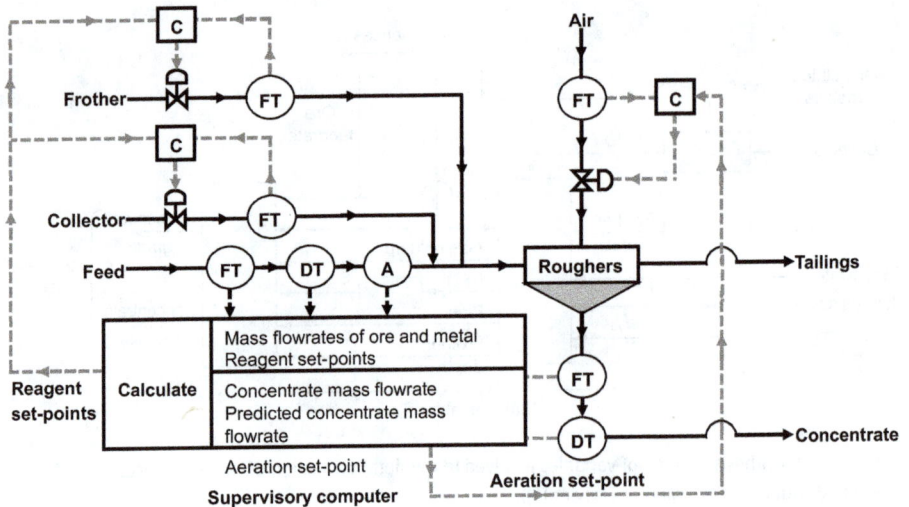

Fig. 15.48: Control scheme for a rougher. The rougher is the first stage in flotation producing a rougher concentrate. Its objective is to remove the maximum amount of valuable mineral at a practical coarse size, to consume less energy. A, metal content; C, controller; DT, density transmitter; and FT, flow transmitter.

The pulp level can be controlled measuring and keeping stable the interface of the liquid–froth phases in a flotation cell and more importantly to increase or decrease this level as dictated by measure concentrate grade and recovery (Fig. 15.49). The control is achieved by opening and closing the tailings control valve.

Grade and recovery in flotation are always inversely proportional, thus increment of one leads to a decrease in the other, and hence the shape of the grade-recovery curves. Operating flotation cells at lower levels is achieved by increase tailings valve opening and it leads to increase in grade but decreases recovery due increase probability of the valuable minerals to detach from the bubbles and fall back into the pulp. The opposite is true. Lower levels also lead to decreased mass pull/concentration ratio.

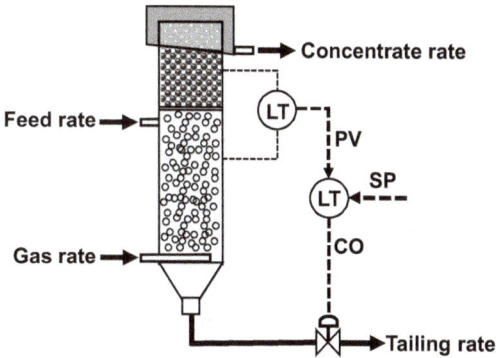

Fig. 15.49: Feedback control of the pulp level in a flotation cell. CO, control signal; LC, level controller; LT, level transmitter; PV, process variable (level); and SP, setpoint (desired level).

Advanced control schemes are devoted to increase the economic efficiency of the process, optimizing it. Real-time optimization using nonlinear programming is an option. Using these methods, as presented in Chapter 4, setpoints calculated according to the optimization procedure (e.g., pulp level, air rate, and reagents' flowrates) are slowly moved from the operational values to the new calculated values.

Bibliography

[1]   Hodouin, D., *Methods for automatic control, observation, and optimization in mineral processing plants*, Journal of Process Control, 21, 211–225, 2011.
[2]   Sushanta, K.P., *Process Control and Instrumentation in Mineral Processing Operations*, Lecture notes, Department of Mineral Engineering Government College of Engineering, Keonjhar, Odisha, http://www.gcekjr.ac.in/pdf/lectures/2020/3014All_7th%20Semester_Mineral%20Engineering.pdf
[3]   Wills, B.A., Napier-Munn T., *Mineral Processing Technology*, Elsevier Science & Technology Books, 2016, chapter 12.
[4]   Hodouin, D., *Automatic Control In Mineral Processing Plants: An Overview*, pp. 14–16 in *Presentation at IFACMMM 2009*, Viña del Mar, Chile, October 2009.
[5]   Nuñez, E., et.al, *Self-Optimizing Grinding Control for Maximizing Throughput While Maintaining Cyclone Overflow Specifications*, pp. 541–555 in *Proc. 41st Annual Meeting of The Canadian Mineral Processors Conference*, Ottawa, Ontario, Canada, 2009.
[6]   Wills, B.A., Napier-Munn T., *Mineral Processing Technology*, Elsevier Science & Technology Books, 2016, chapter 3
[7]   Sbarbaro, D., Ascensio, P., Espinoza, P., Mujica, F., Cortez, G., *Adaptive soft-sensors for online particle size estimation in wet grinding circuits*, Control Engineering Practice, 16(2), 171–178, 2008.
[8]   Agachi, P.S, Cristea, V.M, Makhura, E.P., *Basic Process Engineering Control*, 2nd Edition, De Gruyter Publishing House, Berlin, Boston, 265, 2020.

[9] Eren, H., Gupta, A., *Automation in hydrocyclone operations*, in *IFAC Conference in Mining and Metal Processing*, Buenos Aires, Argentina, 1989.

[10] Savytski, A.I., Timosenko, M.A., *Automated control of classification in a hydrocyclone with incomplete information*, Computer science, Information Technology, Automation, 3(2), 4, 2017.

[11] Quintanilla, P., Neethling, S., Navia, D., Brito-Parada, P.R., *A dynamic flotation model for predictive control incorporating froth physics. Part i: model development*, Minerals Engineering, 173(1), 107192, 2021.

[12] Sun, B., Yang, W., Mingfang, H., Wang, X., *An integrated multi-mode model of froth flotation cell based on fusion of flotation kinetics and froth image features*, Minerals Engineering, 172, 107169, 1 October 2021.

[13] Quintanilla, P., Neethling, S., Navia, D., Brito-Parada, P.R., *Modelling for froth flotation control: A review*, Minerals Engineering, 162, 106718, 1 March 2021.

[14] Bascur, O.A., *Modeling and Computer Control OF A Flotation Cell*, University of Utah, Salt Lake City, 1982

[15] Zaragoza, R., Herbst, J.A., *Model based feed forward control scheme for flotation plants*, Mineral Metallurgy Process, 177–185, 1989.

[16] Wills, B.A. and Napier-Munn T., *Mineral Processing Technology, Chapter 12.15, Control Of Flotation Plants*, Elsevier Science & Technology Books, 345–350, 2016.

15.5 The quality-by-control paradigm in pharmaceutical engineering

Regulatory agencies supervise the pharmaceutical industry, which controls both the product and process-related aspects of manufacturing. For example, there are regulations for the dosage error of an active pharmaceutical ingredient (API) in each tablet, limits for the dissolution rate (i.e. the time at which a well-specified percentage of API must dissolve), etc. Furthermore, limits for different kinds of impurities must be strictly and consistently realized. This includes solvent, reagent, or by-product, and catalysts. The operation that guarantees product quality is regulated as well. Accordingly, the freedom to implement operation or technology improvements without regulatory implications is limited. Hence, pharmaceutical process and operation design are critical, first and foremost, for the patients' safety and because these technologies are expected to operate for a long term without significant operation policy improvement.

The current pharmaceutical design approach is the quality-by-design (QbD) [1], a generic, quality-focused design paradigm. The pharmaceutical projection of QbD entails defining the product property profile composed of critical quality attributes (CQAs) that a manufacturing technology must realize. CQA can be a product's purity, by-product content, particle size distribution, dissolution rate, etc. Then, technology should be designed, and an essential part of the design is the identification of critical process parameters (CPPs). CPPs are those parameters that influence the CQAs. Finally, a so-called design space must be determined – the space of CPPs within which the CQAs are realized. One must recognize that a great advantage of the QbD from a process design and operation perspective is that it allows some flexibility in CPPs. It is not an expectation to set or control CPPs tightly – which on a practical level would hardly be realizable, but to keep them within a reasonable domain. From a control engineering perspective, the QbD could be summarized as design, then control: a process is designed with traditional methods, usually having a relatively high experimental burden (not rarely, design of experiments – DoE), then it is implemented using an appropriate control system.

Quality-by-control (QbC) was proposed recently as an extension of QbD, and a definition was given as follows [2]:

> "QbC consists of the design and operation of a robust manufacturing system that is achieved through an active process control system designed in accordance with hierarchical process automation principles, based on a high degree of quantitative and predictive product and process understanding. QbC in general enables reliable batch and continuous process operations, especially the real-time release in continuous manufacturing of pharmaceutical products."

Hence, QbC proposes the targeted application of process control to reinforce the control of CQAs directly via the CPPs, instead of controlling the CPPs tightly and letting the disturbances propagate through the process and reach the CQAs. The roles of process control in QbD and QbC are compared in Tab. 15.8.

Tab. 15.8: Active process control in the QbD and QbC approaches in pharmaceutical engineering.

Property	QbD	QbC
Highest level controller in terms of CQAs	0: the CPPs are controlled directly, such as temperature and feeding flow rate	1 or 2: the CQAs are controlled directly in a local (level 1), or centralized plantwide supervisory control system (level 2)
Controlled variable of the highest level controller	CPPs (temperature, flow rate, etc.).	CQAs
Manipulated variables of the highest level controller	Technological parameters (coolant flow rates, pump RMPs, etc.).	CPPs
Necessity/role of CQA monitoring or real-time estimation	Not necessary to operate the process but may be used for real-time release	It is necessary to operate the process

The evolution of quality control approaches at a process design and operational level is illustrated in Fig. 15.50. The current regulatory standard is the QbD. However, the QbC brings numerous important improvements.

From a process operation perspective, QbC culminates in fault-tolerant control, which has been applied in other (continuous) manufacturing industries over the last decades. These strategies can accommodate faults among system components automatically while maintaining system stability along with a desired level of overall performance [3]. One approach responds to a failure by reorganizing the remaining system elements in real time to carry out necessary control functions. The other is to make the

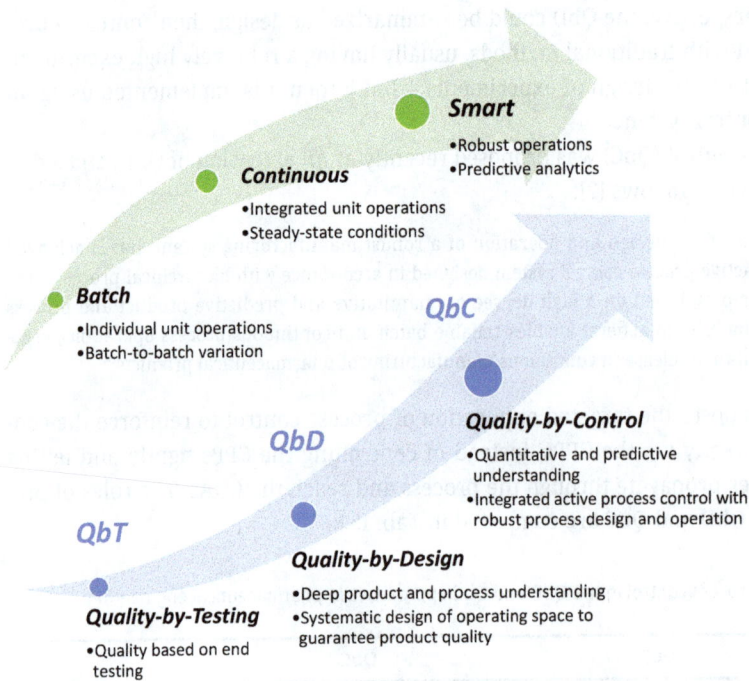

Fig. 15.50: High-level comparison of quality-by-testing, quality-by-design, and quality-by-control approaches [2].

system failure-proof for specific well-defined risk/fault sets at the design stage [4]. Beyond the system stability, product quality under system component failures is even more important under the International Council for Harmonization (ICH) Q9 Quality Risk Management guidance. QbC aims to combine fault-tolerant control practice with product quality concerns for process control design and risk analysis in pharmaceutical manufacturing and for effective alarm management frameworks.

When QbC is implemented as a design tool, it utilizes direct process control strategies, implying the CQA as the controlled variable (or variables) and the CPP as the manipulated variable (or variables). By that, it drives the CPPs to realize the desired CQAs. This is a direct process design as it provides the dynamic trajectory of CPPs that leads to the desired product CQA. Yet, the design space must be determined around the CPPs provided by the QbC experiments. Still, the core solution is obtained quickly, which may be an optimum solution if found with an optimal control strategy. QbC methods can be divided into two groups: model-free (mf), and model-based (mb) QbC [5].

mfQbC relies entirely on real-time CQA measurements or estimations using process analytical technology (PAT) tools. The experimental setup for mfQbC is similar to the setup used for QbD experimentation, and it is available in modern pharmaceutical process development laboratories. Generally, mfQbC consists of hierarchical feedback control strategies: the high-level controller adjusts the CQAs by manipulating the setpoints for low-level CPP (e.g., temperature, flow rate, and concentration) controllers. This latter is often a PID scheme. Meanwhile, the application of QbC is relatively fast and has the potential to find a suitable solution with less experimentation, it struggles with the satisfaction of constraints and gives a suboptimal solution.

mbQbC implements a model-based control strategy. It is not unusual that pharmaceutical processes involve a solid particle phase, translating to nonlinear processes. Hence, mbQbCs are often nonlinear MPCs. These have the advantage of directly accounting for constraints and being used for unmeasurable/inestimable CQAs. Still, the challenges associated with developing a high-fidelity, numerically sufficiently fast model delimit its applications.

Despite its highly regulated nature, QbC can help the pharmaceutical industry exploit the benefits of advanced process control through streamlined process design. Several studies applied QbC, and a few representative titles are listed in Tab. 15.9.

Tab. 15.9: Representative publications implying the concepts of quality-by-control in pharmaceutical engineering (as of early 2023).

Title	Reference
Application of process analytical technology-based feedback control strategies to improve purity and size distribution in biopharmaceutical crystallization	[6]
A perspective on quality-by-control (QbC) in pharmaceutical continuous manufacturing	[2]
Experimental implementation of a QbC framework using a mechanistic PBM-based nonlinear model predictive control involving chord length distribution measurement for the batch cooling crystallization of l-ascorbic acid	[7]
QbCl: toward model predictive control of mammalian cell culture bioprocesses	[8]
Application of model-free and model-based QbC for the efficient design of pharmaceutical crystallization processes	[5]
Quality-by-control of intensified continuous filtration-drying of active pharmaceutical ingredients	[9]
Enzyme cascade reaction monitoring and control	[10]
Evaluation of a combined MHE-NMPC approach to handle plant-model mismatch in a rotary tablet press	[11]
Semimechanistic reduced order model of pharmaceutical tablet dissolution for enabling Industry 4.0 manufacturing systems	[12]
A PSE perspective for the efficient production of monoclonal antibodies: integration of process, cell, and product design aspects	[11]

Note: Population Balance Model (PBM): Process Systems Engineering (PSE).

Besides the potential advantages, there are remarkable engineering challenges associated with the broader spread of QbC across pharmaceutical industries:

The pharmaceutical industry lacks common understanding and systematic framework for process control development in secondary manufacturing. This is particularly true for solid-based unit operations, which require faster response times (order of seconds or minutes) than most fluid-based industries, where process control has been widely applied. Variability in raw materials has a rapid and direct impact on downstream processes, making it difficult to produce consistent quality.

The deployment of real-time process analytical technology (PAT) tools for reliable CQA measurements remains challenging due to sensor calibration complexity, measurement drifts, and bias caused by sensor positioning, sampling concerns, and fouling. QbC must be prepared to use noisy and biased CQA measurements to supervise the control of CPPs, making it necessary to incorporate redundancy in sensor networks and use data reconciliation combined with joint state and parameter estimation to address uncertain measurements.

Nevertheless, to illustrate the capabilities of QbC, two case studies will be presented that demonstrate how process control can be employed as a high-performance pharmaceutical process design and operation tool. The focus will be on secondary manufacturing: from crystallization to final dosage form. The first case study is the process design allowing the size-controlled, thermodynamically guaranteed production of desired polymorphic crystals. The second case study presents the accurate operation of a pilot-scale tablet press machine.

15.5.1 QbC for thermodynamic polymorph selection in crystallization processes

Polymorphic structure is often a CQA of APIs as the undesired polymorph may have reduced bioavailability – that is, insoluble in the body fluids. Furthermore, different polymorphs often exhibit different crystal shapes, which may cause manufacturing-related difficulties through poor flowability, bulk density, etc. Numerous polymorphic systems follow the Ostwald's rule of stages, that is, at high supersaturations, the metastable form (the thermodynamically less favored under the given conditions) may be formed, which opens the door to polymorph selection in kinetic control. This metastable phase is slowly transformed to the thermodynamically stable polymorph through solvent-mediated polymorphic transformation, unless the process is stopped (e.g. the slurry is not filtered). However, the nucleation kinetics that drives the kinetic control is highly sensitive to process conditions, which is a straightforward source of batch-to-batch variation of polymorphic purity, and may make the scale-up and technology transfer challenging. A specific case of polymorphs is the enantiotropic behavior, where the solubility lines cross each other, as illustrated in Fig. 15.51. Hence, the stability of the polymorphs depends on the temperature.

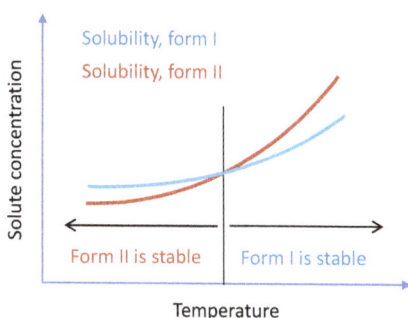

Fig. 15.51: Schematic representation of an enantiotropic polymorphic system in the concentration-temperature phase diagram: the solubility lines cross each other in the transition temperature, resulting in a temperature dependency of the relative stability of the two polymorphs.

The transition point allows operating in the domain where the desired polymorph is stable, hence, no longer relying on kinetic control. Hence, thermodynamic control is feasible in enantiotropic systems. Still, the applicability of traditional batch cooling crystallization may be yield-constrained by the distance between the freezing and transition temperature and the transition point and the solvent's boiling point (or, more precisely, the solubilities corresponding to these three temperatures). Fed-batch systems can help overcome the yield limitation posed by the transition temperature.

The polymorphic content can only be measured with advanced PAT tools. Yet, the polymorphic purity can be guaranteed if the system is supersaturated for the desired and undersaturated for the undesired polymorph under the applied process conditions. In a QbC context, these conditions are ensured through direct feedback control of supersaturation (σ). σ is a function of actual solute concentration (c) and the solubility (c_s, which is subsequently a function of temperature T), and it is defined as

$$\sigma = \frac{c_s(T) - c}{c_s(T)} \qquad (15.31)$$

Hence, the supersaturation is calculated from measured temperature and actual concentration, in the knowledge of temperature dependency of the solubility. This makes the SSC an inferential controller. The batch SSC implies the reactor temperature as the manipulated variable (resolved by the low-level temperature controller). In contrast, for batch SSC the manipulated variable is the flow rate of the concentrated feeding stream. In this latter strategy, independent control loops apply for a robust operation to set the temperature of the crystallizer and the feeding vessel. The control schemes are presented in Fig. 15.52.

The transition temperature is 14 °C for the aqueous *para*-aminobenzoic acid (PABA) solution. This system does not follow the Ostwald's rule of stages: the β-PABA has slow nucleation kinetics, and the α-PABA has fast nucleation kinetics, including at small supersaturations. This makes it challenging to obtain β-PABA with high purity

(a)

(b)

$T_{2,sp} > T_{1,sp}$

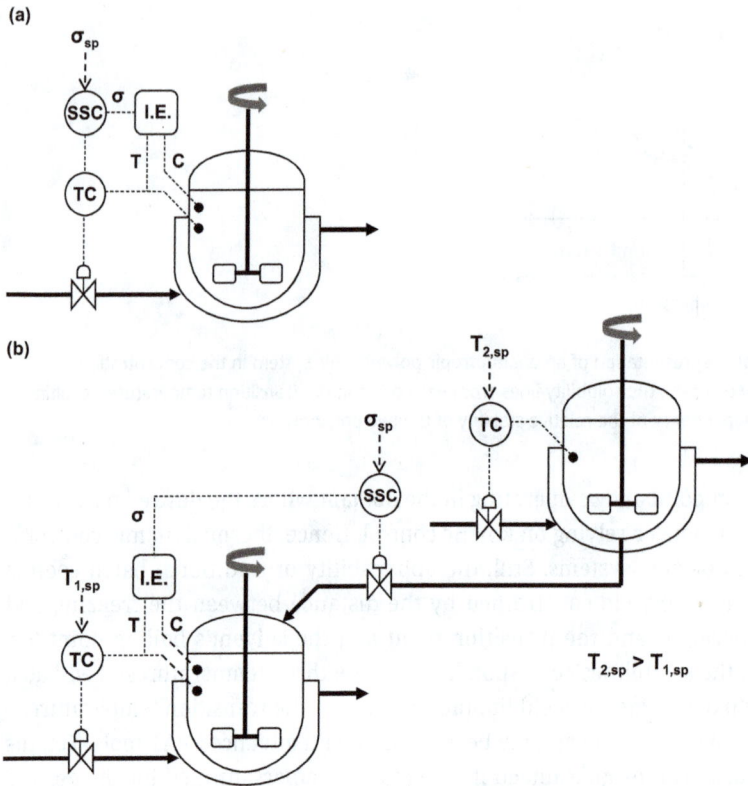

Fig. 15.52: (a) Supersaturation control achieved by manipulating the temperature; (b) supersaturation control achieved by manipulating the flowrate of a concentrated PABA stream (having high temperature to maintain undersaturated conditions despite high concentrations). I.E., inferential estimator.

in a batch-cooling crystallization process with good yield. Below 14 °C the β-PABA is the thermodynamically favored polymorph. Still, the operating window is narrow due to the physical constraints of the system (freezing point). Alternatively, β-PABA can be crystallized in a semibatch process by setting the temperature slightly under 14 °C, and by choosing a supersaturation setpoint that keeps the actual concentration under the solubility of α-PABA. These conditions guarantee thermodynamic polymorph selection by process control for β-PABA. Furthermore, the feeding strategy eliminates the yield limitation of the transition point. Three representative experiments were carried out, as summarized in Fig. 15.53.

1. SSC in batch mode with a relatively good yield in the 22-0 °C domain (Fig. 15.53a). This experiment was started from above the transition point, but where the solubility curves are still relatively close to each other. By seeding with β-PABA and keeping the supersaturation for α-PABA at reasonably low values, one may rely on slow nu-

cleation kinetics of the α-PABA that prohibits its formation and growth, resulting in pure β-PABA at the end of the batch.

Outcome: the α-PABA nucleated shortly after seeding, and the fast growth of α-PABA crystals consumed the supersaturation, leading to the complete dissolution of the β-PABA seeds. In the low-temperature range, despite the α-PABA becoming metastable, the nucleation thereof, and the solvent-mediated polymorphic transformation to β-PABA did not happen. The product consisted of pure α-PABA crystals.

2. SSC in batch mode with modest yield in the 14-0 °C domain (Fig. 15.53b). At the beginning of the batch, the solution is supersaturated for both polymorphs. The supersaturation is maintained for the β-PABA, but it vanishes for α-PABA with the progression of the process.

Outcome: the product consisted of pure β-PABA, but the α-PABA polymorph appeared during the process.

3. SSC in semi-batch mode with relatively good yield in the 22-0 °C domain (Fig. 15.53c). The crystallization temperature is fixed at 0 °C temperature, and the feed solution is saturated at 22 °C (the yield is the same as in the first case).

Outcome: the product consisted of pure β-PABA, and no α-PABA impurity appeared during the process.

The system was seeded with the identical amounts and quality of β-PABA crystals for consistency. The batch time of the semibatch SSC was about half the batch time of the cooling SSC, despite the higher yield. A reason for this may be the emergence of α-PABA impurity in the batch case, which had to undergo solvent-mediated polymorphic transformation to β-PABA at lower temperatures.

To simplify the implementation of SSC at a manufacturing scale, one can apply the principles of QbC: use a feedback control experiment to determine the operating conditions, which lead to desired product properties, then apply the implemented operating profile in an open-loop manner. Another semibatch SSC was executed to demonstrate this concept, using a feeding stream with a concentration corresponding to the solubility at 25 °C. The feeding profile obtained was approximated with linear segments (for the ease of implementation in industrial DCS or SCADA systems), which was then subsequently implemented in an open-loop manner on the supersaturation, but closed-loop way on the flow rate. The supersaturation was measured in the QbC experiment for monitoring purposes only. The results are illustrated in Fig. 15.54. Accordingly, the supersaturation stayed close to the setpoint of the closed-loop experiment, despite the lack of feedback control. From a practical perspective, this operating strategy can be scaled up to the manufacturing scale, which requires extra care as fed-batch crystallization is a mixing-sensitive process [13].

The particle size is often another CQA of APIs. Tuning the size of crystals in SSC is possible through seed strategy design (eq. (14.4)), which works if the supersaturation setpoint is within the metastable zone to reject nucleation.

The case study of this section relies on a research paper [14].

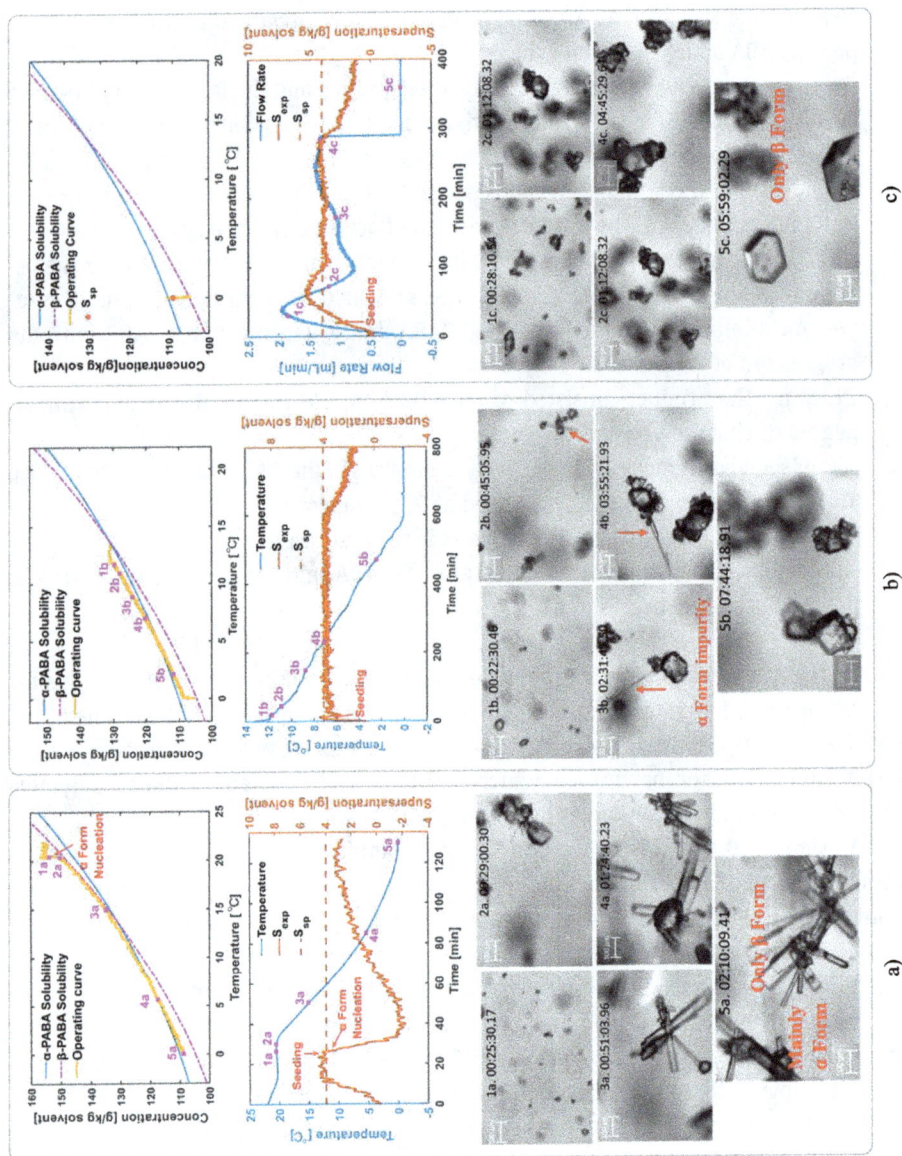

Fig. 15.53: (a) Batch SSC started slightly over the transition point where the solubilities are relatively close to each other (for yield considerations); (b) batch SSC started from the transition temperature; (c) semibatch SSC with the yield domain of the (a) case. Upper row: solubilities and process trajectories in the temperature-concentration phase diagram; mid-row: actual and target supersaturations and the temporal evolution of the manipulated variables; bottom row: representative in-line microscope images captured in the time stamps denoted in the phase diagrams (reprinted with permission from [14]. Copyright 2019 American Chemical Society).

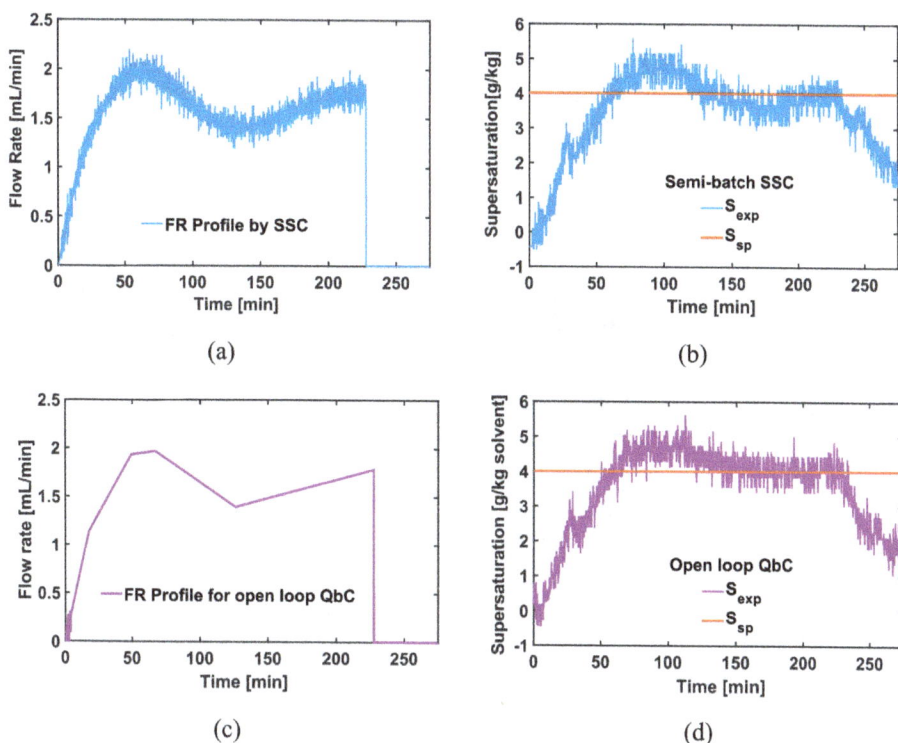

Fig. 15.54: Process operation design by semibatch supersaturation control (mfQbC). The temperature profile ensures that the supersaturation is sufficiently high for good productivity, but it confidently remains under the solubility of the undesired enantiomer ensuring thermodynamic polymorph selection (reprinted with permission from [14]. Copyright 2019 American Chemical Society).

15.5.2 QbC in continuous tablet manufacturing

According to 2021 estimates, tablets dominated the global solid dosage pharmaceutical market segment, with a market share exceeding 55 % over the rest of solid formulations (capsules, powders, and others). Hence, tableting is highly important for the pharmaceutical industry. Tablets are produced in tableting machines, which rely on five repetitive steps (Fig. 15.55b): (i) filling the compression die with powder containing all active ingredients and excipients in the desired quantity and quality; (ii) pre-compression: a compression force is applied to the powder bed to form partial compacts before undergoing the main compression; (iii) compression: the tablet will get its final properties in the second compression step, with exact force and dwell time to be adjusted to the material; and (iv) ejection: the tablet is ejected from the die. The pre-compression force is usually but not necessarily smaller than the main compression. The exact force and compression durations (called dwell time) are adjusted to the material.

The case study presented in this section was conducted in the Continuous Solids Processing Pilot Plant at Purdue University. The manufacturing line includes continuous feeders, blenders, and a rotary tablet press (Fig. 15.55a). The tablet weight can be controlled by adjusting the dosing position, and the punch displacement determines the in-die tablet thickness.

(a)

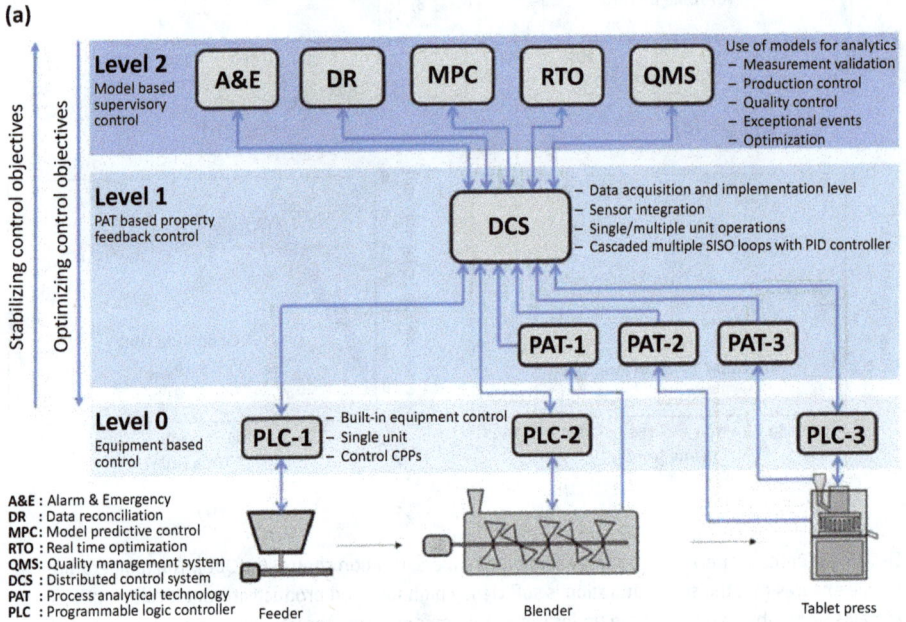

(b)

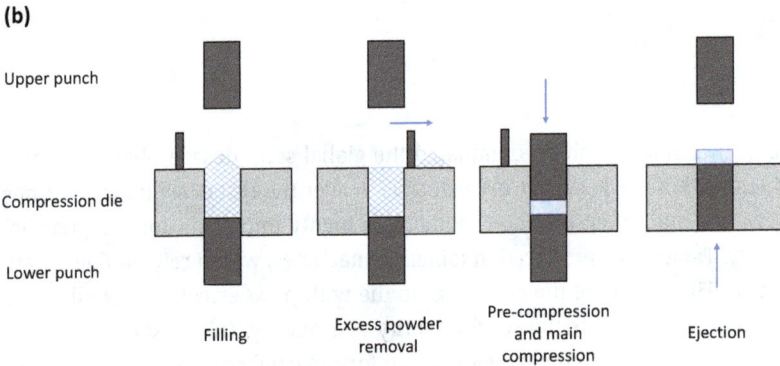

Fig. 15.55: (a) A hierarchical three-level controller for a direct compaction tablet manufacturing line (adapted from the literature with the permission of Elsevier). (b) The main steps involved in tablet pressing (reprinted from [2], with permission from Elsevier).

The tablet must satisfy the regulatory constraints (that includes, API quantity and dissolution rate). Meanwhile, some properties of the tablet will not depend on the tableting process; others may be impacted, such as:

Tablet weight (API amount): The API amount is defined by the filling volume and the bulk density of the powder (which depends further on the size, shape, and other properties). Therefore, any change in the material properties may propagate to the tablets.

Dissolution properties: The effective dissolution of the API from the tablet, beyond the polymorphism and particle size and shape distribution, depends on the tableting settings: higher tableting force leads translate to slower disintegration, thereof, to more prolonged API release.

The aforementioned points project that a tableting machine operated in a feedback manner lets the incoming material disturbances propagate to the product. Sufficiently large disturbances may lead to out-of-specification (OOS) products. Reworking an OOS tablet would consist of recovering the API (or APIs) from the pills and recycling them to the technology in the particle formation (e.g. to the crystallization). This is cost-ineffective and is rarely allowed as the lot number of starting material must be tracked according to the regulations. The application of QbC can prevent these situations.

The QbC strategy of this case study implies near-infrared spectrometry and X-ray-based mass flow meters to continuously measure API mass fraction and powder flow rate, respectively. A real-time tablet weight measurement was based on a Mettler Toledo ME 4001E balance. The role of higher sampling time in tablet weight measurement demonstrated the importance of capturing process dynamics for better process control performance. The critical-to-quality variables in the tablet press were identified as tablet weight, relative density, tensile strength, and main compression force. The Emerson DeltaV 13.3 distributed control system was used to integrate process equipment and develop the automation system. Modular and hierarchical network architecture was implemented following ISA 95 and DeltaV Security Manual recommendations for systematic implementation of QbC. The tools such as KepServerEX, LinkMaster, and Matlab's Instrument Control Toolbox were used to interface the PATs, laptops, and control system.

The continuous measurement of tablets was solved with the Mettler Toledo ME 4001E (Fig. 15.56a). A container on balance collected the tablets continuously, and the cumulative weight was measured. This allowed utilizing eq. (15.33) for approximating the weight of single tablets in real time:

$$W_T = \frac{TPR}{TS \times NS} \tag{15.32}$$

where TPR is the tablet production rate (mg/h), TS is the turret speed (rot/min), and NS is the number of stations (#/rot).

This measurement strategy alone can be misleading. Gross errors can happen when, e.g. tablets are diverted to the Sotax ST4 or when the container is replaced. Hence, a data reconciliation strategy is needed to reduce measurement uncertainty and eliminate gross errors.

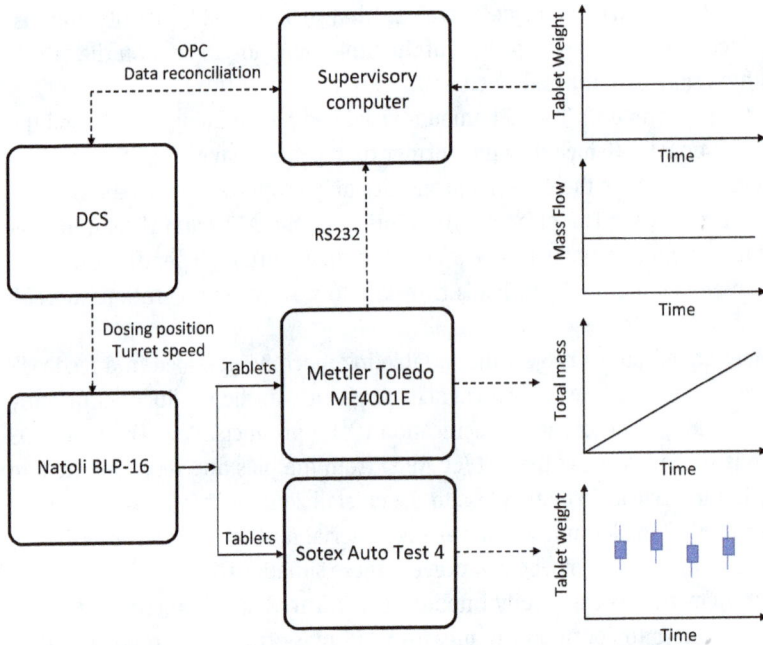

Fig. 15.56: (a) Principle of tablet weight measurement real-time monitoring and control.

First, the powder compressibility is modeled by using the relationship between the main compression force (*CF*) and the resulting in-die tablet relative density ρ_c. Here, the *CF* was captured with the Kawakita model at a given in-die tablet ρ_c:

$$\frac{CF}{1 - \frac{\rho_c}{\rho_r}} = \frac{CF}{a} + \frac{\pi D^2}{4ab} \tag{15.33}$$

where parameters a and b can be interpreted as the maximum degree of compression and the reciprocal of the pressure applied to attain the maximum degree of compression. D is the diameter of the die. The in-die relative density (ρ_r) is computed from the tablet weight as

$$\rho_r = \frac{4W_T}{\pi D^2 \rho_t G} \tag{15.34}$$

where ρ_t is the true density of the powder and H is the in-die tablet thickness.

The ρ_r is an important property that can be used to characterize the material and the compression process, which should be monitored in real time and periodically re-estimated. Data reconciliation was implemented for the tablet weight measurement to correct for the imperfect measurement data and estimate ρ_r based on the process knowledge of material compressibility. The generalized reconciliation problem is posed as an optimization:

$$\min_{z_t} \left(\sum_{t \in T} w_z^T \vartheta(z_t - z_{m,t}) + \sum_{t \in T} w_m^T (z_t - z_{t-1})^2 \right) \tag{15.35}$$

subject to

$$z \subseteq [y, x, u, \theta]$$

$$f(y, x, u, \theta, t) = 0$$

$$g(y, x, u, \theta, t) \le 0$$

where $z_{m,t}$ is the vector of measured process variables z at time t; z_t is the reconciled measurement at time t. x, y, u are process output, state, and input variables. θ is the model parameter in the vector function f and g, which represent the process model.

A robust estimator ϑ was incorporated to eliminate gross errors. w_z is the weight vector for the measurement error, and w_m is the penalty weight for the successive moves of the reconciled variables and model parameters. T is the moving window of calculations. A Welsch robust estimator ϑ_w was used, as shown below:

$$\vartheta_w(\varepsilon) \frac{c_w^2}{2} \left\{ 1 + \exp\left[-\left(\frac{\varepsilon}{c_w}\right)^2 \right] \right\} \tag{15.36}$$

where ε is the standardized residual, z_t is a tuning parameter, $e = (z - z_m)$ is the measurement error, and σ is the standard deviation of measurement error, which was estimated from historical data:

$$\varepsilon = \frac{e}{\sigma} \tag{15.37}$$

To jointly estimate and update the Kawakita model parameter, the ρ_c was included in the reconciled vector variable $z = [CF, W_T, \rho_c]$. A reference density estimated from the historical distribution of ρ_c can be considered. In this manner, both the uncertainty with the tablet weight measurements and the model-plant mismatch or variations in powder compressibility can be tackled using an optimization problem.

Figure 15.57 shows the result of the data reconciliation. The noisy measurement of tablet weight obtained from the balance is shown in Fig. 15.57b). The noisy weight agrees well with the Sotax ST4 measurement, which was executed when the process reached a steady state after each step change. The Kawakita model with a previously determined ρ_c of 0.25 was used to predict the tablet weight from the main compression force (Fig. 16.57a): the data is smooth, but an offset exists between the soft sensor and the (accurate) Sotax ST4 data. When the two measured variables, CF and W_T were reconciled, and the ρ_c was re-estimated, better agreement was achieved between the soft sensor and the actual data (Fig. 15.57c). Figure 15.57d shows the updated values of ρ_c, which showed a slightly increasing trend during the experiment. Notably, the Sotax ST4 measurement is not used in the reconciliation. It is only employed for validation purposes.

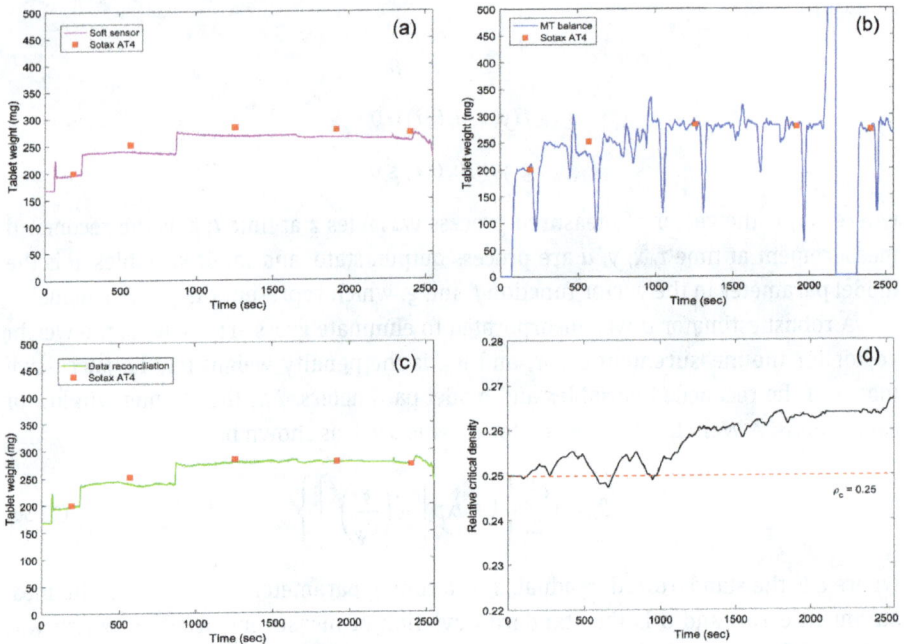

Fig. 15.57: Data reconciliation with level 0 control experiment with at-line Sotax AT4 sampling (reprinted from [2], with permission from Elsevier).

The Natoli BLP-16 tablet press has a built-in PLC panel that can manipulate process parameters of dosing position and turret speed (considered a Level 0 control). A level 1 control with decoupled PID control loops and a level 2 MPC were designed and applied to control tablet weight, tablet production rate, and main compression force. The schemes of the tablet press controlled by level 1 PID controllers and level 2 MPC controller are depicted in Fig. 15.58a and b, respectively.

Continuous tableting experiments were performed in three scenarios to validate online data reconciliation and compare control system performances. The level 0 control operation confirmed that the reconciled tablet weight measurement matched the at-line measurement (Fig. 15.59a). This enabled control of the CQA in real time with high confidence. The level 1 and 2 strategies showed good control performance, with the level 2 MPC showing a more aggressive and promising control performance (Fig. 15.59). The control system design achieved process automation to reach the targeted tablet weight setpoint automatically and steadily. This is particularly important during process startup or operating point switch both by Level 1 and Level 2 control. However, the performance improvement by the level 2 MPC is significant in shortening the period of diversion of

(a)

(b)

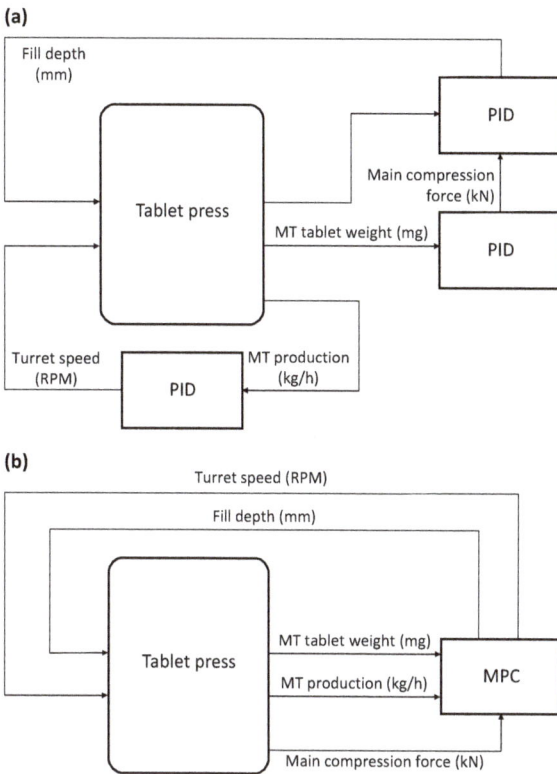

Fig. 15.58: (a) Level 1 (PID) and (b) level 2 (MPC) control of the tablet press machine.

OOS product during setpoint changes or process disturbance (compare the setpoint change dynamics in Fig. 15.59a. Hence, the process control system can maintain the tablet weight under common risks of process disturbances or material property variations, thus attaining a real-time release strategy. During setpoint changes in tablet weight, the tablet production rate was maintained the same to adjust to the campaign production or processing capability upstream. It was assumed that the upstream machinery was being operated steadily. The tablet weight was readjusted to match the new setpoint while maintaining the same overall material throughput (compare Fig. 15.59a and b). On the soft sensor side, the data reconciliation continued updating the model parameter and reached a plateau under level 1 and level 2 control setpoint changes. After a reinitialization of data reconciliation at 3600 s by setting the ρ_c to its initial value of 0.25, offset between reconciled and at-line measurement can be observed, which gradually reduces with the update of ρ_c (Fig. 15.59c).

The case study of this section relies on a research paper [2].

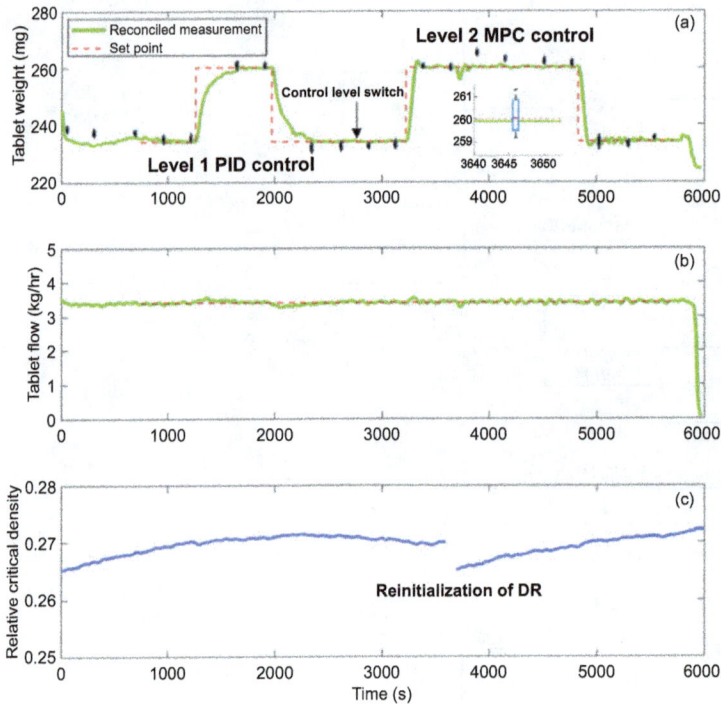

Fig. 15.59: Tablet weight control using level 1 (PID) and level 2 (MPC) control schemes (reprinted from [2], with permission from Elsevier).

References

[1] Juran, M.J., *Juran on Quality By Design: The New Steps For Planning Quality Into Goods And Services*. The Free Press, New York, NY, USA, 1992.

[2] Su, Q., Moreno, M., Bommireddy, Y., Gonzalez, M., Reklaitis, G.V, Nagy, Z.K., A perspective on Quality-by-Control (QbC) in pharmaceutical continuous manufacturing, Computers and Chemical Engineering, 125, 216–231, 2019.

[3] Blanke, M., Izadi-Zamanabadi, R., Bøgh, S.A., Lunau, C.P., *Fault-tolerant control systems – A holistic view*, Control Engineering Practice, 5, 693–702, 1997.

[4] Jiang, J., Yu, X., *Fault-tolerant control systems: A comparative study between active and passive approaches*, Annual Review of Control, 36, 60–72, 2012.

[5] Szilagyi, B., Eren, A., Quon, J.L., Papageorgiou, C.D., Nagy, Z.K., *Application of model-free and model-based quality-by-control (QbC) for the efficient design of pharmaceutical crystallization processes*, Crystal Growth & Design, 20, 3979–3996, 2020.

[6] Simone, E., Zhang, W., Nagy, Z.K., *Application of process analytical technology-based feedback control strategies to improve purity and size distribution in biopharmaceutical crystallization*, Crystal Growth & Design, 15, 2908–2919, 2015.

[7] Szilágyi, B., Borsos, Á., Pal, K., Nagy, Z. K., *Experimental implementation of a Quality-by-Control (QbC) framework using a mechanistic PBM-based nonlinear model predictive control involving chord length*

distribution measurement for the batch cooling crystallization of L-ascorbic acid, Chemical Engineering Science, 195, 335–346, 2019.

[8]   Sommeregger, W., Sissolak, B., Kandra., K., von Stosch, M., Mayer, M., Striedner, G., *Quality by control: Towards model predictive control of mammalian cell culture bioprocesses*, Biotechnology Journal, 12, 1600546, 2017.

[9]   Destro, F., Barolo, M., Nagy, Z.K., *Quality-by-control of intensified continuous filtration-drying of active pharmaceutical ingredients*, AIChE Journal, 69, 17926, 2023.

[10] Hiessl, R., Kleber, J., Liese, A., *Enzyme Cascade Reaction Monitoring and Control*, pp. 141–163 in Kara, S., Rudroff, F., (Eds.), *Enzyme Cascade Design and Modelling*, Springer International Publishing, Cham, 2021.

[11] Badr, S., Sugiyama, H., *A PSE perspective for the efficient production of monoclonal antibodies: integration of process, cell, and product design aspects*, Current Opinion in Chemical Engineering, 27, 121–128, 2020.

[12] Ferdoush, S., Gonzalez, M., *Semi-mechanistic reduced order model of pharmaceutical tablet dissolution for enabling Industry 4.0 manufacturing systems*, International Journal of Pharmaceutical, 631, 122502, 2023.

[13] Ostergaard, I., Szilagyi, B., Nagy, Z.K., Lopez de Diego, H., Qu, H., *Polymorphic control and scale-up strategy for crystallization from a ternary antisolvent system by supersaturation control*, Crystal Growth & Design, 20, 1337–1346, 2020.

[14] Zhang, T., Szilágyi, B., Gong, J., Nagy, Z.K., *Thermodynamic polymorph selection in enantiotropic systems using supersaturation-controlled batch and semibatch cooling crystallization*, Cryst Growth Des, 19, 6715–6726, 2019.

16 Problems and exercises

16.1 Advanced process control

1. A first-order reaction takes place in a cascade of three equal-dimension continuous stirred tank reactors (CSTRs). Concentration transducers are placed after each reactor, and they measure the concentration with a dead time $\tau = 1$ min. An important load disturbance is the feed reactant concentration. The final concentration is modified via input flow to the reactor. Compare the performances (stability and overshoot) of a feedback control system with those of a cascade control with (a_1) inner measurement done after the second reactor and (a_2) inner measurement done after the first reactor.

   Technical data are: $V_1 = V_2 = V_3 = 1$ m^3, $F_1 = F_2 = F_3 = 3$ m^3/h, and $k_1 = k_2 = k_3 = 0.1$ h^{-1}; conversion measurement is done by each transducer with a dead time of 1 min; the transducer gains are of 2 kmol/mA each; the control valve gain and time constant are 12 m^3/h · mA and 0 s, respectively.

2. The automatic control system (ACS) from Fig. 16.1 is tuned for a quarter amplitude decay ratio. Compare in terms of stability and steady-state error, the performances of a simple feedback control system – see controller K_1 in Fig. 16.1, with that of the cascade of two controllers presented in Fig. 16.1.

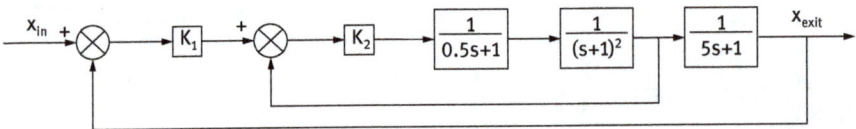

Fig. 16.1: A cascade ACS.

3. The ACS for concentration presented in Fig. 16.2 has the following characteristics: chemical reaction $A \rightarrow B$ is of first order with the rate constant $k = 0.5$ min^{-1}; the residence time in each reactor is of $V/F = 2$ min; and the molar concentrations in steady state are $C_{A1} = 0.4$ kmol/m^3, $C_{A2} = 0.2$ kmol/m^3, $C_{A3} = 0.1$ kmol/m^3, and $C_{A0} = 1.8$ kmol/m^3. The main load disturbance is the variation of input concentration entering through the feed tank placed ahead of the reactors and having a time constant of 5 min. The composition in the tank is measured with a composition disturbance transducer with the gain 1 and dead time 2 min. The actuating device, which controls the composition through the concentrate flow, has the total transfer function $H_{AD}(s) = 1$.

   Synthesize the disturbance controller transfer function by knowing that the manipulating variable is the concentrate flow.

https://doi.org/10.1515/9783110789737-018

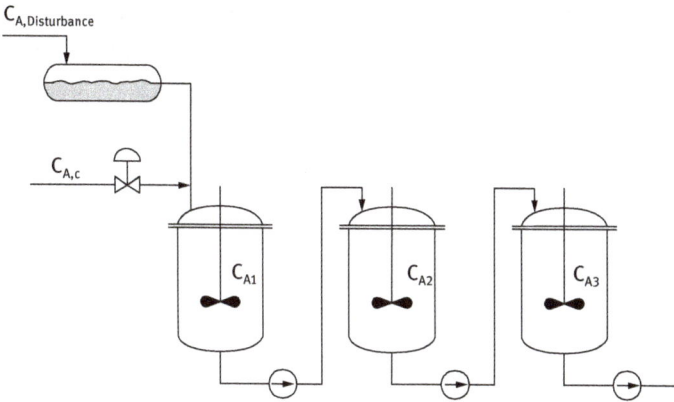

Fig. 16.2: Cascade of three CSTR units.

4. The transfer functions of a binary distillation column are as follows:

$$\frac{X_D(s)}{F(s)} = \frac{K_F e^{-\tau_F s}}{(T_F s + 1)^2}; \quad \frac{X_D(s)}{X_F(s)} = \frac{K_X e^{-\tau_X s}}{(T_X s + 1)^2}; \quad \frac{X_D(s)}{R(s)} = \frac{K_R e^{-\tau_R s}}{(T_R s + 1)^2}$$

where x_D is the output that has to be kept constant, F and x_F are the disturbances, and R is the manipulated variable. Carry out a numerical simulation with the following values: $K_F = 0.51 \frac{kmol/m^3}{kmol/min}$, $K_X = 0.12 \frac{kmol/m^3}{kmol/m^3}$, $K_R = 0.22 \frac{kmol/m^3}{kmol/min}$, $T_F = 5$ min, $T_X = 6$ min; $T_F = 2$ min, $\tau_F = \tau_X = 0.5$ min, and $\tau_R = 0.1$ min. Steady-state values are $F = 1$ kmol/min, $R = 0.4$ kmol/min, $x_F = 0.5$, and the disturbances are 10% of the nominal values. The transfer functions of the transducers and control valves are included in the given transfer functions.

5. Determine the transfer functions that link Y to Y_i within the interacting control systems presented in Fig. 16.3. Calculate the damping factor ζ if $T_1 = 8$ min, $T_2 = 8$ min, $K_1 = K_2 = 10$, $K_3 = 3$, and $K_4 = 5$. Calculate the transfer functions of the decoupling elements.

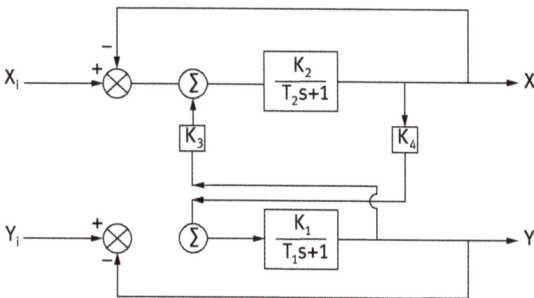

Fig. 16.3: Two interacting control systems.

6. Determine the transfer function of the disturbance controller for the heat-transfer device in Fig. 16.4, when the main disturbance is T_{in}^o. The temperature transducer has the following transfer function: $H(s) = H_{DT}(s) = \frac{0.12\,mA/°C}{0.024s+1}$, whereas of the steam flow-regulating valve is $H_{AD}(s) = \frac{5.26\,L/s/mA}{0.083s+1}$. For the transfer process on the path $F_{ag,in} \rightarrow T_{out}^o$, it is $H_{prm}(s) = \frac{2\,°C/L/s}{(0.017s+1)(0.432s+1)}$, whereas for the $T_{in}^o \rightarrow T_{out}^o$ path, it is $H_{prp}(s) = \frac{0.2e^{-0.1s}}{(0.8s+1)(0.5s+1)}$, respectively. Assess whether the transfer function of the controller is physically feasible and place the control scheme on the figure.

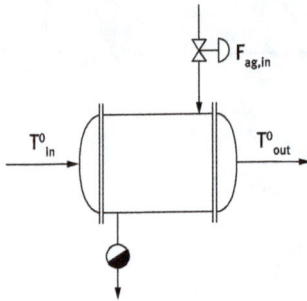

Fig. 16.4: Tubular heat exchanger.

7. Explain how the feedforward control may be included in the model predictive control (MPC) algorithm. Are there any additional tuning considerations to be considered for the feedback-feedforward MPC controller?

8. Can the MPC controller cope with processes showing inverse response? What simplifications may be brought to the MPC performance index for controlling processes with large pure time delay?

9. What are the tuning guidelines for achieving zero steady-state offset for the unconstrained MPC design?

10. How would the multivariable MPC tuning take into account the control of output variables having values of different orders of magnitude?

11. Is it possible to apply inferential control to the MPC control algorithm, i.e., to control nonmeasurable (primary) output variables using other measured (secondary) outputs? Please elaborate.

12. The MPC controller can use a higher number of control variables than the number of controlled variables. What practical circumstances may motivate such control approach?

13. In a mass transfer reactor, gas A is dispersed in aqueous saline solution B. The reactor has a jacket with cooling water. The temperature is measured with a thermoresistor with sheath, with the temperature controller modifying the water flow in the

jacket. The level transducer is a differential pressure gauge and the level controller manipulates the outflow from the reactor. Gas A flow is used to control the conversion in the reactor which has to be at 50%. The pressure is kept at 2 bar by manipulating the output unreacted gas flow and of the vapors at the operating temperature of 110 ° C. Design the overall control block scheme of the system and describe its interactions.

14. Explain the reason for considering the fuzzy controller to have good robust characteristics.

15. How is it possible to include the integral mode in the fuzzy controller design?

16. The fuzzy controller may be designed so as to include adaptive features. Elaborate on the way the form of the membership functions, fuzzy sets selection, and fuzzy rules formulation may be used for accomplishing this adaptive capability.

17. Explain the way a fuzzy controller may be designed to operate with different control performance in the cases of the rising and descending change of the controlled variable to control processes showing integral behavior.

18. What measures can be applied for the fuzzy controller tuning to reduce oscillations of the controlled variable?

19. Explain why the relative gain array (RGA) matrix is scale independent and the sum of its elements on each row or column is equal to unity.

20. How many decouplers are necessary for a 5×5 multiloop control system? If some of the decouplers result as physically unrealizable, what solution still exists to implement the decoupling?

21. How does the nonlinearity affect the loop pairing based on steady-state RGA?

22. Elaborate a methodology for determining the RGA matrix based on an experimental approach.

23. What are the incentives of the centralized control approach, compared to the decentralized one?

24. Which of the layers of the hierarchical structure for controlling the whole plant should consider multiobjective requirements?

25. Explain how you would integrate the following plant-wide control requirements: minimizing the utility, energy, raw material, and environment protection costs and maximizing the throughput.

26. Explain the importance of the active constraint variables and the self-optimizing variables for the plant-wide control structure design methodology.

27. Please motivate why the selection of the primary and secondary controlled variables features a mutual implication relationship.

? 28. How and at what part of the plant-wide control structure design methodology should be constraints taken into account?

? 29. What is the effect of sampling on the design of the discrete controllers? What are the effects of increasing the sampling time on the control performance?

i 30. Explain why the PID velocity algorithm does not need initialization. How is initialization performed for the PID position algorithm?

? 31. How can continuous tuning rules, such as Ziegler-Nichols and Cohen-Coon, be used for tuning discrete controllers? What is the effect of the sampling time value on tuning discrete controllers?

? 32. What are the design solutions for limiting the ringing effect?

? 33. What are the design solutions to avoid getting physically unrealizable controllers, in the case of the controller design imposing the desired closed-loop Z transform response $Y(z)$?

? 34. How can a discrete PID controller be designed to have adaptive features in order to respond to changes (nonlinearity) in the steady-state and dynamic behavior of the controlled process?

16.2 Applied process engineering control

i 1. Write the steady-state model and simulate a CSTR in function for the assumption of three control loops such as those illustrated in Fig. 8.23. Consider the data specified for Example 8.1. Other technical data are: jacket volume $V_{jacket} = 220$ L, heat capacity of heat-transfer agent $C_{P,ag} = 1$ kJ/kg·grd, density $\rho_{ag} = 1000$ kg/m^3, and inlet temperature T°_{ag}, at a value that ensures thermal stability. Feed and discharge occurs on pipes with 50 mm inner diameter. The pressure drops are 0.5 and 0.2 bar on the inlet and outlet pipes, respectively. The transducer time constants are $T_{TT} = 2$ min for temperature, $T_{TF} = 5$ s for flow rate, and $T_{TL} = 2$ s for level control, respectively. The corresponding amplification factors are $K_{TT} = 0.08$ mA/°C, $K_{TF} = 0.8$ mA/m^3·h, and $K_{TL} = 4$ mA/m, respectively.

i 2. Check the thermal stability of a CSTR for an exothermic first-order reaction considered in Example 8.1 and calculate the oscillation period. The reactor has a heat-transfer surface of $A_T = 4$ m^2 characterized by $K_T = 2.10^3$ kJ/m^2·h·°C. The reaction heat is temperature independent and has a value of $\Delta H_r = -12,000$ kJ/kg. The inlet flow has a rate of $F_{in} = 2000$ kg/h and contains $C_{Ao} = 0.2$ kmol/m^3 at $\rho = 10^3$ kg/m^3. Its inlet temperature is $T^\circ_{in} = 95$ °C. The reaction mass does not suffer significant density changes and keeps its heat capacity constant at $C_P = 935$ J/kg·°C. The setpoint is adjusted at 100 °C and 2 h residence time (see Fig. 8.1(b)). The heat-transfer agent's inlet temperature into the jacket is of $T^{\circ}_{ag,\,in} = 85\,^{\circ}C$. Comment on the results. How does operation alter if (a) F_{in} and (b) C_{Ao} drop to half?

3. Recalculate the parameters D/R, V/F, D/F of the distillation column presented in Example 9.2 that allows an $r = 0.95$ recovery factor if the feeding concentration decreases to $x_F = 0.4$.

4. Calculate the optimal D/F ratio that minimizes the losses generated by contamination of distillate and bottom product. The distillate product can be sold at US \$5.3/kg, but the bottom product at US \$2.3/kg. The feeding stream composition is $x_F = 0.5$, the separation factor is $S = 361$, whereas the molar masses of bottom and distillate species are of $M_B = 92$ and $M_D = 78$ kg/kmol, respectively.

5. Calculate the batch time for a discontinuous distillation required to achieve $x_B = 0.1$ when the loaded mixture is of 100 kmol with an $x_F = 0.3$ (initial) concentration. The mean distillate concentration at the end of the batch should be of 0.7. The distillate flow rate is of 10 kmol/h.

6. An absorption column has to ensure absorption of species A. Design a feedforward control system for the following situations:
(a) It attenuates 0.2% variation of y_F.
(b) It attenuates 0.8 bar variation of operating pressure.

The operating pressure is kept constant.
7. Consider a countercurrent L-L packed extraction tower, with light-phase dispersion and in continuous steady-state operation mode at 300 K (see Figs. 11.3 and 11.4). It is characterized by the following data: $F = 100$ m^3/h, $\rho_F = 1000$ kg/m^3, $\rho_S = 870$ kg/m^3, $x_{AF} = 0.15$, $x_{AS} = 0.01$, $K_A = 1.36$ and setpoint $x_{AE} = 0.05$. Design a suitable control system that countervails $\pm30\%$ disturbances in x_{AF} and x_{AS}.

8. Consider a countercurrent L-L packed extraction tower, in continuous steady-state operation mode at 300 K (see Figs. 11.3 and 11.4). Technical data are $F = S = 5$ kmol/h and $K_A = 4.76$. Assess whether the temperature control is necessary.

9. An evaporator concentrates a solution from 20% initial mass concentration to a final concentration of 80%. Calculate the effect on the final concentration of (a) 1% change in steam flow or feed flow rate and (b) 1% change in feed concentration.

10. A single effect evaporator concentrates an aqueous solution of NaOH (sodium hydroxide) from a 0.2 kg/kg solution to a 0.7 kg/kg solution. The feed flow rate of diluted solution is $F = 70.4$ kg/h and the latent heat of vaporization of water is $l_v = 525$ kcal/kg. Calculate the setpoint of a steam flow controller that ensures the vaporization of the unnecessary water.

11. Calculate for the evaporator in Example 12.1 the gains on the transfer paths $V_0 \rightarrow x_o$ and $x_F \rightarrow x_o$.

12. Consider an aspirin powder batch dryer, operating to bring the moisture content from $w_c = 0.1$ to desired $w^* = 0.05$, having the input air temperature $T_i^{\circ} = 93\,°C$ and the wet bulb temperature of $T_{wb}^{\circ} = 34\,°C$. Calculate what additional energy consumption and what additional costs are involved if the drying of 500 kg powder stops at 80 °C. The cost is of US $70/Gcal.

13. A chemical engineer determines the following solubility concentrations for a certain chemical species:

Tab. 16.1: Solubility concentrations at different temperatures for a chemical species.

Temperature (°C)	10	30	60
Solubility (g/g)	0.013	0.046	0.182

If the batch crystallizer is loaded at 40 °C with saturated solution and cooled to 15 °C, what is the expected crystal mass per kilogram suspension?

14. A continuous mixed suspension mixed product removal (CMSMPR) crystallizer is operated at $T = 20\,°C$, and fed with a $T = 40\,°C$ solution. After some structural modification, the crystallizer is fed with a second inlet, saturated stream. Let us calculate the required temperature of the second inlet flux that doubles the outlet stream's specific solid content (g/g). The yield is of $\eta = 90\%$, whereas the solubility is characterized by data in Tab. 16.1.

15. Consider Example 14.3, where a slurry is filtered with a laboratory filter. The engineer performing the calculations made a mistake and used the viscosity of water (0.001 Pa · s) in the calculations. However, the solvent used in the experiment was acetonitrile, which has a viscosity of 0.00038 Pa · s. What is the consequence of this mistake on the calculated specific cake resistance and medium resistance?

16. The filtration time of a compound is excessively long due to the small particle sizes (in the size of micrometers; no particles are in the nanometer range – see the relative trend in Fig. 16.5).

What solutions would you propose for the following scenarios to reduce the filtration time, preferentially, without using another filter or pump?
a. The temperature dependency of the solubility is weak in the operating temperature range.
b. The temperature dependency of the solubility is significant.

What are the process control implications of the decisions? Propose the necessary control loops for your solutions.

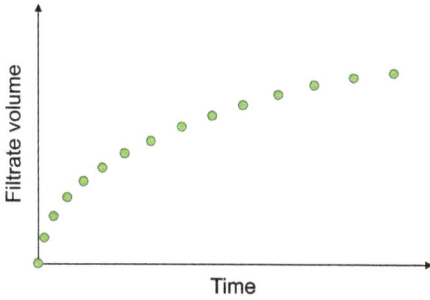

Time

Filtrate volume

17. Determine the number of filtration cycles that minimizes the overall filtration time of a V_t volume of slurry. The duration of a filtration cycle is given by

$$\tau = \tau_f + \tau_w + \tau_a$$

where τ is the total cycle time, τ_f is the filtration time, τ_w is the washing time of the cake, and τ_a denotes the auxiliary time (equipment washing, discharge, charge, etc.).

Let N denote the number of filtration cycles. Then the filtration time can be written as

$$\tau_f = k \left(\frac{V_t}{N} \right)^2$$

where k is a constant. Considering that τ_w is proportional to the filtration time, we can write that

$$\tau_w = \beta k \left(\frac{V_t}{N} \right)^2$$

The auxiliary time does not depend on the filtration and washing times. Therefore, the total duration (τ_t) of the N cycles can be written as

$$\tau_t = \left[k \left(\frac{V_t}{N} \right)^2 + \beta k \left(\frac{V_t}{N} \right)^2 + \tau_t \right] = \frac{k V_t^2 + \beta k V_t^2}{N} + N \tau_t$$

Introducing the notations of $y = \tau_t$, $n = x_1$, and $A = k V_t^2 + \beta k V_t^2$, the following simplified form of the function to be minimized is obtained:

$$y = A x_1^{-1} + \tau_t x_1$$

The problem can be viewed as geometric programming, where $c_1 = A$, $c_2 = \tau_t$, $a_{11} = -1$, $a_{12} = 1$, $n = 1$, $T = 2$. The equation system that determines the weights are:

$$-w_1 + w_2 = 0$$

$$w_1 + w_2 = 1$$

where the solution is $w_1 = w_2 = 0.5$. Therefore, the minimal filtration time is

$$\tau_t^* = y^* = \left(\frac{\tau_a}{0.5}\right)^{0.5} \left(\frac{A}{0.5}\right)^{0.5} = 2(\tau_a A)^{0.5}$$

18. The following reactions are conducted in an isothermal fed-batch system:

$$A + B \rightarrow C$$

$$C + B \rightarrow D$$

$$C \rightarrow E$$

The main product is E, whereas D is a by-product, which is difficult to separate from the main product. According to the reaction kinetic measurements, the formation of by-product D is negligible if the actual concentration of the intermediate product C is lower than a critical value C_{crit}. Initially, the reactor contains a solution of reactant A, whereas reactant B is fed into the reactor with a flow rate, specifying the flow rate is the process design objective.

a. How would you apply the principles of QbC to obtain the feeding flow rate profile that minimizes the concentration of by-product in the final mixture and simultaneously minimizes the operating time? Draw the required control scheme.

b. What extension of the control system would you propose to counteract the highly exothermic reaction rate of the $C \rightarrow E$ chemical reaction? What transfer function must be identified to design a suitable control solution?

Index

https://doi.org/10.1515/9783110789737-019

www.ingramcontent.com/pod-product-compliance
Lightning Source LLC
Chambersburg PA
CBHW080136220326
41598CB00032B/5086

9 783110 789720